INTRODUCTION TO TOPOLOGY AND GEOMETRY

INTRODUCTION TO TOPOLOGY AND GEOMETRY

Second Edition

Saul Stahl

Department of Mathematics
The University of Kansas
Lawrence, KS

Catherine Stenson

Department of Mathematics
Juniata College
Huntington, PA

WILEY

A JOHN WILEY & SONS, INC., PUBLICATION

Library of Congress Cataloging-in-Publication Data:

Stahl, Saul.
  Introduction to topology and geometry. — 2nd edition / Saul Stahl, University of Kansas, Catherine
Stenson, Juniata College.
      pages cm. — (Pure and applied mathematics)
  Includes bibliographical references and index.
  ISBN 978-1-118-10810-9 (hardback)
1. Topology 2. Geometry. I. Stenson, Catherine, 1972– II. Title.
  QA611.S814 2013
  514—dc23   2012040259

To Denise, with love from Saul.

To my family, with love from Cathy.

CONTENTS

PREFACE

This book is intended to serve as a text for a two-semester undergraduate course in topology and modern geometry. It is devoted almost entirely to the geometry of the last two centuries. In fact, some of the subject matter was discovered only within the last two decades. Much of the material presented here has traditionally been part of the realm of graduate mathematics, and its presentation in undergraduate courses necessitates the adoption of certain informalities that would be unacceptable at the more advanced levels. Still, all of these informalities either were used by the mathematicians who created these disciplines or else would have been accepted by them without any qualms.

The first four chapters aim to serve as an introduction to topology. Chapter 1 provides an informal explanation of the notion of homeomorphism. This naive introduction is in fact sufficient for all the subsequent chapters. However, the instructor who prefers a more rigorous treatment of basic topological concepts such as homeomorphisms, topologies, and metric spaces will find it in Chapter 10.

The second chapter emphasizes the topological aspects of graph theory, but is not limited to them. This material was selected for inclusion because the accessible nature of some of its results makes it the pedagogically perfect vehicle for the transition from the metric Euclidean geometry the students encountered in high school to the combinatorial thinking that underlies the topological results of the subsequent chapters. The focal issue here is planarity: Euler's Theorem, coloring theorems, and the Kuratowski Theorem.

Chapter 3 presents the standard classifications of surfaces of both the closed and bordered varieties. The Euler–Poincaré equation is also proved.

Chapter 4 is concerned with the interplay between graphs and surfaces—in other words, graph embeddings. In particular, a procedure is given for settling the question of whether a given graph can be embedded on a given surface. Polygonal (2-cell) embeddings and their rotation systems are discussed. The notion of covering surfaces is introduced via the construction of voltage graphs.

The theory of knots and links has recently received tremendous boosts from the work of John Conway, Vaughan Jones, and others. Much of this work is easily accessible, and some has been included in Chapter 5: the Conway–Gordon–Sachs Theorem regarding the intrinsic linkedness of the graph K_6 in $\mathbb{R}^3$ and the invariance of the Jones polynomial. While this discipline is not, properly speaking, topological, connections to the topology of surfaces are not lacking. Knot theory is used to prove the nonembeddability of nonorientable surfaces in $\mathbb{R}^3$, and surface theory is used to prove the nondecomposability of trivial knots. The more traditional topic of labelings is also presented.

The next three chapters deal with various aspects of differential geometry. The exposition is as elementary as the author could make it and still meet his goals: explanations of Gauss's Total Curvature Theorem and hyperbolic geometry. The geometry of surfaces in $\mathbb{R}^3$ is presented in Chapter 6. The development follows that of Gauss's *General Investigations of Curved Surfaces*. The subtopics include Gaussian curvature, geodesics, sectional curvatures, the first fundamental form, intrinsic geometry, and the Total Curvature Theorem, which is Gauss's version of the famed Gauss–Bonnet formula. Some of the technical lemmas are not proved but are instead supported by informal arguments that come from Gauss's monograph. A considerable amount of attention is given to polyhedral surfaces for the pedagogical purpose of motivating the key theorems of differential geometry.

The elements of Riemannian geometry are presented in Chapter 7: Riemann metrics, geodesics, isometries, and curvature. The numerous examples are also meant to serve as a lead-in to the next chapter.

The eighth chapter deals with hyperbolic geometry. Neutral geometry is defined in terms of Euclid's axiomatization of geometry and is described in terms of Euclid's first 28 propositions. Various equivalent forms of the parallel postulates are proven, as well as the standard results regarding the sum of the angles of a neutral triangle. Hyperbolic geometry is also defined axiomatically. Poincaré's half-plane geometry is developed in some detail as an instance of the Riemann geometries of the previous chapter and is demonstrated to be hyperbolic. The isometries of the half-plane are described both algebraically and geometrically.

The ninth chapter is meant to serve as an introduction to algebraic topology. The requisite group theory is summarized in Appendix B. The focus is on the derivation of fundamental groups, and the development is based on Poincaré's own exposition and makes use of several of his examples. The reader is taught to derive presentations for the fundamental groups of the punctured plane, closed surfaces, 3-manifolds, and knot complements. The chapter concludes with a discussion of the Poincaré Conjecture.

The tenth chapter serves a dual purpose. On the one hand it aims to acquaint the reader with the elegant topic of general topology and the joys of sequence chasing. On the other hand, it contains the rigorous definitions of a variety of fundamental concepts that were only informally defined in the previous chapters. In terms of mathematical maturity, this is probably the most demanding part of the book.

The last chapter is devoted to the study of polytopes. Following an introduction, attention is given to the graphs of polytopes, regular polytopes, and the enumeration of faces of polytopes.

Wherever appropriate, historical notes have been interspersed with the exposition. Care was taken to supply many exercises that range from the routine to the challenging. Middle-level exercises were hard to come by, and the author welcomes all suggestions.

An Instructor's solution manual is available upon request from Wiley.

SAUL STAHL

Lawrence, Kansas
Stahlex@ku.edu

ACKNOWLEDGMENTS

The first author is deeply indebted to Mark Hunacek for reading the manuscript and suggesting many improvements. Jack Porter helped by making his *Graduate Topology Notes* and other materials available to me. He also corrected some errors in the early version of Chapter 10. Encouraging kind words and valuable criticisms were provided by the reviewers Michael J. Kallaher, David Royster, Dan Gottlieb, and David W. Henderson, as well as several others who chose to remain anonymous. Stephen Quigley and Susanne Steitz-Filler at John Wiley and Sons supervised the conversion of the notes into a book. The manuscript was expertly typeset by Larisa Martin and Sandra Reed. Prentice Hall gave me its gracious permission to reprint portions of Chapters 9–12 of my book *Geometry from Euclid to Knots*, which it published in 2003.

S. S.

The second author is grateful to Saul Stahl for the invitation to join this project and for his help throughout. Thanks to Marge Bayer for recommending me and for her comments on a draft, and to Lou Billera for his comments and for his guidance over the years. Thanks also to the Mathematics Institute of Leiden University for their hospitality in the spring of 2012.

C. S.

CHAPTER 1

INFORMAL TOPOLOGY

In this chapter the notion of a topological space is introduced, and informal ad hoc methods for identifying equivalent topological spaces and distinguishing between nonequivalent ones are provided.

The last book of Euclid's opus *Elements* is devoted to the construction of the five Platonic solids pictured in Figure 1.1. A fact that Euclid did not mention is that the counts of the vertices, edges, and faces of these solids satisfy a simple and elegant relation. If these counts are denote by v, e, and f, respectively, then

$$v - e + f = 2. \tag{1}$$

Specifically, for these solids we have:

$$
\begin{aligned}
\text{Cube:} \quad & 8 - 12 + 6 = 2. \\
\text{Octahedron:} \quad & 6 - 12 + 8 = 2. \\
\text{Tetrahedron:} \quad & 4 - 6 + 4 = 2. \\
\text{Dodecahedron:} \quad & 20 - 30 + 12 = 2. \\
\text{Icosahedron:} \quad & 12 - 30 + 20 = 2.
\end{aligned}
$$

Introduction to Topology and Geometry, Second Edition.
By Saul Stahl and Catherine Stenson Copyright © 2013 John Wiley & Sons, Inc.

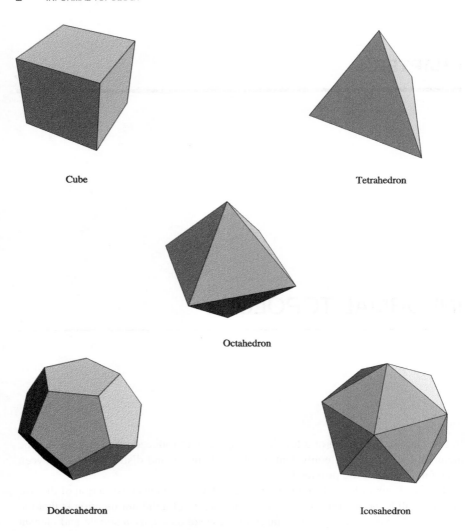

Figure 1.1   The Platonic solids.

A Platonic solid is defined by the specifications that each of its faces is the same regular polygon and that the same number of faces meet at each vertex. An interesting feature of Equation (1) is that while the Platonic solids depend on the notions of length and straightness for their definition, these two aspects are absent from the equation itself. For example, if each of the edges of the cube is either shrunk or extended by some factor, whose value may vary from edge to edge, a lopsided cube is obtained (Fig. 1.2) for which the equation still holds by virtue of the fact that it holds for the (perfect) cube. This is also clearly true for any similar modification of the other four Platonic solids. The fact of the matter is that Equation (1) holds

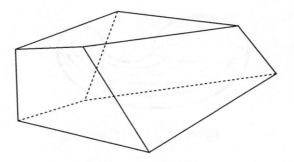

Figure 1.2   A lopsided cube.

not only for distorted Platonic solids, but for all solids as well, provided these solids are carefully defined. Thus, for the three solids of Figure 1.3 we have respectively $5 - 8 + 5 = 2$, $6 - 9 + 5 = 2$, and $7 - 12 + 7 = 2$. The applicability of Equation (1) to all such solids was first noted by Leonhard Euler (1707–1783) in 1758, although some historians contend that this equation was presaged by certain observations of René Descartes (1596–1650).

Euler's equation remains valid even after the solids are subjected to a wider class of distortions which result in the curving of their edges and faces (see Figure 1.4). One need simply relax the definition of edges and faces so as to allow for any non-self intersecting curves and surfaces. Soccer balls and volleyballs, together with the patterns formed by their seams, are examples of such curved solids to which Euler's equation applies. Moreover, it is clear that the equation still holds after the balls are deflated.

Topology is the study of those properties of geometrical figures that remain valid even after the figures are subjected to distortions. This is commonly expressed by saying that topology is *rubber-sheet geometry*. Accordingly, our necessarily informal definition of a *topological space* identifies it as any subset of space from which the notions of straightness and length have been abstracted; only the aspect of contiguity remains. Points, arcs, loops, triangles, solids (both straight and curved), and the surfaces of the latter are all examples of topological spaces. They are, of course, also geometrical objects, but topology is only concerned with those aspects of their ge-

 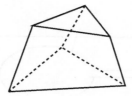 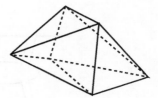

Figure 1.3   Three solids.

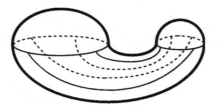

Figure 1.4   A curved cube.

ometry that remain valid despite any translations, elongations, inflations, distortions, or twists.

Another topological problem investigated by Euler, somewhat earlier, in 1736, is known as the *bridges of Koenigsberg*. At that time this Prussian city straddled the two banks of a river and also included two islands, all of which were connected by seven bridges in the pattern indicated in Figure 1.5. On Sunday afternoons the citizens of Koenigsberg entertained themselves by strolling around all of the city's parts, and eventually the question arose as to whether an excursion could be planned which would cross each of the seven bridges exactly once. This is clearly a geometrical problem in that its terms are defined visually, and yet the exact distances traversed in such excursions are immaterial (so long as they are not excessive, of course). Nor are the precise contours of the banks and the islands of any consequence. Hence, this is a topological problem. Theorem 2.2.2 will provide us with a tool for easily resolving this and similar questions.

The notorious Four-Color Problem, which asks whether it is possible to color the countries of every geographical map with four colors so that adjacent countries sharing a border of nonzero length receive distinct colors, is also of a topological nature. Maps are clearly visual objects, and yet the specific shapes and sizes of the countries in such a map are completely irrelevant. Only the adjacency patterns matter.

Every mathematical discipline deals with objects or structures, and most will provide a criterion for determining when two of these are identical, or equivalent. The equality of real numbers can be recognized from their decimal expansions, and two vectors are equal when they have the same direction and magnitude. Topological equivalence is called *homeomorphism*. The surface of a sphere is homeomorphic to

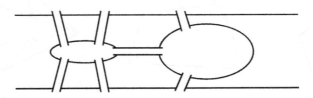

Figure 1.5   The city of Koenigsberg.

Figure 1.6 Homeomorphic open arcs.

those of a cube, a hockey puck, a plate, a bowl, and a drinking glass. The reason for this is that each of these objects can be deformed into any of the others. Similarly, the surface of a doughnut is homeomorphic to those of an inner tube, a tire, and a coffee mug. On the other hand, the surfaces of the sphere and the doughnut are not homeomorphic. Our intuition rejects the possibility of deforming the sphere into a doughnut shape without either tearing a hole in it or else stretching it out and juxtaposing and pasting its two ends together. Tearing, however, destroys some contiguities, whereas juxtaposition introduces new contiguities where there were none before, and so neither of these transformations is topologically admissible. This intuition of the topological difference between the sphere and the doughnut will be confrmed by a more formal argument in Chapter 3.

The easiest way to establish the homeomorphism of two spaces is to describe a deformation of one onto the other that involves no tearing or juxtapositions. Such a deformation is called an *isotopy*. Whenever isotopies are used in the sequel, their existence will be clear and will require no formal justification. Such is the case, for instance, for the isotopies that establish the homeomorphisms of all the open arcs in Figure 1.6, all the loops in Figure 1.7, and all the ankh-like configurations of Figure 1.8. Note that whereas the page on which all these curves are drawn is two-dimensional, the context is definitely three-dimensional. In other words, all our curves (and surfaces) reside in Euclidean 3-space $\mathbb{R}^3$, and the isotopies may make use of all three dimensions.

The concept of isotopy is insufficient to describe all homeomorphisms. There are spaces which are homeomorphic but not isotopic. Such is the case for the two loops in Figure 1.9. It is clear that loop b is isotopic to all the loops of Figure 1.7 above, and it is plausible that loop a is not, a claim that will be justified in Chapter 5. Hence,

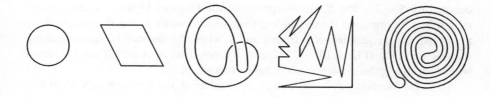

Figure 1.7 Homeomorphic loops.

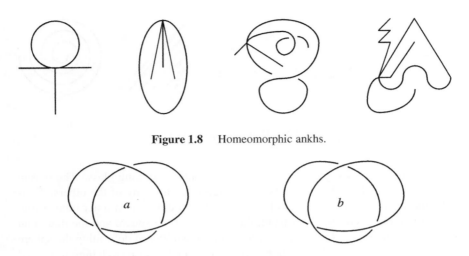

Figure 1.8   Homeomorphic ankhs.

Figure 1.9   Two spaces that are homeomorphic but not isotopic.

the two loops are not isotopic to each other. Nevertheless, they are homeomorphic in the sense that ants crawling along these loops would experience them in identical manners. To express this homeomorphism somewhat more formally it is necessary to resort to the language of functions. First, however, it should be pointed out that the word *function* is used here in the sense of an association, or an assignment, rather than the end result of an algebraic calculation. In other words, a function $f : S \to T$ is simply a rule that associates to every point of S a point of T. In this text most of the functions will be described visually rather than algebraically.

Given two topological spaces S and T, a *homeomorphism* is a function $f : S \to T$ such that

1. f matches all the points of S to all the points of T (distinct points of S are matched with distinct points of T and vice versa);

2. f preserves contiguity.

It is the vagueness of the notion of *contiguity* that prevents this from being a formal definition. Since any two points on a line are separated by an infinitude of other points, this concept is not well defined. The homeomorphism of S and T is denoted by $S \approx T$. The homeomorphism of the loops of Figure 1.9 can now be established by orienting them, labeling their lowest points A and B, and matching points that are at equal distances from A and B, where the distance is measured along the oriented loop (Fig. 1.10). Of course, the positions of A and B can be varied without affecting the existence of the homeomorphism.

A similar function can be defined so as to establish the homeomorphism of any two loops as long as both are devoid of self-intersections. Suppose two such loops c and d, of lengths γ and δ respectively, are given (Fig. 1.11). Again begin by specifying orientations and initial points C and D on the two loops. Then, for every real number $0 \leq r < 1$, match the point at distance $r\gamma$ from C along c with the point at distance $r\delta$ from D along d.

Figure 1.12 contains another instructive example. Each of its three topological spaces consists of a band of, say, width 1 and length 20. They differ in that e is untwisted, f has one twist, and g has two twists. Band f differs from the other two in that its border is in fact one single loop whereas bands e and g have two distinct borders each. It therefore comes as no surprise that band f is not homeomorphic to either e or g. These last two, however, are homeomorphic to each other. To describe this homeomorphism a coordinate system is established on each of the bands as follows. For each number $0 \leq r \leq 1$ let $L_{e,r}$ and $L_{g,r}$ denote the oriented loops of length 20 that run along the band at a constant distance r from the bottom borders of e and g respectively. Choosing start lines as described in Figure 1.13, the coordinate pair (r, s), $0 \leq r \leq 1$, $0 \leq s < 20$, describes those points on the loops $L_{e,r}$ and $L_{g,r}$ at a distance s from the respective starting line. The required homeomorphism simply matches up points of e and g that have the same coordinate pairs. The reason this wouldn't work for band f is that for this band the coordinatization process fails (see Fig. 1.14).

As mentioned above, f is not homeomorphic to e and g, because it has a different number of borders. In general, borders and other extremities are a good place to look for differences between topological spaces. For example, every two of the spaces in Figure 1.15 are nonhomeomorphic because they each have a different number of extremities. The number of components of a space can also serve as a tool for distinguishing between homeomorphism types. All the spaces in Figure 1.16 have the same number of extremities, but they are nevertheless nonhomeomorphic because each has a different number of components: 1, 2, 3, and 4, respectively.

Another method for distinguishing between spaces is to examine what remains when an equal number of properly selected points are deleted from each. For instance, both spaces of Figure 1.17 have one component, and neither has extremities. Nevertheless, they are not homeomorphic, because the removal of the two endpoints of the diameter of the θ-like space results in a space with three components, whereas the removal of any two points of the circle leaves only two components. In general a *topological property* of a space is a property that is shared by all the spaces that are homeomorphic to it. The number of endpoints and the number of components are

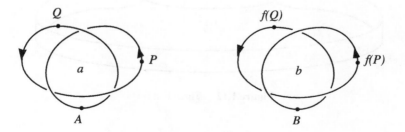

Figure 1.10   A homeomorphism of two loops.

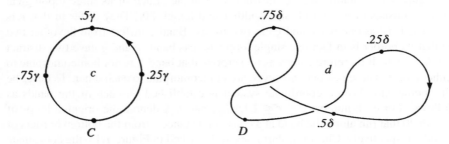

Figure 1.11   A homeomorphism of two loops.

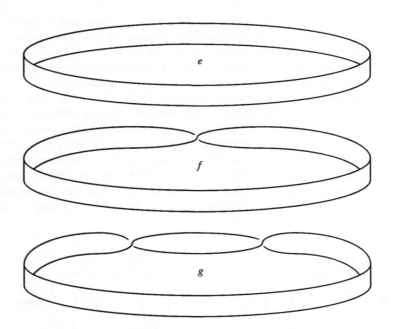

Figure 1.12   Three bands.

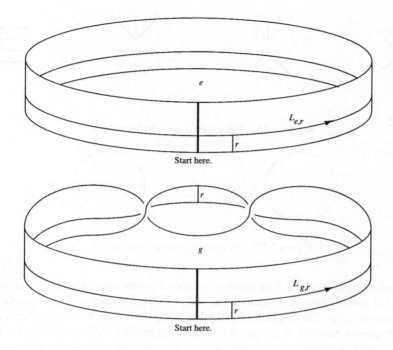

Figure 1.13 The homeomorphism of two bands.

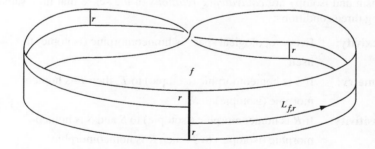

Figure 1.14 A failed homeomorphism.

Figure 1.15 Four nonhomeomorphic spaces.

both such topological properties. On the other hand, neither the length of an interval nor the area of a region is a topological property.

Figure 1.16   Four nonhomeomorphic spaces.

Figure 1.17   Two nonhomeomorphic spaces.

The foregoing discussion of topological spaces, homeomorphisms, and isotopies is an informal working introduction that will serve for the purposes of this text. Experience indicates that this lack of precision will not hamper the comprehension of the subsequent material. Rigorous definitions are provided in Chapter 10, which can be read out of sequence.

In working out the exercises, the readers may find it useful to note that both homeomorphism and isotopy are *equivalence relations* in the sense that they satisfy the following three conditions:

Reflexivity: Every topological space is homeomorphic (isotopic) to itself.

Symmetry: If S is homeomorphic (isotopic) to T, then T is homeomorphic (isotopic) to S.

Transitivity: If R is homeomorphic (isotopic) to S and S is homeomorphic (isotopic) to T, then R is homeomorphic (isotopic) to T.

Exercises 1.1

1. Which of the letters in Figure 1.18 are homeomorphic?
2. Which of the topological spaces in Figure 1.19 are homeomorphic?
3. Which of the topological spaces in Figure 1.20 are homeomorphic?
4. Which of the topological spaces of Figure 1.21 are isotopic?
5. Are the following statements true or false? Justify your answers.

    (a) If two topological spaces are homeomorphic, then they are also isotopic.

(b) If two topological spaces are isotopic, then they are also homeomorphic.

(c) Topological equivalence is synonymous with homeomorphism.

(d) Topological equivalence is synonymous with isotopy.

(e) Every two loops are isotopic.

(f) Every two loops are homeomorphic.

A B C D E F G H I J K L M N O P Q R S T U V W X Y Z

Figure 1.18 Twenty-six topological spaces.

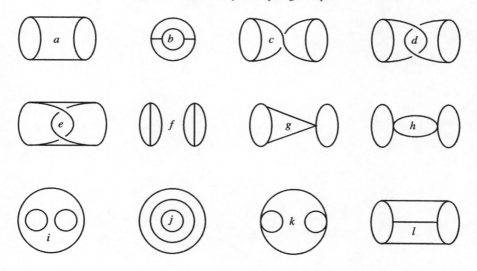

Figure 1.19 Some one-dimensional topological spaces.

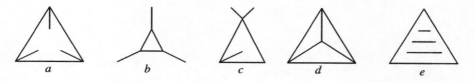

Figure 1.20 Some one-dimensional topological spaces.

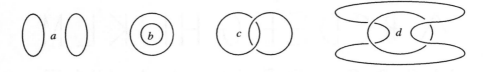

Figure 1.21   Some one-dimensional topological spaces.

CHAPTER 2

GRAPHS

The one-dimensional topological objects are arcs. Graphs are created by the juxtaposition of a finite number of arcs, and they underlie many applications of mathematics as well as popular riddles. Traversability, planarity, colorability, and homeomorphisms of graphs are discussed in this chapter.

2.1 Nodes and Arcs

An *open arc* is any topological space that is homeomorphic to a line segment (Fig. 1.6), and a *loop* is any topological space that is homeomorphic to a circle (Fig. 1.7); both are collectively referred to as *arcs*. A graph G consists of a set of points $N(G) = \{v_1, v_2, \ldots, v_p\}$, called the *nodes* of G, and a set of arcs $A(G) = \{a_1, a_2, \ldots, a_q\}$. The endpoints of each open arc are nodes, and every loop is assumed to begin and end at a node; other than that, arcs contain no nodes. Two nodes are said to be *adjacent* in G if they are the endpoints of a common arc of G. The adjacency of u and v is denoted by $u \sim v$. The number of arcs of which a node v is an endpoint, with each loop counted twice, is called the *degree* of v in G and is denoted by $\deg_G v$, or simply $\deg v$. For example, in the graph G of Figure 2.1 we have $\deg u = 4$, $\deg v = 4$,

Introduction to Topology and Geometry, Second Edition.
By Saul Stahl and Catherine Stenson Copyright © 2013 John Wiley & Sons, Inc.

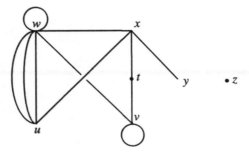

Figure 2.1   A graph.

$\deg w = 7$, $\deg x = 4$, $\deg y = 1$, $\deg t = 2$, $\deg z = 0$. The location of nodes of degree at least 3 is, of course, indicated by the conjunction of the arcs at that node. Nodes of degree 1 are equally conspicuous. Nodes of degrees 2 and 0 have their location marked by a solid dot. If $v_1, v_2, \ldots, v_p$ is a listing of the nodes of G such that

$$\deg v_1 \geq \deg v_2 \geq \cdots \geq \deg v_p,$$

then $(\deg v_1, \deg v_2, \ldots, \deg v_p)$ is the *degree sequence* of G. Thus, the degree sequence of the graph of Figure 2.1 is $(7, 4, 4, 4, 2, 1, 0)$. The numbers of nodes and arcs of the graph G are denoted by $p(G)$ and $q(G)$, or simply p and q, respectively.

The first proposition describes a simple and useful relation between the degrees of the nodes of a graph and the number of its arcs.

Proposition 2.1.1 *In any graph G, $\sum_{v \in N(G)} \deg v = 2q(G)$.*

PROOF: Each open arc of G contributes 1 to the degrees of each of its two endpoints, and each loop contributes 2 to the degree of its only node. Hence each arc contributes 2 to $\sum_{v \in N(G)} \deg v$. The desired equation now follows immediately.   Q.E.D.

This proposition immediately eliminates $(3, 3, 2, 2, 1)$ as a possible degree sequence, since the sum of the degrees of a graph must be even.

Distinct arcs that join the same endpoints are said to be *parallel*, and a graph that contains neither loops nor parallel arcs is said to be *simple*. Of the three graphs in Figure 2.2 only the middle one is simple.

If n is any positive integer, then the *complete graph* K_n is the simple graph with n nodes in which every pair of distinct nodes is joined by an arc. It is clear that each of the n nodes of K_n has degree $n-1$ and that K_n has $\binom{n}{2} = n(n-1)/2$ arcs. The graph K_1, which consists of a single node and no arcs, is also known as the *trivial graph*. If $m \geq n$ are two positive integers, then the *complete bipartite graph* $K_{m,n}$ is formed as follows. The node set $N(K_{m,n})$ is the union of two disjoint sets $A = \{u_1, u_2, \ldots, u_m\}$ and $B = \{v_1, v_2, \ldots, v_n\}$, and the arc set consists of mn open arcs that join each u_i to each v_j. In general, given positive integers $n_1 \geq n_2 \geq \cdots \geq n_k$, the *complete n-partite (simple) graph* $K_{n_1, n_2, \ldots, n_k}$ has the union $A_1 \cup A_2 \cup \cdots \cup A_k$ as its node set, where the sets $A_1, A_2, \ldots, A_k$ have cardinalities $n_1, n_2, \ldots, n_k$ respectively and are pairwise disjoint. The arcs of $K_{n_1, n_2, \ldots, n_k}$ join all node pairs u, v where u and v belong to distinct A_i's. Three examples of such complete n-partite graphs appear in Figure 2.3.

If G and G' are graphs such that $N(G) \supset N(G')$ and $A(G) \supset A(G')$, then G' is said to be a subgraph of G. If G' is a subgraph of G, then $G - G'$ is the subgraph of G with node set $N(G)$ and arc set $A(G) - A(G')$ (i.e., all the arcs of G that are not arcs of G').

Exercises 2.1

1. Compute the degree sequences of the following graphs.

   (a) K_5 (b) K_n (c) $K_{3,4}$ (d) $K_{m,n}$

   (e) $K_{3,2,1}$ (f) $K_{l,m,n}$ (g) $K_{7,6,5,4}$ (h) $K_{n_1, n_2, \ldots, n_k}$

2. Which of the following sequences are degree sequences of some graph? Justify your answer.

   (a) $(1, 1)$ (b) $(2, 1)$ (c) $(2, 2)$ (d) $(3, 1)$

   (e) $(3, 2)$ (f) $(3, 3)$ (g) $(1, 1, 1)$ (h) $(2, 1, 1)$

   (i) $(2, 2, 1)$ (j) $(2, 2, 2)$ (k) $(4, 2, 0)$ (l) $(8, 0)$

3. Which of the following sequences are degree sequences of some simple graph? Justify your answer.

   (a) $(1, 1)$ (b) $(2, 1)$ (c) $(2, 2)$ (d) $(3, 1)$

   (e) $(3, 2)$ (f) $(3, 3)$ (g) $(1, 1, 1)$ (h) $(2, 1, 1)$

   (i) $(2, 2, 1)$ (j) $(2, 2, 2)$ (k) $(4, 2, 0)$ (l) $(8, 0)$

4. For which values of n are the following sequences degree sequences of some graph? Justify your answer.

   (a) $(n, n-1, \ldots, 1)$ (b) $(n, n, n-1, n-2, \ldots, 1)$

   (c) $(n, n-1, n-2, \ldots, 0)$ (d) $(n, n, n-1, \ldots, 0)$

   (e) $(n, n, n-1, n-1, \ldots, 0, 0)$ (f) $(k, k, \ldots, k)$

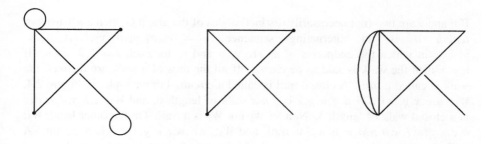

Figure 2.2   Simple and nonsimple graphs.

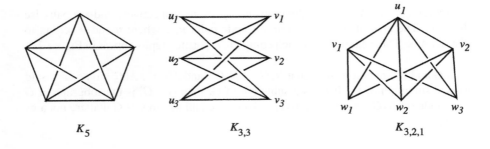

Figure 2.3   Three simple graphs.

5. For which values of n are the following sequences degree sequences of some simple graph? Justify your answer.

(a) $(n, n-1, \ldots, 1)$ (b) $(n, n, n-1, n-2, \ldots, 1)$

(c) $(n, n-1, n-2, \ldots, 0)$ (d) $(n, n, n-1, \ldots, 0)$

(e) $(n, n, n-1, n-1, \ldots, 0, 0)$ (f) $(k, k, \ldots, k)$

6. Prove that in any graph G the number of nodes with odd degree is even.

7. Compute the number of arcs of $K_{n_1, n_2, \ldots, n_k}$.

8. Prove that if $a_1 \geq a_2 \geq \cdots \geq a_p \geq 0$ are all integers, then $(a_1, a_2, \ldots, a_p)$ is the degree sequence of some graph if and only if $\sum_{i=1}^{p} a_i$ is even.

9*. Prove that if $a_1 \geq a_2 \geq \cdots \geq a_p \geq 0$ are all integers, then $(a_1, a_2, \ldots, a_p)$ is the degree sequence of some loopless graph if and only if $\sum_{i=1}^{p} a_i$ is even and $a_1 \leq a_2 + a_3 + \cdots + a_p$.

10**. Prove that if $a_1 \geq a_2 \geq \cdots \geq a_p \geq 0$ are all integers, then $(a_1, a_2, \ldots, a_p)$ is the degree sequence of some graph if and only if $\sum_{i=1}^{p} a_i$ is even and $\sum_{i=1}^{k} a_i \leq k(k-1) + \sum_{i=k+1}^{p} \min\{k, a_i\}$ for $1 \leq k \leq p-1$.

11*. Prove that every simple graph can be placed in $\mathbb{R}^3$ so that its arcs are in fact straight line segments. (Hint: If $N(G) = \{v_1, v_2, \ldots, v_p\}$, place v_k at the point (k, k^2, k^3).)

2.2 Traversability

If u and v are two (not necessarily distinct) nodes of the graph G, then a *u-v walk* of *length* n is an alternating sequence $u = v_0, a_1, v_1, a_2, v_2, \ldots, v_{n-1}, a_n$, $v_n = v$ in which the endpoints of a_k are v_{k-1} and v_k for each $k = 1, 2, \ldots, n$. If $u = v$ then the walk is said to be *closed*. If all the arcs of a walk are distinct, the walk is called a *trail*. A closed trail is called a *circuit*. For example, in Figure 2.4, W_1: $u, c, x, g, w, i, w, b, u, a, w, g, x$ is a u-x walk of length 6, and W_2: v, h, v, d, w, d, v is a closed walk of length 3. Neither W_1 nor W_2 is a trail. On the other hand, W_3: $u, c, x, g, w, i, w, b, u, a, w$ is a u-w trail, and W_4: $v, d, w, i, w, g, x, e, v$ is a circuit. A trail $v_0, a_1, v_1, a_2, v_2, \ldots, v_{n-1}, a_n, v_n$ all of whose nodes are distinct is said to be a *path*. A circuit $v_0, a_1, v_1, a_2, v_2, \ldots, v_{n-1}, a_n, v_0$ in which $v_0, v_1, \ldots, v_{n-1}$ are all distinct is called a *cycle*. The above trail W_3 is not a path, nor is W_4 a cycle. However,

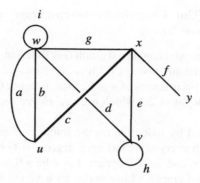

Figure 2.4   A graph with labeled arcs and nodes.

u, c, x, g, w, d, v is a path, and u, c, x, g, w, b, u is a cycle. When no ambiguities can arise, walks will be abbreviated by listing only their vertices. Thus, the short form of the walk W_5: $y, f, x, e, v, d, w, i, w, g, x, c, u$ of Figure 2.4 is y, x, v, w, w, x, u.

A graph in which every two nodes can be joined by a trail is said to be *connected*. The maximal connected subgraphs of a graph are its *components*. A circuit that contains all the arcs of the graph G is an *Eulerian circuit* of G. A graph that possesses an Eulerian circuit is said to be an *Eulerian graph*. Thus, K_3 and K_5 are clearly Eulerian, whereas trial and error shows that K_4 is not. It turns out that Eulerian graphs are easily recognized by the fact that all their nodes have even degrees. This characterization of Eulerian graphs is one of the earliest theorems of graph theory. Its statement is preceded by a lemma.

Lemma 2.2.1 *Suppose G is a graph all of whose nodes have positive even degree. Then there is a sequence $C_1, C_2, \ldots, C_m$ of cycles of G such that every arc of G is in exactly one of these cycles.*

PROOF: We proceed by induction on the number of arcs of G. The lemma clearly holds if G has one arc. Let n be a fixed integer, let G have n arcs, and suppose the lemma holds for all graphs that satisfy the evenness hypothesis and have less than n arcs. Let P: $v_0, a_1, v_1, a_2, v_2, \ldots, a_k, v_k$ be a path of G that has maximum length. Since $\deg_G v_k$ is at least 2, there is another arc a at v_k that is distinct from a_k. It follows from the maximality of P that this arc joins v_k to v_i for some $i = 1, 2, \ldots, k$, thus creating a cycle C: $v_i, a_{i+1}, v_{i+1}, \ldots, a_k, v_k, a, v_i$. Now let H denote the graph obtained from G by deleting all the arcs of C as well as any resulting isolated nodes. Note that

$$\deg_H v = \begin{cases} \deg_G v - 2 & \text{for } v = v_i, v_{i+1}, \ldots, v_k, \\ \deg_G v & \text{otherwise.} \end{cases}$$

By the induction hypothesis, the arc set of H can be decomposed into cycles $C_1, C_2, \ldots, C_m$, and it follows that $C_1, C_2, \ldots, C_m, C$ constitutes the desired partition of the arcs of G.   Q.E.D.

Theorem 2.2.2 (Euler 1736) *A nontrivial connected graph is Eulerian if and only if each of its nodes has even degree.*

PROOF: Suppose the nontrivial connected graph G is Eulerian, so that it possesses an Eulerian circuit C. Let v be any node of G. If $\ldots a_{i_1}, v, a_{i_2}, \ldots, a_{i_3}, v, a_{i_4}, \ldots, a_{i_{2n-1}}, v,$ $a_{i_{2n}}, \ldots$ constitute all the occurrences of v in C, then, because C contains each of the arcs of G exactly once, $\deg_G v = 2n$, which is even. Hence each node of G has even degree.

The converse is proved by induction on the number of arcs in G. Let G be a connected graph in which every node has even degree. If G has only one arc, then it consists of a single loop and so is Eulerian. Let n be a fixed positive integer, and suppose that all connected graphs whose nodes all have even degrees are Eulerian whenever they have fewer than n arcs. Suppose next that G is such a graph with n arcs. By the lemma the arcs of G can be partitioned into a set of disjoint cycles $C_1, C_2, \ldots, C_m$. If H is the graph obtained from G by deleting the arcs in C_m, then H is a possibly disconnected graph with components, say, $H_1, H_2, \ldots, H_c$ (some of these components may be trivial). For all the nodes v of G

$$\deg_H v = \begin{cases} \deg_G v - 2 & \text{if } v \text{ is on } C_m, \\ \deg_G v & \text{otherwise.} \end{cases}$$

It follows that the nodes of H all have even degrees. By the induction hypothesis each of the nontrivial components H_i has an Eulerian circuit. These Eulerian circuits and the trivial components can be combined by means of C_m into an Eulerian circuit of G. Hence, G is Eulerian.   Q.E.D.

The resolution of the problem of the bridges of Koenigsberg of Chapter 1 is now an easy matter (see Exercise 8). The simplicity and efficacy of Euler's theorem may mislead the beginner into the expectation that all graph-theoretical problems lend themselves to such easy resolutions. To counter this misconception, a superficially similar question is posed whose satisfactory resolution is still wanting, and not for lack of trying.

A *Hamiltonian cycle* of the graph G is a cycle that contains all the nodes of G. A graph that possesses a Hamiltonian cycle is said to be a *Hamiltonian graph*. It is clear that K_n is Hamiltonian for all $n \geq 3$ whereas K_2 is not. Other interesting Hamiltonian and non-Hamiltonian graphs appear in the exercises. Hamiltonianness is not a topological property of graphs. Figure 2.5 displays two homeomorphic graphs only one of which is Hamiltonian. Given a specific graph, it is of course possible in principle to check on all of the permutations of its nodes to see whether one of them yields a Hamiltonian cycle or not. In practice, however, this method is impractical as it is much too time-consuming. While better and more sophisticated algorithms have been found, they too are essentially impractical. Efficient methods for deciding on the Hamiltonianness of a graph would also help resolve many other problems in both pure and applied mathematics and have been the subject of much research.

A characterization of Hamiltonian graphs that is analogous to that of Eulerian graphs is as yet unavailable and probably nonexistent. Instead, a sufficient condition for a graph to be Hamiltonian is offered. If u and v are nodes of a graph G then $G + uv$ denotes the graph obtained by adding to G an arc that joins u and v (but is otherwise disjoint from G).

Proposition 2.2.3 (O. Ore 1960) *If G is a simple graph with at least three nodes such that for all nonadjacent nodes u and v,*

$$\deg u + \deg v \geq p,$$

then G is Hamiltonian.

PROOF: We proceed by contradiction. Suppose the theorem is false. Of all the simple non-Hamiltonian graphs with p nodes that satisfy the hypothesis of the theorem, let G be one with the maximum number of arcs. Since $p \geq 3$ and every complete graph K_p with $p \geq 3$ is Hamiltonian, it follows that G is not complete. Let u and v be some two nonadjacent nodes of G. The aforementioned maximality of G implies that $G + uv$ contains a Hamiltonian cycle C, which, since G is not Hamiltonian, contains the new arc joining u and v. Thus, G has a u-v path P: $u = v_1, v_2, \ldots, v_p = v$ which contains every node of G.

For every $i = 2, \ldots, p$, either $v_1 \nsim v_i$ or $v_{i-1} \nsim v_p$, since otherwise

$$v_1, v_i, v_{i+1}, \ldots, v_p, v_{i-1}, v_{i-2}, \ldots, v_1$$

would constitute a Hamiltonian cycle of G (see Fig. 2.6). Hence, if $v_{i_1}, v_{i_2}, \ldots, v_{i_d}$ are the nodes of G that are adjacent to $u = v_1$, then $v_{i_1-1}, v_{i_2-1}, \ldots, v_{i_d-1}$ are all not adjacent to $v = v_p$. Hence

$$\deg v \leq p - 1 - d = p - 1 - \deg u,$$

which contradicts the assumption that $\deg u + \deg v \geq p$. It follows that G is Hamiltonian. Q.E.D.

Exercises 2.2

1. Which of the diagrams in Figure 2.7 can be drawn without lifting the pencil, without retracing any lines, and so that the start and finish points are the same?

2. Which of the floor plans of Figure 2.8 can be traversed so that each door is used exactly once and the tour both starts and ends in the yard?

Figure 2.5   A Hamiltonian and a non-Hamiltonian graph.

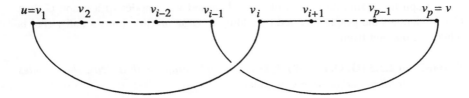

Figure 2.6 A would-be Hamiltonian cycle of G.

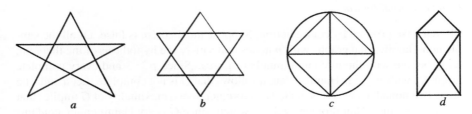

Figure 2.7 Drawing figures.

3. Which of the following graphs are Eulerian?

    (a) K_7 (b) K_8 (c) $K_{3,4}$ (d) $K_{4,4}$

    (e) $K_{3,5}$ (f) $K_{4,5}$ (g) $K_{1,2,2}$ (h) $K_{1,2,3}$

    (i) $K_{2,2,2}$ (j) $K_{1,2,3,4}$ (k) $K_{2,2,2,2}$ (l) $K_{3,3,3,3}$

4. For which values of n are K_n Eulerian?

5. For which values of n_1, n_2 is K_{n_1,n_2} Eulerian?

6. For which values of n_1, n_2, n_3 is K_{n_1,n_2,n_3} Eulerian?

7*. For which values of $n_1, n_2, \ldots, n_m$ is $K_{n_1,n_2,\ldots,n_m}$ Eulerian?

8. Solve the Koenigsberg bridges problem of Chapter 1.

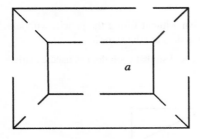
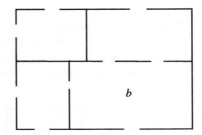

Figure 2.8 Floor plans.

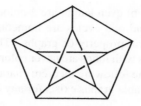

Figure 2.9   The Petersen graph.

9. A graph is said to be *traversable* if it has a walk that contains each edge exactly once. Prove that a connected graph is traversable if and only if it has either two or no nodes of odd degree.

10. Which of the diagrams in Figure 2.7 can be drawn without lifting the pencil and without retracing any lines?

11. Prove that if G is a graph with $p \geq 3$ nodes such that $\deg v \geq p/2$ for each node v, then G is Hamiltonian.

12. Which of the following graphs are Hamiltonian?

    (a) K_7 (b) K_8 (c) $K_{3,4}$ (d) $K_{4,4}$

    (e) $K_{3,5}$ (f) $K_{4,5}$ (g) $K_{1,2,2}$ (h) $K_{1,2,3}$

    (i) $K_{2,2,2}$ (j) $K_{1,2,3,4}$ (k) $K_{2,2,2,2}$ (l) $K_{3,3,3,3}$

13. For which values of n_1, n_2 is K_{n_1,n_2} Hamiltonian?

14. For which values of n_1, n_2, n_3 is K_{n_1,n_2,n_3} Hamiltonian?

15*. For which values of $n_1, n_2, \ldots, n_m$ is $K_{n_1,n_2,\ldots,n_m}$ Hamiltonian?

16. Prove that the Petersen graph of Figure 2.9 is not Hamiltonian.

17**. The graph P_n has node set $\{u_1, u_2, \ldots, u_n, v_1, v_2, \ldots, v_n\}$. For each $i = 1, 2, \ldots, n$, u_i is adjacent to u_{i-1}, u_{i+1}, v_i, and v_i is adjacent to v_{i-2}, v_{i+2}, u_i (addition modulo n). For which values of n is the graph P_n Hamiltonian? Prove your answer.

18**. Prove that the Tutte graph of Figure 2.10 is not Hamiltonian.

19. Prove that if a is an arc of the connected graph G, then $G - a$ has either one or two components.

20. The *subdivision graph* $S(G)$ of the graph G is obtained by placing a node of degree 2 on each arc of G. For example, if C_n denotes the cycle of length n, then $S(C_n) = C_{2n}$. Characterize those graphs G for which $S(G)$ is

    (a) Eulerian

    (b) Hamiltonian.

21. Is the graph of Figure 2.11 Hamiltonian? Justify your answer.

2.3   Colorings

A graph G is said to be *k-colorable* if each of its nodes can be labeled with one of the numbers (also called *colors*) $1, 2, \ldots, k$ in such a manner that adjacent nodes receive distinct colors (Fig. 2.12).

It is easy to see that cycles of even length are 2-colorable whereas cycles of odd length are not. Since even and odd cycles are homeomorphic, this means that k-colorability is *not* a topological property of graphs. The complete graph K_n is k-colorable only if $k \geq n$. Graphs with loops are never colorable. A graph is 1-colorable if and only if it has no arcs. The following theorem characterizes 2-colorable graphs. No characterization of k-colorable graphs exists for any $k > 2$.

Theorem 2.3.1 *The graph G is 2-colorable if and only if it has no odd cycles.*

PROOF: Suppose G has been 2-colored and C: $v_1, v_2, \ldots, v_n$ is a cycle of G. It may be assumed without loss of generality that the nodes v_1 and v_2 are colored 1 and 2 respectively. It follows that for each $1 \leq i \leq n$ the node v_i is colored 1 or 2 according as i is odd or even. Since the node v_n is adjacent to v_1, it is necessarily colored 2, and hence n is even. Thus, every cycle of G is even.

Conversely, suppose G contains no odd cycles. It clearly suffices to show that each component of G is 2-colorable, and so it may be assumed that G is connected. Let v be any node of G, and set $V_1 = \{v\}$. Assuming that the node set V_n, $n \geq 1$, has been defined, V_{n+1} is defined to be the set of all those nodes of G that are not in $V_1 \cup V_2 \cup \cdots \cup V_n$ and are adjacent to some nodes of V_n. It follows from this definition that a node of V_n can be adjacent only to nodes of $V_{n-1} \cup V_n \cup V_{n+1}$. It is next demonstrated by contradiction that for each n, nodes of V_n cannot be adjacent to each other. For suppose u and w are adjacent nodes of some V_n. Let P: $v = u_1, u_2, \ldots, u_n = u$ and Q: $v = w_1, w_2, \ldots, w_n = w$ be v-u and v-w paths of length n, whose existence is guaranteed by the definition of V_n. Let $x = u_m = w_m$ be the last node shared by these paths. Then $x = u_m, u_{m+1}, \ldots, u_n = u, w = w_n, w_{n-1}, \ldots, x = w_m$ is an odd cycle (of length $2(n - m) + 1$) of G. This contradicts the hypothesis, and hence for each n the nodes of V_n are adjacent only to nodes of $V_{n-1} \cup V_{n+1}$. A 2-coloring of the nodes of

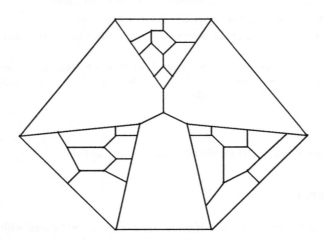

Figure 2.10   The Tutte graph.

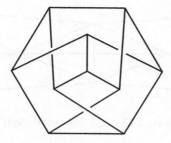

Figure 2.11

G is now obtained by assigning to the nodes of each V_n either 1 or 2 according as n is either odd or even.   Q.E.D.

Given any graph G and any positive integer k, it is, in principle, easy to determine whether or not G is k-colorable. Every k-*coloring* can be viewed as a function that assigns to each node of G a number in $\{1, 2, \ldots, k\}$. Hence one need merely methodically examine all the k^p functions from $V(G)$ to $\{1, 2, \ldots, k\}$ and see whether there is at least one such function which assigns distinct colors to adjacent nodes. In practice, however, this method is so time-consuming, even for fairly small graphs, that it is ineffective. Two theoretical partial answers to this question are now provided. The first is a sufficient condition guaranteeing colorability under certain circumstances, and the second is a necessary condition.

Let k be a positive integer. A graph G is said to be k-*degenerate* if there is a listing $v_1, v_2, \ldots, v_p$ of its nodes such that for each $i = 1, 2, \ldots, p$, the node v_i is adjacent to at most k of the nodes $v_1, v_2, \ldots, v_{i-1}$. For example, the graph of Figure 2.13 is 3-degenerate, even though it has several nodes of degree greater than three. Any listing of the nodes that begins with the four central nodes constitutes a proof of this fact.

Proposition 2.3.2 *Every loopless k-degenerate graph is $(k+1)$-colorable.*

PROOF: Let G be a k-degenerate graph, and $v_1, v_2, \ldots, v_p$ a listing of its nodes such that each node is adjacent to at most k of its predecessors. The nodes can now be colored by an inductive process. Assign the color 1 to v_1. Assume that $v_1, v_2, \ldots, v_i$

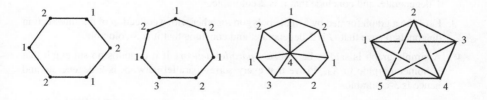

Figure 2.12   Four graph colorings.

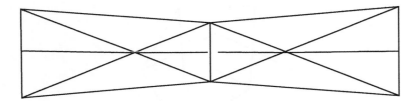

Figure 2.13   A 3-degenerate graph.

have been assigned colors in $\{1, 2, \ldots, k+1\}$ so that adjacent nodes have different colors. Since v_{i+1} is adjacent to at most k of the previous nodes, it can be assigned a color from $\{1, 2, \ldots, k+1\}$ that is distinct from its neighbors' colors. This will eventually result in a $(k+1)$-coloring of G.   Q.E.D.

It follows from this proposition that the graph of Figure 2.13 is 4-colorable. It also follows that if $\Delta(G)$ denotes the *maximum degree* of the graph G, then every loopless graph G is $(\Delta(G) + 1)$-colorable. Necessary conditions for colorability are even harder to come by than sufficient conditions. Only the following obvious criterion can be offered here.

Proposition 2.3.3 *If the graph G contains the complete graph K_{k+1} as a subgraph then G is not k-colorable.* $\square$

Since the graph of Figure 2.13 contains a K_4 in its center, it is not 3-colorable.

Exercises 2.3

1. For which values of k are the following graphs k-colorable? Justify your answers.

    (a) K_7 (b) $K_{3,4}$ (c) $K_{m,n}$ (d) $K_{m,n,r}$ (e) $K_{n_1,n_2,\ldots,n_m}$

    (f) The Petersen graph of Figure 2.9.

    (g) The Tutte graph of Figure 2.10.

    (h) The 3-degenerate graph of Figure 2.13.

2*. A graph is said to be *acyclic* if it contains no cycles. Prove that every acyclic graph is 1-degenerate, and conclude that it is 2-colorable.

3. Let G be a graph formed by a plane polygon together with a collection of nonintersecting diagonals. Prove that G is 2-degenerate, and conclude that it is 3-colorable.

4*. A simple graph is said to be a *series–parallel network* if it contains no subgraph that is homeomorphic to K_4. Prove that every series–parallel network is 2-degenerate and hence is 3-colorable.

5. Prove that if the loopless graph G has less than $k(k+1)/2$ arcs, then G is k-colorable.

6. The wheel-like graph W_n is formed by joining every node of the cycle C_n to a new node w. For each positive integer n determine those k's for which W_n is k-colorable.

Two plane graphs.   Two (planar) nonplane graphs. K_5: a nonplanar graph.

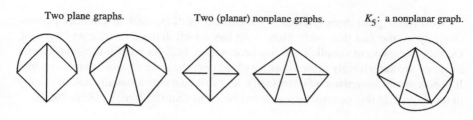

Figure 2.14   Plane, planar, and nonplanar graphs.

7**. Prove that if G is neither a complete graph nor an odd cycle, then G is $\Delta(G)$-colorable.

8. Prove that every graph is homeomorphic to a 2-colorable graph.

9. Prove that the graph G is 2-colorable if and only if it is a subgraph of some complete bipartite graph.

2.4 Planarity

A *plane graph* is one that is drawn in the plane in its entirety, without any spurious intersections. In other words, any two of the arcs in the drawing intersect only in their common endpoints. A *planar graph* is a graph that can be drawn as a plane graph. A *nonplanar graph* is one that cannot be drawn in the plane (Fig. 2.14). The graph K_5 is in a sense the smallest nonplanar graph (Exercise 11).

This section deals with the characterization of planar graphs. The discussion begins with a closer examination of plane graphs. The arcs of a plane graph divide the ambient plane into two-dimensional pieces called *regions*. Note that each plane graph has one region of infinite extent, which is called the *exterior region*. The two plane graphs of Figure 2.14 have four and five regions respectively. The set of arcs of G that delimit a region is called its *perimeter*. In the two plane graphs of Figure 2.14 every arc is on the perimeter of two regions. That need not always be the case. Both the arcs a and b in Figure 2.15 are on the perimeter of a single region each. When the perimeters of two distinct regions share an arc, they are said to *abut* along that arc. The regions R and R' in Figure 2.15 are said to *self-abut* along a and b respectively.

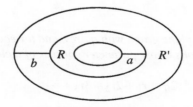

Figure 2.15   Two self-abutting regions.

Most topological investigations of the plane as well as much of Euclidean geometry rely on the fact that every plane loop has a well-defined interior and exterior. Despite the apparent simplicity of this proposition, both its precise formulation and its proof are surprisingly difficult and evolved over the span of more than a century. It is now stated somewhat more formally, and will be used repeatedly and implicitly in the sequel. In this general form it is attributed to Camille Jordan (1838–1922).

Theorem 2.4.1 (Jordan Curve Theorem) *Every plane loop divides the plane into two regions such that any arc joining a point of one region to a point of the other must intersect the dividing loop.* $\square$

The following is one of the earliest of all topological theorems. It is as central to topology as the Theorem of Pythagoras is to Euclidean geometry. It was first discovered by Euler as Equation (1) of Chapter 1, and the exact relationship between the two versions is the subject of Exercise 16.

Theorem 2.4.2 (Euler's equation, 1758) *If G is a connected plane graph with p nodes, q arcs, and r regions, then $p - q + r = 2$.*

PROOF: We proceed by induction on p. If $p = 1$, then every arc of G is a loop, and it follows from the Jordan curve theorem that $r = q + 1$. Hence,

$$p - q + r = 1 - q + (q + 1) = 2.$$

If $p > 1$, assume that the equation holds for all connected plane graphs with less than p nodes. Since G is connected, it has an open arc a. Let G' be the plane graph obtained from G by contracting the arc a to a point (see Fig. 2.16). Then G' is a connected plane graph such that $p(G') = p - 1$, $q(G') = q - 1$, and $r(G') = r$, because all the regions of G are passed on to G', albeit one or two with a slightly shorter perimeter. It follows that

$$\begin{aligned} p - q + r &= (p(G') + 1) - (q(G') + 1) + r(G') \\ &= p(G') - q(G') + r(G') = 2. \end{aligned}$$

This completes the induction process.   Q.E.D.

The following lemma and corollaries provide some necessary conditions for planarity.

Lemma 2.4.3 *Let G be a plane graph with q arcs, and for each $k = 1, 2, 3, \ldots$ let r_k denote the number of regions whose perimeter consists of a closed walk of length k. Then*

$$2q = r_1 + 2r_2 + 3r_3 + \cdots.$$

PROOF: Since G is a plane graph, every arc of G has two sides in the drawing of G. The total number of such sides is clearly $2q$. For every fixed positive integer k, the r_k

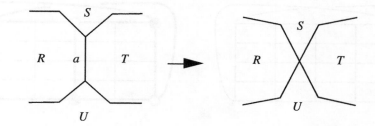

Figure 2.16 The contraction of the arc a.

regions whose perimeters have length k contribute kr_k sides to $2q$. Hence the desired equation.   Q.E.D.

Corollary 2.4.4 *Let G be a connected simple planar graph with p nodes and q arcs. Then:*

1. $q \le 3p - 6$.

2. $q \le 2p - 4$ *if G is 2-colorable.*

PROOF: Suppose G is drawn in the plane and the number of regions whose perimeter has length k is r_k. Since G is simple, $r_1 = r_2 = 0$. Hence, by Lemma 2.4.3,

$$2q = 3r_3 + 4r_4 + 5r_5 + \cdots \ge 3(r_3 + r_4 + r_5 + \cdots) = 3r.$$

By Euler's equation,

$$2 = p - q + r \le p - q + 2q/3 = p - q/3,$$

and the first inequality follows.

If G is 2-colorable, then it contains no 3-cycles, so that $r_3 = 0$. Hence, by Lemma 2.4.3,

$$2q = 4r_4 + 5r_5 + 6r_6 + \cdots \ge 4(r_4 + r_5 + r_6 + \cdots) = 4r.$$

By Euler's equation,

$$2 = p - q + r \le p - q + q/2 = p - q/2,$$

and the second inequality follows immediately.   Q.E.D.

Proposition 2.4.5 *The graphs K_5 and $K_{3,3}$ are not planar.*

PROOF: For K_5, we have $p = 5$ and $q = 10$. Since

$$10 \nleq 3 \cdot 5 - 6$$

it follows from Corollary 2.4.4.1 that K_5 is not planar.

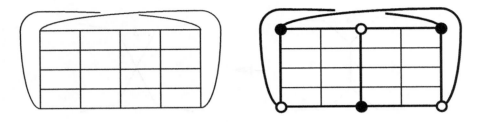

Figure 2.17 A nonplanar graph.

For the 2-colorable $K_{3,3}$, $p = 6$ and $q = 9$. Since

$$9 \not\leq 2 \cdot 6 - 4,$$

it follows from Corollary 2.4.4.2 that $K_{3,3}$ is not planar. Q.E.D.

It turns out that K_5 and $K_{3,3}$ can be used to formulate a succinct description of the difference between planar and nonplanar graphs. The proof of this theorem, while minuscule in comparison with some of the graph-theoretical proofs produced in the last three decades, is still too long to be included in this text. The theorem itself is attributed to Kasimierz Kuratowski (1896–1980), who wrote one of the first books on topology.

Theorem 2.4.6 (Kuratowski 1930) *A graph is nonplanar if and only if it contains a subgraph that is homeomorphic to either K_5 or $K_{3,3}$.* □

Example 2.4.7 The left graph in Figure 2.17 is nonplanar because, as is demonstrated on the right, it contains a subgraph homeomorphic to $K_{3,3}$ (drawn with a bold line).

As a rule of thumb, a nonplanar graph is more likely to contain a copy of $K_{3,3}$ than K_5. Efficient methods for finding such subgraphs are known. One good informal strategy for finding a copy of $K_{3,3}$ in a graph is to locate a long cycle six of whose nodes can be connected by either arcs or paths in the pattern indicated by Figure 2.17.

In 1852 Francis Guthrie, a graduate student at University College, London, asked whether every plane map could be colored with four colors so that adjacent countries received different colors. A *plane map* differs from a plane graph only in that attention is focused on the regions, which are called *countries*. It is implicit in Guthrie's question that each country consists of a single contiguous region, that the exterior region is also to be colored, and that two countries are adjacent if and only if they share a border of positive length. This question, which became known as the *Four-Color Problem*, turned out to be very difficult, but, because of the simplicity of its statement, it attracted the attention of many mathematicians, both the professional and amateur. More erroneous proofs were produced in pursuit of its solution than in that of any other mathematical problem. In 1976, Kenneth Appel and Wolfgang Haken used a localized version of Euler's equation, developed by Henri L. Lebesgue (1875–1941) and H. Heesch, to prove that four colors do indeed suffice. Their intricate reasoning was supplemented by over 1200 hours of computerized calculations. Another, somewhat shorter proof was produced recently by Neil Robertson, Paul D. Seymour, and Robin Thomas. No proof that is suitable for inclusion in a textbook has appeared so far. A proof of the 6-colorability of all planar maps is offered next.

Lemma 2.4.8 *Every simple plane graph is 6-colorable.*

PROOF: It follows from Proposition 2.1.1 and Corollary 2.4.4.1 that if G is a simple plane graph, then

$$\sum_{v \in V(G)} \deg v = 2q \le 6p - 12,$$

and hence every simple plane graph contains a node of degree at most 5. Since every subgraph of a plane graph is also plane, it may be concluded by a straightforward induction that every simple plane graph is 5-degenerate. Hence, by Proposition 2.3.2, every simple plane graph is 6-colorable.   Q.E.D.

Proposition 2.4.9 *Every plane map is 6-colorable.*

PROOF: Let M be a plane map. Associate to it a simple graph G_M by selecting a node inside each country, two distinct new nodes being joined by a new arc if and only if their countries are adjacent (in the example of Fig. 2.18 the arcs of G_M are dashed). The graph G_M is, by its construction, simple and plane. It follows from Lemma 2.4.8 that G_M is 6-colorable, and any such 6-coloring can clearly be construed as a 6-coloring of the map M.   Q.E.D.

A plane map is said to be Eulerian if the graph that consists of the map's perimeters is Eulerian. The following proposition will prove useful in Section 5.6. Its converse is also true (Exercise 10).

Proposition 2.4.10 *Every Eulerian plane map is 2-colorable.*

PROOF: Let M be an Eulerian map, so that the graph G that consists of the perimeters of M is an Eulerian graph. It follows from Lemma 2.2.1 that there exist arc-disjoint cycles $C_1, C_2, \ldots, C_m$ of G that contain all the arcs of G. For each country R of M (i.e., each region of G) let n_R denote the number of these cycles that contain R in their interior. Assign to each country the color 1 or 2 according as n_R is odd or even.

Suppose the two countries R_1 and R_2 abut along an arc a of G, and suppose further that a is contained in the (unique) cycle C_{i_0}. With the single exception of C_{i_0}, every one of the cycles $C_1, C_2, \ldots, C_m$ contains either both R_1 and R_2 in its interior or else neither of them. The cycle C_{i_0}, on the other hand, contains exactly one of R_1 and R_2 in its interior. Hence

$$n_{R_1} - n_{R_2} = \pm 1,$$

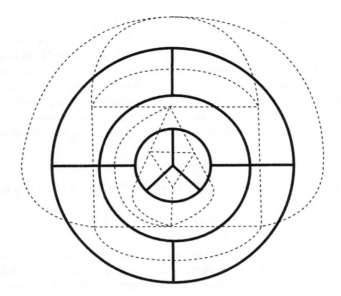

Figure 2.18   Associating a map to a plane graph.

so that n_{R_1} and n_{R_2} have different parities. Consequently R_1 and R_2 have been assigned different colors.   Q.E.D.

Exercises 2.4

1. Which of the graphs of Figure 2.19 are planar and which are nonplanar? Justify your answers.

2. Which of the graphs of Figure 2.20 are planar and which are nonplanar? Justify your answers.

3. Prove that the regions of a Hamiltonian plane graph form a 4-colorable plane map.

4. A plane graph G is formed by drawing a polygon in the plane and adding a set of nonintersecting diagonals in its interior. Prove that both G and the map formed by its regions are 3-colorable.

5. Prove that the Tutte graph is 3-colorable and the map formed by its regions is 4-colorable. Is the map 3-colorable?

6. Find a 2-coloring of the maps in Figure 2.21.

7. For which values of n_1, n_2 is K_{n_1,n_2} planar? Justify your answer.

8. For which values of n_1, n_2, n_3 is K_{n_1,n_2,n_3} planar? Justify your answer.

9. For which values of n_1, n_2, n_3, n_4 is K_{n_1,n_2,n_3,n_4} planar? Justify your answer.

10. Prove that if a plane map is 2-colorable then it is Eulerian.

11. Prove that K_5 is the only simple nonplanar graph with five or fewer nodes.

12. Prove that the plane map formed by a finite number of infinitely extended straight lines is 2-colorable.

13. Is it true that every loopless plane graph is 6-colorable?

14. Prove that the plane map formed by a finite number of circles is 2-colorable.

15. Formulate and prove a generalization of Theorem 2.4.2 that holds for disconnected graphs.

16. Use Theorem 2.4.2 to prove that Equation (1) of Chapter 1 holds for solids.

2.5   Graph Homeomorphisms

This section addresses the following natural question:

How does one determine whether or not two graphs are homeomorphic?

We begin with the description of some necessary conditions for two graphs to be homeomorphic. It is clear that connectedness, number of components, and planarity are all topologically distinguishing characteristics. A connected graph and a disconnected graph cannot be homeomorphic to each other, and neither can a planar and a nonplanar graph be homeomorphic. The degree sequence can also serve the

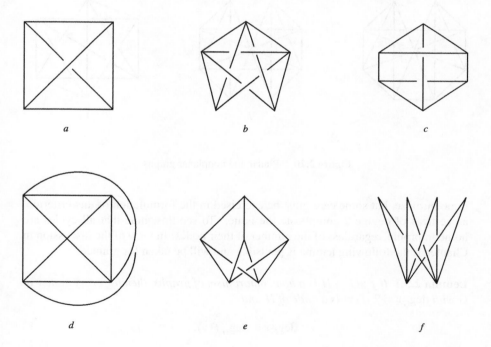

a   *b*   *c*

d   *e*   *f*

Figure 2.19

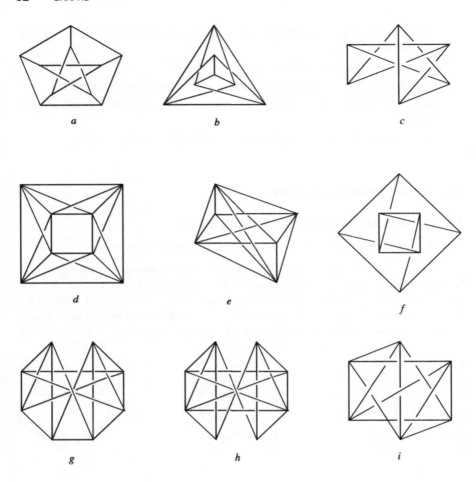

Figure 2.20 Planar and nonplanar graphs.

same purpose, but some care must be exercised in the formulation of this criterion, as vertices of degree 2 complicate the issue. To see this, note that all cycles are homeomorphic, regardless of the number of their nodes. In view of the discussion in Chapter 1, the following lemma is plausible and will be taken for granted.

Lemma 2.5.1 *If* $f : G \rightarrow H$ *is a homeomorphism of graphs, then for every node* v *of* G *with* $\deg_G v \neq 2$, $f(v)$ *is a node of* H *and*

$$\deg_G v = \deg_H f(v). \qquad \square$$

The *reduced degree sequence* of the graph G consists of those terms of its degree sequence that differ from 2. Thus, if three graphs have degree sequences (4, 3, 3, 2,

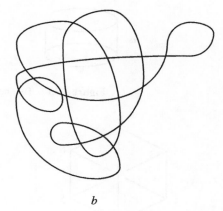

a

b

Figure 2.21   2-Colorable maps.

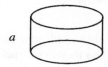

a

b

Figure 2.22   Are these graphs homeomorphic?

2, 2, 1, 1, 0), (4, 3, 3, 2, 1, 1, 0), and (4, 3, 3, 1, 1, 0) respectively, then they all have (4, 3, 3, 1, 1, 0) as their reduced degree sequence. The reduced degree sequence of every cycle is empty. Lemma 2.5.1 yields the following proposition.

Proposition 2.5.2 *Homeomorphic graphs have identical reduced degree sequences.*
□

Since both degree sequences and reduced degree sequences are easily computed, this proposition is very useful in proving the nonhomeomorphism of graphs. Two other techniques for demonstrating topological differences are illustrated in the next example.

Example 2.5.3 The two graphs of Figure 2.22 are both connected and planar and have the same reduced degree sequences. Nevertheless, they are not homeomorphic. One difference is that *a* has two distinct arcs joining the same two vertices, which is not the case for *b*. Alternatively, if one of the nodes of each graph is deleted, the remainders pictured in Figure 2.23 are clearly not homeomorphic.

One way to prove the homeomorphism of two graphs is to find a labeling of their nodes that makes the matching of points obvious, as has been done for the two graphs in Figure 2.24. The homeomorphism can then be proved by (somewhat laboriously)

a

  b

Figure 2.23   Two nonhomeomorphic graphs.

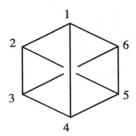

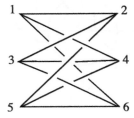

Figure 2.24   Two homeomorphic graphs.

verifying that similarly labeled pairs of vertices are either both adjacent or both non-adjacent in the two graphs. Unfortunately, even when they exist, such corresponding labelings are not easily found, and all that can be done here is to describe one method that works in principle but is not practical. We begin with a plausible proposition which asserts that it is possible to augment homeomorphic graphs to larger homeomorphic graphs by joining corresponding pairs of points with new arcs.

Proposition 2.5.4 *Let $f : G \to H$ be a homeomorphism of graphs. If u and v are two points of G, then $G + uv$ and $H + f(u)f(v)$ are also homeomorphic graphs.*   □

Any node of degree 2 may be deleted from or restored to the node set of a graph G without affecting that graph's homeomorphism type, so long as that node is not the sole node of a component of G. Hence, it is reasonable to define the *topologically reduced form* of the graph G as that graph obtained from G by eliminating as many degree 2 nodes as possible from its node set, short of changing its topological type (see Fig. 2.25 for an example). Note that the degree sequence of a topologically reduced graph is not necessarily reduced. Such graphs can have loops as components, each of which will contribute a 2 to the degree sequence.

An answer to the graph homeomorphism problem can be phrased in terms of matrices that describe the combinatorial structure of graphs. Given a graph G with ordered node set $(v_1, v_2, \ldots, v_p)$, the *associated adjacency matrix* is the square array $A = [a_{i,j}]$ where $a_{i,j}$, the entry in the ith row and jth column, denotes the number of distinct open arcs that join v_i and v_j when $i \neq j$ and the number of loops at v_i when $i = j$ (see Fig. 2.26 and Table 2.1 for an example). It is clear that a graph is completely reconstructible (up to homeomorphism) from any of its adjacency tables. One need merely assign a node v_i to every row r_i of the matrix and connect the nodes v_i and v_j with $a_{i,j}$ parallel arcs. Of course, each permutation of the ordered node

Table 2.1   An adjacency matrix.

1	1	2	1	0	0	0	0
1	0	2	1	0	0	0	0
2	2	0	1	0	0	0	0
1	1	1	0	0	0	0	0
0	0	0	0	0	1	0	0
0	0	0	0	1	0	0	0
0	0	0	0	0	0	1	0
0	0	0	0	0	0	0	0

set of G defines its own adjacency matrix, and so the number of different adjacency matrices of a graph with p nodes is $p!$, which is rather large in general. Still, these matrices do provide an answer to the graph homeomorphism problem.

Theorem 2.5.5 *Two graphs are homeomorphic if and only if their topologically reduced forms share an adjacency matrix.*

PROOF: Suppose two graphs with the respective ordered node sets $(u_1, u_2, \ldots, u_p)$ and $(v_1, v_2, \ldots, v_p)$ have the same adjacency matrices A and B. This means that for

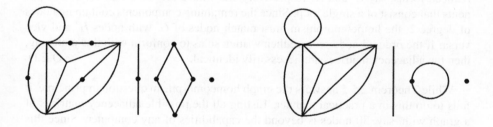

Figure 2.25   A graph and its topologically reduced form.

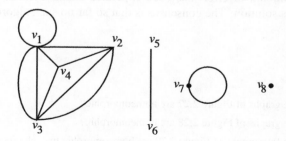

Figure 2.26   A graph.

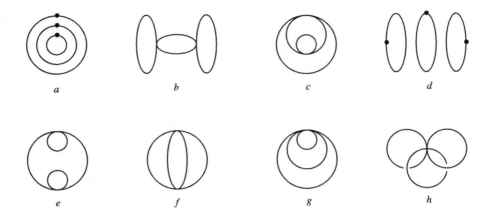

Figure 2.27 Some graphs.

every pair i, j of indices the number of arcs that join u_i with u_j (i.e., $a_{i,j}$) equals the number of arcs that join v_i to v_j (i.e., $b_{i,j}$). Hence the matching of u_i to v_i for each i can be extended, by repeated applications of Proposition 2.5.4, to a homeomorphism of the entire graphs.

Conversely, if two graphs G and H are homeomorphic, then their topologically reduced forms, say G' and H' respectively, must have the same number of components that consist of a single loop. Since the remaining components contain no nodes of degree 2, the homeomorphism must match nodes of G' with nodes H' and vice versa. If the ordered node sets are then written so as to conform with this matching, then the adjacency matrices are necessarily identical.   Q.E.D.

While Theorem 2.5.5 answers the graph homeomorphism question in principle, it fails to do this in a practical manner. Listing all the possible adjacency matrices of a graph with, say, 30 nodes is beyond the capabilities of any computer. Since this homeomorphism problem (better known as the graph isomorphism problem, which is its combinatorial formulation) has many implications for applied mathematics, a considerable amount of effort has been expended on the derivation of efficient algorithms for its solution. The consensus is that so far no such algorithm has been found.

Exercises 2.5

1. Which of the graphs of Figure 2.27 are homeomorphic?
2. Which of the graphs of Figure 2.28 are homeomorphic?
3. Which of the four graphs of Figure 2.29 are homeomorphic to each other?
4. Which of the graphs of Figure 2.30 are homeomorphic to each other?

5. Compute all the six adjacency matrices of the graph of Figure 2.31.

6. Suppose the adjacency matrices of the graph G of the ordered node set $(v_1, v_2, \ldots, v_p)$ is A. Prove that if $A^n = [a_{i,j}^{(n)}]$, then $a_{i,j}^{(n)}$ is the number of distinct walks from v_i to v_j.

Chapter Review Exercises

1. Let G be the graph formed by the edges of the truncated cube of Figure 2.32. Answer the following questions and justify your answers.

   (a) Is G Eulerian?

   (b) Is G Hamiltonian?

   (c) For which k is G k-colorable?

   (d) Is G planar?

2. Let G be the graph formed by the edges of the truncated cube of Figure 2.33. Answer the following questions and justify your answers.

   (a) Is G Eulerian?

   (b) Is G Hamiltonian?

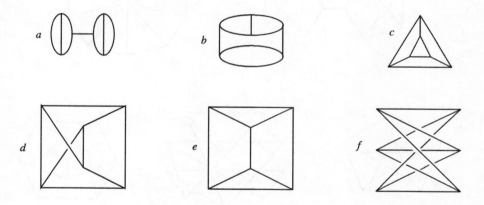

Figure 2.28   Some homeomorphic and nonhomeomorphic graphs.

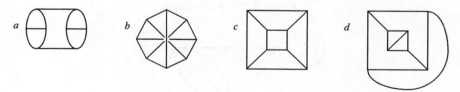

Figure 2.29   Some graphs.

(c) For which k is G k-colorable?

(d) Is G planar?

3. Let G be the graph formed by the edges of the tunneled solid of Figure 2.34. Answer the following questions and justify your answers.

(a) Is G Eulerian?

(b) Is G Hamiltonian?

(c) For which k is G k-colorable?

(d) Is G planar?

4. Are the following statements true or false?

(a) Every graph is k-colorable for some integer k.

(b) There is an integer k such that every graph is k-colorable.

(c) Every complete graph with at least three vertices is Eulerian.

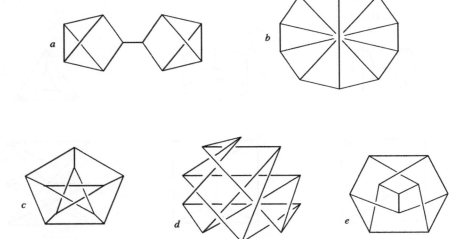

Figure 2.30   Some graphs.

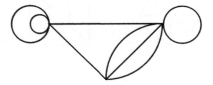

Figure 2.31   A graph.

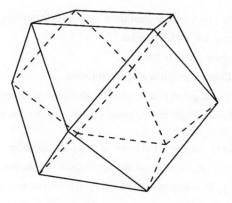

Figure 2.32   A truncated cube.

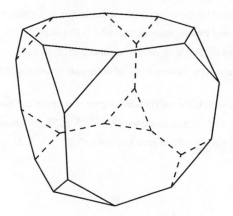

Figure 2.33   A truncated cube.

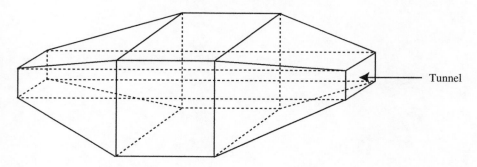

Figure 2.34   A tunneled solid.

(d) Every complete graph with at least three vertices is Hamiltonian.

(e) Every Eulerian graph is Hamiltonian.

(f) Every Hamiltonian graph is Eulerian.

(g) Every planar Eulerian graph is also Hamiltonian.

(h) Every simple planar graph that is both Hamiltonian and Eulerian is also 3-colorable.

(i) If a connected plane graph has the same number of nodes and arcs, then it has exactly two regions.

(j) If a 2-colorable planar graph has twice as many arcs as nodes, then it is Hamiltonian.

(k) The graph formed by the edges of the tetrahedron is homeomorphic to K_4.

(l) The graph $K_{2,2,2}$ is homeomorphic to the graph formed by the edges of the octahedron.

(m) The graph formed by the edges of the dodecahedron is Hamiltonian.

(n) The graph formed by the edges of the icosahedron is 3-colorable.

(o) The graph formed by the edges of the cube is Eulerian.

(p) There is a simple graph with degree sequence $(5, 5, 4, 4, 3)$.

(q) The graph obtained by joining a pair of antipodal vertices in the graph of Figure 2.32 is planar.

(r) Simple graphs with identical reduced degree sequences are homeomorphic.

(s) All the graphs with degree sequence $(5, 5, 5, 5, 5, 5)$ are homeomorphic.

(t) All the simple graphs with degree sequence $(5, 5, 5, 5, 5, 5)$ are homeomorphic.

CHAPTER 3

SURFACES

The two-dimensional topological objects are surfaces. Though the study of the metric properties of surfaces goes back to the ancient Greeks, their topological aspects only came under scrutiny in the eighteen-hundreds. In the early part of that century Niels H. Abel (1802–1829), Carl G. J. Jacobi (1804–1851), and later Georg F. B. Riemann (1826–1866) pointed out that *elliptic integrals*, then at the forefront of mathematical research, as well as more complicated integrals, became more tractable if their independent variables were assumed to be complex rather than real. Moreover, the proper domain of each integrand should consist of a stack of several copies of the complex plane which are so interconnected as to form new mathematical objects, which later became known as *Riemann surfaces*. The topological classification of these surfaces turned out to be crucial for the proper understanding of the aforementioned integrals. While some earlier work of Euler, Carl F. Gauss (1777–1855) and others could also be viewed as topological in nature, it was these concerns with analytic issues that brought topology into the mainstream of mathematics.

Introduction to Topology and Geometry, Second Edition.
By Saul Stahl and Catherine Stenson Copyright © 2013 John Wiley & Sons, Inc.

Figure 3.1 Disks.

3.1 Polygonal Presentations

The prototypical surface is the Euclidean plane. Today it is customary to understand by this a flat surface that extends infinitely far in all directions. Euclid, however, thought of it as a flat surface of finite, though arbitrarily large, extent. For our purposes here it is convenient to adopt this earlier point of view. A *disk* is a plane circle with its interior, or any topological space homeomorphic to it, such as a triangle (with its interior) or indeed any polygon whatsoever. It should be stressed that the polygon's perimeter must be devoid of self intersections, i.e., it is a loop. The context being topology, curved isotopes of polygons are also disks, and the requirement that the edges be straight line segments is relaxed to the stipulation that they can be any open arcs (Fig. 3.1).

The Cartesian graph of a continuous single-valued function $z = f(x,y)$, $a \leq x \leq b$, $c \leq y \leq d$, is also a disk, because it is homeomorphic to a rectangle. This homeomorphism is established by matching every point (x,y) in the given rectangular domain with the point $(x,y,f(x,y))$ on the surface.

The *sphere*, i.e., surface of a ball, is an example of a surface that is not homeomorphic to the disk. The latter has a rim, or *border*, which is lacking in the sphere. Surfaces, such as the sphere, that are completely lacking in such borders are said to be *closed*. Now, while the sphere is homeomorphic to neither the plane nor the disk, it can be cut into two hemispheres each of which is homeomorphic to a disk. This decomposability into polygons has been chosen to be the defining characteristic of all surfaces. The open-ended cylinder of Figure 3.2 can be "unrolled" into the plane if it is first cut lengthwise along the arc a. The direction of the cut is recorded by means of an arrowhead just in case the original surface needs to be reconstructed from its flattened version (this will come up later). The sphere can also be flattened

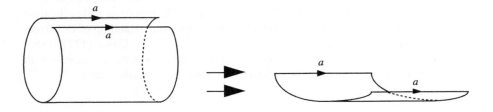

Figure 3.2 Flattening the cylinder.

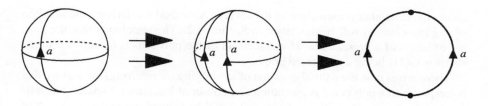

Figure 3.3   Flattening the sphere.

into a single polygon by means of a cut (see Fig. 3.3). The *torus*, i.e., the surface of a doughnut, is transformed in Figure 3.4 into a rectangle by means of two cuts a and b or else into two rectangles (Fig. 3.5) by means of the four cuts a, b, c, d. The *double torus* of Figure 3.6 can be transformed into two octagons by means of the seven cuts a, b, c, d, e, f, g.

Note that in all of the foregoing "flattenings" each cut leaves a trace consisting of two directed edges on the perimeters of the resultant polygons. Thus, if the unit square $\{(x,y) \mid 0 \le x \le 1, 0 \le y \le 1\}$ is cut along the vertical straight line segment joining (0.5, 0) to (0.5, 1) then the result consists of the two rectangles $\{(x,y) \mid 0 \le x \le 0.5, 0 \le y \le 1\}$ and $\{(x,y) \mid 0.5 \le x \le 1, 0 \le y \le 1\}$.

The definition of closed surfaces reverses the above flattening process. A *pasting* is the reversal of a cut. The pasted arcs will always be directed, and the pasting must be consistent with these directions. In other words, if the directed edge AB is to be pasted to the directed edge CD, then A and B are pasted to C and D respectively, with the rest of the points of the edges following in a continuous manner. This (directed) pasting operation can also be visualized by means of an imaginary zipper that joins the two edges.

A *polygonal presentation* is a set of plane polygons together with a pairing of their directed edges, such that the pasting of paired edges connects all the polygons. The edges of these polygons are labeled so that paired edges carry the same letter (see Figs. 3.7 and 3.8 for examples). The topological space obtained by performing the indicated pastings is a *closed surface*. (Wherever possible the term *closed surface* will be abbreviated to *surface*.) The set of labeled polygons obtained by the flattening of surfaces depicted in Figures 3.3–3.6 is a polygonal presentation of the given surface. The open-ended cylinder of Figure 3.1 is not a closed surface. Rather, it belongs to the class of bordered surfaces that will be discussed in Section 3.4.

It is this chapter's goal to describe the topological classification of surfaces. By this is meant that the reader will be provided with a procedure for determining when two surfaces are the same (i.e., homeomorphic).

The vertices and (pasted) edges of a polygonal presentation Π constitute a graph called the *skeleton* of Π and denoted by $G(\Pi)$. For example, the graph G_1 of Figure 3.9 is the skeleton of the presentation of the sphere (Fig. 3.3), and G_2 is the skeleton of the presentation of the torus in Figure 3.4. The graph G_3 of the same figure is the skeleton of the rightmost polygon of Figure 3.7, which, as will be demonstrated momentarily, is another presentation of the sphere. A method for deriving the skeleton of any presentation will be described in Section 3.2. The stipulation that the pairing of labels of a presentation Π connects all of the polygons is tantamount to the skeleton $G(\Pi)$ being a connected graph.

There arises now the natural question of identifying (or realizing) the surface that corresponds to a polygonal presentation as a subset of Euclidean 3-space. The leftmost of the three presentations of Figure 3.7 will be recognized as the end result of the flattening of the sphere illustrated above (Fig. 3.3). The middle presentation of Figure 3.7 looks very much like the flattening of the torus depicted in Figure 3.4 except that the labels and directions are different. The different labeling is, of course, immaterial, nor do the differing directions matter. As a subset of Euclidean space every arc is identical to its reverse, and so the reversal of two arrows that carry the same label does not affect the presented surface. Consequently, the surface that corresponds to the middle presentation is indeed the torus. The rightmost presentation of Figure 3.7 also bears some resemblance to the flattening of the torus, but this similarity is misleading. If the union of the edges a and b is replaced by a single arc c, we obtain the presentation if Figure 3.10, which is clearly identical to the leftmost presentation of Figure 3.7. The rightmost presentation therefore also yields a sphere.

But what about the presentation of Figure 3.11, which greatly resembles that of the torus? One could begin reconstructing the surface by first pasting the two edges labeled a so as to obtain a cylinder. This would then be followed by a pasting of the b's, perhaps by keeping the cylinder's left end fixed in place while bending the rest of the cylinder so as to bring the right end flush against the fixed left end (Figure 3.12).

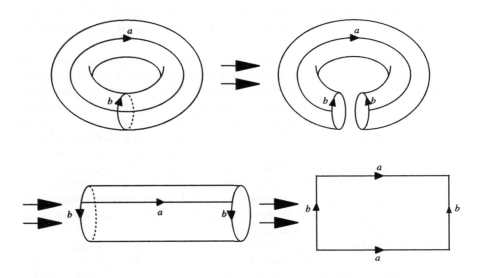

Figure 3.4   Flattening the torus.

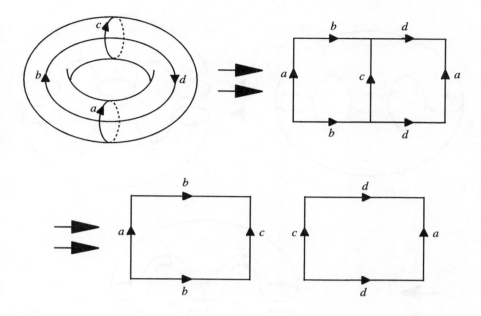

Figure 3.5   Flattening the torus into two polygons.

Unfortunately, as is indicated in the diagram, the directions of the two versions of *b* are such as to make the required pasting impossible. This difficulty is inescapable, and the two *b*'s must wind up facing each other with inconsistent orientations no matter how the cylinder is twisted through space. To see this, note that when the cylinder was straight, each of the *b*'s was endowed with the counterclockwise orientation from the point of view of an observer who is positioned at the geometrical center of the border loop and is looking into the cylinder from the outside. No amount of bending or twisting will alter these relative orientations, and when the two *b*'s are eventually brought together, the observers must stand back to back, so that the orientations of the *b*'s are inconsistent.

The impossibility of carrying out the pasting of Figure 3.12 in 3-space notwithstanding, there is a surface, called the *Klein bottle*, which corresponds to it and which can be realized in four-dimensional Euclidean space. To demonstrate this it is helpful to think of time as the fourth dimension. Begin as before by converting the presentation into a cylinder, fix its left end, and, bending the cylinder, pass the right end *through* the cylinder's surface and bring this end up to the left end through the cylinder's *interior* (Fig. 3.13). Note that the orientations of the two *b*'s are now consistent, so that there is no obstacle to pasting them together. Of course, this pasting involves a "cheat"—the self-crossing *c* of the cylinder, which is not allowed for in the polygonal form. However, this self-crossing can be eliminated if each point on the bent cylinder is endowed with an additional time coordinate above and beyond

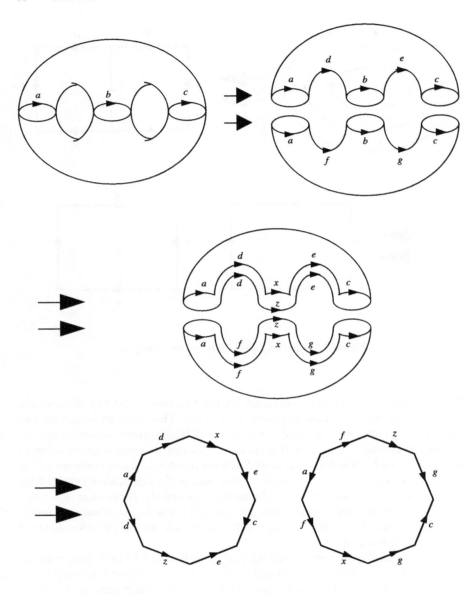

Figure 3.6 Flattening the double torus.

its usual three spatial coordinates, which time coordinate is assumed to be smoothly distributed in conformance with the pattern indicated by the Now, Yesterday, Now, and Tomorrow labels in the diagram. The two parts of the cylinder do not really cross each other in the loop c, because they pass through it at different times.

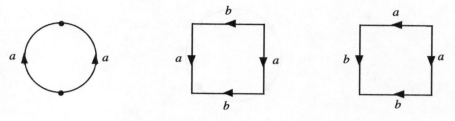

Figure 3.7   Three polygonal presentations.

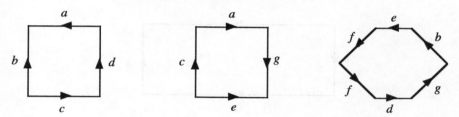

Figure 3.8   A presentation with three polygons.

Four-dimensional Euclidean space can be formally represented by the abstract mathematical object

$$\mathbb{R}^4 = \{(x, y, z, w) \mid x, y, z, w \text{ are real numbers}\}.$$

Nevertheless, the informal space–time continuum point of view is quite useful and will arise again.

Another surface that is unrealizable in $\mathbb{R}^3$ is the *projective plane*, whose presentation appears in Figure 3.14. Just like the Klein bottle, this surface is realizable in $\mathbb{R}^4$. The realization process is depicted in Figure 3.15 and also makes use of an illusory self-crossing. The top left-hand circle describes some points on the "perimeter" of the projective plane which need to be pasted. This disk is first deepened into a bowl, which is further distorted as in the central diagram. Next, a time coordinate is attached to every point on the surface of the distorted bowl. This time coordinate is assumed to vary smoothly with values between -4 and 4. Note that each of the two occurrences of each of the points A, B, C, D, E, F, G, H has been assigned the same

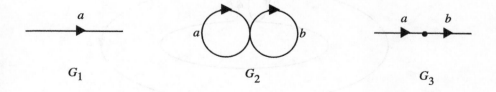

Figure 3.9   Some skeletons.

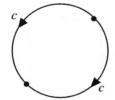

Figure 3.10   A polygonal presentation.

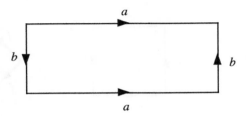

Figure 3.11   A mystery polygonal presentation.

time coordinate: 0 to A, -2 to B, -4 to C, -2 to D, 0 to E, 2 to F, 4 to G, and 2 to H. Consequently the two halves of the distorted lip can be identified as in the bottom diagram, where what looks like a self-crossing is no such thing at all—the points B and H occupy the same location at different times -2 and 2, etc.

There are many other such surfaces that cannot be realized in $\mathbb{R}^3$, and they will be discussed below. These observations call for some new terminology. A topolog-

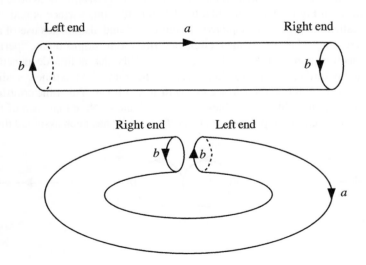

Figure 3.12   A seemingly impossible pasting.

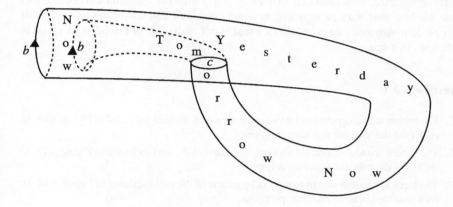

Figure 3.13   Constructing a 4-dimensional realization of the Klein bottle.

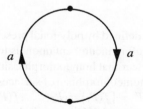

Figure 3.14   The projective plane.

ical space S is said to be *embeddable* or *realizable* in the topological space T if S is homeomorphic to a subspace of T. Thus, a graph is planar if and only if it is embeddable in a disk. Every graph is embeddable in $\mathbb{R}^3$, and the projective plane and the Klein bottle are not embeddable in $\mathbb{R}^3$. They are, however, embeddable in $\mathbb{R}^4$. In fact, every closed surface is embeddable in $\mathbb{R}^4$.

To summarize, it is assumed that *every polygonal presentation corresponds to some surface that may or may not be embeddable in Euclidean 3-space.* It will later be demonstrated that every surface that is not realizable in Euclidean 3-space is realizable in 4-space.

Exercises 3.1

1. The sphere is decomposed in two ways by means of the cuts indicated in Figure 3.16. In each case describe the resulting polygons.

2. The sphere is decomposed in two ways by means of the cuts indicated in Figure 3.17. In each case describe the resulting polygons.

3. The torus is decomposed in two ways by means of the cuts indicated in Figure 3.18. In each case describe the resulting polygons.

4. The torus is decomposed in two ways by means of the cuts indicated in Figure 3.19. In each case describe the resulting polygons.

5. Explain how the surface A of Figure 3.20 can be flattened into one polygon. Describe the resulting polygonal presentation.

6. Explain how the surface B of Figure 3.20 can be flattened into one polygon. Describe the resulting polygonal presentation.

7. Explain how the surface C of Figure 3.20 can be flattened into one polygon. Describe the resulting polygonal presentation.

8. Explain how the surface D of Figure 3.20 can be flattened into one polygon. Describe the resulting polygonal presentation.

3.2   Closed Surfaces

In this section all the surfaces defined by polygonal presentations will be classified up to homeomorphism. It is therefore incumbent upon us to clarify the notion of *homeomorphism of surfaces*. It is clear that homeomorphic surfaces have identical presentations. For, if $f : S \to S'$ is a homeomorphism of surfaces, and if $\Pi = \{P_1, P_2, \ldots, P_n\}$ is a presentation of S, then $\Pi' = \{f(P_1), f(P_2), \ldots, f(P_n)\}$ is an identical presentation of S', assuming that the arcs of $S(\Pi')$ inherit their labels from Π. It therefore makes sense to define two surfaces to be *homeomorphic* provided they have identical presentations.

The classification project begins with an examination of the vertices of the constituent polygons. As is indicated by the surfaces of Figures 3.3–3.6 and their polygonal presentations, different vertices of the constituent polygons may in fact designate *occurrences* of the same node of the skeleton $G(\Pi)$—in other words, they may

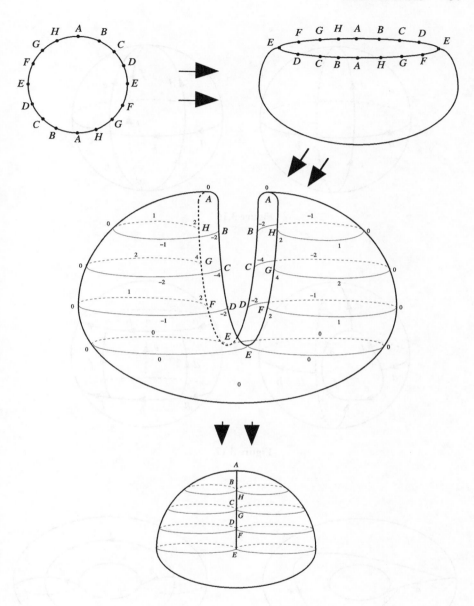

Figure 3.15   A realization of the projective plane in $\mathbb{R}^4$.

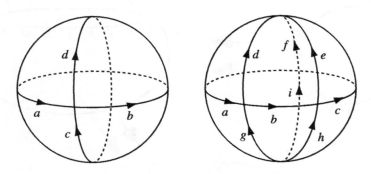

Figure 3.16

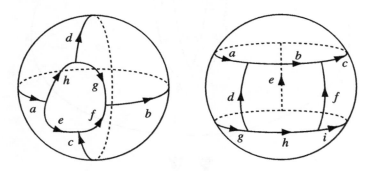

Figure 3.17

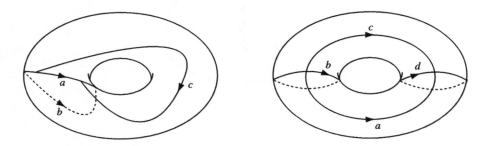

Figure 3.18

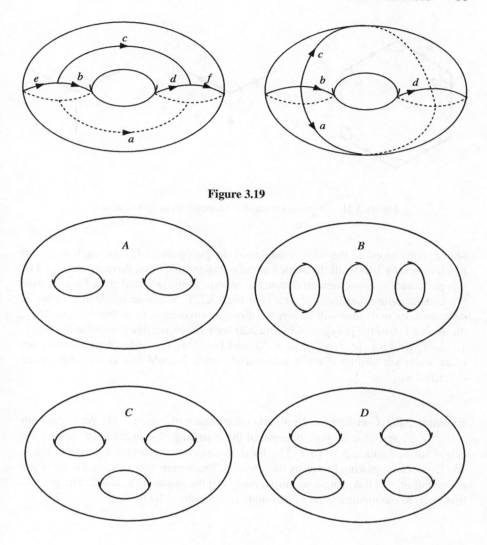

Figure 3.19

Figure 3.20

correspond to the same point on the surface that is being presented. If the vertex of a polygon is an occurrence of a node, then the vertex is also said to *belong* to the node. Two vertices that belong to the same node are said to be *equivalent*.

It will be necessary to ascertain the actual number of distinct nodes of $G(\Pi)$ that appear on the perimeters of the polygons of Π, and a method for deriving this number can be based on the examination of the surface in the vicinity of its nodes. If Π is a presentation of the surface S, then the vicinity of any node u of $G(\Pi)$ on S consists of a disk that is divided by the arcs emanating from u into sectors (Fig. 3.21). These

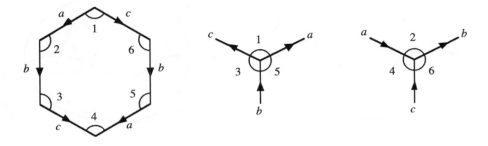

Figure 3.21   A presentation and the vicinities of its two nodes.

sectors correspond to the interior angles of the polygons of Π, and each node can be identified by listing all the angles of the polygons of Π that form its vicinity. For this purpose it is convenient to denote the *reverse* of the oriented arc a by a^{-1}, with the understanding that $(a^{-1})^{-1} = a$ (see Fig. 3.22). When an *angle* at a vertex is described, its two sides will be written down as emanating from that vertex. Thus, the angles 1–6 of the polygon of Figure 3.21 have the respective pairs of sides $\{c,a\}$, $\{a^{-1},b\}$, $\{b^{-1},c\}$, $\{c^{-1},a^{-1}\}$, $\{a,b^{-1}\}$, and $\{b,c^{-1}\}$. The order in which the sides of an angle are written down is immaterial; angle 2 could just as well have been described as $\{b,a^{-1}\}$.

Example 3.2.1 Consider the 1-polygon presentation of Figure 3.21. We begin with $\angle 1 = \{c,a\}$ and observe that, because of the a-pasting, this angle must abut on the actual surface with $\angle 5 = \{a,b^{-1}\}$. Next, because of the b-pasting, comes $\angle 3 = \{b^{-1},c\}$. The c-pasting brings us back to $\angle 1$. This means that there is a node of the presentation, call it A, whose vicinity consists of the angles 1, 3, and 5. The process that leads to the identification of this node is symbolized by the list

$$A: \qquad c,a,b^{-1},c.$$

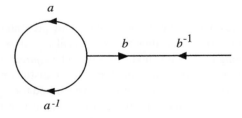

Figure 3.22   Arcs and their inverses.

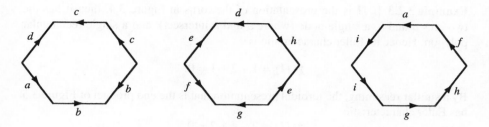

Figure 3.23 A polygonal presentation.

The other three angles constitute the vicinity of another node

$$B: \quad a^{-1}, b, c^{-1}, a^{-1}.$$

Hence the presentation of Figure 3.21 has two nodes.

Example 3.2.2 The polygonal presentation of Figure 3.23 has three nodes u, v, w that are given by the following lists:

$$u: \quad d, a, f^{-1}, g^{-1}, h, d.$$
$$v: \quad c, c^{-1}, d^{-1}, e^{-1}, h^{-1}, f, e, g, i^{-1}, i, a^{-1}, b, c.$$
$$w: \quad b^{-1}, b^{-1}.$$

The skeleton of this presentation is therefore the graph of Figure 3.24.

Presentations are classified by two attributes: the numerical Euler characteristic and the geometric orientability character.

Euler characteristic: Let the presentation Π have r polygons, and suppose $G(\Pi)$ has p nodes and q arcs. Then the quantity $p - q + r$ is called the *Euler characteristic* of Π and is denoted by $\chi(\Pi)$.

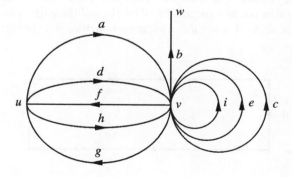

Figure 3.24 A skeleton.

Example 3.2.3 If Π is the presentation of the torus in Figure 3.4, then it has the two arcs a and b, a single node (where a and b intersect), and a single rectangular polygon. Hence its Euler characteristic is

$$\chi(\Pi) = 1 - 2 + 1 = 0.$$

By a similar reasoning, the toroidal presentation that is the end product of Figure 3.5 has Euler characteristic

$$\chi(\Pi) = 2 - 4 + 2 = 0.$$

If Π is the presentation of the Klein bottle given in Figure 3.11, then, since the underlying surface is not given, it is necessary to determine the number of nodes by the method of Example 3.2.2. Accordingly, there is a unique node given by the list

$$b, a, b^{-1}, a^{-1}, b,$$

and so the presentation has Euler characteristic

$$\chi(\Pi) = 1 - 2 + 1 = 0.$$

On the other hand, if Π is the polygonal presentation of the surface of Figure 3.23, then it has three nodes (see Example 3.2.2), nine arcs, and three polygons. It therefore has Euler characteristic

$$\chi(\Pi) = 3 - 9 + 3 = -3.$$

Orientability character: An *orientation* in a single polygon consists of a sense of traversal of its perimeter. Every polygon therefore has two orientations, which, when the polygon is visualized as drawn on a page, are referred to as clockwise and counterclockwise (from the point of view of the reader). In the case of the rectangle of Figure 3.25 these orientations are $a^{-1}bcd$ and $ad^{-1}c^{-1}b^{-1}$ respectively. A directed edge a on the perimeter of a polygon P is said to be *consistent* with an orientation of P provided that its direction agrees with the orientation. Otherwise a is *inconsistent* with the orientation. Equivalently, the directed edge a is *consistent* with an orientation when it appears with exponent 1 (i.e., no exponent) when that orientation is written out as a succession of labels, and the edge is *inconsistent* if it appears with exponent -1 when the orientation is written out. In Figure 3.25, edge

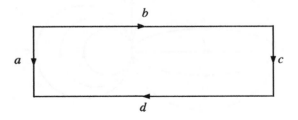

Figure 3.25   A polygon to be oriented.

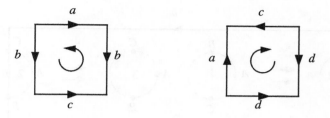

Figure 3.26   A coherently oriented presentation.

a is consistent with the counterclockwise orientation $ad^{-1}c^{-1}b^{-1}$, whereas the other edges are consistent with the clockwise orientation $a^{-1}bcd$. Orientations are usually described by a circular arc in the center of the oriented polygon (see Fig. 3.26).

An *orientation* of a polygonal presentation Π consists of an assignment of an orientation to each of its polygons. Such an orientation of Π is said to be *coherent* if, of the two occurrences of every arc on the oriented polygon perimeters, one is consistent and the other is inconsistent. The orientation of the polygonal presentation of Figure 3.26 is coherent, since the indicated traversals are $a^{-1}bcb^{-1}$ and $ac^{-1}dd^{-1}$. It is clear that the coherence of an orientation of a presentation is tantamount to every arc label appearing with both a^1 and a^{-1} as exponents. A presentation is said to be *orientable* if it possesses a coherent orientation. The presentations of Figures 3.21 and 3.26 are both orientable. On the other hand, a minimal amount of trial and error demonstrates that the rectangular presentation of the Klein bottle (Fig. 3.11) is not orientable, nor can the partially labeled presentation of Figure 3.27 be completed to a coherent orientation. Such polygonal presentations are said to be *nonorientable*. It is clear that the reversal of a coherent orientation is also coherent.

To explain the significance of orientability we begin by demonstrating some of the undesirable consequences of nonorientability. Imagine that in Figure 3.28 the circle 1 drawn on a presentation of the nonorientable Klein bottle glides to the left so that it crosses the edge a (position 2) and then reappears at the right end. The apparent change in the positions of B and C is dictated by the arrows on the two edges marked a. Then, still moving to the left, bring the circle to position 3, close to its starting point. Note that notwithstanding the fact that the circle has been subjected

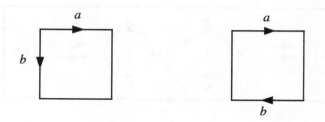

Figure 3.27   A nonorientable presentation.

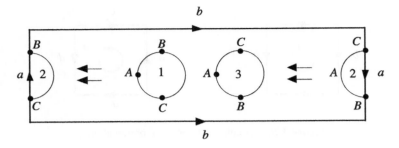

Figure 3.28 Movement on a nonorientable surface.

to a smooth, or *continuous*, sliding motion, its sense has been reversed. The cyclic order of A, B, C is clockwise in position 1 but counterclockwise in position 3. Such a reversal is impossible on the sphere. For suppose the three points A, B, C on a circle on the sphere have a clockwise cyclic order from the point of view of an observer who is stationed at a distance of, say, one unit outside the sphere. No matter where this circle glides, as long as the observer moves with it, he will remain outside the sphere and so A, B, and C will always stay in a clockwise cyclic order from his point of view. Thus, no reversal of cyclic orders can occur on the sphere. The same argument clearly applies to any surface that is placed in $\mathbb{R}^3$ with a clearly defined interior and exterior, such as the torus and the double torus.

This makes it possible to interpret orientability in terms of the surface rather than any of its presentations. A surface is orientable when the ambiguity displayed in Figure 3.28 cannot arise. In other words, it is possible to specify a rotational sense on the surface at each point so that if an oriented circle glides arbitrarily on the surface, the orientation of the circle always agrees with the rotational sense of the surface in its location (see Fig. 3.29). Consequently, it is possible to visualize an oriented surface as one with several arched arrows such that whenever one arrow is slid towards another, their rotational senses agree (see Fig. 3.30). Note that when two such arcs are placed next to each other they must pass through the point of contact in

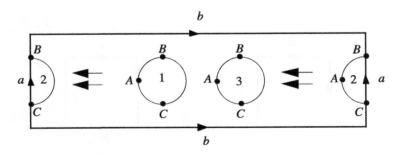

Figure 3.29 Movement on an orientable surface.

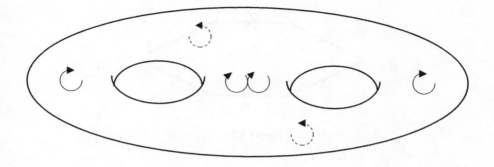

Figure 3.30   An oriented double torus (dashed lines lie on the back side).

opposite directions. That is the analogous concept to the coherence of an orientation of a presentation.

We digress here to discuss some alternative terminology. It is also customary to refer to orientable surfaces as *two-sided* and nonorientable ones as *one-sided*. This terminology can be misleading. Unlike a sheet of paper, a polygon consists of a single layer of points. If one were to write on a polygon with a blue pen, the color and the writing would be equally visible no matter which of the two sides of the polygon was examined. Nevertheless, it is not altogether unreasonable to speak of the sphere and the torus as two-sided in that each clearly has an *interior* and an *exterior*. This *separation* property is a feature of all closed surfaces embedded in $\mathbb{R}^3$. Moreover, it will be demonstrated in Corollary 3.3.3 that all orientable surfaces are embeddable in $\mathbb{R}^3$ with well-defined interiors and exteriors and hence they can all be said to be two-sided. Nonorientable surfaces, on the other hand, can be at best embedded in $\mathbb{R}^4$. There they fail to separate the ambient space, much as any loop fails to separate $\mathbb{R}^3$.

It should be mentioned here that the equivalence of orientability to two-sidedness holds only in the context of embeddings in $\mathbb{R}^3$. In higher dimensions the two notions are independent of each other (see Section 9.3).

The orientability character of a presentation is easily determined by trying to choose orientations of the constituent polygons one at a time in such a manner that each choice is coherent with the previous choices.

Example 3.2.4   Neither the counterclockwise $abca^{-1}b^{-1}c$ nor the clockwise $c^{-1}bac^{-1}b^{-1}a^{-1}$ orientation of the polygon of Figure 3.31 is coherent (c is the troublesome edge). The presentation is therefore nonorientable.

Example 3.2.5   The presentation of Figure 3.32 is orientable, because we can choose the indicated counterclockwise abc and clockwise $c^{-1}a^{-1}b^{-1}$ traversals.

Example 3.2.6   For the presentation of Figure 3.33 we begin by arbitrarily choosing the traversal abc for the first triangle. The requirement of coherence (applied to both

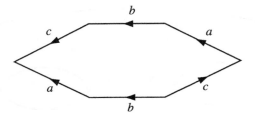

Figure 3.31

b and c) forces the choice of the traversal $d^{-1}b^{-1}c^{-1}$ for the second triangle. Once again, coherence requires the choice of $a^{-1}d$ for the third polygon. The presentation has thus been coherently oriented and is therefore orientable.

Example 3.2.7 For the presentation of Figure 3.34 we again begin by arbitrarily choosing the traversal abc for the first triangle. The requirement of coherence (applied to both b and c) forces the choice of the traversal $d^{-1}b^{-1}c^{-1}$ for the second triangle. The alternatives for the third polygon are da and $a^{-1}d^{-1}$. The first of these fails to be coherent with the first triangle along a, whereas the second of these fails to be coherent with the second triangle along d. The presentation is therefore not orientable.

Six *modifications* of presentations are now defined that clearly have no effect on the corresponding surfaces and also change neither the Euler characteristic nor the orientability character of the presentations.

1. *A transverse cut.* One of the polygons of a presentation is cut in two along an interior arc that joins two (distinct and nonconsecutive) vertices of the polygon, and

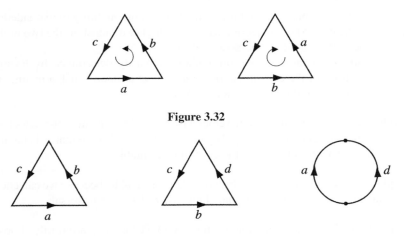

Figure 3.32

Figure 3.33

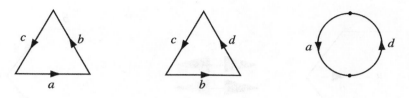

Figure 3.34

the two new edges created by this cutting are paired in the derived presentation. In Figure 3.35 this operation transforms the polygon on the left into the two polygons on the right. Note that while the vertex u gives rise to two new vertices u_1 and u_2, these necessarily belong to the same node of the skeleton of Π'.

2. *A disjoint pasting*. This is the reverse of a transverse cut—two different polygons are pasted along two paired edges, and the corresponding arc is of course no longer present in the skeleton of the derived presentation. In Figure 3.35, this operation transforms the two polygons on the right into the single polygon on the left.

3. *A partial cut*. One of the polygons of the presentation is cut along an arc that joins one of the vertices to an interior point, and the two new edges created by this cut are paired in the derived presentation. In Figure 3.36 this operation transforms the polygon on the left into the one on the right.

4. *A self-pasting*. This is the reversal of a partial cut—two paired directed edges of a polygon that happen to share their terminal vertices are pasted. In Figure 3.36 this operation transforms the polygon on the right into the one on the left.

5. *An arc subdivision*. Both occurrences of an edge are subdivided into a pair of consecutive edges. In Figure 3.37 the pair of polygons on the left is transformed into the pair on the right.

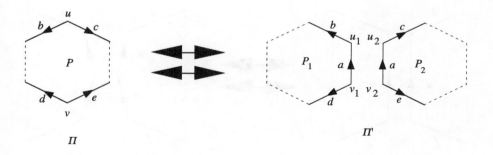

Figure 3.35   A transverse cut and a disjoint paste.

Figure 3.36 A partial cut and a self-pasting.

6. *An arc unification.* This is the reversal of an arc subdivision. If the directed edges b, c occur consecutively twice along the perimeters, then they can be unified into a single edge. In Figure 3.37 the pair of polygons on the right is transformed into the pair on the left.

Two presentations Π and Π' are said to be *cut-and-paste equivalent* if there is a sequence of modifications of types 1–6 that transforms one to the other. Since none of these modifications actually change the underlying surface, it follows that cut-and-paste equivalent surfaces are homeomorphic. Conversely, if the two surfaces S and S' are homeomorphic, then the homeomorphism transforms a polygonal presentation of one into a polygonal presentation of the other. Hence we have the following proposition.

Proposition 3.2.8 *Two surfaces are homeomorphic if and only if they have cut-and-paste equivalent presentations.*

In addition to leaving the topological nature of a presentation unchanged, the above modifications have no effect on its connectedness, Euler characteristic, and orientability character.

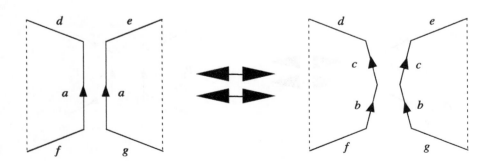

Figure 3.37 A subdivision and a unification.

Lemma 3.2.9 *Modifications 1–6 preserve the connectedness, orientability charac-ter, and the Euler characteristic of a presentation.*

PROOF: The lemma is demonstrated for transverse cuts and disjoint pastings only. Partial cuts, self-pastings, subdivisions, and unifications are relegated to Exercises 19 and 20.

For the modifications in question (Fig. 3.35), the skeletons $G(\Pi)$ and $G(\Pi')$ differ only in that the former has one fewer arc, namely a. Consequently, the con-nectedness of $G(\Pi)$ implies that of $G(\Pi')$. Conversely, suppose $G(\Pi')$ is connected. If W is a w_1-w_2 walk in $G(\Pi')$, then the replacement of each occurrence of a in W by the rest of the perimeter of P_1 (or P_2) converts W to a w_1-w_2 walk in $G(\Pi)$. Hence $G(\Pi)$ is also connected.

Turning to the issue of orientability, note that if the presentation Π of Figure 3.35 is coherently oriented, then, if the orientation of the split polygon is transferred to its two pieces, a coherent orientation of Π' is obtained. Conversely, if the presentation Π' is coherently oriented, then the presence of the common edge a in both of the polygons P_1 and P_2 implies that in this orientation, both of the polygons must receive the same orientation relative to the page. This orientation can therefore be passed on to the presentation Π in an unambiguous manner. Thus Π and Π' have the same orientability character.

It is clear that $r(\Pi') = r(\Pi) - 1$ and $q(\Pi') = q(\Pi) - 1$. Moreover, there is a one-to-one correspondence between the angle chains that define the nodes of Π and those chains that define the nodes of Π'. This correspondence is obtained by replacing every occurrence of the angles $\{b, c\}$ and $\{d, e\}$ in Π' with the angle pairs $\{b, a^{-1}, c\}$ and $\{d, a, e\}$ in Π. It follows that $p(\Pi') = p(\Pi)$. Hence,

$$\chi(\Pi') = p(\Pi') - q(\Pi') + r(\Pi') = p(\Pi) - (q(\Pi) - 1) + r(\Pi) - 1$$
$$= p(\Pi) - q(\Pi) + r(\Pi) = \chi(\Pi).$$
Q.E.D.

In the interest of conciseness, polygons will be encoded by the cyclical sequence of their edges relative to either of their orientations. Thus, using the counterclock-wise sense, the three polygons of Figure 3.7 can be denoted as $aa^{-1} = a^{-1}a$, $ab^{-1}a^{-1}b = b^{-1}a^{-1}ba$, and $abb^{-1}a^{-1} = a^{-1}abb^{-1}$ respectively. Using the clockwise sense, the same three polygons would be denoted by aa^{-1}, $b^{-1}aba^{-1}$, and $abb^{-1}a^{-1}$. The presentation of the Klein bottle in Figure 3.11 is denoted by either $aba^{-1}b$ or $ab^{-1}a^{-1}b^{-1}$.

It is now possible to state and prove this chapter's main theorem.

Theorem 3.2.10 *Every polygonal presentation Π is cut-and-paste equivalent to ex-actly one of the 1-polygon normal presentations below:*

$$\Pi_0: \quad aa^{-1};$$
$$\Pi_n: \quad a_1 b_1 a_1^{-1} b_1^{-1} a_2 b_2 a_2^{-2} b_2^{-2} \cdots a_n b_n a_n^{-1} b_n^{-1},$$
$$\text{where } n = \frac{2 - \chi(\Pi)}{2} = 1, 2, 3, \ldots;$$

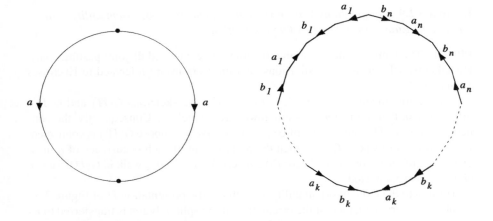

Figure 3.38 The normal presentations Π_0 and Π_n, $n \geq 1$.

$$\tilde{\Pi}_n: \qquad a_1 a_1 a_2 a_2 \cdots a_n a_n, \qquad \text{where } n = 2 - \chi(\Pi) = 1, 2, 3, \ldots.$$

PROOF: Let Π be a polygonal presentation. We shall produce a sequence of presentations $\Pi = \Pi^1, \Pi^2, \ldots, \Pi^m$ where Π^m is one of the polygonal presentations in Figures 3.38–3.39, and for each $i = 1, 2, \ldots, m - 1$, Π^{i+1} is obtained from Π^i by one or several of modifications 1–6. Of necessity, then, Π and Π^m represent the same surface. It also follows from Lemma 3.2.9 that these presentations all have the same Euler characteristics, orientability characters, and connectedness. The construction of this sequence of presentations has five *phases*, each of which consists of the repeated application of several *reductions*. Each of the reductions consists of a finite number of modifications.

Phase 1 (One polygon). This phase produces a presentation that consists of a single polygon. If Π^i has more than one polygon, then, because $G(\Pi^i)$ is connected, some

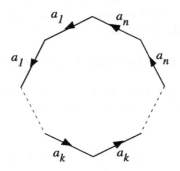

Figure 3.39 The normal presentation $\tilde{\Pi}_n$.

two polygons contain the same labeled edge, say a. Reduction A is then applied to these two polygons:

Reduction A. Paste the two polygons along the common arc a as indicated in Figure 3.35.

Reduction A is repeated until a polygonal presentation Π^i is obtained that consists of only one polygon, which will be denoted by P^i. All the subsequent steps will be such as to convert this single polygon into another single polygon. In other words, from this point on, $r(\Pi^i) = 1$. In each case the single polygon will be assumed to be oriented in the counterclockwise sense of the page.

Phase 2 (One node or Π_0). The goal of this phase is to reduce the number of nodes of $G(\Pi^i)$ to 1. For the entire duration of this phase, a node v of $G(\Pi^i)$ is fixed and it is supposed that P^i has at least one vertex that does not belong to v (otherwise, we proceed to the next phase). Consequently P^i has an edge labeled a whose initial vertex is an occurrence of v and whose terminal vertex belongs to a node $u \neq v$. Let b be the label that follows a along the perimeter of P^i. Since the initial vertices of a and b are the nonequivalent vertices v and u respectively, it follows that $b \neq a$. There are now three cases to be considered:

Reduction B. If $b = a^{-1}$, then u occurs along the perimeter of P^i only once. If $ab = aa^{-1}$ constitutes the entire perimeter of P^i, then the first normal presentation has been obtained and we are done. Otherwise, perform the self-pasting of Figure 3.36. This reduces the number of occurrences of u.

Reduction C. If $b \neq a^{-1}$ and the perimeter of P^i contains $\cdots ab \cdots b^{-1} \cdots$, perform the reduction of Figure 3.40: cut the polygon along the indicated arc c, and paste the two pieces along b. This augments the number of occurrences of v by 1 and diminishes the number of occurrences of u by 1.

Reduction D. If $b \neq a^{-1}$ and the perimeter of P^i contains $\cdots ab \cdots b \cdots$, perform the reduction of Figure 3.41: cut the polygon along the indicated arc c, and paste the two pieces along b. This too augments the number of vertices belonging to v by 1 and diminishes those belonging to u by 1.

Reductions B, C, and D are to be repeated until v is the only node of the presentation.

Figure 3.40   Diminishing the occurrences of u.

Figure 3.41 Diminishing the occurrences of u.

If Phase 2 does not produce the normal presentation Π_0, then it results in a polygon P^i all of whose vertices belong to the same node v of $G(\Pi^i)$. If the perimeter of this polygon does not contain a repetition of the form $\cdots a \cdots a \cdots$, we go on to Phase 4 below. Otherwise Phase 3 is required.

This is the place to note that in all of the remaining phases the number of arcs is unchanged and the number of polygons remains 1. As, by Lemma 3.2.9, the Euler characteristic is also unchanged, it follows that these phases all result in a presentation with one node only.

Phase 3 (Cross-cap normalization). The purpose of this phase is to juxtapose the occurrences of any two labels that have the same exponent.

Reduction E. Given a polygon P^i with perimeter $\cdots a \cdots a \cdots$, cut it along c and paste along a as indicated in Figure 3.42. This results in a consecutive repeated pair $\cdots cc \cdots$ called a *cross-cap*. If the perimeter $\text{per}(P^{i+1})$ contains another repetition of the form $\cdots b \cdots b \cdots$, Reduction E is repeated, and this repetition does not affect any previously obtained crosscaps. Hence this process is bound to eventually lead to a polygon P^i all of whose repeated labels are assembled into cross-caps (Figure 3.43).

If $\text{per}(P^{i+1})$ consists entirely of cross-caps, then we have arrived at the third normal presentation $\tilde{\Pi}_n$. Otherwise, it is necessary to resort to Phase 4.

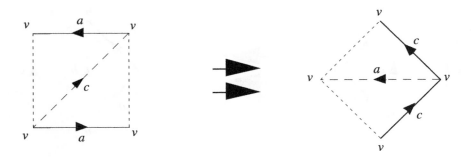

Figure 3.42 Cross-cap normalization.

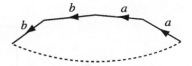

Figure 3.43 Normalized cross-caps.

Phase 4 (Handle normalization). If a is not part of a cross-cap, then a^{-1} is also on the perimeter. Of all such a's, choose that one such that the section $\alpha = a \cdots a^{-1}$ of $\operatorname{per}(P^i)$ is minimal. It follows that if b is in the interior of α, then b^{-1} is not there. If each b on this segment is repeated there, then

$$\alpha = abbcc \cdots xxa^{-1},$$

and an angle-chain argument demonstrates that the vertices of $b, c, \cdots, x$ constitute *all* the appearances of a node of Π^i, which node is of necessity different from the initial node of a. This would contradict the fact that Π^i has only one node. Hence there must exist an arc b such that

$$\operatorname{per}(P^i) = \cdots a \cdots b \cdots a^{-1} \cdots b^{-1} \cdots .$$

This phase's Reduction F replaces a perimeter of the form $\cdots a \cdots b \cdots a^{-1} \cdots b^{-1} \cdots$ with one of the form $\cdots xyx^{-1}y^{-1} \cdots$. A perimeter portion of the form $xyx^{-1}y^{-1}$ is called a *handle*, which accounts for the title of this phase. The term *handle* is explained by Proposition 3.3.7.

Reduction F. As indicated in Figure 3.44, cut P^i along y and paste the two pieces along y. Then cut along x and paste along c. Note that no previously obtained cross-caps and/or handles are destroyed by this process.

Reduction F is iterated until a polygon P^i is obtained whose counterclockwise perimeter consists entirely of cross-caps and handles. If P^i contains no cross-caps, we have obtained a normal presentation of the form $\Pi_n, n = 0, 1, 2, \ldots$. If P^i contains no handles, we have obtained a canonical presentation of the form $\tilde{\Pi}_n, n = 1, 2, 3, \ldots$.

Phase 5 (Handles to cross-caps). Successive iterations of Reduction G will alter a polygon with both handles and cross-caps into one with cross-caps only.

Reduction G. The sequence of cuts and pastings of Figure 3.45 alters a perimeter $cc \cdots aba^{-1}b^{-1} \cdots$ into the perimeter $xxyyzz \cdots$. First cut along d and paste along c. Next cut along x and paste along b. Again, cut along y and paste along a. Finally, cut along z and paste along d. This reduction does not affect the connectedness, the Euler characteristic, the number of edges, or the number of regions. Consequently, the derived presentation still has only one node.

The above arguments prove that every polygonal presentation is cut-and-paste equivalent to one of the normal presentations of Figures 3.38–3.39. It now remains to show that none of the standard presentations of these figures are equivalent to each other.

Note that Π_0 and Π_n are orientable, whereas $\tilde{\Pi}_n$ is nonorientable for each $n = 1, 2, 3, \ldots$ and it follows from Lemma 3.2.9 that orientable presentations cannot be

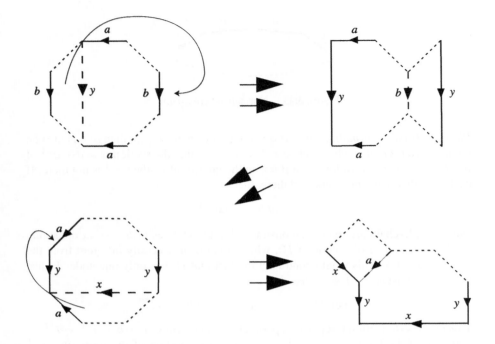

Figure 3.44   Handle normalization.

cut-and-paste equivalent to nonorientable ones. Moreover (see Exercise 3.2.13),

$$\chi(\Pi_0) = 2 - 1 + 1 = 2,$$
$$\chi(\Pi_n) = 1 - 2n + 1 = 2 - 2n,$$
$$\chi(\tilde{\Pi}_n) = 1 - n + 1 = 2 - n.$$

Since, again by Lemma 3.2.9, equivalent presentations have the same Euler characteristic, it follows that no two of the standard polygons are equivalent.   Q.E.D.

Let S_n and $\tilde{S}_n$ be the surfaces that correspond to Π_n and $\tilde{\Pi}_n$ respectively. The following corollaries elucidate the meaning of Theorem 3.2.10.

Corollary 3.2.11 *Every surface is homeomorphic to exactly one of the surfaces S_0, S_n, $\tilde{S}_n$, $n = 1, 2, 3, \ldots$.*

PROOF: By definition, every surface has some presentation which, by Theorem 3.2.11, is cut-and-paste equivalent to a normal presentation. Since equivalent presentations have homeomorphic surfaces, every surface is homeomorphic to one of S_0, S_n, $\tilde{S}_n$, $n = 1, 2, 3, \ldots$.

Moreover, homeomorphic surfaces have identical presentations, and so it follows that no two of the normal surfaces are homeomorphic.   Q.E.D.

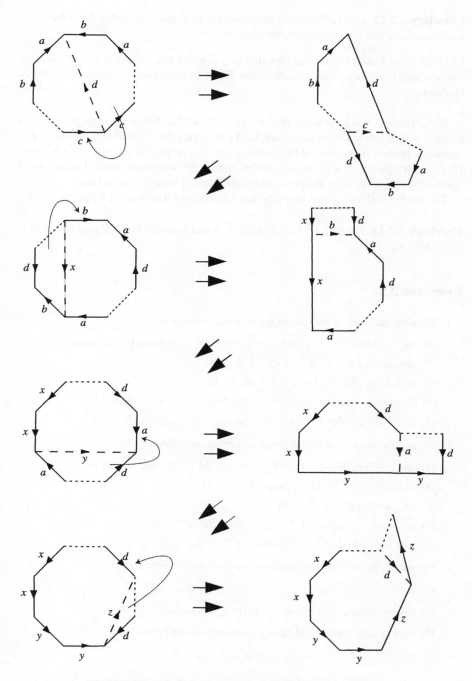

Figure 3.45 The conversion of a handle into two cross-caps.

Corollary 3.2.12 *Two surfaces are homeomorphic if and only if they have the same Euler characteristic and orientability character.*

PROOF: This follows from the fact that any distinct pair of the surfaces listed in the previous corollary differ in either their Euler characteristic or their orientability character.   Q.E.D.

If the surface S is homeomorphic to either S_n or $\tilde{S}_n$, then the *genus* $\gamma(S)$ of S is defined as n. It follows from Corollary 3.2.11 that once the orientability character of a surface is known, its genus and Euler characteristic completely determine each other. However, the genus, as will be seen in Corollary 3.3.3, has a very natural geometrical interpretation and is often the preferred parameter for describing surfaces.

The next corollary follows from the last few lines of the proof of Theorem 3.2.10.

Corollary 3.2.13 *For any surface S, $\chi(S) \leq 2$ and equality holds if and only if S is the sphere S_0.*   □

Exercises 3.2

1. Draw the skeletons of the following polygonal presentations:

   (a) aa (b) aa^{-1} (c) $aabb$ (d) $aa^{-1}bb$ (e) $abc, b^{-1}ed, dc, ae$

   (f) $abc, defc, fgi^{-1}a^{-1}h^{-1}, jhdg^{-1}b^{-1}, je^{-1}i$

   (g) $abc, defc, fgd^{-1}a^{-1}h^{-1}, bgj^{-1}, ei^{-1}j, hai$

   (h) $abcabc$ (i) $abca^{-1}b^{-1}c^{-1}$ (j) $abcdabcd$ (k) $aabbcc$

   (l) $aba^{-1}b^{-1}cdc^{-1}d^{-1}$ (m) $aba^{-1}b^{-1}cdc^{-1}d^{-1}efe^{-1}f^{-1}$

2. Identify the surfaces of the following polygonal presentations:

   (a) aa (b) aa^{-1} (c) $aabb$ (d) $aa^{-1}bb$ (e) $abc, b^{-1}ed, dc, ae$

   (f) $abc, defc, fgi^{-1}a^{-1}h^{-1}, jhdg^{-1}b^{-1}, je^{-1}i$

   (g) $abc, defc, fgd^{-1}a^{-1}h^{-1}, bgj^{-1}, ei^{-1}j, hi$

   (h) $abcabc$ (i) $abca^{-1}b^{-1}c^{-1}$ (j) $abcdabcd$ (k) $aabbcc$

   (l) $aba^{-1}b^{-1}cdc^{-1}d^{-1}$ (m) $aba^{-1}b^{-1}cdc^{-1}d^{-1}efe^{-1}f^{-1}$

3. For each positive integer n identify the surface with the presentation

   (a) $a_1a_2\cdots a_na_1a_2\cdots a_n$ (b) $a_1a_2\cdots a_na_1^{-1}a_2^{-1}\cdots a_n^{-1}$

   (c) $a_1a_2\cdots a_na_na_{n-1}a_{n-2}\cdots a_1$ (d) $a_1a_2\cdots a_na_n^{-1}a_{n-1}^{-1}a_{n-2}^{-1}\cdots a_1^{-1}$

4. For each positive integer n identify the surface with the presentation

   $$a_1a_1^{-1}a_2a_2^{-1}\cdots a_na_n^{-1}.$$

5. A closed orientable surface has a polygonal presentation with the same number of nodes and arcs. Identify this surface.

6. A closed nonorientable surface has a polygonal presentation with the same number of nodes and arcs. Explain why this surface must be homeomorphic to the projective plane.

7. Show that for every positive integer n there is a polygonal presentation of the sphere with 2 nodes, n arcs, and n polygons.

8. Show that for every positive integer n there is a polygonal presentation of the sphere with n nodes, n arcs, and 2 polygons.

9*. A polygonal presentation Π is said to be *regular* if each of its polygons has the same number (≥ 3) of arcs, each node of $G(\Pi)$ has the same degree (≥ 3), and no arc appears twice on the perimeter of a polygon. Prove that if Π is a regular presentation of the sphere, then $(p(\Pi), q(\Pi), r(\Pi))$ is one of the triples (4, 6, 4), (6, 12, 8), (8, 12, 6), (12, 30, 20), or (20, 30, 12).

10. Show that the torus has regular presentations with an arbitrarily large number of nodes.

11. Show that the Klein bottle has regular presentations with an arbitrarily large number of nodes.

12*. Does the projective plane have any regular presentations? If so, can the number of nodes in such a presentation be arbitrarily large?

13. Determine the Euler characteristic and orientability character of each of the presentations in Figures 3.38–3.39.

14. A *triangulation* of a surface S is a presentation Π of S in which each polygon is a triangle and the skeleton is a simple graph. Prove that if Π is a triangulation with parameters p, q, r, then $3r = 2q$, $q = 3(p - \chi(S))$.

15. What are the minimum values of $p(\Pi)$, $q(\Pi)$, $r(\Pi)$ for any triangulation of the sphere?

16*. Prove that the minimum value of $p(\Pi)$ for any triangulation of the torus is 7.

17*. Prove that the minimum value of $p(\Pi)$ for any triangulation of the projective plane is 6.

18. Prove Corollary 3.2.13.

19. Prove Lemma 3.2.9 for self-pastings and partial cuts.

20. Prove Lemma 3.2.9 for arc subdivisions and unifications.

3.3 Operations on Surfaces

So far, the normal surfaces have been displayed only as presentations. We now set out to describe them as surfaces in $\mathbb{R}^3$ whenever possible, or in $\mathbb{R}^4$ otherwise.

An *excision* of a surface is a cut whose effect is permanent in the sense that it is not meant to be reversed—the borders it creates are not meant to be repasted to each other. These borders, however, may be pasted to other borders. Such will always be the case in this section, so that the end product of the operations will always be a surface without borders. For example, given two disjoint surfaces S^1 and S^2, their *connected sum* $S^1 + S^2$ is the surface obtained by excising a disk from each and pasting the resulting borders. This is illustrated in Figure 3.46, where the excised disks leave borders c^1 and c^2, which are then pasted to form the loop c on the sum $S^1 + S^2$.

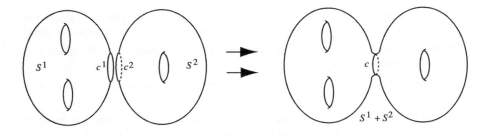

Figure 3.46   The addition of surfaces.

It may seem at first that the sum might depend on the specific orientations that are assigned to the excisions and on their locations, but that, surprisingly, is not the case.

Proposition 3.3.1 *If S^1 and S^2 are closed surfaces, then*

$$\chi(S^1 + S^2) = \chi(S^1) + \chi(S^2) - 2,$$

and $S^1 + S^2$ is orientable if and only if both S^1 and S^2 are orientable.

PROOF: Consider any presentation Π of $S^1 + S^2$ such that the excision path along which S^1 and S^2 have been pasted is a cycle c of Π with n nodes and n arcs. For each $i = 1, 2$ let Π^i be the polygonal presentation of S^i obtained by restricting Π to S^i and adding a polygon P^i that seals the hole surrounded by c^i. Then, because the nodes and arcs of c appear in both Π^1 and Π^2,

$$\begin{aligned}
\chi(S^1 + S^2) = \chi(\Pi) &= p(\Pi) - q(\Pi) + r(\Pi) \\
&= \left(p(\Pi^1) + p(\Pi^2) - n\right) - \left(q(\Pi^1) + q(\Pi^2) - n\right) \\
&\quad + \left(r(\Pi^1) - 1 + r(\Pi^2) - 1\right) \\
&= \left(p(\Pi^1) - q(\Pi^1) + r(\Pi^1)\right) + \left(p(\Pi^2) - q(\Pi^2) + r(\Pi^2)\right) - 2 \\
&= \chi(\Pi^1) + \chi(\Pi^2) - 2 = \chi(S^1) + \chi(S^2) - 2.
\end{aligned}$$

Turning to the issue of orientability, suppose c consists of the consecutive directed arcs $a_1, a_2, \ldots, a_n$ which bound the polygons P^1 and P^2 in Π^1 and Π^2 respectively. If S^1 and S^2 are both orientable, it may be assumed that Π^1 and Π^2 have been coherently oriented so that

$$(\text{orientation of } P^1) = a_1 a_2 \cdots a_n,$$
$$(\text{orientation of } P^2) = a_n^{-1} a_{n-1}^{-1} \cdots a_1^{-1}.$$

Then $\Pi = \Pi^1 \cup \Pi^2 - \{P^1, P^2\}$ is coherently oriented, so that $S^1 + S^2$ is orientable. Conversely, suppose $S^1 + S^2$ is orientable and that Π has in fact been coherently oriented. Of necessity the arcs $a_1, a_2, \ldots a_n$ are either all inconsistent with the orientations of the polygons of $\Pi^1 - \{P^1\}$ and consistent with the orientations of the

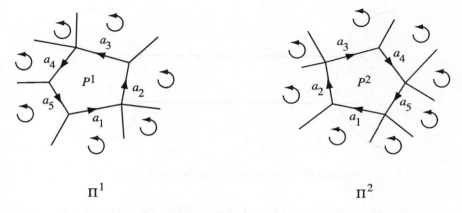

Π^1 Π^2

Figure 3.47 The addition of surfaces.

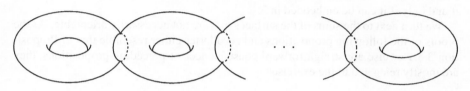

Figure 3.48 S_n as the sum of n tori.

polygons of the second presentation $\Pi^2 - \{P^2\}$ (see Figure 3.47), or vice versa. In the first case coherent orientations of Π^1 and Π^2 are obtained by assigning to P^1 and P^2 the respective orientations $a_1 a_2 \cdots a_n$ and $a_n^{-1} a_{n-1}^{-1} \cdots a_1^{-1}$. In the second case the assignments are reversed. In either case we may conclude that the surfaces S^1 and S^2 are orientable.   Q.E.D.

Corollary 3.3.2 *If m, $n \geq 0$ then $S_m + S_n \approx S_{m+n}$.*

PROOF: By Proposition 3.3.1 both $S_m + S_n$ and S_{m+n} are orientable. Moreover,

$$\chi(S_m + S_n) = \chi(S_m) + \chi(S_n) - 2 = (2 - 2m) + (2 - 2n) - 2$$
$$= 2 - 2(m+n) = \chi(S_{m+n}).$$

The desired result now follows from Corollary 3.2.12.   Q.E.D.

     This corollary now provides a simple realization of the closed orientable surfaces in $\mathbb{R}^3$.

Corollary 3.3.3 *For $n = 1, 2, 3, \ldots$, the closed orientable surface S_n can be realized in $\mathbb{R}^3$ as the sum of n tori (Fig. 3.48).*   □

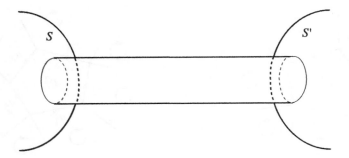

Figure 3.49   Connecting two surfaces with a tube.

Nonorientable surfaces cannot be embedded in $\mathbb{R}^3$, and hence the above corollary characterizes orientability of surfaces. In other words, a closed surface is orientable if and only if it can be embedded in $\mathbb{R}^3$.

We turn next to the sums of the surfaces that are not necessarily orientable. As the proofs of the following propositions and corollaries either resemble that of Proposition 3.3.1 or else are straightforward consequences of preceding propositions, they are mostly relegated to the exercises.

Corollary 3.3.4 *If $m \geq 0$ and $n \geq 1$ then $S_m + \tilde{S}_n \approx \tilde{S}_{2m+n}$.*

PROOF: Exercise 18.   □

Corollary 3.3.5 *Every closed nonorientable surface can be realized in $\mathbb{R}^4$.*

PROOF: It was demonstrated in Section 3.1 that the projective plane $\tilde{S}_1$ and the Klein bottle $\tilde{S}_2$ can be embedded in $\mathbb{R}^4$. It follows from Corollary 3.3.4 that

$$\tilde{S}_n \approx \begin{cases} S_{(n-1)/2} + \tilde{S}_1 & \text{for } n = 1, 3, 5, \ldots, \\ S_{(n-2)/2} + \tilde{S}_2 & \text{for } n = 2, 4, 6, \ldots . \end{cases}$$

In either case the four-dimensional realizations of $\tilde{S}_1$ and $\tilde{S}_2$ depicted in Figures 3.15 and 3.13 can be added to the three-dimensional realizations of $S_{(n-1)/2}$ and $S_{(n-2)/2}$ given in Figure 3.48 so as to produce a four-dimensional realization of $\tilde{S}_n$.   Q.E.D.

Corollary 3.3.6 *If $m, n \geq 1$ then $\tilde{S}_m + \tilde{S}_n \approx \tilde{S}_{m+n}$.*

PROOF: Exercise 19.   □

Surface addition can also be described as *tube connection*, an operation wherein the excised borders are joined by a tube (Fig. 3.49) instead of being pasted together. The resulting surface is homeomorphic to the sum of the two surfaces. To see this one need only imagine the length of the connecting cylinder to be shrinking to zero.

The foregoing operations created a new surface out of two distinct surfaces. Next, operations on a single surface are examined.

Suppose first that two disjoint disks are removed from a surface S and each of the resulting borders is given a direction and labeled a. Suppose further that the two borders are now pasted to each other in a manner that is consistent with their directions. Unlike the connected sum of two surfaces, in this case the nature of the resulting surface may depend on the choice of directions for the borders. If the surface S is orientable and the two borders have been assigned directions of which one is consistent and the other is inconsistent with some orientation of S, then the pasting process is illustrated in Figure 3.50, where the two borders labeled a are "pulled out" and joined. As suggested by this figure, this operation is called *handle addition*, and the result of adding a handle to the surface S is denoted by $S + h$. It is clear that this process can also be visualized by simply joining the two borders a by a tube or an open-ended cylinder (Fig. 3.51).

Proposition 3.3.7 *If S is an orientable surface, then $S + h$ is also orientable and* $\gamma(S + h) = \gamma(S) + 1$.

PROOF: See Figure 3.52. $\qquad\qquad\qquad\qquad\qquad\qquad\qquad\qquad\qquad\qquad\qquad$ $\square$

It follows by a straightforward induction that S_n is also realizable in $\mathbb{R}^3$ as a sphere S_0 with n attached handles (see Fig. 3.53 for an example).

Example 3.3.8 The solid of Figure 3.54 is obtained from K_4 by a "fattening" process. Without the three outer tubes this is a three-spiked sphere whose surface has

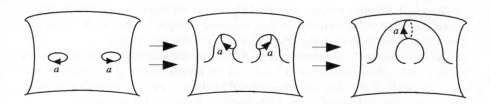

Figure 3.50   Adding a handle to a surface.

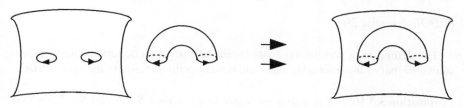

Figure 3.51   Adding a handle to a surface.

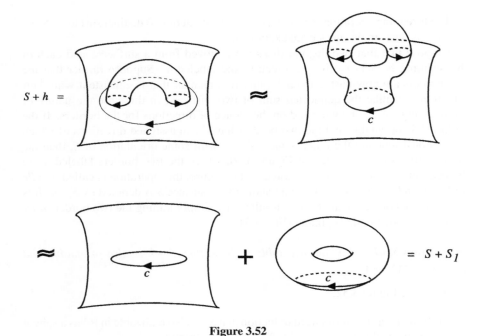

$S + h =$

$\approx$

$\approx$ $+$ $= S + S_1$

Figure 3.52

genus 0. It therefore follows from Proposition 3.3.7 that the original surface has genus 3.

The handle addition process described above was applied to an *orientable* surface S, and the two borders were directed as indicated in Figure 3.50, i.e., they had different directions relative to the orientation of S. If the two borders have the same direction relative to this orientation (Fig. 3.55), then the identification of the borders yields a different surface, denoted $S + \tilde{h}$, and the process is called *cross-handle addition*. It can be realized in four-dimensional Euclidean space by means of the same device that was used to realize the Klein bottle in Section 3.1 (Exercise 17).

Proposition 3.3.9 *If S is an orientable surface, then $S + \tilde{h}$ is a nonorientable surface and $\gamma(S + \tilde{h}) = 2\gamma(S) + 2$.*

PROOF: Exercise 20.   □

The definition of handle and cross-handle addition to a nonorientable surface is similar to that of the orientable case, and both operations result in the same surface.

Proposition 3.3.10 *If S is a nonorientable surface, then $S + h$ and $S + \tilde{h}$ are both nonorientable surfaces and $\gamma(S + h) = \gamma(S + \tilde{h}) = \gamma(S) + 2$.*

PROOF: Exercise 21.   □

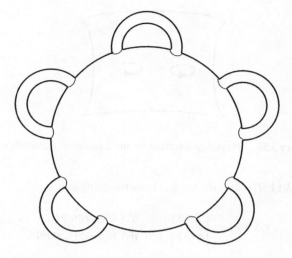

Figure 3.53 S_5 as a sphere with five handles.

Finally, suppose a single disk is excised from a surface and the remaining border is divided into two arcs, each of which is labeled b and both of which are given the same direction relative to S (Fig. 3.56). If the two b's are then pasted to each other, a new surface $S + c$ is obtained, and the process is called a *cross-cap addition*.

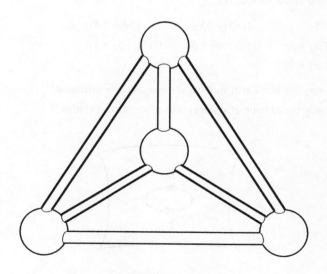

Figure 3.54   A fat K_4.

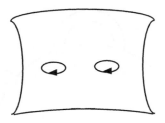

Figure 3.55   Preparing a surface for the addition of a cross-handle.

Proposition 3.3.11 *The surface $S + c$ is nonorientable and*

$$\gamma(S+c) = \begin{cases} 2\gamma(S)+1 & \text{if } S \text{ is orientable,} \\ \gamma(S)+1 & \text{if } S \text{ is nonorientable.} \end{cases}$$

PROOF: Exercise 22.   □

Corollary 3.3.12 *The nonorientable surface of genus n is homeomorphic to a sphere with n cross-caps.*

PROOF: Exercise 23.   □

Exercises 3.3

1. Identify the following surfaces:

   (a) $S_1 + S_1$ (b) $S_{12} + \tilde{S}_{21}$ (c) $S_{21} + \tilde{S}_{12}$

   (d) $S_1 + S_2 + S_3$ (e) $\tilde{S}_1 + S_2 + S_3$ (f) $S_1 + \tilde{S}_2 + \tilde{S}_3$

   (g) $\tilde{S}_1 + \tilde{S}_2 + \tilde{S}_3$

2. Explain why the addition of surfaces is a commutative operation.

3. Explain why the addition of surfaces is an associative operation.

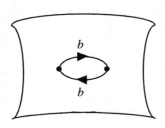

Figure 3.56   Preparing a surface for a cross-cap addition.

4. Two tori are connected by means of three tubes. Identify the resulting surface.

5. The surfaces S^1, S^2, S^3 are all tori. Each is connected to the other two by means of a tube. Identify the resulting surface.

6. The surfaces S^1, S^2, S^3 are all Klein bottles. Each is connected to the other two by means of a tube. Identify the resulting surface.

7. The surfaces S^1, S^2, S^3 are all tori. Each is connected to the other two by means of two tubes. Identify the resulting surface.

8. The surfaces S^1, S^2, S^3 are all Klein bottles. Each is connected to the other two by means of two tubes. Identify the resulting surface.

9. The surfaces $S^1, S^2, \ldots, S^n$ are all tori. For each $i = 2, 3, \ldots, n-1$, S^i is connected by means of a tube to both S^{i-1} and S^{i+1}. Identify the resulting surface.

10. The surfaces $S^1, S^2, \ldots, S^n$ are all Klein bottles. For each $i = 2, 3, \ldots, n-1$, S^i is connected by means of a tube to both S^{i-1} and S^{i+1}. Identify the resulting surface.

11. The surfaces $S^1, S^2, \ldots, S^n$ are all tori. For each $i = 2, 3, \ldots, n-1$, S^i is connected by means of two tubes to S^{i-1} and two other tubes to S^{i+1}. Identify the resulting surface.

12. The surfaces $S^1, S^2, \ldots, S^n$ are all Klein bottles. For each $i = 2, 3, \ldots, n-1$, S^i is connected by means of two tubes to S^{i-1} and two other tubes to S^{i+1}. Identify the resulting surface.

See Example 3.3.8 for an illustration of a fat graph.

13. Show that surface of the fat K_5 has genus 6.

14. Show that the surface of the fat version of K_n has genus $(n-1)(n-2)/2$.

15. Show that if the connected graph G has p nodes and q arcs, then the surface of the fat version of G has genus $q - p + 1$.

16. Two distinct surfaces S and S' are connected by means of n disjoint tubes to form a new surface S''. Identify the surface S''.

17. Show by means of a diagram how the addition of a cross-handle can be realized in $\mathbb{R}^4$.

18. Prove Corollary 3.3.4.

19. Prove Corollary 3.3.6.

20. Prove Proposition 3.3.9.

21. Prove Proposition 3.3.10.

22. Prove Proposition 3.3.11.

23. Prove Corollary 3.3.12.

3.4 Bordered Surfaces

Surfaces were defined as the topological spaces obtained by pasting the edges of the polygons of a presentation. Just so, *bordered surfaces* are the topological spaces obtained by pasting *some* of the edges of a bordered polygonal presentation. A *bordered polygonal presentation*, in turn, consists of a set of polygons with labeled edges, such that each label occurs *at most* twice and the pasting recipe connects all of the polygons. In the bordered presentation of Figure 3.57 each of the labels a, b, c, d appears twice, whereas the labels x, y, z, w appear only once. The arcs whose labels appear only once are the *border arcs* of Π, and their nodes are the *border nodes*. The skeleton $G(\Pi)$ of the bordered presentation Π is again the graph formed by all the labeled arcs of Π. The Euler characteristic and the orientability character of bordered presentations are defined in the same manner as was used in the previous section. When the pasting process is applied to a bordered presentation, the border arcs remain exposed and form border cycles which can be identified by the node-computing technique of the previous section.

Example 3.4.1 Consider the bordered presentation of Figure 3.57. It has the following angle chains:

$$
\begin{aligned}
A: &\quad x, a^{-1}, w \\
B: &\quad y, c^{-1}, z^{-1} \\
C: &\quad x^{-1}, b^{-1}, y^{-1} \\
D: &\quad w^{-1}, d^{-1}, z \\
E: &\quad a, b, c, d, a
\end{aligned}
$$

Note that the angle chains of border nodes must begin and end with border edges. The border nodes are A, B, C, and D, and their vicinities are depicted in Figure 3.58, where the lined angles are understood to be missing.

Figure 3.59 shows how the missing angles of Figure 3.58 can be assembled so as to form a polygon which is missing from Π. The page's counterclockwise orientation is a coherent orientation of the triangles of Figure 3.57, so that Π is orientable. Moreover, $\chi(\Pi) = 5 - 8 + 4 = 1$.

Example 3.4.2 Consider the presentation Π of Figure 3.60.

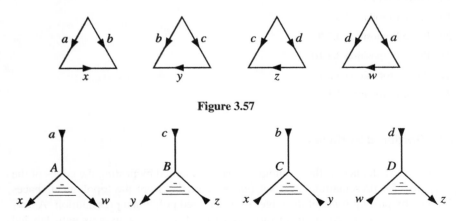

Figure 3.57

Figure 3.58

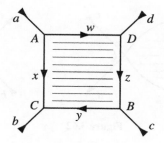

Figure 3.59

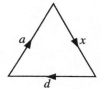 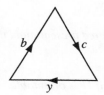

Figure 3.60

The nodes are given by the following angle chains:

$$A: \quad x,a^{-1},b,y^{-1}$$
$$B: \quad y,c^{-1},d,x^{-1}$$
$$C: \quad z,b^{-1},c,z^{-1}$$
$$D: \quad a,d^{-1}$$

The border nodes are A, B, and C, and their vicinities are depicted in Figure 3.61. The missing angles are assembled into borders in Figure 3.62.

A coherent orientation of Π is obtained when the two triangles of Figure 3.60 are given the page's counterclockwise orientation whereas the pentagon is oppositely oriented. The characteristic of this surface is $4 - 7 + 3 = 0$.

It is clear from the above examples that every border arc of a bordered presentation Π belongs to a border cycle. The number of such border cycles is $\beta(\Pi)$. The *closure* Π^c of Π is obtained by adding to Π a polygon with perimeter $abc\cdots$ for every border

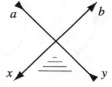  

Figure 3.61

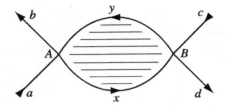 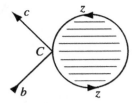

Figure 3.62

cycle $abc\cdots$ of Π. Figures 3.63 and 3.64 contain the closures of the presentations in Figures 3.57 and 3.60 respectively.

Proposition 3.4.3 *If Π is a bordered presentation, then*

1. $\chi(\Pi^c) = \chi(\Pi) + \beta(\Pi)$;

2. Π^c *is orientable if and only if Π is orientable.*

PROOF: Note that

$$\chi(\Pi^c) = p(\Pi^c) - q(\Pi^c) + r(\Pi^c) = p(\Pi) - q(\Pi) + (r(\Pi) + \beta(\Pi))$$
$$= \chi(\Pi) + \beta(\Pi).$$

Because every polygon of Π is also a polygon of Π^c, it follows that every coherent orientation of Π^c also constitutes a coherent orientation of Π. Conversely, suppose the polygons of Π have been coherently oriented. Let b be a border of Π whose

Figure 3.63

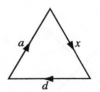

Figure 3.64

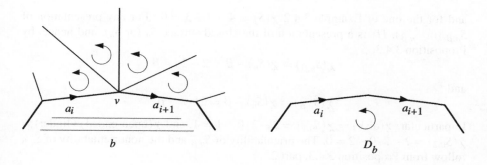

Figure 3.65 Extending a coherent orientation.

perimeter is $a_1^{\pm 1} a_2^{\pm 1} \cdots a_n^{\pm 1}$. Whenever necessary, reverse the directions of the a_i's so that they are all consistent with the orientations of the polygons of Π. It follows from the coherence of the orientation of Π (Figure 3.65) that if a_i is directed into the vertex v, then a_{i+1} is directed out of v. Consequently, the depicted orientation of the polygon D_b is coherent with the given orientation of Π. Thus, Π is coherently orientable if and only if Π^c is coherently orientable. Q.E.D.

The topological space obtained by the prescribed gluing of the polygons of a bordered presentation Π is a *bordered surface*, denoted by $S(\Pi)$. One large class of bordered surfaces is obtained by excising a finite number of disjoint disks from a surface embedded in $\mathbb{R}^3$. Thus, one can think of the open-ended cylinder as the surface obtained from the sphere S^0 by excising (and throwing away) two disks. The rim of the opening left by the excision of a disk is, of course, a border. Since a minute sliding of such a border along the surface results in a new surface that is still homeomorphic to the original bordered surface, the same holds when any number of such minute slides are performed. In other words, the bordered surface obtained by the removal of one or more disks from a given surface is the same regardless of the exact location of the borders so created on the surface, so long as these borders do not intersect. Each such disk removal is also called a *perforation*. The surfaces obtained from S_n and $\tilde{S}_n$ by β perforations are denoted by $S_{n,\beta}$ and $\tilde{S}_{n,\beta}$ respectively (see Fig. 3.66). There are, however, other bordered surfaces which cannot be described quite so easily by such surgery on closed surfaces, and examples of these appear in Figure 3.67 and in the exercises below. In general we define the *borders* of $S(\Pi)$ to be those of Π. The closure of $S = S(\Pi)$ is $S^c = S(\Pi^c)$. The following assumptions are plausible and will be taken for granted.

Assumption 3.4.4 *If $S(\Pi) \approx S(\Pi')$ then $\beta(\Pi) = \beta(\Pi')$ and $S(\Pi^c) \approx S(\Pi'^c)$.*

Note that these assumptions obviously hold for the excised surfaces $S_{n,\beta}$ and $\tilde{S}_{n,\beta}$.

The *Euler characteristic* of $S = S(\Pi)$ is $\chi(S) = \chi(\Pi)$, and S is *orientable* if Π is orientable. For example, for the presentation of Example 3.4.1 $\chi(S) = 5 - 8 + 4 = 1$, and for the one of Example 3.4.2 $\chi(S) = 4 - 7 + 3 = 0$. For any presentation of $S_{n,\beta}$ (or $\tilde{S}_{n,\beta}$), Π^c is a presentation of the closed surface S_n (or $\tilde{S}_n$), and hence, by Proposition 3.4.3,

$$\chi(S_{n,\beta}) = \chi(S_n) - \beta = 2 - 2n - \beta$$

and

$$\chi(\tilde{S}_{n,\beta}) = \chi(\tilde{S}_n) - \beta = 2 - n - \beta.$$

In particular, $\chi(\text{disk}) = \chi(S_{0,1}) = 2 - 2 \cdot 0 - 1 = 1$, and $\chi(\text{open-ended cylinder}) = \chi(S_{0,2}) = 2 - 2 \cdot 0 - 2 = 0$. The orientability of $S_{n,\beta}$ and the nonorientability of $\tilde{S}_{n,\beta}$ follow from Proposition 3.4.3, part 2.

Proposition 3.4.5 *If Π and Π' are bordered presentations such that $S(\Pi) \approx S(\Pi')$, then*

1. $\chi(\Pi) = \chi(\Pi')$;

2. Π and Π' have the same orientability characteristic.

PROOF: By Proposition 3.4.3

$$\chi(\Pi) = \chi(\Pi^c) - \beta(\Pi) \quad \text{and} \quad \chi(\Pi') = \chi(\Pi'^c) - \beta(\Pi').$$

However, by Assumption 3.4.4, $S(\Pi^c) \approx S(\Pi'^c)$, so that, by Proposition 3.2.8,

$$\chi(\Pi^c) = \chi(\Pi'^c).$$

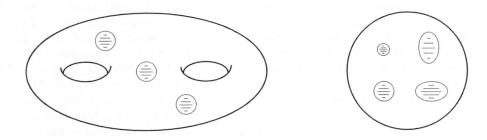

Figure 3.66   The bordered surfaces $S^{2,3}$ and $S^{0,5}$.

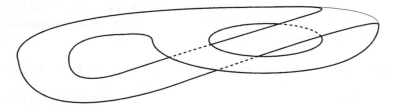

Figure 3.67

It was also assumed in 3.4.4 that $\beta(\Pi) = \beta(\Pi')$. It therefore follows that $\chi(\Pi) = \chi(\Pi')$.

Moreover, by Proposition 3.4.3,

$$\Pi \text{ is orientable iff } \Pi^c \text{ is orientable;}$$
$$\Pi' \text{ is orientable iff } \Pi'^c \text{ is orientable.}$$

Since $S(\Pi^c) \approx S(\Pi'^c)$, it follows from Proposition 3.2.8 that Π and Π' have the same orientability characteristic. Q.E.D.

Theorem 3.4.6 *Two bordered surfaces are homeomorphic if and only if they have the same Euler characteristic, orientability character, and number of border loops.*

PROOF: If $S = S(\Pi)$ and $S' = S(\Pi')$ are homeomorphic, then this theorem merely reiterates the conclusions of Assumption 3.4.4 and Proposition 3.4.5.

Conversely, suppose S and S' have the same Euler characteristic, orientability character, and number of border loops. It follows from Proposition 3.4.3 that the same holds for the closures $S(\Pi^c)$ and $S(\Pi'^c)$, and so Corollary 3.2.12 implies that S and S' have homeomorphic closures. Since S and S' are obtained from their respective homeomorphic closures by the excision of the same number of disks, they too must be homeomorphic. Q.E.D.

The *genus* $\gamma(S)$ of the bordered surface S is that of its closure, i.e., $\gamma(S^c)$. In particular $\gamma(S_{n,\beta}) = \gamma(\tilde{S}_{n,\beta}) = n$. The following corollary will prove useful in the proof of Proposition 4.3.3.

Corollary 3.4.7 *If the bordered surface S has $\gamma(S) = 0$ and $\beta(S) = 1$, then S is homeomorphic to a disk.* □

Example 3.4.8 Identify the surface B_n obtained by introducing n similarly directed twists into a band. Figure 3.68 depicts B_3 and B_4.

The two-polygon presentation produced by the cuts a_1, a_{n+1} has four nodes and six arcs. It therefore has Euler characteristic $4 - 6 + 2 = 0$. When n is even, this presentation is orientable and has two borders. Consequently the closure of B_n has characteristic $0 + 2 = 2$ and genus 0, and so, for even n, $B_n \approx S_{0,2}$. When n is odd this presentation is nonorientable and has a single border. Consequently the closure of B_n has characteristic 1 and genus 1, and so, for odd n, $B_n \approx \tilde{S}_{1,1}$. These computations are of course consistent with the discussion in Chapter 1.

Bands with twists can be used to identify nonorientable surfaces.

Proposition 3.4.9 *Every bordered surface that is embedded in $\mathbb{R}^3$ and contains a band with an odd number of twists is nonorientable.*

PROOF: Note that consecutive but oppositely oriented twists cancel each other out. Consequently it may be assumed that the twists of the given band are all similarly

The non-orientable B^3.

The orientable B^4.

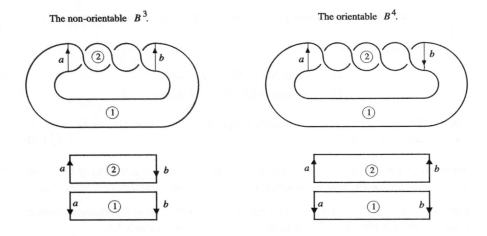

Figure 3.68   Two twisted bands.

oriented. Let Π be a presentation of such a surface in which the given band is the union of two quadrilaterals of which one contains all the twists and the other contains none. An argument similar to that employed in Example 3.4.8 leads to the conclusion that this presentation cannot be coherently oriented. Hence the given surface is nonorientable.   Q.E.D.

Example 3.4.10 To identify the bordered surface S on the left of Figure 3.69, introduce the ten cuts indicated on the right. The Euler characteristic of this presentation is $20 - 30 + 9 = -1$, and there are two borders. By Lemma 3.4.3, $\chi(S^c) = \chi(S) + \beta(S) = -1 + 2 = 1$. A narrow band passing through all of the four circu-

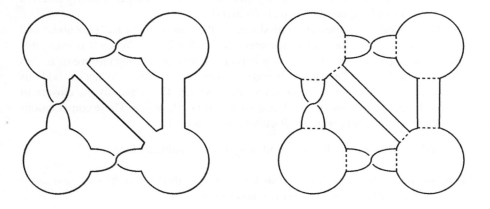

Figure 3.69   The identification of a bordered surface.

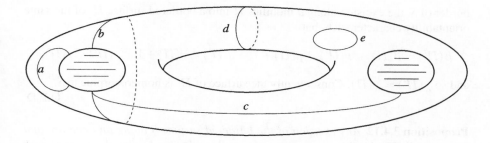

Figure 3.70 Separating and nonseparating excisions of the perforated torus $S^{0,2}$.

lar portions of the surface in succession, has three twists. The surface is therefore nonorientable. Hence S^c is the nonorientable surface of Euler characteristic 1, or genus $2 - 1 = 1$. It follows that S is homeomorphic to the twice perforated projective plane $\tilde{S}_{1,2}$.

It is useful to understand the effect of all excisions on a bordered surface. An excision of a bordered surface S along an arc is said to be *separating* if it disconnects the surface. It is *border-to-border* if both the endpoints of the path lie on borders. In Figure 3.70 the excisions along a and e are separating, whereas those along c and d are not. The excisions along a, b, and c are border-to-border.

Lemma 3.4.11 *An excision of a bordered surface along an open arc that has exactly one border point does not change the homeomorphism type of the surface.*

PROOF: Let Π be a presentation of the bordered surface S, and let a be an open arc on S. Using modification 5, subdivide a so finely that each of its subarcs either is an arc of the skeleton $G(\Pi)$ or else joins two points on the perimeter of a polygon through its interior. In the latter case, use modification 1 to subdivide all such polygons (see Fig. 3.71), so that it may be assumed without loss of generality that a is in fact an arc of $G(\Pi)$. Bearing in mind that one of the endpoints of a is on the

Figure 3.71 The subdivision of a polygon.

border of S, the excision along a modifies Π into a bordered surface Π' of the same orientability character such that

$$\beta(\Pi') = \beta(\Pi), \quad p(\Pi') = p(\Pi) + 1, \quad q(\Pi') = q(\Pi) + 1, \quad r(\Pi') = r(\Pi),$$

and so $\chi(\Pi') = \chi(\Pi)$. Consequently the surface of Π' is homeomorphic to S.

$$\text{Q.E.D.}$$

Proposition 3.4.12 *Any excision of the bordered surface S along an open arc that joins two distinct borders is nonseparating and results in a bordered surface S' such that $\chi(S') = \chi(S) + 1$. If S is orientable, so is S', and $\gamma(S') = \gamma(S)$.*

PROOF: Think of this excision as the end result of a succession of small excisions of the type described in Lemma 3.4.11 followed by a single small border-to-border excision. Since the excisions of 3.4.11 do not change the homeomorphism type of the surface, it may be assumed that the excision arc is an arc of the skeleton of a presentation Π of S. If the excision results in the presentation Π' of a surface S', then let b denote the border of Π' which consists of the lips of the excision itself together with the two borders of Π that are joined by it. Note that if u and v are two nodes of $G(\Pi)$, then every $u - v$ walk in $G(\Pi)$ that has been disconnected by the excision can be augmented by portions of b into a u-v walk in $G(\Pi')$. Hence the connectedness of S implies the connectedness of the excised surface S'.

As for the characteristics,

$$p(\Pi') = p(\Pi) + 2, \quad q(\Pi') = q(\Pi) + 1, \quad r(\Pi') = r(\Pi),$$

and hence

$$\begin{aligned}
\chi(S') &= p(\Pi') - q(\Pi') + r(\Pi') \\
&= (p(\Pi) + 2) - (q(\Pi) + 1) - r(\Pi) = \chi(S) + 1.
\end{aligned}$$

If S is orientable, then every coherent orientation of S yields a coherent orientation of S', so that the latter is also orientable. In that case, since $\beta(S') = \beta(S) - 1$, it follows that

$$\begin{aligned}
\gamma(S') &= \gamma(S'^c) = \frac{1}{2}(2 - \chi(S'^c)) = \frac{1}{2}(2 - \chi(S') - \beta(S')) \\
&= \frac{1}{2}(2 - \chi(S) - 1 - \beta(S) + 1) = \frac{1}{2}(2 - \chi(S^c)) \\
&= \gamma(S^c) = \gamma(S).
\end{aligned}$$

$$\text{Q.E.D.}$$

Proposition 3.4.13 *If the border-to-border excision of the bordered surface S along an open arc separates it into the surfaces S' and S'', then the excision path begins and ends at the same border, and $\chi(S) = \chi(S') + \chi(S'') - 1$. If S is orientable, so are S' and S'' and $\gamma(S) = \gamma(S') + \gamma(S'')$.*

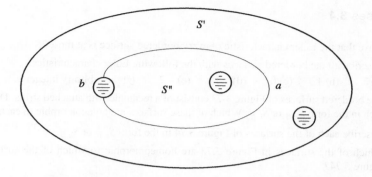

Figure 3.72   A separating border-to-border excision.

PROOF: The fact that the excision path must begin and end in the same border, say b, follows from the previous proposition. Hence, if a denotes the excision path, then Figure 3.72 illustrates a typical separation of S into S' ad S''. As was the case before, it may be assumed that the excision arc is in fact an arc of the skeleton of a presentation Π of S. Then

$$p(\Pi') + p(\Pi'') = p(\Pi) + 2,$$
$$q(\Pi') + q(\Pi'') = q(\Pi) + 1,$$
$$r(\Pi') + r(\Pi'') = r(\Pi),$$
$$\beta(\Pi') + \beta(\Pi'') = \beta(\Pi) + 1.$$

Hence,

$$\chi(S') + \chi(S'') = p(\Pi') - q(\Pi') + r(\Pi') + p(\Pi'') - q(\Pi'') + r(\Pi'')$$
$$= (p(\Pi) + 2) - (q(\Pi) + 1) + r(\Pi) = \chi(S) + 1.$$

If S is orientable, then any coherent orientation of S is also a coherent orientation of S' and S''. Hence, by Theorem 3.2.10,

$$\gamma(S) = \gamma(S^c) = \frac{2 - \chi(S^c)}{2} = \frac{2 - \chi(S) - \beta(S)}{2}$$
$$= \frac{2 - \chi(S') - \chi(S'') + 1 - \beta(S') - \beta(S'') + 1}{2}$$
$$= \frac{2 - \chi(S') - \beta(S')}{2} + \frac{2 - \chi(S'') - \beta(S'')}{2}$$
$$= \frac{2 - \chi(S'^c)}{2} + \frac{2 - x(S''^c)}{2} = \gamma(S'^c) + \gamma(S''^c)$$
$$= \gamma(S') + \gamma(S''). \qquad \text{Q.E.D.}$$

Exercises 3.4

1. Prove that the Euler characteristic of every bordered surface is at most 2.

2. Describe all the bordered surfaces with the following Euler characteristics:
   (a) 2   (b) 1   (c) 0   (d) −1   (e) −2   (f) the arbitrary integer n.

3. The bordered surfaces of Figure 3.73 consist of a rectangle with attached strips. Describe each in the form $S_{n,\beta}$ or $\tilde{S}_{n,\beta}$. Which of these surfaces are homeomorphic to each other?

4. Describe each of the surfaces of Figure 3.74 in the form $S_{n,\beta}$ or $\tilde{S}_{n,\beta}$.

5. Which of the surfaces in Figure 3.73 are homeomorphic to which of the surfaces in Figure 3.74?

6. Does the relation $2 - 2\gamma(S) = \chi(S)$ hold for orientable bordered surfaces?

7. Are the two bordered surfaces in Figure 3.75 homeomorphic? Justify your answer.

8. Which of the surfaces in Figure 3.76 are homeomorphic? Justify your answer.

9. Identify the closures of the surfaces in the following exercises:
   (a) 7   (b) 8.

10. Describe the topological space obtained when each of the Möbius bands in Figure 3.77 is cut along the dashed lines.

11.* Suppose c is a loop of the bordered surface S that contains no border points. Prove that if the excision along c separates S into two bordered surfaces S' and S'', then $\chi(S) = \chi(S') + \chi(S'')$.

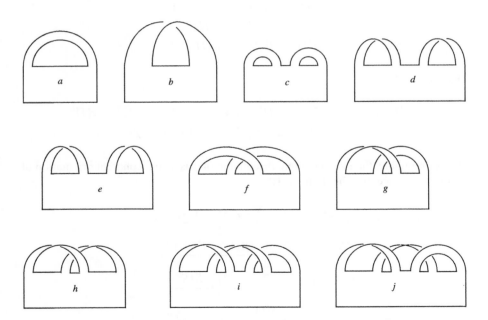

Figure 3.73   Rectangles with strips.

Figure 3.74   Wheel-like surfaces.

12. Suppose c is a loop of the bordered surface S that contains no border points. If the excision along c results in a single bordered surface S', find the relation between $\chi(S)$ and $\chi(S')$.

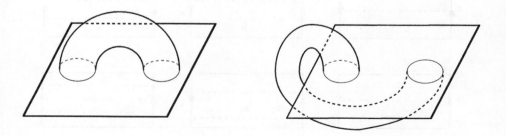

Figure 3.75   Are these surfaces homeomorphic?

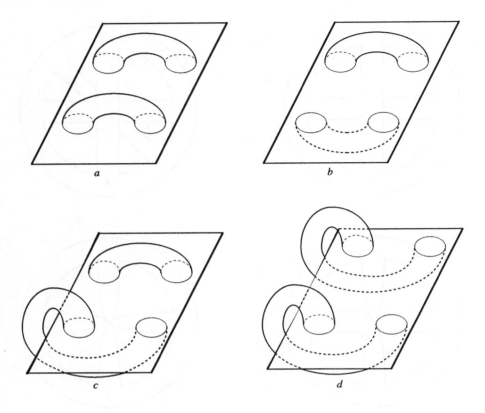

Figure 3.76   Which of these surfaces are homeomorphic?

13. Is the surface of Figure 3.78 homeomorphic to the Möbius band?

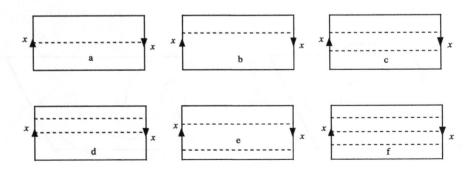

Figure 3.77   Cutting a Möbius band.

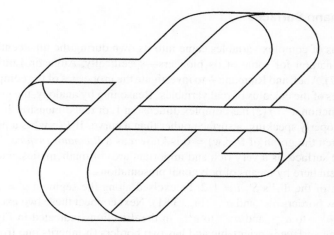

Figure 3.78   Is this surface homeomorphic to the Möbius band?

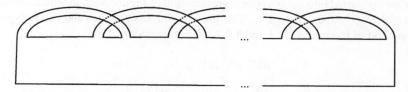

Figure 3.79

14. The bordered surface of Figure 3.79 consists of a rectangle to which n untwisted bands have been attached. Prove that this surface is homeomorphic to $S_{n,1}$.

15. The bordered surface of Figure 3.80 consists of a rectangle to which n bands have been attached, of which only the first is twisted. Prove that this surface is homeomorphic to $\tilde{S}_{n,1}$.

16. Assuming that $\tilde{S}_n$ cannot be embedded in $\mathbb{R}^3$ (see Proposition 5.3.6), use the previous exercise to prove that a nonorientable surface is embeddable in $\mathbb{R}^3$ if and only if it has at least one border.

17. Prove that a bordered surface is nonorientable if and only if it contains a subset that is homeomorphic to the Möbius band.

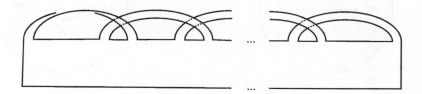

Figure 3.80

3.5 Riemann Surfaces

The calculus of complex variables came into its own during the nineteenth century, and it was natural for some of its pioneers—specifically, Augustin-Louis Cauchy (1789–1857), Abel, and Riemann—to investigate the properties of the complex analogs of the curves of the calculus of real variables. Reasoning by analogy, the graph of the complex function $w = f(z)$ has complex dimension 1, or real dimension 2. Thus, this graph is, properly speaking, a surface rather than a curve. If $P(z, w)$ is a polynomial function, then the graph of $P(z, w) = 0$ is known as a *Riemann surface*. The study of Riemann surfaces is a very rich and important area of mathematics, and we shall describe them here by means of polygonal presentations.

Let each of the disks S^i, $i = 1, 2$, be excised along the segment $a^i = A^i B^i$, thus creating new borders $a^{i,+}$ and $a^{i,-}$ (Fig. 3.81). Next connect these two excised disks by pasting $a^{1,+}$ to $a^{2,-}$ and $a^{2,+}$ to $a^{1,-}$ in the directions indicated in Figure 3.81. The resulting surface is orientable and has two borders (it inherits one from each of the original disks). To compute its genus, cap each of the borders so that the two disks are converted into the polygonal presentation of Figure 3.82, in which one is to paste $a^{1,+}$ to $a^{2,-}$ and $a^{2,+}$ to $a^{1,-}$. This presentation has two arcs and two polygons. The method of Examples 3.2.1 and 3.2.2 yields two nodes:

$$A: \quad a^{1,+}, a^{2,+}, a^{1,+},$$
$$B: \quad \left(a^{1,+}\right)^{-1}, \left(a^{2,+}\right)^{-1}, \left(a^{1,+}\right)^{-1}.$$

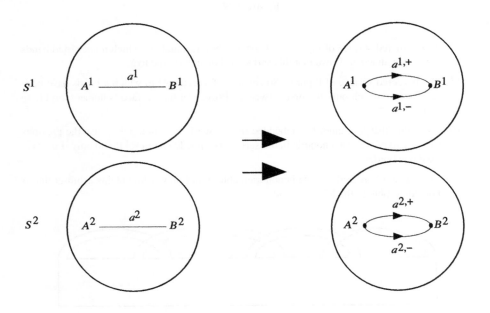

Figure 3.81

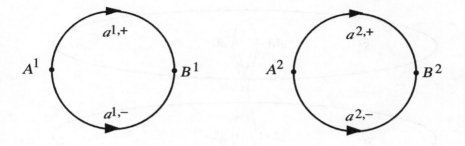

Figure 3.82

The capped surface therefore has characteristic $2 - 2 + 2 = 2$ and genus 0. The uncapped surface obtained in Figure 3.81 is therefore $S_{0,2}$.

Because of the historical origins of this construction, it is customary to visualize the two disks with S^1 stacked up on top of S^2 and with the points A^1, B^1 identified with the points A^2, B^2 respectively. This identification is reasonable because these points get pasted in the construction process anyway. The excision path a^1 is imagined as lying on top of a^2, and their location is marked by an unindexed a (see Fig. 3.83).

Figure 3.84 offers an exaggerated side view of the stacked version of the aforesaid Riemann surface. The troublesome crossing of the two slanted sheets is meant to be disregarded. It can be removed by employing the same device as was used to dispose of the self-crossings of the projective plane and the Klein bottle in Section 3.1. One need only stipulate that the two sheets pass through the crossing line at different times.

Suppose three disks are used instead of two. It is then necessary to imagine Figure 3.83 as a stack of three disks and to specify a pasting rule. One such rule would paste

$$a^{1,+} \text{ to } a^{2,-}, \qquad a^{2,+} \text{ to } a^{3,-}, \qquad a^{3,+} \text{ to } a^{1,-}.$$

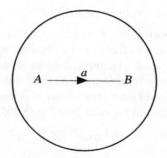

Figure 3.83

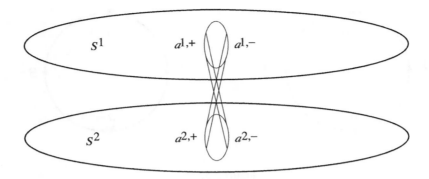

Figure 3.84   A two-sheeted Riemann surface.

In this case the three disks are pasted to form an orientable surface with three borders. The numbers of polygons and arcs in the presentation implicit in Figure 3.83 are three each. There are again two nodes:

$$A: \quad a^{1,+}, a^{3,+}, a^{2,+}, a^{1,+}$$
$$B: \quad \left(a^{1,+}\right)^{-1}, \left(a^{3,+}\right)^{-1}, \left(a^{2,+}\right)^{-1}, \left(a^{1,+}\right)^{-1}.$$

The capped surface therefore has characteristic $2 - 3 + 3 = 2$ and genus 0. The uncapped surface obtained from Figure 3.83 is therefore $S_{0,3}$. A further generalization could use a stack of n disks with the pasting rule

$$a^{1,+} \text{ to } a^{2,-}, \qquad a^{2,+} \text{ to } a^{3,-}, \ldots, a^{n-1,+} \text{ to } a^{n,-}, \qquad a^{n,+} \text{ to } a^{1,-}.$$

The resulting orientable surface has n borders. It has a presentation similar to that of Figure 3.82, except that there are n polygons. Chains similar to those above establish that this presentation has two nodes. The closure of the surface therefore has characteristic $2 - n + n = 2$, yielding a genus of 0. The unclosed surface is therefore $S_{0,n}$.

It is customary to encode the pasting rule as a permutation π with the understanding that

$$a^{j,+} \text{ is to be pasted to } a^{\pi(j),-}.$$

In the three examples discussed so far, π is $(1\ 2)$, $(1\ 2\ 3)$, and $(1\ 2\ \ldots\ n)$ respectively.

More complicated Riemann surfaces can be obtained by increasing the complexity of the excisions. Suppose, for example, that three stacked disks are cut along the three arcs a, b, c emanating from O and with endpoints A, B, C respectively (Fig. 3.85). The "+" and "−" sides of each excision are also displayed in the diagram and they are to be pasted according to the rules (see Fig. 3.86)

$$a^{j,+} \text{ to } a^{j+1,-}, \qquad b^{j,+} \text{ to } b^{j+1,-}, \qquad c^{j,+} \text{ to } c^{j+1,-},$$

where the index addition is done modulo 3. The resulting surface, call it S, has three borders and is orientable. As was the case before, the pasting rules result in all the

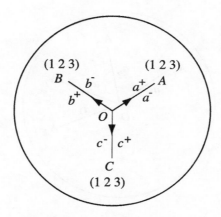

Figure 3.85   The code for a Riemann surface.

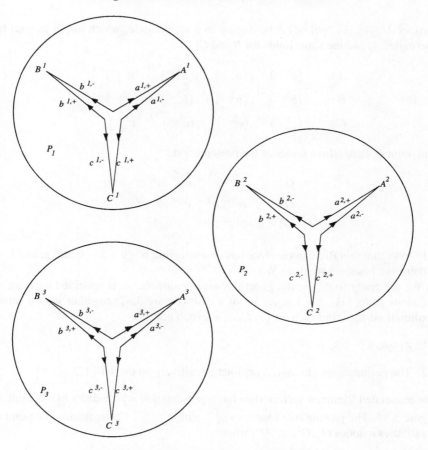

Figure 3.86   A presentation of a Riemann surface.

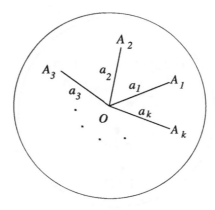

Figure 3.87   The specification of a general Riemann surface.

vertices A^1, A^2, A^3 "below" A belonging to a single node, which might as well be also called A, and the same holds for B and C:

$$A: \quad \left(a^{1,-}\right)^{-1}, \left(a^{2,-}\right)^{-1}, \left(a^{3,-}\right)^{-1}, \left(a^{1,-}\right)^{-1};$$
$$B: \quad \left(b^{1,-}\right)^{-1}, \left(b^{2,-}\right)^{-1}, \left(b^{3,-}\right)^{-1}, \left(b^{1,-}\right)^{-1};$$
$$C: \quad \left(c^{1,-}\right)^{-1}, \left(c^{2,-}\right)^{-1}, \left(c^{3,-}\right)^{-1}, \left(c^{1,-}\right)^{-1}.$$

The point O yields three nodes of the presentation:

$$O^1: \quad a^{1,-}, c^{2,-}, b^{3,-}, a^{1,-};$$
$$O^2: \quad b^{1,-}, a^{2,-}, c^{3,-}, b^{1,-};$$
$$O^3: \quad c^{1,-}, b^{2,-}, a^{3,-}, c^{1,-}.$$

It follows that this Riemann surface has characteristic $6 - 9 + 3 = 0$ and genus 1. It is therefore homeomorphic to $S_{1,3}$.

We are ready to define the general *Riemann surface*. It is specified by a set of excisions paths OA_i, $i = 1, 2, \ldots, m$, of a disk (Figure 3.87) together with a set of permutations π_i acting each on $\{1, 2, \ldots, n\}$, such that:

1. $\pi_1 \circ \pi_2 \circ \cdots \circ \pi_k = \text{Id}$.

2. The permuations $\{\pi_1, \pi_2, \ldots, \pi_k\}$ act transitively on the set $\{1, 2, \ldots, n\}$.

The associated Riemann surface then has a presentation with n disks like the one in Figure 3.88. The pasting rule identifies $a_i^{j,+}$ with $a_i^{\pi_i(j),-}$. Consequently the point O contributes n nodes $O^1, O^2, \ldots, O^n$, where

$$O^j: \quad a_1^{j,-}, a_k^{\pi_k(j),-}, a_{k-1}^{\pi_{k-1}\pi_k(j),-}, \ldots, a_1^{\pi_1 \cdots \pi_{k-1}\pi_k(j),-} = a_1^{j,-}.$$

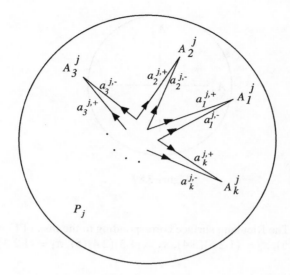

Figure 3.88   A typical disk in the presentation of a Riemann surface.

On the other hand, the point A_i contributes one point for each of its cyclic factors. Specifically, if $\sigma = (j_1 \ j_2 \ \cdots \ j_s)$ is any cyclic factor of π_i, then it contributes the node

$$A_{i,\sigma}: \qquad \left(a_i^{j,-}\right)^{-1}, \left(a_i^{\pi_i(j),-}\right)^{-1}, \left(a_i^{\pi_i^2(j),-}\right)^{-1}, \ldots, \left(a_i^{\pi_i^s(j),-}\right)^{-1} = \left(a_i^{j,-}\right)^{-1},$$

where j is any of the indices $j_1, j_2, \ldots, j_s$. Note that this Riemann surface has n borders.

Theorem 3.5.1 *The general Riemann surface is orientable and has genus*

$$\frac{2 + (k-2)n - \sum_{i=1}^{k} \|\pi_i\|}{2}.$$

PROOF: Condition 2 of the definition of the general Riemann surface implies that it is possible to move along the surface from any sheet to any other sheet. It follows that the surface is connected. The node calculations above imply that the capped surface has characteristic

$$\chi(S) = n + \sum_{i=1}^{k} \|\pi_i\| - kn + n = \sum_{i=1}^{k} \|\pi_i\| - (k-2)n.$$

As the orientability of the surface follows directly from the orientability of the disk $S_{1,0}$ and the pasting rule, we conclude that the surface has the desired genus. Q.E.D.

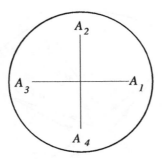

Figure 3.89

Example 3.5.2 The Riemann surface corresponding to the disk of Figure 3.89 with $\pi_1 = (1\ 4\ 3)(2)(5)$, $\pi_2 = (1\ 2\ 5)(3\ 4)$, $\pi_3 = (1\ 3)(2\ 4)(5)$, $\pi_4 = (1\ 2\ 3)(4\ 5)$ has the following nodes:

$$O^1: \quad a_1^{1,-}, a_4^{2,-}, a_3^{4,-}, a_2^{3,-}, a_1^{1,-}$$

$$O^2: \quad a_1^{2,-}, a_4^{3,-}, a_3^{1,-}, a_2^{2,-}, a_1^{2,-}$$

$$O^3: \quad a_1^{3,-}, a_4^{1,-}, a_3^{3,-}, a_2^{4,-}, a_1^{3,-}$$

$$O^4: \quad a_1^{4,-}, a_4^{5,-}, a_3^{5,-}, a_2^{1,-}, a_1^{4,-}$$

$$O^5: \quad a_1^{5,-}, a_4^{4,-}, a_3^{2,-}, a_2^{5,-}, a_1^{5,-}$$

$$A_{1,(1\ 4\ 3)}: \quad \left(a_1^{1,-}\right)^{-1}, \left(a_1^{4,-}\right)^{-1}, \left(a_1^{3,-}\right)^{-1}, \left(a_1^{1,-}\right)^{-1}$$

$$A_{1,(2)}: \quad \left(a_1^{2,-}\right)^{-1}, \left(a_1^{2,-}\right)^{-1}$$

$$A_{1,(5)}: \quad \left(a_1^{5,-}\right)^{-1}, \left(a_1^{5,-}\right)^{-1}$$

$$A_{2,(1\ 2\ 5)}: \quad \left(a_2^{1,-}\right)^{-1}, \left(a_2^{2,-}\right)^{-1}, \left(a_2^{5,-}\right)^{-1}, \left(a_2^{1,-}\right)^{-1}$$

$$A_{2,(3\ 4)}: \quad \left(a_2^{3,-}\right)^{-1}, \left(a_2^{4,-}\right)^{-1}, \left(a_2^{3,-}\right)^{-1}$$

$$A_{3,(1\ 3)}: \quad \left(a_3^{1,-}\right)^{-1}, \left(a_3^{3,-}\right)^{-1}, \left(a_3^{1,-}\right)^{-1}$$

$$A_{3,(2\ 4)}: \quad \left(a_3^{2,-}\right)^{-1}, \left(a_3^{4,-}\right)^{-1}, \left(a_3^{2,-}\right)^{-1}$$

$$A_{3,(5)}: \quad \left(a_3^{5,-}\right)^{-1}, \left(a_3^{5,-}\right)^{-1}$$

$$A_{4,(1\ 2\ 3)}: \quad \left(a_4^{1,-}\right)^{-1}, \left(a_4^{2,-}\right)^{-1}, \left(a_4^{3,-}\right)^{-1}, \left(a_4^{1,-}\right)^{-1}$$

$$A_{4,(4\ 5)}: \quad \left(a_4^{4,-}\right)^{-1}, \left(a_4^{5,-}\right)^{-1}, \left(a_4^{4,-}\right)^{-1}$$

The genus of this surface is $(2+10-10)/2 = 1$, and so its closure is homeomorphic to the torus.

The constituent polygons of a Riemann surface are called its *sheets*, and the nodes $A_{i,\sigma}$ are its *branch points*.

Exercises 3.5

1. Describe the nodes of the Riemann surface with $\pi_1 = \pi_2 = \pi_3 = \pi_4 = (1\ 2\ 3\ 4)$, and compute its genus.

2. Describe the nodes of the Riemann surface with $\pi_1 = \pi_2 = \cdots = \pi_n = (1\ 2\ \cdots\ n)$, and compute its genus.

3. Describe the nodes of the Riemann surface with $\pi_i = (i\ i+1)$ for $i = 1, 2, \ldots, n-1$ and $\pi_n = (n\ n-1\ \cdots\ 2\ 1)$, and compute its genus.

4. Which of the following defines a Riemann surface?

    (a) $\pi_1 = (1\ 3)(2\ 4)$, $\pi_2 = (1)(2\ 4)(3)$, $\pi_3 = (1\ 3)(2)(4)$;

    (b) $\pi_1 = \pi_2 = \pi_3 = (1\ 2\ 3\ 4)$;

    (c) $\pi_1 = (1\ 3)(2\ 4)$, $\pi_2 = (1\ 4)(2)(3)$, $\pi_3 = (1\ 2\ 4\ 3)$.

5. Prove that if $\pi_1, \pi_2, \ldots, \pi_k$ define an n-sheeted Riemann surface, then kn and $\sum_{i=1}^{k} \|\pi_i\|$ have the same parity.

Chapter Review Exercises

1. Identify the following surfaces:

    (a) One with the polygonal presentation $abcd, a^{-1}b^{-1}c^{-1}d^{-1}$.

    (b) One with the polygonal presentation $abcd, a^{-1}b^{-1}c^{-1}d$.

    (c) Surface a of Figure 3.90.

    (d) Surface b of Figure 3.90.

    (e) Surface c of Figure 3.90.

    (f) $S_3 + S_4$.

    (g) $\tilde{S}_3 + S_4$.

    (h) $S_3 + \tilde{S}_4$.

    (i) $\tilde{S}^3 + \tilde{S}_4$.

    (j) $S_3 + h$.

    (k) $S_3 + \tilde{h}$.

    (l) $S_3 + c$.

    (m) $\tilde{S}_3 + h$.

    (n) $\tilde{S}_3 + \tilde{h}$.

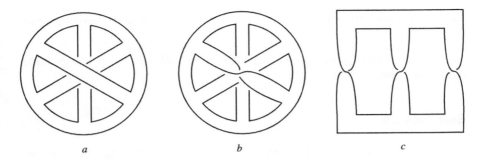

Figure 3.90

(o) $\tilde{S}_3 + c$.

2. Are the following statements true or false? Justify your answers.

(a) Every two closed surfaces are homeomorphic.

(b) Every two bordered surfaces with the same number of borders are homeomorphic.

(c) If two closed surfaces have the same Euler characteristic, then they are homeomorphic.

(d) If two orientable closed surfaces have the same Euler characteristic, then they are homeomorphic.

(e) If two bordered surfaces have the same Euler characteristic and orientability character, then they are homeomorphic.

(f) Every closed surface has a presentation with exactly one polygon.

(g) Every closed surface has a presentation with 100 polygons.

(h) Given 200 closed surfaces each of which has characteristic greater than -100, some two of them are homeomorphic to each other.

(i) Given 300 closed surfaces each of which has characteristic greater than -100, some two of them are homeomorphic to each other.

(j) Every bordered surface has the same Euler characteristic as its closure.

(k) Every bordered surface has the same genus as its closure.

(l) Every nonorientable bordered surface is homeomorphic to some surface of the form $\tilde{S}^{n,\beta}$.

(m) $S_7 + \tilde{S}_{15} \approx \tilde{S}_{22}$.

(n) $S_{15} + \tilde{S}_7 \approx \tilde{S}_{29}$.

(o) $S_{15} + \tilde{S}_{15} \approx S_{45}$.

CHAPTER 4

GRAPHS AND SURFACES

This chapter is concerned with the interplay between the concepts of graphs and surfaces. Given a graph and a surface, it is reasonable to ask whether the graph can be embedded on the surface. This question is answered in full for orientable surfaces.

4.1 Embeddings and Their Regions

A graph G is *embedded* on a surface S if it is homeomorphic to a subset of S. Less formally, G can be embedded on S provided that it can be drawn on the surface so that any two arcs intersect only in their common endpoints. By definition, every planar graph can be embedded on the disk $S_{0,1}$ and vice versa. This class of graphs is also identical with those that are embeddable on the sphere S_0.

Proposition 4.1.1 *A graph G is embeddable on the sphere S_0 if and only if it is planar.*

PROOF: Figure 4.1 shows how the stereographic projection can be used to convert spherical embeddings to plane embeddings and vice versa. One need merely make sure that when a spherical embedding is given, the sphere is so positioned on the

Introduction to Topology and Geometry, Second Edition.
By Saul Stahl and Catherine Stenson Copyright © 2013 John Wiley & Sons, Inc.

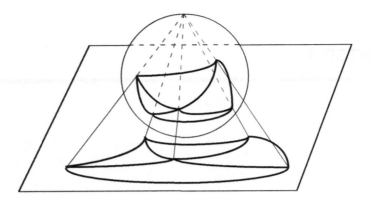

Figure 4.1 Shuttling embeddings between the sphere and the plane.

target plane that the graph on the sphere does not contain the north pole. This is clearly always possible. Q.E.D.

This chapter is concerned with the following natural question:

Given a graph and a surface, can the graph be embedded on the surface?

There is a well-defined procedure which answers this question in theory. In practice, the computational aspects of this procedure are such as to render it impractical for all but the simplest graphs. The existence of this method will be demonstrated below for all orientable surfaces.

As was noted in the previous chapter, every orientable surface can be realized in $\mathbb{R}^3$ as a sphere with handles (or a sum of tori). Consequently such a surface may be assumed to have a well-defined exterior as well as an orientation that will always be counterclockwise from the point of view of an observer stationed in that exterior. The resolution of the above question, which takes up the bulk of this chapter, begins with showing that every graph can be embedded on some orientable surface.

Proposition 4.1.2 *Every graph G embeds on some orientable surface.*

PROOF: Position the given graph G in $\mathbb{R}^3$ so that its arcs do not intersect (except at their endpoints, of course). The feasibility of such a placement is quite plausible, but a formal argument is indicated in Exercise 5. Next, replace each node of G with a small sphere, and each arc with a narrow tube that connects the spheres that have replaced that arc's endpoints (see Figure 4.2). The girths of the spheres and tubes should be such as to avoid any spurious intersections. It is clear that the graph G can be embedded on the resulting collection of surfaces, with each node and arc being drawn on the sphere or tube that replaced it. If G is disconnected, the various embeddings of its components can be added in the sense of Proposition 3.3.1 so as to obtain a connected surface on which G is embedded in its entirety. Q.E.D.

It is next demonstrated that the set of integers n such that G can be embedded on S_n consists of all the integers that are greater than some fixed integer.

Proposition 4.1.3 *For every graph G there is a nonnegative integer $\gamma(G)$ such that G embeds on the orientable surface S_n if and only if $n \geq \gamma(G)$.*

PROOF: Let G be a fixed graph. In view of Proposition 4.1.2, it suffices to show that if $m \geq n$ and G embeds on S_n, then G also embeds on S_m. Suppose the graph G is embedded on S_n for some nonnegative integer n. Choose two disjoint loops on S_n that do not intersect G, and connect them by a handle. This results in an embedding of G on S_{n+1}. It follows by induction that G embeds on S_m whenever $m \geq n$. Q.E.D.

The parameter $\gamma(G)$ defined in the statement of Proposition 4.1.3 is called the *(orientable) genus* of the graph G. Thus, a graph G is planar if and only if its genus is 0. Since both K_5 and $K_{3,3}$ can be embedded on the torus S_1, and since neither of them is planar, it follows that $\gamma(K_5) = \gamma(K_{3,3}) = 1$. It will be demonstrated in Section 4.3 that for every integer n there exists a graph G_n such that $\gamma(G_n) = n$.

Suppose a graph G is embedded on a surface S, and suppose further that this surface is excised along all the arcs of G. This decomposes the surface into a collection of bordered surfaces each of whose borders consists of one or more arcs of G. These surfaces are called the *regions* of the embedding.

Example 4.1.4 The spherical embedding of K_4 in Figure 4.3 has four triangular regions marked A, B, C, D. Assuming that each region is oriented in the counterclockwise sense, their perimeters are described as

$$A : bed, \qquad B : ac^{-1}b^{-1}, \qquad C : fe^{-1}c, \qquad D : f^{-1}a^{-1}d^{-1}.$$

Example 4.1.5 In the toroidal embedding of Figure 4.4 the region R has perimeter $cbc^{-1}a$, whereas the perimeter of the region T consists of the disjoint union of the arcs a and b.

The visualization of regions and their perimeters is of course more difficult on surfaces of higher genus. The identification of the (counterclockwise) perimeters of the regions is facilitated by the following observation (Fig. 4.5):

Figure 4.2   An embedding of K_4 on an orientable surface.

When a <u>counterclockwise</u> perimeter enters a node u along an arc x, it leaves u along that arc y which is first encountered when we turn away from x^{-1} in the <u>clockwise</u> direction.

Example 4.1.6 The toroidal embedding of K_4 in Figure 4.6 has two regions A and B whose perimeters are

$$A : afe^{-1}b^{-1}, \qquad B : f^{-1}c^{-1}edbca^{-1}d^{-1}.$$

Given an embedded graph, it is both convenient and natural to visualize its regions as parts of the ambient surface. This visualization, however, can be misleading as it is easy to forget that some of the apparent self-abutments of a region on the ambient surface are in fact not really there. Such is the case for the aforementioned region B in Figure 4.6, which, on the surface, looks as if it abuted itself along c and d, whereas in reality it has been excised along these two arcs, so that the region B is actually an octagon whose sides are labeled f^{-1}, c^{-1}, e, d, b, c, a^{-1}, and d^{-1}.

There is another important difference between regions and their visualization on the ambient surfaces. In addition to spuriously identifying distinct arcs, this visualization may also do the same to distinct nodes on the perimeters of the regions. Note, for instance, the points labeled u on the borders of the actual regions in Figure 4.7. They all correspond to the same point on the surface, but become distinct points once the region has been excised from the surface. Moreover, when visualized on the ambient surface, the region may not look quite like a bordered surface. The version of the region A visualized on the surface in Figure 4.7 has its two borders intersecting in the point u, thus violating the definition, which stipulates that the borders of a surface do *not* intersect. The actual regions, on the other hand, are always bordered surfaces.

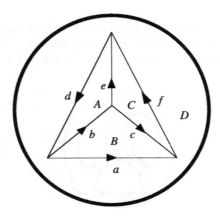

Figure 4.3   A spherical embedding of K_4.

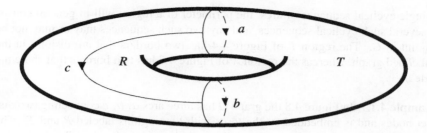

Figure 4.4 A toroidal embedding

Note also that the perimeter of a region need not be connected. The perimeter of the actual region A in Figure 4.7 has two components, whereas that of B has three. In general

> *Each border of a region is a subset of its perimeter, and the union of the region's borders constitutes its perimeter.*

The collection of regions of an embedding of a graph on a surface S resembles a polygonal presentation of S in that all the arcs on the perimeters are paired and the surface can be reconstructed by pasting the paired arcs. There are important differences, however. In the first place, the regions need not be polygons; unlike the polygons, they can be surfaces with positive genus. Such is the case for the region R of the toroidal embedding of Figure 4.8, which is homeomorphic to $S_{1,2}$. Moreover, while the oriented perimeter of a polygon in a presentation consists of

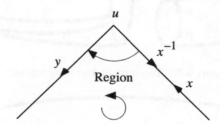

Figure 4.5 Computing the perimeter of a region.

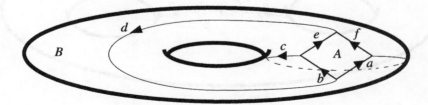

Figure 4.6 A toroidal embedding of K_4.

a single cyclical sequence of arcs, the perimeter of a region will in general consist of several such cyclical sequences, and any two such sequences may or may not be disjoint in G. The region T of Figure 4.4 has two borders that are disjoint in the embedded graph, whereas the region A of Figure 4.7 has two borders that share the node u.

Example 4.1.7 In Figure 4.8 the graph G has three arcs a, b, c connecting two distinct nodes and is embedded on the torus with two regions labeled R and T. The region T consists of a polygon with perimeter $b^{-1}c$, which also describes its border.

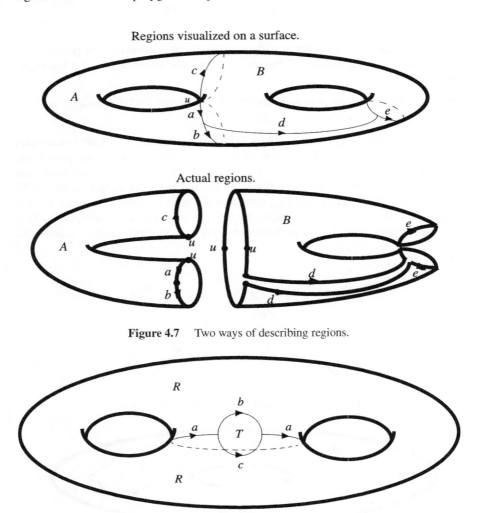

Regions visualized on a surface.

Actual regions.

Figure 4.7 Two ways of describing regions.

Figure 4.8 A graph embedding on a torus with two regions.

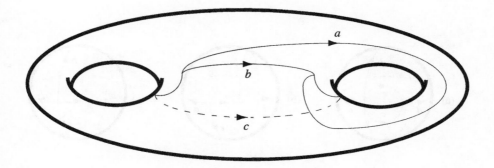

Figure 4.9   A graph embedded on a double torus with a single region.

The region R is a twice perforated torus whose borders are $c^{-1}a^{-1}$ and ab respectively.

Example 4.1.8 In Figure 4.9 the graph G has three arcs a, b, c connecting two distinct nodes and is embedded on the double torus with a single region. One way to see that this embedding has only one region is to note that it is possible to connect any surface point off the graph to any other surface point off the graph by means of an arc that, while lying on the surface, does not intersect the graph. Alternatively, an application of the process described in Examples 4.1.5 and 4.1.6 also yields a single perimeter $ab^{-1}ca^{-1}bc^{-1}$. The region is actually a perforated torus. To see this observe that the excision along b leaves a once perforated double torus $S_{2,1}$ with border bb^{-1}. The further excision along c results in a twice perforated torus with borders bc^{-1} and $b^{-1}c$. The final excision along a converts this to a once perforated torus with border $ab^{-1}ca^{-1}bc^{-1}$.

Exercises 4.1

1. Describe the perimeter, borders, and genus of each of the regions of the graph embeddings in Figure 4.10.

2. Describe the perimeter, borders, and genus of each of the regions of the graph embeddings in Figure 4.11.

3. Describe the perimeter, borders, and genus of each of the regions of the embedding of the one-node graph G with loops $\{e, f, g, h\}$ in Figure 4.12. What is the genus of the ambient surface?

4. Describe the perimeter, borders, and genus of each of the regions of the graph embeddings in Figure 4.13.

5. Prove that if R is a region of the embedding of some graph on an orientable surface S, then $\gamma(R) \leq \gamma(S)$.

6. Prove that every graph embeds on some nonorientable surface.

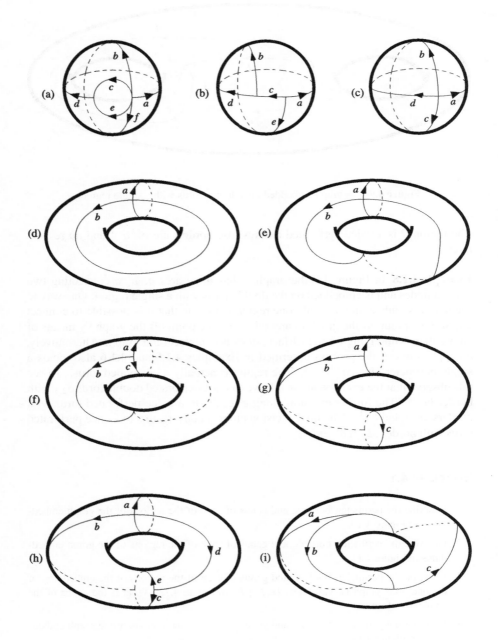

Figure 4.10 Planar and torodial embeddings.

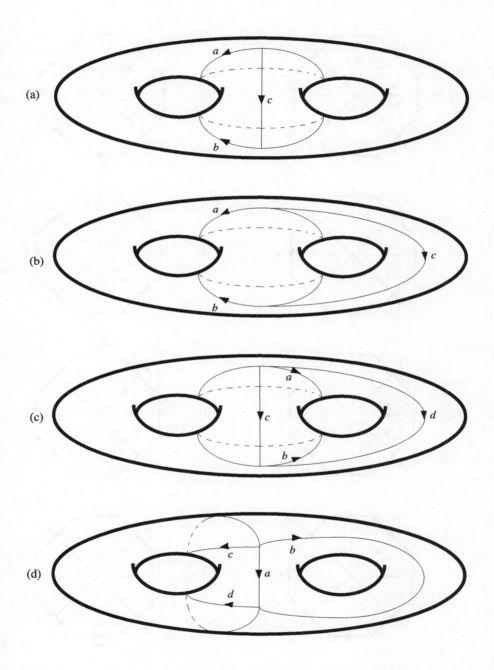

Figure 4.11   Embeddings on the double torus.

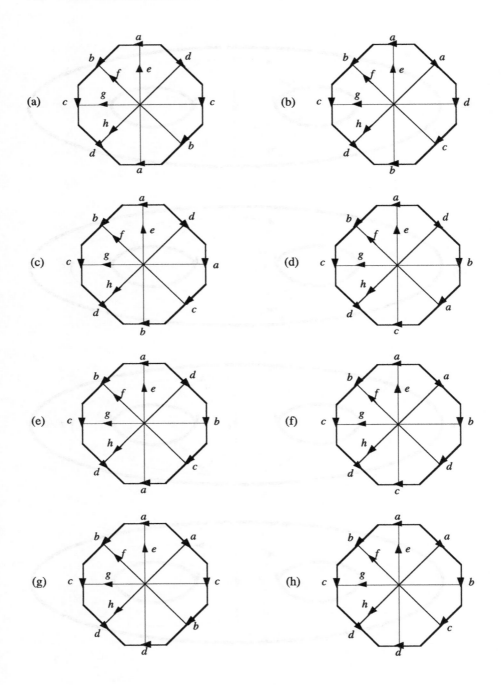

Figure 4.12   Embeddings on polygonal presentations.

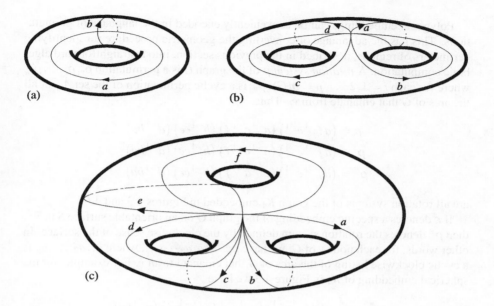

Figure 4.13   Embedded graphs.

7. Prove that if the graph G embeds on the nonorientable surface $\tilde{S}_n$, then it embeds on $\tilde{S}_m$ for each $m \geq n$.

4.2 Polygonal Embeddings

Polygonal embeddings are those graph embeddings whose regions are polygons. They are of special interest because their regions constitute a polygonal presentation of the ambient surface. Since the skeletons of these presentations are the embedded graphs themselves, the following important equation is an easy consequence of the classification of surfaces in Theorem 3.2.10

Theorem 4.2.1 (Euler–Poincaré Equation) *If the connected graph G has a polygonal embedding with r regions on the orientable surface S_γ, then*

$$p - q + r = 2 - 2\gamma,$$

where $p = p(G)$ and $q = q(G)$.

PROOF: Let Π denote the polygonal presentation of S_γ defined by the regions of the embedding of G, so that $p(\Pi) = p(G), q(\Pi) = q(G)$, and $r(\Pi) = r$. It then follows from Theorem 3.2.6 that

$$p - q + r = p(\Pi) - q(\Pi) + r(\Pi) = \chi(\Pi) = 2 - 2\gamma. \qquad \text{Q.E.D.}$$

Polygonal embeddings can be conveniently encoded in the language of permutations. This has the advantage of converting the geometric procedure for identifying perimeters of regions, described in the previous section, into a straightforward algebraic computation. A *rotation system* ρ of the graph G is a permutation $\rho_{v_1} \rho_{v_2} \cdots \rho_{v_p}$ where for each $i = 1, 2, \ldots, p = p(G)$, ρ_{v_i} is a cyclic permutation of the set A_{v_i} of all the arcs of G that emanate from v_i. Thus,

$$\rho = \left(df^{-1}e^{-1}\right)\left(a^{-1}c^{-1}f\right)\left(b^{-1}ce\right)\left(d^{-1}ba\right)$$
$$\rho' = \left(df^{-1}e^{-1}\right)\left(a^{-1}c^{-1}f\right)\left(b^{-1}ec\right)\left(d^{-1}ba\right)$$
$$\rho'' = \left(df^{-1}e^{-1}\right)\left(c^{-1}a^{-1}f\right)\left(b^{-1}ce\right)\left(d^{-1}ba\right)$$

are all rotation systems of the graph K_4 embedded in Figures 4.3 and 4.6.

If ε denotes a specific embedding of the graph G on an orientable surface S in $\mathbb{R}^3$, then ρ^ε denotes the rotation system defined by the clockwise sense of the surface. In other words, for each node v of G, we have $\rho_v^\varepsilon = (a_1 a_2 \cdots a_d)$, where $a_1, a_2, \ldots, a_d$ is a cyclic clockwise listing of the arcs of G that emanate from v. For example, for the spherical embedding of K_4 in Figure 4.3 we have

$$\rho^\varepsilon = \rho' = \left(df^{-1}e^{-1}\right)\left(a^{-1}c^{-1}f\right)\left(b^{-1}ec\right)\left(d^{-1}ba\right),$$

whereas for the toroidal embedding of K_4 in Figure 4.6 we have

$$\rho^\varepsilon = \rho'' = \left(df^{-1}e^{-1}\right)\left(c^{-1}a^{-1}f\right)\left(b^{-1}ce\right)\left(d^{-1}ba\right).$$

For every graph G the *arc involution* I_G is that permutation such that $I_G(a) = a^{-1}$ for every directed arc a of G. In other words,

$$I_G = \left(a_1 a_1^{-1}\right)\left(a_2 a_2^{-1}\right)\ldots\left(a_i a_i^{-1}\right)\ldots\left(a_q a_q^{-1}\right),$$

where $\left(a_i a_i^{-1}\right)$ runs through all the arcs of G. For example, for the above K_4 we have

$$I_{K_4} = \left(aa^{-1}\right)\left(bb^{-1}\right)\left(cc^{-1}\right)\left(dd^{-1}\right)\left(ee^{-1}\right)\left(ff^{-1}\right).$$

Proposition 4.2.2 *Let ε denote a polygonal embedding of the connected graph G into the orientable surface S. Let $I = I_G$ be the arc involution of G. Then the perimeters of the regions of the embedding ε coincide with the cycles of the composition $\rho^\varepsilon \circ I_G$.*

PROOF: Let R be any region of the embedding ε, and let $\cdots abc \cdots$ be a portion of the counterclockwise perimeter of R (Fig. 4.14). Then $\rho^\varepsilon \circ I_G(a) = \rho^\varepsilon(a^{-1}) = b$ and $\rho^\varepsilon \circ I_G(b) = \rho^\varepsilon(b^{-1}) = c$. In other words, if $(abc \cdots)$ is the cycle of $\rho^\varepsilon \circ I_G$ that contains the arc a, then a, b, $c, \ldots$ also constitute the counterclockwise traversal of the perimeter of R. This is the desired correspondence. Note that the coherence of the orientation guarantees that for every arc a, both a and a^{-1} are accounted for in this process.   Q.E.D.

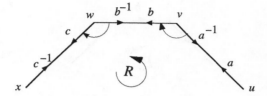

Figure 4.14   Computing the perimeter of a region.

The restriction of the above proposition to connected graphs is necessitated by the fact that disconnected graphs do not have polygonal embeddings (Exercise 16). Of course, as noted in Proposition 4.1.2, all graphs, connected or not, do have orientable embeddings.

Example 4.2.3 For the spherical embedding of K_4 in Figure 4.3 we have

$$\rho' \circ I_{K_4} = \left(df^{-1}e^{-1}\right)\left(a^{-1}c^{-1}f\right)\left(b^{-1}ec\right)\left(d^{-1}ba\right)$$
$$\circ \left(aa^{-1}\right)\left(bb^{-1}\right)\left(cc^{-1}\right)\left(dd^{-1}\right)\left(ee^{-1}\right)\left(ff^{-1}\right)$$
$$= \left(ac^{-1}b^-1\right)\left(bed\right)\left(cfe^{-1}\right)\left(d^{-1}f^{-1}a^{-1}\right),$$

whose cycles constitute the perimeters of the regions B, A, C, D respectively.

Example 4.2.4 For the toroidal embedding of K_4 in Figure 4.6 we have

$$\rho'' \circ I_{K_4} = \left(df^{-1}e^{-1}\right)\left(c^{-1}a^{-1}f\right)\left(b^{-1}ce\right)\left(d^{-1}ba\right)$$
$$\circ \left(aa^{-1}\right)\left(bb^{-1}\right)\left(cc^{-1}\right)\left(dd^{-1}\right)\left(ee^{-1}\right)\left(ff^{-1}\right)$$
$$= \left(afe^{-1}b^{-1}\right)\left(a^{-1}d^{-1}f^{-1}c^{-1}edbc\right),$$

whose cycles constitute the perimeters of the regions A and B respectively.

It is next demonstrated that every rotation system of a graph G actually describes one of its orientable embeddings.

Proposition 4.2.5 *Given any rotation system ρ of the connected graph G, there exists an orientable polygonal embedding ε on some closed orientable surface S such that $\rho^\varepsilon = \rho$.*

PROOF: For each cycle $\sigma = (abc \cdots)$ of $\rho \circ I_G$, let P_σ be a plane polygon whose sides are labeled, in their counterclockwise traversal, $abc \cdots$ (Fig. 4.15). For any arc a, the arc a^{-1} also belongs to some polygon P_τ whose counterclockwise perimeter is consistent with a^{-1} and so inconsistent with a. It follows that the set of polygons $\Pi = \{P_\sigma\}$ constitutes an orientable polygonal presentation such that the skeleton $G(\Pi)$ has the same arc set as G. If the ambient orientable surface S is placed in $\mathbb{R}^3$, then this constitutes an embedding ε of $G(\Pi)$ on S with a corresponding rotation system ρ^ε. By Proposition 4.2.2, the counterclockwise perimeters of the regions of

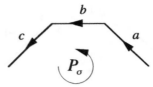

Figure 4.15   The perimeter of a region.

this embedding are given by the cycles of $\rho^{\varepsilon} \circ I_G$. By construction, however, the counterclockwise perimeters of this embedding are given by the cycles of $\rho \circ I_G$. It follows that $\rho^{\varepsilon} \circ I_G = \rho \circ I_G$ and hence $\rho^{\varepsilon} = \rho$. Since the cycles of ρ^{ε} and ρ completely determine the nodes of $G(\Pi)$ and G respectively, it follows that $G(\Pi)$ and G constitute the same graph on their common arc set.   Q.E.D.

Example 4.2.6 Let $\rho = (df^{-1}e^{-1})(a^{-1}c^{-1}f)(b^{-1}ce)(d^{-1}ba)$ be a rotation system of K_4 (whose arcs are labeled as in Figures 4.3 and 4.6). Then

$$\rho \circ I_G = \left(df^{-1}e^{-1}\right)\left(a^{-1}c^{-1}f\right)\left(b^{-1}ce\right)\left(d^{-1}ba\right)$$
$$\circ \left(aa^{-1}\right)\left(bb^{-1}\right)\left(cc^{-1}\right)\left(dd^{-1}\right)\left(ee^{-1}\right)\left(ff^{-1}\right)$$
$$= \left(ac^{-1}edbcfe^{-1}b^{-1}\right)\left(a^{-1}d^{-1}f^{-1}\right).$$

Hence the corresponding embedding has the regions in Figure 4.16.

Since in this polygonal embedding $p(K_4) = 4$, $q(K_4) = 6$, and $r = 2$, the genus of the ambient surface is given by the equation

$$4 - 6 + 2 = 2 - 2\gamma(S),$$

and so it must be the torus.

Exercises 4.2

1. List all the rotation systems of the graph G_1 of Figure 4.17 and compute the genera of the associated embeddings.

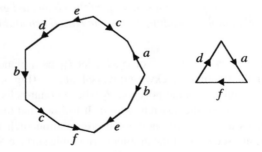

Figure 4.16   An embedding of K_4.

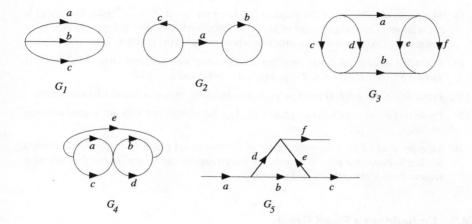

Figure 4.17   Some small graphs.

2. List all the rotation systems of the graph G_2 of Figure 4.17, and compute the genera of the associated embeddings.

3. List all the rotation systems of the graph G_3 of Figure 4.17, and compute the genera of the associated embeddings.

4. For each nonnegative integer n, how many of the rotation systems of the graph G_4 of Figure 4.17, define embeddings of genus n?

5. For each nonnegative integer n, how many of the rotation systems of the graph G_5 of Figure 4.17 define embeddings of genus n?

6. For each nonnegative integer n, how many of the rotation systems of the graph K_4 define embeddings of genus n?

7. What is the rotation system associated with the embeddings in Figure 4.10 (a), (b), (c), (d)?

8. What is the rotation system associated with the embeddings in Figure 4.11 (c), (d)?

9. What is the rotation system associated with the embeddings in Figures 4.12 (c), (d)?

10. Prove that if the graph G has degree sequence $d_1, d_2, \ldots, d_p$, then it has $\prod_{i=1}^{p}(d_i - 1)!$ distinct rotation systems.

11. Prove that if G has an orientable polygonal embedding of genus m, then

$$m \leq \frac{q(G) - p(G) + 1}{2}.$$

12*. Prove that if H has orientable polygonal embeddings of genera $m < n$, then G has an orientable polygonal embedding of genus k whenever $m \leq k \leq n$.

13**. A graph G is said to have embedding range 1 if all of its orientable polygonal embeddings have the same genus. Prove that every graph G with embedding range 1 is planar.

14. The genus distribution of the graph G is a function $f_G : \mathbb{Z}^{\geq 0} \to \mathbb{Z}^{\geq 0}$ such that $f_G(g)$ is the number of rotation systems of G whose associated orientable polygonal embedding has genus g. Write a computer program whose output is the function f_G.

15. Prove that if the graph G has a polygonal embedding with r regions on the nonorientable surface $\tilde{S}_\gamma$, then $p - q + r = 2 - \gamma$, where $p = p(G)$ and $q = q(G)$.

16*. Prove that if the graph G has a polygonal embedding, then it is necessarily connected.

17*. Prove that every connected graph has a polygonal embedding with one region on some surface.

18. Suppose ε and i are two embeddings of the connected graph G on the same oriented surface S. Prove that $\rho^\varepsilon = \rho^i$ if and only if their regions can be matched so that matching regions have identically labeled perimeters.

4.3 Embedding a Fixed Graph

This section describes all the orientable bordered surfaces on which a given graph G can be embedded. In particular it also provides a clear-cut answer to the question:

Given a graph G and an orientable bordered surface S, does G embed on S?

This question can immediately be simplified by its restriction to unbordered surfaces. The reason for this is that every embedding of G on S_n can be modified by b excisions into an embedding on $S_{n,b}$ and, vice versa, every embedding of G on $S_{n,b}$ can be modified by means of b caps into an embedding of G on S_n. Moreover, as noted in Proposition 4.1.3, every embedding of G on S_n can be modified by means of a handle addition into an embedding of G on S_{n+1}. Thus, we have the following proposition.

Proposition 4.3.1 *The graph G embeds in the orientable surface $S_{n,b}$ if and only if $n \geq \gamma(G)$.*

The evaluation of $\gamma(G)$ is in general difficult. It will be demonstrated below that for a specific graph G this evaluation can be reduced to a finite algorithm and hence it is possible, in principle at least, to always resolve the question of whether a given graph can be embedded on a given bordered surface.

An embedding of the graph G on the surface $S = S_{\gamma(G)}$ is called a *minimal embedding*. It will next be shown that minimal embeddings are necessarily polygonal and hence they are governed by the Euler–Poincaré equation. Our strategy will be to demonstrate that if a region R of a minimal embedding of G on S were not a polygon, then it could be "simplified," thereby producing an embedding of G on a surface with smaller genus, which would contradict the minimality of the genus of the original embedding.

The simplification process begins with the excision of the nonpolygonal region R from the ambient surface, leaving a bordered surface $S - R$. If the region R has no self-abutments, then such an excision leaves G embedded on $S - R$ with all of the arcs

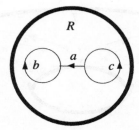

Figure 4.18   A minimal embedding.

of G that lie in R's perimeter appearing on the borders of $S - R$. The closure $(S - R)^c$ can then be shown to have smaller genus than G (see proof of Proposition 4.3.3), thus producing the requisite contradiction. Unfortunately, whenever the region R self-abuts along an arc of G, that arc leaves no trace on $S - R$, so that $(S - R)^c$ need no longer contain an embedding of G. Such, for example, is the case for the minimal (spherical) embedding of Figure 4.18, where the excision of R from the ambient sphere results in the disappearance of the arc a. In fact, the excision of R in this figure also leaves a disconnected $S - R$ (consisting of two disks bounded by b and c respectively), which, strictly speaking, is not even a bordered surface. Our way out of this bind is to excise a slightly trimmed version of the region R so as to endow $S - R$ with a "lip," or "collar," which ensures that G remains embedded on this slightly augmented version of $S - R$. This modification of the excision process will now be defined somewhat more formally.

Let R be a region of some embedding of a graph G on an orientable surface that is placed in $\mathbb{R}^3$. The *t-buffer zone* of the region R consists of all the points of the region that are within distance t of its perimeter. The *trimming R^t* of the region R is the bordered surface obtained from it by the excision of its t-buffer zone (Fig. 4.19). If R is a polygon, then R^t is a slightly shrunk version of R with no identifications along its perimeter. Of course, if t is too large then R^t is empty, but it is the trimmings that correspond to small values of t that are of interest here. It is clear that components of the perimeter of R give rise to the borders of R^t. However, any single component of the perimeter may give rise to any number of borders of R^t. Figure 4.20 shows a graph G consisting of two loops a and b embedded on the torus with regions R and

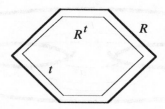

Figure 4.19   A trimming of a polygonal region.

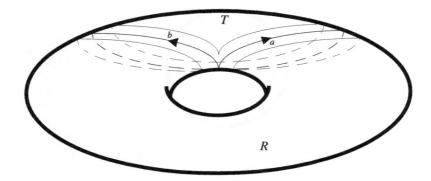

Figure 4.20   Trimmings on a torus.

T. When these regions are trimmed, R^t, being an open cylinder, has two borders, whereas T^t is simply a disk. The trimming of the region R of the minimal spherical embedding of Figure 4.18 is also a disk. In Figure 4.21, where the embedded graph G consists of the single loop a, the trimmed region R^t is a twice perforated torus $S_{1,2}$.

If R is a region of an embedding of the graph G on the surface S, then $S - R^t$ denotes the bordered surface that remains after R^t has been excised from S along its perimeter. It is clear that each border of $S - R^t$ is also a border of R^t and vice versa. In the toroidal embedding of Figure 4.20 above, $S - R^t$ is an open cylinder whereas $S - T^t$ is a once perforated torus. Note that the t-buffer zone of the region R in the minimal spherical embedding of Figure 4.18 runs on both sides of the arc a, and hence $S - R^t$ is a disk. Such is also the case in Figure 4.21 above, where $S - R^t$ is an open cylinder $S^{0,2}$. The next lemma says that sufficiently narrow trimmings do not affect the topological nature of the region.

Lemma 4.3.2 *Let R be a region of an embedding of the graph G on the orientable surface S. Then there exists a trimming R^t of R such that $\gamma(R^t) = \gamma(R)$ and $\beta(R^t) = \beta(R)$.*

PROOF: If the region R has genus 0, then it can be embedded in the plane as a disk with $\beta(R)$ borders. The desired trimming is then obtained by a slight shrinking of the exterior border and a slight expansion of the interior borders as indicated in Figure

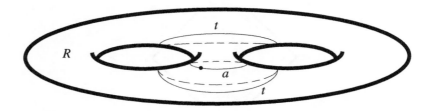

Figure 4.21   A trimmed region of genus 1.

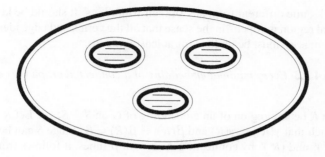

Figure 4.22   A slight trimming of a region of genus 0.

4.22. If R has a positive genus, set $\gamma = \gamma(R) = \gamma(R^c)$ and suppose that R^c has been placed in Euclidean 3-space $\mathbb{R}^3$ as a sphere with $\gamma \geq 1$ handles. Since each of the borders $b_1, b_2, b_3, \ldots$ of R bounds a disk in R^c, namely, the cap that was used to close it up, it follows that these borders can be shrunk, if necessary, and slid until they are lined up in a row on one handle of R^c. If R^c is now excised along loops $c_1, c_2, c_3, \ldots$ that run around $b_1, b_2, b_3, \ldots$ respectively and still do not intersect (see Fig. 4.23), then a trimming R^t of R is obtained such that $\beta(R^t) = \beta(R)$ and $(R^t)^c = R^c$, so that

$$\gamma(R^t) = \gamma\big((R^t)^c\big) = \gamma(R^c) = \gamma(R). \qquad \text{Q.E.D.}$$

The reader is reminded here of the difference between a region and its visualization on the surface. If this distinction is ignored, then the regions R of Figures 4.20

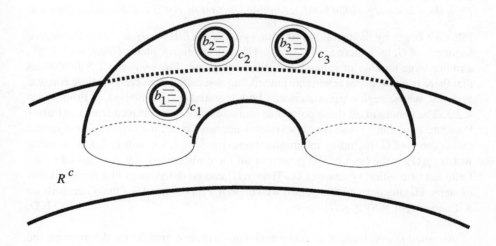

Figure 4.23   A slight trimming of a region with positive genus.

and 4.21 look quite different from any of their trimmings. It should be kept in mind that the actual region is "open" in the sense that all the vertex and edge identifications along its boundaries must be undone (or at least ignored).

Proposition 4.3.3 *Every minimal embedding of a connected graph G is also polygonal.*

PROOF: Let R be any region of an embedding of G on $S = S_{\gamma(G)}$. Let R^t be a trimming of R such that $\gamma(R^t) = \gamma(R)$ and $\beta(R^t) = \beta(R) = \beta$. Since S can be recovered from $(S - R^t)^c$ and $(R^t)^c$ by connecting them with β tubes, it follows from Propositions 3.3.1 and 3.3.9 that

$$\gamma(S) = \gamma\big((S - R^t)^c\big) + \gamma\big((R^t)^c\big) + \beta - 1.$$

However, because G is embedded on $(S - R^t)^c$, it follows from the minimality of $\gamma(G)$ that

$$\gamma\big((S - R^t)^c\big) \geq \gamma(G) = \gamma(S),$$

so that

$$\gamma\big((R^t)^c\big) + \beta - 1 \leq 0.$$

Hence $\beta = 1$ and $\gamma(R^t) = \gamma((R^t)^c) = 0$. In other words, R^t has genus 0 and a single border. It follows that R too has genus 0 and one border. Thus, by Corollary 3.4.3, R is a polygon.   Q.E.D.

There is now enough information available to resolve this chapter's main issue for connected graphs.

Theorem 4.3.4 *Let G be a connected graph, and S an orientable bordered surface. Then there is an algorithm for determining whether or not G can be embedded on S.*

PROOF: Begin by listing all the rotation systems of G. If $d_1, d_2, \ldots, d_p$ is the degree sequence of G, then there are $\prod_{i=1}^{p}(d_i - 1)!$ such rotation systems (Exercise 4.2.10), a rather large number in general, but still a finite one. Proposition 4.2.5 guarantees that there is a polygonal orientable embedding associated with each of these rotation systems, whose region perimeters can be computed by the method of Proposition 4.2.2. The genera of all these polygonal embeddings can be derived from the Euler–Poincaré equation. These rotation systems account for all the orientable polygonal embeddings of G, including the minimal ones on $S_{\gamma(G)}$ (Proposition 4.3.3). In other words, $\gamma(G)$ is the least of the genera of all the embeddings that correspond to the finite list of rotation systems of G. Thus $\gamma(G)$ can be determined in a finite number of steps. Having derived this genus, we know by Proposition 4.3.1 that G embeds on S if and only if $\gamma(S) \geq \gamma(G)$.   Q.E.D.

As noted above, there is no computationally efficient method for determining the genus of a graph. In a technical sense that cannot be clarified here, this problem is at least as difficult as that of deciding whether a graph is Hamiltonian or not. Even the determination of the genera of the complete graphs K_n has turned out to be a major project (see Gerhard Ringel's book). Such computations usually fall into two parts. Having conjectured a value k for $\gamma(G)$, one then usually proves separately that $k \leq \gamma(G)$ and that $k \geq \gamma(G)$. The only known general method for finding a lower bound on $\gamma(G)$ calls for an application of the Euler–Poincaré equation of Theorem 4.2.1.

Example 4.3.5 We show that $\gamma(K_n) \geq \{(n-3)(n-4)/12\}$, where $\{x\}$ denotes the least integer that is at least as large as x. For suppose K_n is embedded on $S_{\gamma(G)}$ with r regions. Since the embedding is necessarily polygonal (Proposition 4.3.3) it follows from the Euler–Poincaré equation that

$$p(K_n) - q(K_n) + r = 2 - 2\gamma(K_n).$$

Here $p(K_n) = n$ and $q(K_n) = n(n-1)/2$. As K_n has neither loops nor multiple arcs, the perimeter of every region has at least three arcs. Moreover, every arc appears on the perimeters twice. Hence $3r \leq 2q(K_n) = n(n-1)$ (see the proof of Corollary 2.4.4), and consequently

$$
\begin{aligned}
\gamma(K_n) &= \frac{2 - p(K_n) + q(K_n) - r}{2} \\
&\geq \frac{2 - n + n(n-1)/2 - n(n-1)/3}{2} \\
&= \frac{n^2 - 7n + 12}{12} = \frac{(n-3)(n-4)}{12}.
\end{aligned}
$$

The desired result follows from the integrality of $\gamma(G_n)$.

Having obtained a lower bound for $\gamma(G)$, one is then faced with the much harder task of demonstrating the reverse inequality. The proof that

$$\gamma(K_n) \leq \{(n-3)(n-4)/12\}$$

was completed by Gerhard Ringel and J. W. T. Youngs, with some help from William Gustin, Charles M. Terry, and Lloyd R. Welch, in 1969. It marked a milestone in the evolution of topological graph theory, and its details fill out the best part of Ringel's book. This part of the computation of $\gamma(G)$ usually calls for the construction of an embedding of G on some surface. For example, Figure 4.24 displays an embedding of K_8 on the double torus S_2. This demonstrates that $\gamma(K_8) \leq 2 = \{(8-3)(8-4)/12\}$.

Very little theoretical information is available about the genus parameter. One area that has been explored with some success is the question of how the genus of a graph depends on the genera of its parts. The earliest of the results of this nature is now presented.

For each $i = 1, 2$, let v_i be a node of the graph G_i. If the nodes v_1 and v_2 are identified (pasted) to a single node v, the resulting graph is denoted by $G_1 +_v G_2$ and is called the *amalgamation* of G_1 and G_2 at v (or at v_1 and v_2; see Fig. 4.25).

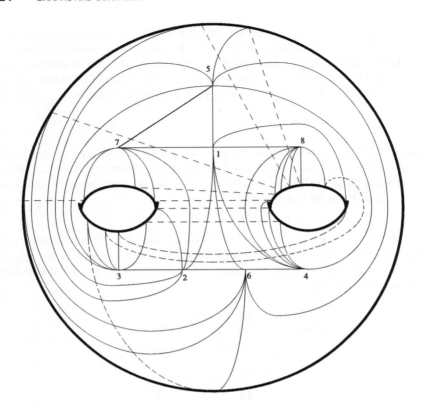

Figure 4.24   An embedding of K_8 on the double torus S_2.

Proposition 4.3.6 *For any two connected graphs G_1 and G_2,*

$$\gamma(G_1 +_v G_2) = \gamma(G_1) + \gamma(G_2).$$

PROOF: Suppose v results from the identification of the nodes v_i of G_i, where G_i is embedded on the surface $S^i \approx S_{\gamma(G_i)}$, $i = 1, 2$. Suppose further that R_i is a region of G_i on S^i that contains v_i in its perimeter. Let S_*^i be the bordered surface obtained

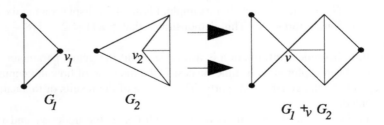

Figure 4.25   The amalgamation of two graphs.

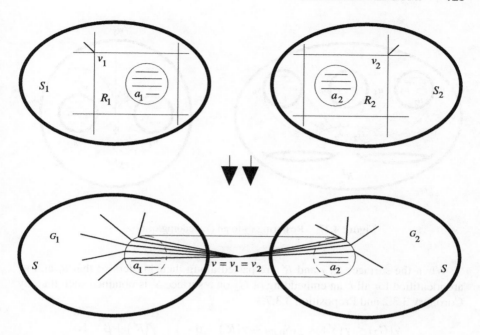

Figure 4.26   Combining embeddings of G_1 and G_2 into an embedding of $G_1 +_v G_2$.

by excising out from R_i a disk whose border is labeled a_i. Note that S_*^i has a_i as its single border and that $\gamma(S_*^i) = \gamma(S^i)$. If S is the closed surface that results from the connection of a_1 on S_1 with a_2 on S_2 by means of a tube, then the embeddings of G_1 and G_2 can be combined, as depicted in Figure 4.26, into an embedding of $G_1 +_v G_2$ on S. Hence, by Proposition 3.3.2,

$$\gamma(G_1 +_v G_2) \le \gamma(S) = \gamma(S_1) + \gamma(S_2) = \gamma(G_1) + \gamma(G_2).$$

To obtain the reverse inequality, assume first $G_2 - v_2$ is connected, and let the graph $G_1 +_v G_2$ be embedded on some surface S of genus $\gamma(G_1 +_v G_2)$. If we think of this as an embedding of G_1 on S, then, since $G_2 - v_2$ is connected, it is contained in a single region R of this embedding of G_1 on S. Let R^t be a trimming of R such that $\gamma(R^t) = \gamma(R)$ and $\beta(R^t) = \beta(R) = \beta$ (see Lemma 4.3.2). Since G_1 is embedded on $(S - R^t)^c$, it follows that

$$\gamma(G_1) \le \gamma\big((S - R^t)^c\big).$$

Turning to R^t, note that it contains all the arcs of G_2, except those arcs that are incident to v_2, say $a_1, a_2, \ldots, a_d$. These have been trimmed, say at the points $u_1, u_2, \ldots, u_d$ respectively. Choose points $w_1, w_2, \ldots, w_n$ on the β borders of the surface $S_{0,\beta}$ so that their positions on these borders match those of the points $u_1, u_2, \ldots, u_d$ on the β borders of R^t, and connect them all to one point v_2 that lies in the interior of $S_{0,\beta}$ (see Fig. 4.27).

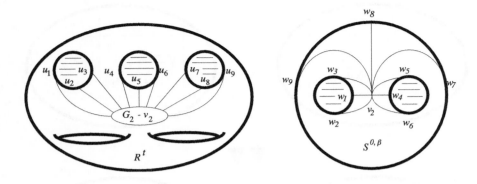

Figure 4.27   Restoring v_2 to an embedding of $G_2 - v_2$.

When the surfaces $S_{0,\beta}$ and R^t are pasted along their borders so that u_i and w_i are identified for all i, an embedding of G_2 on a surface S' is obtained such that, by Corollary 3.3.2 and Proposition 3.3.7,

$$\gamma(G_2) \leq \gamma(S') = \gamma(S_{0,\beta}) + \gamma(R^t) + \beta - 1 = \gamma(R^t) + \beta - 1.$$

Hence,

$$\gamma(G_1 +_v G_2) \leq \gamma(G_1) + \gamma(G_2) \leq \gamma\left((S - R^t)^c\right) + \gamma(S')$$
$$= \gamma\left((S - R^t)^c\right) + \gamma\left((R^t)^c\right) + \beta - 1 = \gamma(S) = \gamma(G_1 +_v G_2),$$

since S can be viewed as the connection of $(S - R^t)^c$ and $(R^t)^c$ by β tubes. This concludes the proof when $G_2 - v_2$ is connected.

If $G_2 - v_2$ is not connected, let $C_1, C_2, \ldots, C_k$ be its components, and for each $i = 1, 2, \ldots, k$ let H_i be the graph obtained by joining v_2 to those vertices of C_i to which it is adjacent in G_2. Each of the graphs H_i has the property that $H_i - v_2$ is connected. Since

$$G_1 +_v G_2 \approx G_1 +_v H_1 +_v H_2 +_v \cdots +_v H_k,$$

the general result now follows by a straightforward induction.   Q.E.D.

Corollary 4.3.7 *Let H_n be the graph obtained by amalgamating n copies of K_5 at one node. Then $\gamma(H_n) = n$.*

PROOF: Since $\gamma(K_5) = 1$, this follows from Proposition 4.3.6 by induction.   Q.E.D.

The important Theorem 4.3.4 is now extended to cover disconnected graphs as well. First, a proposition.

Proposition 4.3.8 *If* $C_1, C_2, \ldots, C_k$ *are the connected components of the graph G, then* $\gamma(G) = \gamma(C_1) + \gamma(C_2) + \cdots + \gamma(C_k)$.

Sketch of proof: That $\gamma(G) \leq \gamma(C_1) + \gamma(C_2) + \cdots + \gamma(C_k)$ follows from a tube connection of minimal embeddings of the components. The reverse inequality follows from an inductive excision argument similar to the one employed in the proof of Proposition 4.3.6, with G_1 (whose connectedness is not used in that proof) replaced by $C_1 \cup C_2 \cup \cdots \cup C_{k-1}$ and G_2 by C_k. □

The proof of the next theorem follows directly from the above lemma and Theorem 4.3.4.

Theorem 4.3.9 *Let G be a graph, and S an orientable bordered surface. It is possible to determine in a finite amount of time whether or not G can be embedded on S.* □

EXERCISES 4.3

1*. Prove that $\gamma(K_5) = \gamma(K_6) = \gamma(K_7) = 1$.

2**. Prove that $\gamma(K_9) = 3$.

3. Prove that if the simple graph G is 2-colorable, then every region of every embedding of G has at least four sides in its perimeter.

4. Prove that $\gamma(K_{m,n}) \geq \{(m-2)(n-2)/4\}$. (Hint: Use Exercise 3.)

5*. Prove that $\gamma(K_{3,4}) = \gamma(K_{4,4}) = 1$.

6*. Prove that $\gamma(K_{3,5}) = 1$.

7*. Prove that $\gamma(K_{4,5}) = 2$.

Given two graphs G_1 and G_2, the result of pasting an arc a of one to an arc of the other is denoted by $G_1 +_a G_2$.

8. Prove that $\gamma(K_5 +_a K_5) = 1$.

9. Prove that $\gamma(K_{3,3} +_a K_{3,3}) = 1$.

10. Prove that $\gamma(G_1 +_a G_2) \leq \gamma(G_1) + \gamma(G_2)$.

11*. Prove that if the graph G with p nodes and q arcs embeds on an orientable surface of genus γ, then $p - q + r \geq 2 - 2\gamma$.

The nonorientable genus $\tilde{\gamma}(G)$ of the graph G is the smallest positive integer n such that G embeds on $\tilde{S}_n$.

12. Prove that for every graph G, $\tilde{\gamma}(G) \leq 2\gamma(G) + 1$.

13. Prove that the graph G embeds in the nonorientable surface $\tilde{S}^{n,b}$ if and only if $n \geq \tilde{\gamma}(G)$.

14. Prove that $\tilde{\gamma}(K_5) = 1$.

15. Prove that $\tilde{\gamma}(K_{3,3}) = 1$.

16. Prove that $\tilde{\gamma}(K_n) \geq \{(n-3)(n-4)/6\}$.

Figure 4.28 A voltage graph.

17. Prove that $\tilde{\gamma}(K_6) = 1$.

18**. Prove that $\tilde{\gamma}(K_7) = 3$ (yes, 3).

19. Prove that $\tilde{\gamma}(K_{m,n}) \geq \{(m-2)(n-2)/2\}$.

20. Determine $\tilde{\gamma}(K_{3,4})$.

21. Determine $\tilde{\gamma}(K_{3,5})$.

22**. Determine $\tilde{\gamma}(K_{4,4})$.

23*. Show that $\tilde{\gamma}(K_5 +_v K_7) \neq \tilde{\gamma}(K_5) + \tilde{\gamma}(K_7)$. (Hint: Use Exercises 1 and 18.)

24*. Prove that if the graph G has an embedding with r regions on the nonorientable surface $\tilde{S}_\gamma$, then $p - q + r \geq 2 - \gamma$, where $p = p(G)$ and $q = q(G)$.

25. Draw a graph embedding with a region R whose perimeter is connected in the graph, but such that R^t has three borders.

26. State and prove an analog of Lemma 2.4.3 for embeddings of graphs on arbitrary closed surfaces.

27*. A simple graph G is said to *triangulate* a surface S provided G is the skeleton of some triangulation of S. Prove that if G triangulates the orientable surface S, then $\gamma(G) = \gamma(S)$.

28*. Suppose the simple bipartite graph G can be embedded on the orientable surface S so that the perimeter of each region has length 4. Prove that $\gamma(G) = \gamma(S)$.

4.4 Voltage Graphs and Their Coverings

This section assumes familiarity with elementary group theory up to and including the concept of a coset decomposition. The group operation will be written additively for all commutative groups. Otherwise, it will be written multiplicatively, in which case the group identity will be denoted by Id. The additive group of integers modulo n is denoted by $\mathbb{Z}_n$.

Given a connected graph G and a group Γ, a *voltage assignment* (*with values in* Γ) on the graph G is a function φ that assigns an element of Γ to every directed arc of G with the stipulation that for every directed arc a of G

$$\varphi(a^{-1}) = [\varphi(a)]^{-1}.$$

The triple (G, φ, Γ) is called a *voltage graph*. Voltage graphs should be pictured as graphs whose directed arcs are labeled by the group elements. For example, Figure 4.28 displays a voltage graph in which $G = K_2$, $\varphi(a) = 1$, $\varphi(a^{-1}) = -1$. In this particular case the group can be any group of which 1 is an element, such as the

Figure 4.29   Two identical voltage graphs

cyclic group $\mathbb{Z}_n, n = 1, 2, 3, \ldots$, and so Figure 4.28 actually denotes an infinitude of different voltage graphs. Figure 4.29 displays two identical voltage graphs whose underlying graph G has three nodes and four arcs. Here, too, Γ can be any of the cyclic groups $\mathbb{Z}_n$. For every voltage assignment φ there is a corresponding *covering graph* $\tilde{G}$ (the φ and Γ are hidden). The nodes of $\tilde{G}$ consist of all the elements u_g where u is an arbitrary node of G and g is an arbitrary element of Γ. Moreover, if a is a directed arc of G from u to v, then for every g in Γ there is a directed arc a_g of $\tilde{G}$ from u_g to $v_{g\varphi(a)}$. For example, Figure 4.30 displays the two covering graphs of the voltage graph of Figure 4.28 that result from the specifications $\Gamma = \mathbb{Z}_2$ and $\Gamma = \mathbb{Z}_3$ respectively. Figure 4.31 contains the covering graph associated with the voltage graph of Figure 4.29 with $n = 5$. Figure 4.32 makes it clear that in general

$$(a_g)^{-1} = (a^{-1})_{g\varphi(a)} \, .$$

The objects $u_g, a_g, \tilde{G}$ are said to be *lifts* (or *covers*) of u, a, G, respectively. If $a_g, b_h, c_i, \ldots$ is a walk of $\tilde{G}$, then it too is said to be a *lift*, or a *cover*, of the walk $a, b, c, \ldots$ of G. If $w : a, b, c, \ldots$ is a walk of G, then $\varphi(w) = \varphi(a)\varphi(b)\varphi(c) \cdots$. Note that if w is a u-v walk, then $\tilde{w}$ is a u_g-$v_{g\varphi(w)}$ walk for some g. Finally, we define the *isotropy group* at the node v:

$$\Gamma_v = \{\varphi(w) \mid w \text{ is a closed walk of } G \text{ at } v\}.$$

The voltage graph (G, φ, Γ) is said to be *full* provided $\Gamma_v = \Gamma$ for each (or, equivalently, for some) vertex v.

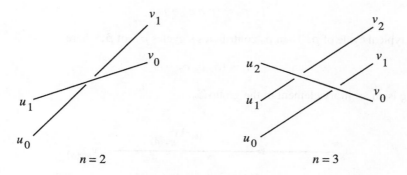

Figure 4.30   Two covering graphs.

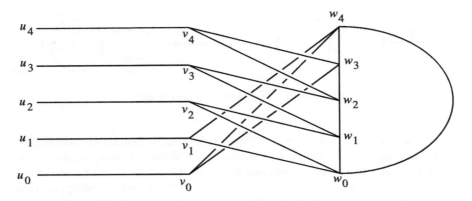

Figure 4.31 A covering graph with $n = 5$.

Theorem 4.4.1 *Let (G, φ, Γ) be a voltage graph where G is connected. Then the number of components of the covering graph $\tilde{G}$ is $[\Gamma : \Gamma_u]$, where u is any vertex of G.*

PROOF: Let u and v be any two nodes of G. Since G is connected, there is a u-v walk w in G. It follows that for any $g \in \Gamma$ there is a covering walk $\tilde{w}$ of w from u_g to $v_{g\varphi(w)}$. Thus, every component of $\tilde{G}$ contains lifts of all the vertices of G. It follows from the definition of the arcs of $\tilde{G}$ that the nodes u_g and u_h belong to the same component of $\tilde{G}$ if and only if there is a closed walk w of G at u such that $g\varphi(w) = h$. Consequently, the number of components of $\tilde{G}$ is $[\Gamma : \Gamma_u]$. Q.E.D.

Note that in the voltage graph of Figure 4.28 $\Gamma_u = \Gamma_v = 0$. Hence $[\Gamma : \Gamma_u] = [\Gamma : \Gamma_v] = n$. In Figure 4.33, on the other hand, it is clear that, because $\varphi(d) = 1$, $\Gamma_w = \Gamma = \mathbb{Z}_3$. Hence $\tilde{G}$ must be connected, as is also indicated by the diagram.

Suppose next that the graph G is embedded on some orientable surface with rotation system $\rho = \rho_u \circ \rho_v \circ \cdots$. Then $\tilde{\rho}$ is a rotation system of the covering graph $\tilde{G}$, which consists of $np(G)$ cycles which are obtained in the following manner. Let

$$\rho_u = (a\ b\ c\ \cdots)$$

be a typical cycle of ρ. Then ρ_u contributes n cycles ρ_{u_g} of $\tilde{\rho}$, where

$$\rho_{u_g} = (a_g\ b_g\ c_g\ \cdots)$$

and g is an arbitrary element of the group $\mathbb{Z}_n$.

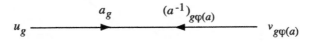

Figure 4.32 An arc and its inverse in the covering graph.

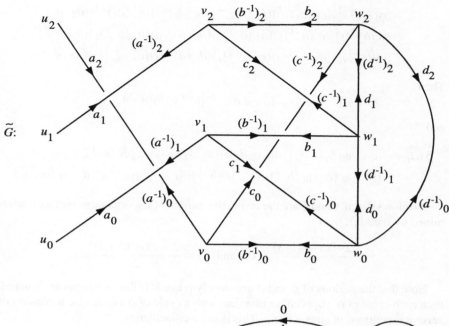

Figure 4.33   A voltage graph and its covering graph.

Example 4.4.2 Let

$$\rho = (a)(a^{-1}\, c\, b^{-1})(b\, c^{-1}\, d\, d^{-1})$$

be a rotation system of the voltage graph of Figure 4.33. Then ρ lifts to

$$\tilde{\rho} = (a_0)(a_1)(a_2)((a^{-1})_0\, c_0\, (b^{-1})_0)((a^{-1})_1\, c_1\, (b^{-1})_1)((a^{-1})_2\, c_2\, (b^{-1})_2)$$
$$\cdot (b_0\, (c^{-1})_0\, d_0\, (d^{-1})_0)(b_1\, (c^{-1})_1\, d_1\, (d^{-1})_1)(b_2\, (c^{-1})_2\, d_2\, (d^{-1})_2).$$

To find the regions of the embeddings determined by both ρ and $\tilde{\rho}$, we set

$$I_G = (a\, a^{-1})(b\, b^{-1})(c\, c^{-1})(d\, d^{-1})$$

and

$$I_{\tilde{G}} = (a_0\, (a_0)^{-1})(a_1\, (a_1)^{-1})(a_2\, (a_2)^{-1})(b_0\, (b_0)^{-1})(b_1\, (b_1)^{-1})(b_2\, (b_2)^{-1})$$

$$\cdot (c_0 \, (c_0)^{-1})(c_1 \, (c_1)^{-1})(c_2 \, (c_2)^{-1})(d_0 \, (d_0)^{-1})(d_1 \, (d_1)^{-1})(d_2 \, (d_2)^{-1})$$
$$= (a_0 \, (a^{-1})_1)(a_1 \, (a^{-1})_2)(a_2 \, (a^{-1})_0)(b_0 \, (b^{-1})_0)(b_1 \, (b^{-1})_1)(b_2 \, (b^{-1})_2)$$
$$\cdot (c_0 \, (c^{-1})_2)(c_1 \, (c^{-1})_0)(c_2 \, (c^{-1})_1)(d_0 \, (d^{-1})_1)(d_1 \, (d^{-1})_2)(d_2 \, (d^{-1})_0).$$

Then

$$\rho \circ I_G = (a \, c \, d \, b \, a^{-1})(b^{-1} \, c^{-1})(d^{-1})$$

and

$$\tilde{\rho} \circ I_{\tilde{G}} = (a_0 \, c_1 \, d_0 \, b_1 (a^{-1})_1)(a_1 \, c_2 \, d_1 \, b_2 (a^{-1})_2)(a_2 \, c_0 \, d_2 \, b_0 (a^{-1})_0)$$
$$\cdot ((b^{-1})_0 \, (c^{-1})_0 \, (b^{-1})_1 \, (c^{-1})_1 \, (b^{-1})_2 \, (c^{-1})_2 \,)((d^{-1})_0 \, (d^{-1})_2 \, (d^{-1})_1).$$

It follows that if γ and $\tilde{\gamma}$ are the respective genera of the orientable surfaces determined by ρ and $\tilde{\rho}$, then

$$\gamma = \frac{2 - (3 - 4 + 3)}{2} = 0 \quad \text{and} \quad \tilde{\gamma} = \frac{2 - (9 - 12 + 5)}{2} = 0.$$

Note that the regions of ρ and $\tilde{\rho}$ are closely related. If the subscripts are ignored, then each cycle of $\tilde{\rho} \circ I_{\tilde{G}}$ is either identical with a cycle of $\rho \circ I_G$ or else it consists of several repetitions of such a cycle. This is not a coincidence.

Proposition 4.4.3 *Let (G, φ, Γ) be a voltage graph, and suppose the rotation system ρ of G is lifted to the rotation system $\tilde{\rho}$ of the covering graph $\tilde{G}$. If the region R of ρ that contains the directed arc a on its boundary has boundary cycle*

$$\partial R = (a \, b \, c \, \cdots),$$

then, for any element g of Γ, the boundary cycle of $\tilde{\rho}$ that contains the lift a_g has the form

$$(a_g \, b_{g\varphi(a)} \, c_{g\varphi(a)\varphi(b)} \, \cdots \, a_{g\varphi(R)} \, b_{g\varphi(R)\varphi(a)} \, c_{g\varphi(R)\varphi(a)\varphi(b)} \, \cdots$$
$$a_{g\varphi(R)^2} \, b_{g\varphi(R)^2\varphi(a)} \, c_{g\varphi(R)^2\varphi(a)\varphi(b)} \, \cdots a_{g\varphi(R)^{k-1}} \, b_{g\varphi(R)^{k-1}\varphi(a)} \, c_{g\varphi(R)^{k-1}\varphi(a)\varphi(b)} \, \cdots \,),$$

where $\varphi(R) = \varphi(a)\varphi(b)\varphi(c) \cdots$ and k is the order of $\varphi(R)$ in Γ.

PROOF: Suppose $I_G = (a \, a^{-1}) \cdots$ and $I_{\tilde{G}} = (a_0 \, (a_0)^{-1}) \cdots$ are the arc involutions of G and $\tilde{G}$ respectively. Then

$$\rho \circ I_G = (a \, b \, c \, \cdots) \cdots,$$

so that necessarily

$$b = \rho(a^{-1})$$

and

$$b_g = \tilde{\rho}((a^{-1})_g) \qquad \text{for all } g \in \mathbb{Z}_n.$$

Then,

$$\tilde{\rho} \circ I_{\tilde{G}}(a_g) = \tilde{\rho}(I_{\tilde{G}}(a_g)) = \tilde{\rho}((a_g)^{-1}) = \tilde{\rho}((a^{-1})_{g\varphi(a)}) = b_{g\varphi(a)} \,.$$

This means that

$$\tilde{\rho} \circ I_{\tilde{G}} = (a_g \, b_{g\varphi(a)} \, \cdots) \cdots .$$

If a_g is replaced by $b_{g\varphi(a)}$, then the above argument yields

$$\tilde{\rho} \circ I_{\tilde{G}} = (a_g \, b_{g\varphi(a)} \, c_{g\varphi(a)\varphi(b)} \, \cdots) \cdots .$$

A sufficient number of iterations of this argument yields the desired expression for the cycle of $\tilde{\rho} \circ I_{\tilde{G}}$ that contains a_g. The reason the cycle terminates where it does is that $\varphi(R)^k = \text{Id}$. $\qquad$ Q.E.D.

Theorem 4.4.4 *Let the rotation system ρ of the voltage graph (G, φ, Γ) be lifted to the rotation system $\tilde{\rho}$ of the covering graph $\tilde{G} = G \times_\varphi \Gamma$. Suppose the region R of ρ has perimeter of length m and the order of $\phi(R)$ is k. Then R lifts to n/k regions of $\tilde{\rho}$, each of which has a perimeter of length mk.*

PROOF: Let $\Gamma_1, \Gamma_2, \ldots, \Gamma_h$ be the cosets of the cyclic subgroup of Γ generated by $\varphi(R)$, and select representatives $g_i \in \Gamma_i$ for $i = 1, 2, \ldots, h$. Let a be a directed arc of G, and suppose it is contained in the boundary cycle $\partial R = (a \cdots)$ (of length m) of $\rho \circ I_G$. Then it follows from Proposition 4.4.3 and the properties of coset decompositions that the boundary cycles of $\tilde{\rho} \circ I_{\tilde{G}}$ that contain lifts of a are

$$(a_{g_1} \cdots a_{g_1\varphi(R)} \cdots a_{g_1\varphi(R)^2} \cdots a_{g_1\varphi(R)^{k-1}} \cdots),$$

$$(a_{g_2} \cdots a_{g_2\varphi(R)} \cdots a_{g_2\varphi(R)^2} \cdots a_{g_2\varphi(R)^{k-1}} \cdots),$$

$$(a_{g_3} \cdots a_{g_3\varphi(R)} \cdots a_{g_3\varphi(R)^2} \cdots a_{g_3\varphi(R)^{k-1}} \cdots),$$

$$\vdots$$

$$(a_{g_h} \cdots a_{g_h\varphi(R)} \cdots a_{g_h\varphi(R)^2} \cdots a_{g_h\varphi(R)^{k-1}} \cdots). \qquad \text{Q.E.D.}$$

Corollary 4.4.5 *Let (G, φ, Γ) be a full voltage graph on the graph G, and suppose the rotation system ρ of G is lifted to the rotation system $\tilde{\rho}$ of the covering graph $\tilde{G}$. If the regions of ρ are $R_1, R_2, \ldots, R_r$, and k_i is the order of $\varphi(R_i)$ for $i = 1, 2, \ldots, r$, then the ambient orientable surface $\tilde{S}$ of the embedding of $\tilde{G}$ determined by $\tilde{\rho}$ is the closed orientable surface of genus*

$$\tilde{\gamma} = 1 + n(\gamma - 1) + \frac{n}{2} \sum_{i=1}^{r} \left(1 - \frac{1}{k_i}\right).$$

PROOF: That $\tilde{S}$ is closed and orientable follows from Proposition 4.2.5. Let its genus be $\tilde{\gamma}$. The covering graph $\tilde{G}$ has $np(G)$ nodes and $nq(G)$ arcs, and by Theorem 4.4.4 the embedding of $\tilde{G}$ on $\tilde{S}$ has $\sum_{i=1}^{r} (n/k_i)$ regions. It follows that

$$2 - 2\tilde{\gamma} = n\left(p - q + \sum_{i=1}^{r} \frac{1}{k_i}\right) = n\left(p - q + r + \sum_{i=1}^{r} \left(\frac{1}{k_i} - 1\right)\right)$$

$$= n \left(2 - 2\gamma + \sum_{i=1}^{r} \left(\frac{1}{k_i} - 1 \right) \right),$$

from which it follows that

$$2 - 2n \left(1 - \gamma + \frac{1}{2} \sum_{i=1}^{r} \left(\frac{1}{k_i} - 1 \right) \right) = 2\tilde{\gamma},$$

or

$$\tilde{\gamma} = 1 + n(\gamma - 1) + \frac{n}{2} \sum_{i=1}^{r} \left(1 - \frac{1}{k_i} \right). \qquad \text{Q.E.D.}$$

Example 4.4.6 The embedding of the voltage graph in Example 2 (Fig. 4.33) has the three region cycles

$$\partial R_1 = (a \, c \, d \, b \, a^{-1}), \qquad \partial R_2 = (b^{-1} \, c^{-1}), \qquad \partial R_3 = (d^{-1}).$$

It follows that ρ determines an embedding of G on the orientable surface of genus

$$\gamma = \frac{2 - (3 - 4 + 3)}{2} = 0.$$

Moreover,

$$\varphi(R_1) = \varphi(a) + \varphi(c) + \varphi(d) + \varphi(b) - \varphi(a) = 1 + 2 + 1 + 0 - 1 \equiv 0 \pmod{3},$$
$$\varphi(R_2) = -\varphi(b) - \varphi(c) = -0 - 2 \equiv 1 \pmod{3},$$
$$\varphi(R_3) = -\varphi(d) = -1 \equiv 2 \pmod{3}.$$

It follows that

$$k_1 = 1, \quad k_2 = k_3 = 3,$$

and hence

$$\tilde{\gamma} = 1 + 3(0 - 1) + \frac{3}{2} \left(1 - \frac{1}{1} + 1 - \frac{1}{3} + 1 - \frac{1}{3} \right) = 0.$$

Example 4.4.7 Suppose the rotation system $\rho = (a \, b \, c \, d)(d^{-1} \, c^{-1} \, b^{-1} \, a^{-1})$ is applied to the voltage graph of Figure 4.34, and $I_G = (a \, a^{-1})(b \, b^{-1})(c \, c^{-1})(d \, d^{-1})$ is its arc involution. Then

$$\rho \circ I_G = (a \, d^{-1})(b \, a^{-1})(c \, b^{-1})(d \, c^{-1}) = R_1 R_2 R_3 R_4.$$

It follows that ρ determines an embedding of the voltage graph on the orientable surface of genus

$$\gamma = \frac{2 - (2 - 4 + 4)}{2} = 0.$$

Since

$$\varphi(R_1) = 1 - 4 = -3,$$

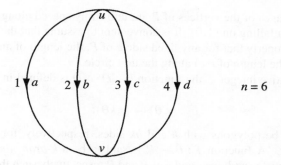

Figure 4.34   A voltage graph.

$$\varphi(R_2) = 2 - 1 = 1,$$
$$\varphi(R_3) = 3 - 2 = 1,$$
$$\varphi(R_4) = 4 - 3 = 1,$$

it follows that $k_1 = k_2 = k_3 = k_4 = 6$ and hence the covering graph is embedded on the orientable surface of genus

$$\tilde{\gamma} = 1 + 6(0 - 1) + \frac{6}{2}\left(4\left(1 - \frac{1}{6}\right)\right) = 5.$$

Example 4.4.8 Suppose the rotation system $\rho = (a\ b\ c\ d)(a^{-1}\ c^{-1}\ b^{-1}\ d^{-1})$ is applied to the voltage graph of Figure 4.34. Then

$$\rho \circ I_G = (a\ c^{-1}\ d\ a^{-1}\ b\ d^{-1})(c\ b^{-1}) = R_1 R_2.$$

It follows that the voltage graph is embedded on the orientable surface of genus

$$\gamma = \frac{2 - (2 - 4 + 2)}{2} = 1.$$

Since

$$\varphi(R_1) = 1 - 3 + 4 - 1 + 2 - 4 = -1 \quad \text{and} \quad \varphi(R_2) = 3 - 2 = 1,$$

it follows that $k_1 = k_2 = 6$, and hence the covering graph $\tilde{G}$ is embedded on the orientable suface of genus

$$\tilde{\gamma} = 1 + 6(1 - 1) + \frac{6}{2}\left(2\left(1 - \frac{1}{6}\right)\right) = 6.$$

Let D denote the disk of radius 1 centered at the origin $O = (0,0)$. A *rounding* of a polygon P is a homeomorphism

$$r : P \to D$$

such that the images of the vertices of P are uniformly spaced along the unit circle with one of them falling on $(1,0)$. It is convenient to assume that the function r has the additional property that for any fixed side e of P, the length of any arc a on e is proportional to the length of $r(e)$ along the unit circle.

For each positive integer k, the function $b_k : D \to D$ is defined in terms of polar coordinates by

$$b_k : (r, \theta) \mapsto (r, k\theta).$$

Let P and $\tilde{P}$ be polygons with h and hk sides, respectively, for some positive integers h and k. A function $f : \tilde{P} \to P$ is said to be a *k-branching* of $\tilde{P}$ onto P provided there exist roundings r and s of P and $\tilde{P}$ respectively such that

$$f = r^{-1} \circ b_k \circ s.$$

It is clear from the definition that every k-branching is a k-to-1 function except at the *center* $r^{-1}(O)$ of $\tilde{P}$ where it is one-to-one.

Given two surfaces S and S', a function $f : S \to T$ is said to be a *branched covering* provided there exist polygonal presentations Π and Π' of S and S' such that the restriction of f to any polygon P' of Π' is a k-branching over some polygon P of Π.

Let S be the orientable surface determined by the embedding scheme ρ of a voltage graph (G, φ, Γ). Let $\tilde{S}$ be the orientable surface determined by the lift $\tilde{\rho}$ of ρ to the covering graph $\tilde{G}$. If $\partial R = (a\ b\ c\ \cdots)$ is a typical boundary cycle of ρ, then, by Proposition 4.4.3, there exists a k-branching of every lift of R onto R, where k is the order of $\varphi(R)$. Because of the proportionality stipulation built into the definition of roundings, the effect on a_g of the two branchings of the two regions that abut along a_g are identical. In other words, if the two regions of $\tilde{G}$ that abut along a_g determine the respective branchings f_1 and f_2, then

$$f_1(A) = f_2(A) \in a \quad \text{for every } A \in a_g.$$

It follows that these k-branchings can then be pieced together to form a function $f : \tilde{S} \to S$ that is a branched covering. These observations are summarized below.

Theorem 4.4.9 *Let the rotation system ρ of the voltage graph (G, φ, Γ) determine an embedding of G on the surface S_γ, and let its lift $\tilde{\rho}$ determine an embedding of the covering graph $\tilde{G}$ on the surface $S_{\tilde{\gamma}}$. Then there is a branched covering*

$$f : S_{\tilde{\gamma}} \to S_\gamma$$

such that

$$f(\tilde{G}) = G$$

and for every lift $\tilde{R}$ of R the restriction

$$f : \tilde{R} \to R$$

is a k-branching, where k is the order of $\varphi(R)$. $\qquad\square$

If every k-branching of the branched covering $f : \tilde{S} \to S$ happens to be a 1-branching, then f is a *covering map*.

Example 4.4.10 Let G be the graph with one node u and two loops a and b, let $\Gamma = \mathbb{Z}_6$, and set $\varphi(a) = 2$ and $\varphi(b) = 3$. Suppose further that G is embedded on the torus as illustrated twice in Figure 4.35. On the left, the graph is shown on a bona fide torus, and on the right the torus is cut along a and b so that the four-sided single region R of the embedding becomes a polygonal presentation of the torus. Since $\varphi(R) = 0$, it follows that the lifted embedding of $\tilde{G}$ has six quadrilaterals as its regions. The lifted embedding is illustrated in two ways in Figure 4.36.

Example 4.4.11 Suppose the graph of Figure 4.35 is endowed with the voltage assignment $\varphi(a) = (1\ 2)$ and $\varphi(b) = (2\ 3)$ with $\Gamma = \Sigma_3$, the group of permutations of the set $\{1, 2, 3\}$. Then $\varphi(R) = (2\ 3)(1\ 2)(2\ 3)(1\ 2) = (1\ 2\ 3)$, which has order 3. It follows that $\tilde{\gamma} = 3$, and so this voltage graph displays S_3 as a three-branched covering of the torus.

Example 4.4.12 Let G be the bouquet with three loops a, b, c embedded on the torus with two triangular regions as shown in Figure 4.37. Suppose further that these loops are assigned the voltages $\varphi(a) = 1, \varphi(b) = 2, \varphi(c) = 3$ in $\mathbb{Z}_6$. The total voltage around the boundary of each of the regions is 0, and hence the lifted embedding of $\tilde{G}$ is also triangular. The shading illustrates which regions of the covering are lifts of which regions of the bouquet.

Example 4.4.13 Figure 4.38 displays an embedding of the dipole D_5 on the sphere S_0 with voltages in $\mathbb{Z}_5$. The covering graph $\tilde{G}$ is homeomorphic to $K_{5,5}$. Note that if the boundary of any region R_i of G is traversed in the clockwise sense, then $\varphi(R_i) = 1$. It follows from Theorem 4.4.4 that R_i lifts to a unique region $\tilde{R}_i$ that is five-branched over R_i and whose boundary is a Hamiltonian cycle of $\tilde{G}$. The covering surface is S_6. For each i, let w_i be a point in the interior of the region R_i, and $a_{i,j}(b_{i,j})$ be an arc that joins w_i to $u_j(v_j)$ in the region R_i in such a manner that these arcs intersect only in their endpoints, if at all. If these new nodes and arcs are added to those of $\tilde{G}$, they together constitute a copy of $K_{5,5,5}$, which is now embedded on S_6

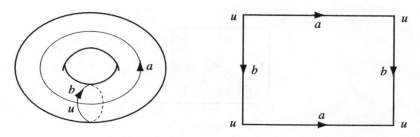

Figure 4.35   The two-loop bouquet on the torus.

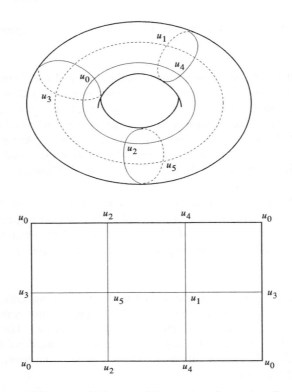

Figure 4.36   A toroidal cover of the two-loop bouquet on the torus.

$\widetilde{G} = K_7$ on the torus

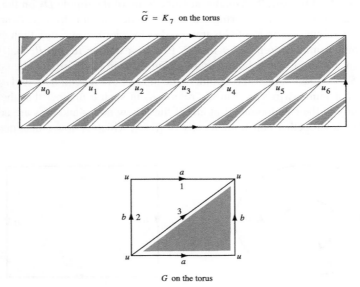

G on the torus

Figure 4.37   A toroidal covering of the three-loop bouquet on the torus.

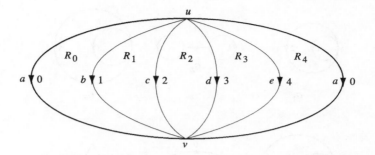

Figure 4.38   A voltage graph embedded on the sphere S_0.

so that each of the regions is a triangle. By Exercise 4.3.27, this is a minimum-genus embedding of $K_{5,5,5}$, and hence $\gamma(K_{5,5,5}) = 6$.

Exercises 4.4

1. Assuming that the voltage assignments depicted in Figure 4.39 take their values in $\mathbb{Z}_n$, for which values of $n \geq 2$ are the covering graphs of the following graphs connected?

   (a) G_1

   (b) G_2

   (c) G_3

2. Assuming $\Gamma = \mathbb{Z}_n$, describe the covering graphs of the following voltage graphs of Figure 4.40:

(a)	$G_1, \Gamma = \mathbb{Z}_4$	(b)	$G_1, \Gamma = \mathbb{Z}_5$	(c)	$G_1, \Gamma = \mathbb{Z}_6$
(d)	$G_1, \Gamma = \mathbb{Z}_7$	(e)	$G_1, \Gamma = \mathbb{Z}_8$	(f)	$G_2, \Gamma = \mathbb{Z}_3$
(g)	$G_2, \Gamma = \mathbb{Z}_4$	(h)	$G_2, \Gamma = \mathbb{Z}_5$	(i)	$G_2, \Gamma = \mathbb{Z}_6$
(j)	$G_3, \Gamma = \mathbb{Z}_{15}$	(k)	$G_4, \Gamma = \mathbb{Z}_{221}$	(l)	$G_5, \Gamma = \mathbb{Z}_3$
(m)	$G_5, \Gamma = \mathbb{Z}_4$	(n)	$G_5, \Gamma = \mathbb{Z}_5$	(o)	$G_5, \Gamma = \mathbb{Z}_6$
(p)	$G_6, \Gamma = \mathbb{Z}_3$	(q)	$G_6, \Gamma = \mathbb{Z}_4$		

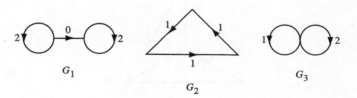

Figure 4.39   Some voltage graphs.

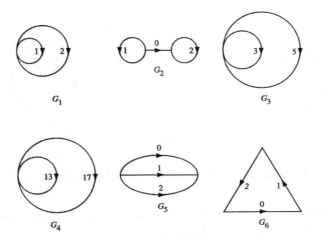

Figure 4.40   Some voltage graphs.

3. Let $\rho = (a\,c\,b)(b^{-1}\,e\,d)(f\,a^{-1}\,d^{-1})(c^{-1}\,f^{-1}\,e^{-1})$ be a rotation system of the voltage graph of Figure 4.41. Compute the genus and describe the regions of the embedding determined by the lift $\bar{\rho}$ if $\Gamma =$

(a) $\mathbb{Z}_3$

(b) $\mathbb{Z}_4$

(c) $\mathbb{Z}_n$

4. Repeat Exercise 3 using $\rho = (c\,b\,a)(b^{-1}\,d\,e)(f\,d^{-1}\,a^{-1})(c^{-1}\,e^{-1}\,f^{-1})$.

5. Let $\rho = (a\,a^{-1}\,b\,b^{-1}\,c\,c^{-1}d\,d^{-1})$ be a rotation system of the current graph in Figure 4.42. If $\Gamma = \mathbb{Z}_4$, compute the genus and describe the regions of the lifted embedding. Use the method described in Example 13 to convert this embedding into an embedding of $K_{8,4}$. Use Exercise 4.3.4 to conclude that $\gamma(K_{8,4}) = 3$.

6. Prove that $\gamma(K_{2n,n}) = (n-1)(n-2)/2$.

7. Use the method of Example 4.4.13 to prove that $\gamma(K_{n,n,n}) = (n-1)(n-2)/2$.

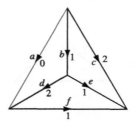

Figure 4.41   A voltage graph.

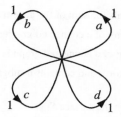

Figure 4.42   A voltage graph.

Appendix

It is natural to inquire whether the Kuratowski Theorem 2.4.6, which characterizes planar graphs by means of two forbidden subgraphs, has analogs for all the other surfaces. In other words:

> *Given a surface S, does there exist a finite list of graphs $\lambda(S)$ such that a graph G can be embedded on S if and only if it does not contain a member of $\lambda(S)$ as a subgraph?*

Such a list of 103 graphs was proposed for the projective plane $\tilde{S}_1$ by H. Glover, J. P. Huneke, and C. S. Wang in 1979 (N. Robertson contributed the 103rd member), and its validity was verified by D. Archdeacon two years later. In 1989 Archdeacon and Huneke proved the existence of such a list for every nonorientable surface $\tilde{S}_n$, while R. Bodendiek and K. Wagner settled the question for orientable surfaces. The following year a unified proof was offered by P. D. Seymour and N. Robertson. All these proofs are so intricate and lengthy that to date no upper bounds on the lengths of any of these lists have been extracted, except, of course, for those of the sphere (2) and the projective plane (103). Attempts to simplify these proofs are still ongoing, and two new proofs have very recently been supplied by Seymour and B. Mohar.

Chapter Review Exercises

1. Are the following statements true or false? Justify your answers.

   (a) The genus of every planar graph is zero.

   (b) Every graph can be embedded on some closed orientable surface.

   (c) Every graph can be embedded on some closed nonorientable surface.

   (d) The graph that consists of three disjoint copies of K_5 can be embedded on the surface S_2.

   (e) A region of a graph embedding may have two perimeters and five borders.

   (f) A region of a graph embedding may have five borders.

   (g) A region of a polygonal graph embedding may have five borders.

   (h) The Euler–Poincaré equation holds for all embeddings of a connected graph.

(i) The genus of a graph equals the sum of the genera of its components.

(j) Every polygonal embedding of a connected graph is minimal.

(k) Every minimal embedding of a connected graph is polygonal.

(l) If a graph embeds on the torus and contains a subgraph homeomorphic to $K_{3,3}$, then it has genus 1.

(m) The graph K_{100} embeds on the surface S_{100}.

(n) There is a voltage graph (G, φ, Γ) such that every node of G lifts to 17 nodes of the covering graph $\tilde{G}$.

(o) There is a voltage graph (G, φ, Γ) such that every arc of G lifts to 71 arcs of the covering graph $\tilde{G}$.

(p) There is a voltage graph (G, φ, Γ) such that every node of G lifts to 17 nodes of the covering graph $\tilde{G}$ and every arc lifts to 71 arcs.

CHAPTER 5

KNOTS AND LINKS

Many of the scientists of the nineteenth century believed that what is now considered to be empty space was actually occupied by a substance called *ether*, which was the medium through which light waves traveled. The famous physicist Lord Kelvin (William Thomson, 1824–1907) thought that atoms might actually be vortices, or knots, in the ether. He therefore asked the mathematicians Peter G. Tait (1831–1901), Reverend Thomas P. Kirkman (1806–1895), and C. N. Little to create a table of knots with the expectation that this would eventually turn into a table of atoms.

The ether hypothesis was abandoned in the early part of the twentieth century, and with it went the possibility of representing atoms as knots. Still, the questions raised in the process of organizing this knot table caught the interest of the mathematical world and gave birth to the new discipline of *knot theory*. Several of these questions have been answered, some very recently, whereas others still await resolution. While the original idea of representing atoms as vortices in the void has gone the way of the geocentric and phlogiston theories, new applications of this theory to biology, chemistry, and physics, have been found.

A variety of techniques are offered here for representing the winding of a knot through the ambient space in an algebraic language. Two fruitful connections with the topology of surfaces are explored.

Introduction to Topology and Geometry, Second Edition.
By Saul Stahl and Catherine Stenson Copyright © 2013 John Wiley & Sons, Inc.

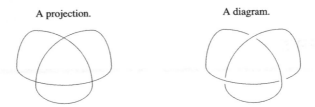

A projection.   A diagram.

Figure 5.1   Projections versus diagrams.

5.1 Preliminaries

A *knot* is a loop in $\mathbb{R}^3$, and portions of the knot are referred to as its *arcs*. A *link* is a finite set of disjoint knots, each of which is said to be a *component* of the link. Knots and links are generally represented by their projections to a plane. When each of the crossings in such a projection is modified so as to indicate which projected arc is in reality closer to the viewer, such a drawing is called a *diagram* of the link (see Fig. 5.1). It will always be assumed that the link and its projection plane are so positioned that no three of the link's arcs project to concurrent lines in the plane and the projections of any two arcs intersect unequivocally (see Fig. 5.2).

Any knot that is isotopic to a plane loop is called an *unknot*. Any link that is isotopic to a set of disjoint plane loops is called an *unlink*. An *(ambient) isotopy* is a deformation of a link in $\mathbb{R}^3$ without self-crossings of the link. In other words, it is no longer permissible to use the fourth dimension as a device for bypassing self-crossings (see Section 3.1). Two links are said to be *equivalent* if there is an isotopy that deforms one link into the other. Equivalent links are considered to be identical and receive the same name. Thus, all the links that are isotopic to the unlink with n components are denoted by 0^n. The main goal of the theory of knots is to provide effective criteria for deciding which links are equivalent. A more restricted goal is to determine which links are in fact unlinks. A general procedure for answering this last question for knots was found by Wolfgang Haken in 1961. However, this procedure is so complicated that to date it has not been implemented as a computer program.

Allowed crossings   Forbidden crossings

Figure 5.2   Allowed versus forbidden crossings.

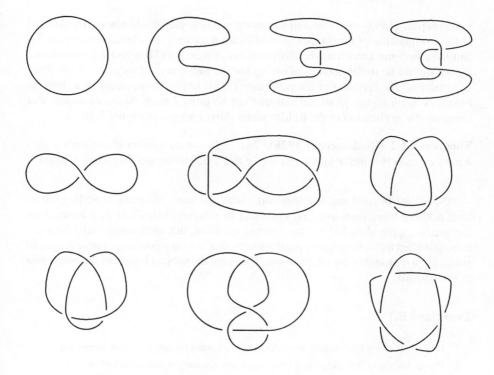

Figure 5.3   Some knot diagrams.

Different diagrams may represent equivalent links. The first six knot diagrams of Figure 5.3 all represent the unknot 0, and the first two link diagrams in Figure 5.4 represent the unlink 0^2. The following observation is plausible and will be taken for granted in the sequel.

Proposition 5.1.1 *If two links have the same diagram, then they are necessarily equivalent.*   □

The subtlety of the issue of equivalence is brought to light by Figure 5.5, in which are exhibited two *trefoil* knots which are mirror images of each other (in a sense that will be made precise later) as well as two *figure 8* knots that are also mirror images of each other. As Figure 5.6 indicates, the mirror-image figure 8 knots are equivalent to each other. The mirror image trefoil knots, however, are not equivalent (a fact that will be established in Section 4).

Figures 5.7 and 5.8 display some standard knots and links.

In general, an isotopy of a link will materially modify its diagram and crossings. An isotopy that consists of a mere deformation of the projection plane and therefore does not materially affect the crossings of the diagram is called a *plane isotopy*. Isotopies *a*, *c*, *d* of Figure 5.6 are plane isotopies, whereas *b* is not. In 1926 Kurt

Reidemeister (1893–1971) proved that nonplane isotopies could always be effected as the composition of three types of modifications, which are depicted in Figure 5.9 and have become known as the *Reidemeister moves.*   (The type III Reidemeister moves should be understood as allowing for the movement of any one of the three arcs "across" the crossing of the other two.) This observation, stated as a theorem below, is again highly plausible and will not be proved here. Some examples that illustrate the application of the Reidemeister moves appear in Figure 5.10.

Theorem 5.1.2 (Reidemeister 1926). *Two links are equivalent if and only if they have diagrams that differ by a sequence of Reidemeister moves and plane isotopies.*
□

Above and beyond the fact that links with the same diagram must be equivalent, nothing more than trial and error can be offered in this text as a method for recognizing equivalent links. The converse question, that of distinguishing between nonequivalent links, has some partial answers that are very powerful, interesting, and subject to a reasonably easy exposition. These are the subject matter of the remainder of this chapter.

Exercises 5.1

1. Prove that every knot which has a diagram with one crossing is in fact an unknot.
2. Prove that every link diagram with at most one crossing is that of an unlink.
3. Prove that every knot which has a diagram with two crossings is in fact an unknot.
4. Is it true that every link with at most two crossings is an unlink?
5. Prove that the top three knots in Figure 5.11 are all equivalent.
6. To which of the knots in Figure 5.5 is the knot at the bottom of Figure 5.11 equivalent?

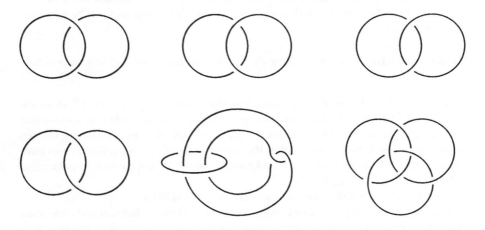

Figure 5.4   Some link diagrams.

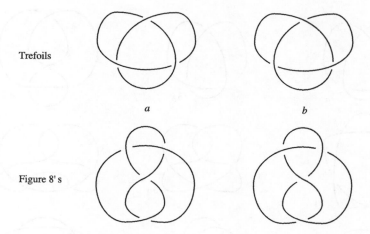

Trefoils

a   b

Figure 8' s

Figure 5.5   Mirror image knots.

5.2 Labelings

The most elementary procedure for distinguishing between links associates a system of modular linear equations to a diagram, one equation to each crossing. The nature of the solutions can sometimes be used to differentiate between diagrams that describe inequivalent links.

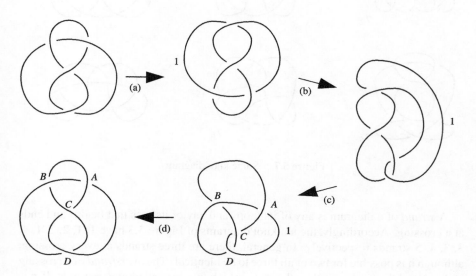

Figure 5.6   The equivalence of the figure 8 knots.

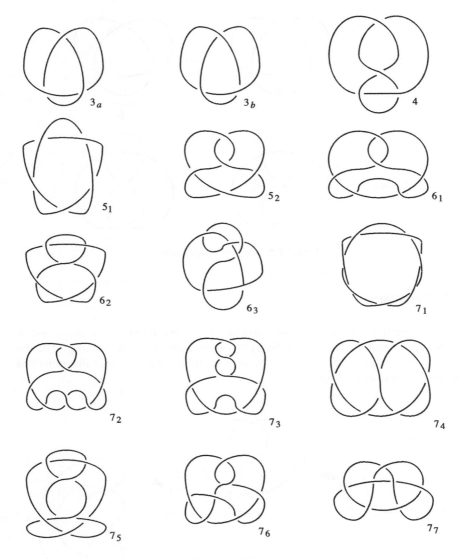

Figure 5.7   Some knot diagrams.

A *strand* of a diagram is any of its loops and any of its arcs that begins and ends at a crossing. Accordingly, the ten knot diagrams of Figure 5.3 have 1, 1, 2, 2, 1, 4, 3, 3, 4, 5 strands respectively. In general, there are three strands at every crossing, although it is possible for two or all three to be identical. The *overstrand* of a crossing is the strand that separates the other two, which two are called *understrands*. If p is any odd prime number, then a *p-labeling* of a knot diagram assigns to every strand of the diagram a number from $\{0, 1, 2, \ldots, p - 1\}$ so that at least two of these numbers

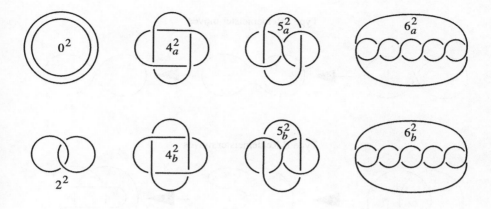

Figure 5.8   Some link diagrams.

are used and, at each crossing, if the overstrand is labeled x and the understrands are labeled y and z, the *labeling equation*

$$2x - y - z \equiv 0 \ (\text{mod } p) \tag{1}$$

holds (see Fig. 5.12). For example, the diagram of the trefoil knot of Figure 5.13 is 3-labeled, and the three associated equations are:

top crossing: $2 \cdot 1 - 2 - 0 \equiv 0 \ (\text{mod } 3);$

I

II

III

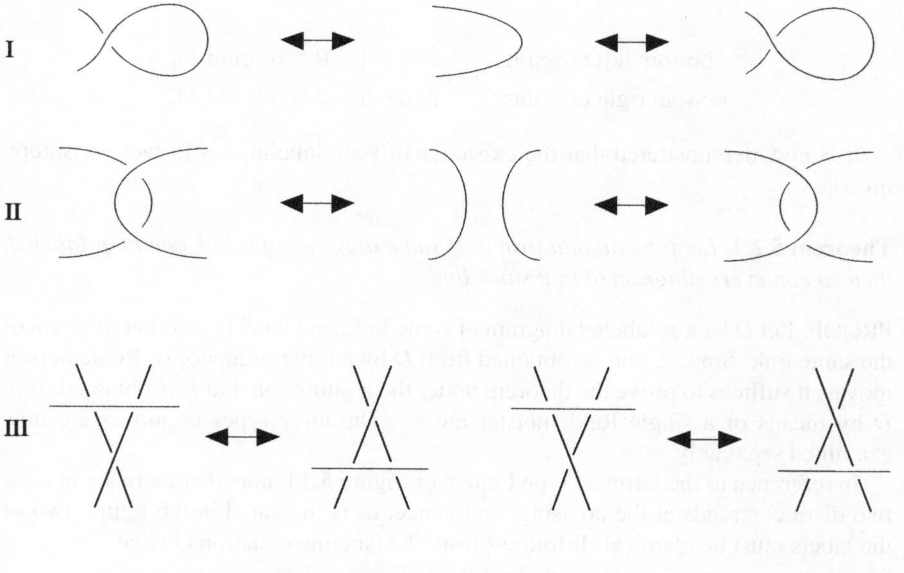

Figure 5.9   The three types of Reidemeister moves.

Type I Reidemeister moves

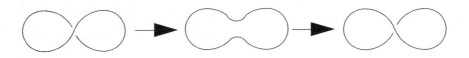

Type II Reidemeister moves

Type III Reidemeister moves

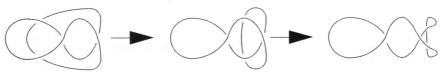

Figure 5.10   Examples of Reidemeister moves.

bottom left crossing: $2 \cdot 2 - 1 - 0 \equiv 0 \pmod 3$;

bottom right crossing: $2 \cdot 0 - 1 - 2 \equiv 0 \pmod 3$.

It is now demonstrated that the existence of such labelings is in fact an isotopy invariant.

Theorem 5.2.1 *Let p be an odd prime. If some diagram of a link can be p-labeled, then so can every diagram of that same link.*

PROOF: Let D be a p-labeled diagram of some link, and let E be another diagram of the same link. Since E can be obtained from D by a finite sequence of Reidemeister moves, it suffices to prove the theorem under the assumption that E is obtained from D by means of a single Reidemeister move. The three types of moves are now examined separately.

In reference to the leftmost type I move of Figure 5.14, note that there are at most two distinct strands at the crossing, and hence, as is indicated in the figure, two of the labels must be identical. It follows from the labeling equation (1) that

$$2x - x - y \equiv 0 \quad \text{or} \quad x \equiv y \pmod{p},$$

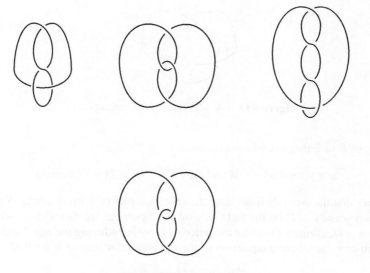

Figure 5.11   Four knots.

and hence all three of the labels at this crossing of D must be identical. Their common value x can now be used to label the strand that replaces them in E, and all the other strands of E can inherit their labels from D without affecting the validity of any of the remaining (unillustrated) labeling equations. Moreover, the total numbers of distinct labels used is unchanged. Hence E can also be p-labeled.

For the Reidemeister move that is depicted in the rightmost half of Figure 5.14, the label of the single depicted strand of D can be transferred to E as indicated, and the labeling equation at the new crossing of E becomes

$$2x - x - x \equiv 0.$$

Hence, the p-labeling of D can be augmented into a p-labeling of E, and the total numbers of distinct labels used is again unchanged. The consideration of the other two Reidemeister moves of type I is relegated to Exercise 11.

Turning to type II moves, suppose D has labels $x, y, z,$ and w as depicted in the left move in Figure 5.15. Then, because of the labeling equations, it follows that $w \equiv z \equiv 2x - y \pmod{p}$, and hence the p-labeling of D can be modified into a p-

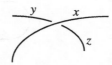

Figure 5.12   Labeling a typical crossing.

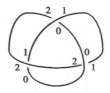

Figure 5.13   A 3-labeling of the trefoil knot.

labeling of E as indicated. Moreover, since

$$x \equiv y \pmod{p} \qquad \text{if and only if} \qquad x \equiv 2x - y \pmod{p},$$

it follows that the two labelings have the same number of distinct labels. Similarly, if the two strands of D on the right side of this figure are labeled with x and y, then the given p-labeling of D can be extended to E by introducing the new label $2x - y$, in which case the labeling equations of the two new crossings of E are both

$$2x - y - (2x - y) \equiv 0.$$

The remaining two type II moves are relegated to Exercise 12.

Finally, we turn to the type III moves. Beginning with the labels x, y, and z in D in Figure 5.16, the labeling equations at the two top crossings of D force the labels $2z - x$ and $2z - y$, and these, coupled with the labeling equation of the bottom crossing, yield the further label

$$2(2z - x) - (2z - y) \equiv 2z - 2x + y$$

of D. This p-labeling of D can be transferred to E as indicated. The labeling equations of the three new crossings of E are

$$\begin{aligned} \text{top:} \qquad & 2x - (2x - y) - y \equiv 0; \\ \text{bottom left:} \qquad & 2z - (2x - y) - (2z - 2x + y) \equiv 0; \\ \text{bottom right:} \qquad & 2z - (2z - x) - x \equiv 0. \end{aligned}$$

If the resulting p-labeling of E had only one label, then we would have in E

$$x \equiv y \equiv z \equiv 2z - x \equiv 2z - 2x + y \equiv 2x - y,$$

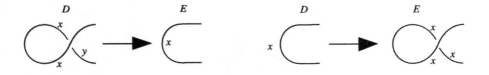

Figure 5.14   Type I labeled Reidemeister moves.

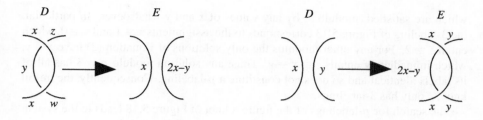

Figure 5.15   Type II labeled Reidemeister moves.

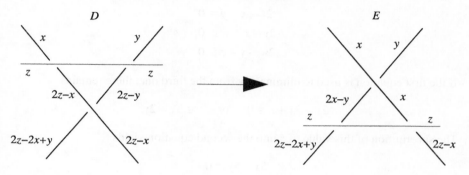

Figure 5.16   Type III labeled Reidemeister moves.

from which it would follow that D could contain only one label, which is not the case. Hence E can also be p-labeled. The three remaining type III moves are relegated to Exercise 13.   Q.E.D.

Corollary 5.2.2   *The trefoil knots 3_a and 3_b are not unknots.*

PROOF: A 3-labeling of the trefoil knot 3_a is exhibited in Figure 5.13. Since the trivial knot possesses no labelings whatsoever, the corollary follows from the above theorem. (See Exercise 13 regarding the trefoil knot 3_b.)   Q.E.D.

In general, finding such labelings is an exercise in the solution of simultaneous equations modulo a prime number. For example, the aforementioned 3-labeling of the trefoil knot 3_a was found by first assigning the unknowns x, y, and z to the strands in its diagram (Fig. 5.17) and writing out the labeling equations:

$$\text{top crossing:} \qquad 2x - y - z \equiv 0;$$
$$\text{bottom left crossing:} \qquad 2z - x - y \equiv 0;$$
$$\text{bottom right crossing:} \qquad 2y - z - x \equiv 0.$$

The elimination of z yields the two equations

$$3x - 3y \equiv 0 \equiv -3x + 3y, \tag{2}$$

which are satisfied, modulo 3, by any values of x and y whatsoever. In particular, the 3-labeling of Figure 5.13 corresponds to the assignments $x \equiv 1$ and $y \equiv 0$, which entail $z \equiv 2$. For any other modulus the only solutions of Equation (2) have $x \equiv y$, which immediately entails $z \equiv x \equiv y$. Thus, any solution modulo $p \neq 3$ has all of its labels identical and so does not constitute a p-labeling. Consequently, the trefoil knot 3_a only has 3-labelings.

The search for p-labelings of the figure 8 knot of Figure 5.18 leads to the system of equations

$$2x - z - w \equiv 0,$$
$$2z - x - y \equiv 0,$$
$$2y - x - w \equiv 0,$$
$$2w - y - z \equiv 0.$$

If the first equation is used to eliminate w from the third one, there remains

$$2y - 3x + z \equiv 0, \quad \text{or} \quad z \equiv 3x - 2y.$$

The substitution of this value of z into the second equation yields

$$5x - 5y \equiv 0.$$

This implies that $x \equiv y \equiv z \equiv w$ for any modulus $p \neq 5$, allowing for no p-labelings. For $p = 5$ there is, amongst others, the labeling $x \equiv 1$, $y \equiv 2$, $z \equiv 4$, $w \equiv 3$.

It may be concluded from the above examples that the figure 8 knot (labeled 4 in Fig. 5.7), the trefoil knot 3_a, and the unknot are all inequivalent to each other. Unfortunately, this elegant method does not contain the whole story. Inequivalent knots which have p-labelings for the same p's are known (see Exercise 4).

Links with several components are also subject to p-labelings. Surprisingly, however, the information gleaned from such labelings is qualitatively different from that obtained for knots. This discussion requires a definition. A link diagram is said to be *split* if there is a straight line in the diagram's plane such that portions of the diagram lie on both sides of the straight line, but the diagram does not intersect the straight line (see Figure 5.19). A link is said to be *splittable* if it is equivalent to a link which has a split diagram. It is easy to see that every split diagram has a p-labeling for

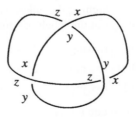

Figure 5.17   Finding a labeling of a trefoil knot.

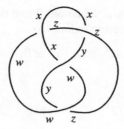

Figure 5.18   Finding a labeling for the figure 8 knot.

A split diagram.   A splittable link.

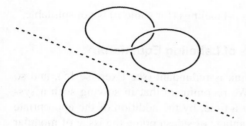   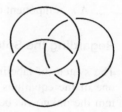

Figure 5.19   Splittings.

every p. One need simply label all the strands on one side of the separating line 1 and all the others 2. All the p-labeling equations then have the form

$$2x - x - x \equiv 0, \qquad x = 1, 2,$$

and are clearly satisfied. Hence the following conclusion can be drawn from Theorem 5.2.1.

Proposition 5.2.3 *If a link with at least two components does not have a p-labeling for even one p, then it is not splittable.*   □

It will next be shown that the *Borromean rings* of Figure 5.20 do not have a 3-labeling and are therefore not splittable. Since the modulus is 3, the labeling equation $2x - y - z \equiv 0$ is equivalent to the equation $x + y + z \equiv 0$. Hence the crossings of this diagram yield the equations

(1) $a + b + c \equiv 0$,   (2) $c + d + f \equiv 0$,   (3) $b + e + f \equiv 0$,
(4) $a + e + f \equiv 0$,   (5) $a + b + d \equiv 0$,   (6) $c + d + e \equiv 0$.

Equations 1 and 5 imply that $c \equiv d$, equations 2 and 6 that $e \equiv f$, and equations 3 and 4 that $a \equiv b$. The system therefore reduces to

$$(1') \quad 2a + c \equiv 0, \qquad (2') \quad 2c + e \equiv 0, \qquad (3') \quad 2e + a \equiv 0,$$

or

$$(1'') \quad -a + c \equiv 0, \qquad (2'') \quad -c + e \equiv 0, \qquad (3'') \quad -e + a \equiv 0.$$

It is now clear that $a \equiv b \equiv c \equiv d \equiv e \equiv f$, so that the condition that at least two different labels are employed cannot hold. Hence the diagram of the Borromean rings does not have a 3-labeling and is therefore not splittable.

It is worthwhile to reiterate here the difference between the two uses of labelings:

1. A knot with a p-labeling is nontrivial.

2. A k-component link ($k > 1$) with no p-labelings (for some p) is not splittable.

Regarding the Solution of Systems of Labeling Equations

The system of labeling equations of a link is redundant (see Exercise 15), and so one of the equations can be ignored. We recommend that in solving such a system the unknowns be eliminated, one at a time, by the addition of the appropriate multiples of equations. This has the advantage of sidestepping the issue of modular inverses. Moreover, it is also suggested that the elimination process begin with the three unknowns that appear in the discarded equation. The elimination process will eventually result in an equation of the form

$$Ax - Ay \equiv 0,$$

and every odd prime divisor p of A has a corresponding p-labeling that comes from setting $x = 0$ and $y = 1$.

Exercises 5.2

1. Verify directly, without using Theorem 5.2.1, that the diagrams in Figure 5.21 do not have any p-labelings.

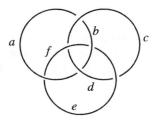

Figure 5.20 The Borromean rings cannot be 3-labeled.

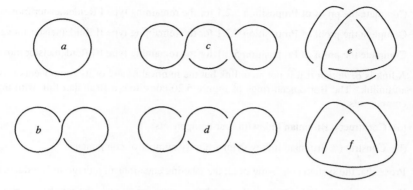

Figure 5.21

2. For each of the diagrams in Figure 5.22 determine directly, without using Theorem 5.2.1, all the primes p for which that diagram has a p-labeling.

3. Show that the knot 3_b has only 3-labelings.

4. For which values of p, if any, do the following diagrams in Figure 5.7 have p-labelings? (a) 5_1, (b) 5_2, (c) 6_1, (d) 6_2, (e) 6_3, (f) 7_1, (g) 7_2, (h) 7_3, (i) 7_4, (j) 7_5, (k) 7_6, (l) 7_7.

5. The diagram of $L_{1,n}$ in Figure 5.23 has a total of n crossings. For which values of n and p does $L_{1,n}$ have a p-labeling?

6. The diagram of $L_{2,n}$ in Figure 5.23 has a total of n crossings. For which values of n and p does $L_{2,n}$ have a p-labeling?

7. The diagram of $L_{3,n}$ in Figure 5.23 has a total of n crossings. For which values of n and p does $L_{3,n}$ have a p-labeling?

8. The diagram of $L_{4,n}$ in Figure 5.23 has a total of n crossings. For which values of n and p does $L_{4,n}$ have a p-labeling?

9. Prove that the Whitehead link of Figure 5.24 has no p-labelings.

10. Prove that no p-labeling of a knot uses exactly two labels. Is the same true for links?

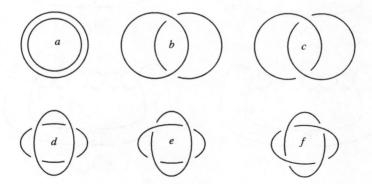

Figure 5.22

11. Complete the proof of Proposition 5.2.1 for the remaining type I Reidemeister moves.

12. Complete the proof of Proposition 5.2.1 for the remaining type II Reidemeister moves.

13. Complete the proof of Proposition 5.2.1 for the remaining type III Reidemeister moves.

14. A link is *Brunnian* if it is not an unlink but the removal of any of its components leaves an unlink. The Borromean rings of Figure 5.20 constitute a Brunnian link with three components.

    (a) Construct a Brunnian link with four components.

    (b) Construct a Brunnian link with an arbitrary number of components.

15*. Prove that the system consisting of all the labeling equations of a diagram is redundant.

16. Prove that in any link diagram, the number of crossings equals the number of strands that are not loops.

5.3 From Graphs to Links and on to Surfaces

It stands to reason that if a graph in $\mathbb{R}^3$ has a high density of arcs (relative to the number of nodes), then some of these arcs form nontrivial knots and/or links. This plausible intuition was corroborated in 1983, when it was demonstrated that in any placement of K_6 in $\mathbb{R}^3$ some of the arcs form a nontrivial link, and in any placement of K_7 in $\mathbb{R}^3$ some of the arcs form a nontrivial knot. This section is devoted to the unexpectedly easy proof of the first of these facts and to its surprising implications regarding surfaces.

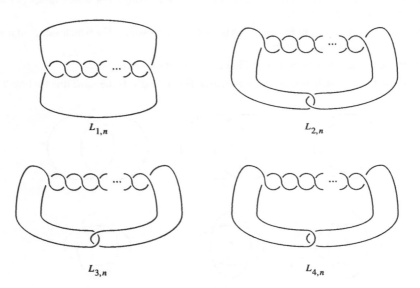

Figure 5.23 Some links in need of labelings.

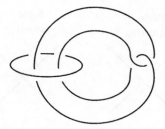

Figure 5.24   The Whitehead link.

An *orientation* of a knot K (or its diagram D) is a choice of one of the two senses in which it can be traversed, and is denoted by $\overrightarrow{K}$ (or $\overrightarrow{D}$). A link is said to be *oriented* if each of its components is oriented. Since each of the link's components has two possible orientations, it follows that a link with k components has 2^k possible orientations. Figure 5.25 exhibits all four orientations of the link 0^2.

The study of knots and links is enhanced by the widening of its scope to include oriented links as well, because the orientations lead to several parameters which also furnish useful information about the underlying unoriented links. This is one of many instances in mathematics where a seemingly artificial broadening of the context results in new tools that can be brought to bear on old problems.

Two oriented links are said to be *equivalent* if there is an ambient isotopy that transforms one to the other so that the orientations of corresponding components agree. It is easy to see that the leftmost and rightmost oriented links in Figure 5.25 are equivalent to each other, as are the two middle ones. The requisite isotopies are provided by a 180° flip around a horizontal axis. It is plausible that these two groupings are *not* equivalent to each other and a rigorous proof of this fact will be provided shortly.

If $\overrightarrow{L}$ is an oriented link, then the link obtained by reversing the orientations of each of the component loops of $\overrightarrow{L}$ is denoted by $\overrightarrow{L^r}$ and is called the *reverse* of $\overrightarrow{L}$. As it happens, all the oriented links displayed in this text are isotopic to their reverses, but that is not always the case (see Exercise 22).

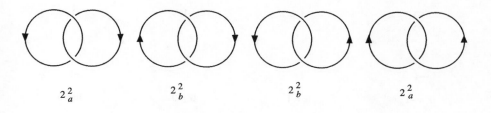

2^2_a 2^2_b 2^2_b 2^2_a

Figure 5.25   Four orientations of a single link.

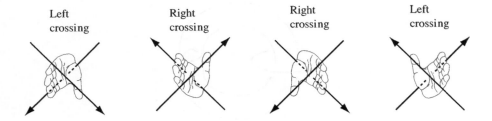

Figure 5.26   Left and right crossings.

In a diagram of an oriented link, a crossing is said to be *right* (*left*) *handed* if, when the right (left) hand is placed between the arcs so as to separate them, with the thumb pointing in the direction of the overstrand's orientation and the palm facing the observer, the other four digits point in the direction of the lower arc's orientation (Fig. 5.26). If C is a crossing of a diagram, then value$(C) = 1, -1$ according as C is right or left handed (Fig. 5.27). If $\overrightarrow{K}_1$ and $\overrightarrow{K}_2$ are two disjoint oriented knots and, in some diagram of the link $\overrightarrow{K}_1 \cup \overrightarrow{K}_2$, the list $C_1, C_2, \ldots, C_m$ constitutes all the crossings that involve arcs from both $\overrightarrow{K}_1$ and $\overrightarrow{K}_2$, then

$$\mathrm{lk}\left(\overrightarrow{K}_1, \overrightarrow{K}_2\right) = \frac{1}{2} \sum_{i=1}^{m} \mathrm{value}(C_i)$$

is called the *linking number* of $\overrightarrow{K}_1$ and $\overrightarrow{K}_2$ (see Fig. 5.27). The definition of $\mathrm{lk}(\overrightarrow{K}_1, \overrightarrow{K}_2)$ depends on the diagrams used to represent $\overrightarrow{K}_1 \cup \overrightarrow{K}_2$, but its numerical value does not.

Theorem 5.3.1 *If $\overrightarrow{K}_1$ and $\overrightarrow{K}_2$ are disjoint oriented knots, then* $\mathrm{lk}(\overrightarrow{K}_1, \overrightarrow{K}_2)$ *is an isotopy invariant of the link* $\overrightarrow{K}_1 \cup \overrightarrow{K}_2$.

PROOF: By Theorem 5.1.2 it suffices to prove that the Reidemeister moves do not change the value of the linking number. Since type I moves involve only self-crossings of a knot, and since such self-crossings do not contribute to the linking number, it follows that these moves have no effect on the linking number.

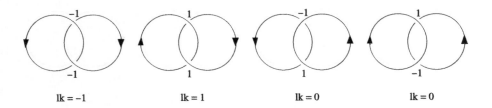

Figure 5.27   Some linking numbers.

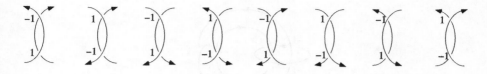

Figure 5.28   The linking number is invariant under type II Reidemeister moves.

Figure 5.29   The linking number is invariant under type III Reidemeister moves.

Turning to type II moves, the two crossings involved fall into eight categories, depending on the orientations of the arcs and on which loop contributes the overstrand (Fig. 5.28). In each case the sum of the crossing values is 0, and so, when the move either eliminates or adds these crossings, the linking number is unaffected.

Finally, if a, b, c, a', b', c' are the values of the crossings involved in a type III Reidemeister move, as indicated in Figure 5.29, then, regardless of the orientations of the arcs that form these crossings,

$$a = a', \quad b = b', \quad \text{and} \quad c = c'.$$

Hence the type III moves do not affect the linking number either, and so the linking number is an isotopy invariant.   Q.E.D.

Returning to the oriented links discussed at the beginning of this section (see Fig. 5.30), note that the two equivalent outside links have linking number -1, whereas the two equivalent inside links have linking number 1. It follows that these two groupings are inequivalent to each other. The oriented Whitehead link of Figure 5.31 has linking number 0. Nevertheless, by Exercise 5.2.9 above, it is not equivalent to the unlink 0^2.

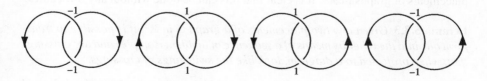

Figure 5.30   Four orientations of a single link.

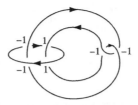

Figure 5.31   The Whitehead link.

Notwithstanding the presence of the 1/2 in its definition, the linking number is always an integer, and the proof of this fact is based on the following observation. If C is a crossing in the diagram of some link L, then the operation of making the lower arc of C into an overstrand and vice versa is called a *switching* of C (this operation is *not* an isotopy). If a crossing C of the knots $\overrightarrow{K_1}$ and $\overrightarrow{K_2}$ is switched, then the value of C is changed either from 1 to -1 or from -1 to 1, and hence its contribution to

$$\text{lk}\left(\overrightarrow{K_1}, \overrightarrow{K_2}\right) = \frac{1}{2} \sum_{i=1}^{m} \text{value}(C_i)$$

is altered by

$$\frac{1}{2}(\pm 2) = \pm 1,$$

which is an integer.

Proposition 5.3.2 *If $\overrightarrow{K_1}$ and $\overrightarrow{K_2}$ are disjoint oriented knots, then $lk(\overrightarrow{K_1}, \overrightarrow{K_2})$ is an integer.*

PROOF: In any diagram of $\overrightarrow{K_1} \cup \overrightarrow{K_2}$, switch all the crossings of $\overrightarrow{K_1}$ and $\overrightarrow{K_2}$, if necessary, so that the arc of $\overrightarrow{K_1}$ is always the overstrand. The resulting link $\overrightarrow{L}$ is clearly splittable and so is equivalent to the unlink 0^2. Consequently $\overrightarrow{L}$ has linking number 0. Since $\text{lk}(\overrightarrow{K_1}, \overrightarrow{K_2})$ differs from the linking number of $\overrightarrow{L}$ by an integer, it follows that $\text{lk}(\overrightarrow{K_1}, \overrightarrow{K_2})$ is also an integer.   Q.E.D.

One more lemma is needed before the linking number can be used to demonstrate the existence of nontrivial links in arbitrary placements of K_6 in $\mathbb{R}^3$. For the purposes of this lemma it is necessary to extend the notions of *diagrams* and *projections* to placements of graphs in $\mathbb{R}^3$. It is clear that this can be done without any difficulties.

Lemma 5.3.3 *Given any two placements of a graph G in $\mathbb{R}^3$, it is possible to transform one into the other by means of a sequence of ambient isotopies and arc crossings that, when projected to a diagram, look like the switchings of crossings.*

PROOF: Given two placements of G, say P and P', it is clear that P' can be deformed by means of an ambient isotopy into a placement P'' such that the nodes of P'' coincide with the nodes of P. It may therefore be assumed without loss of generality that the nodes of P coincide with those of P'. It now suffices to show that any arc a of P can be transformed by means of ambient isotopies and switchings of crossings into the corresponding arc a' of P', without disturbing the position of any other arc.

Let a and a' be corresponding arcs of P and P'. Switch every crossing of the diagram of P in which the understrands are parts of a so that a becomes the overstrand at each of its crossings. Then deform a so that its position in the projection of G coincides with that of a'. This can be done because a is always the overstrand. Finally, switch all the necessary crossings until a, in its new position, is the overstrand at exactly the same crossings as a'. Q.E.D.

Let K_1 and K_2 be two disjoint unoriented knots, and let $\overrightarrow{K_1}$ and $\overrightarrow{K_2}$ be an orientation for each. The (unoriented) *linking number* of K_1 and K_2 is defined as $\mathrm{lk}(K_1, K_2) = |\mathrm{lk}(\overrightarrow{K_1}, \overrightarrow{K_2})|$, and it follows from Theorem 5.3.1, Proposition 5.3.2, and Exercise 16 that this quantity is a well defined nonnegative integer.

A pair of disjoint cycles of a graph placed in $\mathbb{R}^3$ is said to be *linked* if they form a nonsplittable link.

Theorem 5.3.4 (H. Sachs 1982, J. H. Conway and C. M. Gordon, 1983). *Every placement of the graph K_6 in $\mathbb{R}^3$ contains a linked pair of 3-cycles.*

PROOF: The graph K_6 contains $\frac{1}{2}\binom{6}{3} = 10$ distinct pairs of disjoint cycles, which may be supposed to be labeled (C_{2i}, C_{2i+1}), $i = 1, 2, \ldots, 10$, where C_{2i} and C_{2i+1} are disjoint. It will be shown that for any placement of K_6 in $\mathbb{R}^3$, the quantity

$$U = \sum_{i=1}^{10} \mathrm{lk}(C_{2i}, C_{2i+1})$$

is not zero, which implies the desired conclusion. This will be accomplished by demonstrating that this quantity is in fact odd.

Such is the case for the placement of Figure 5.32. The triangles 123, 125, 134, 135, 145, 235, 345, 356 are clearly unlinked to their complements, leaving us with

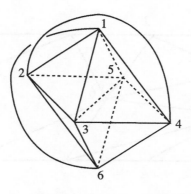

Figure 5.32 A placement of K_6.

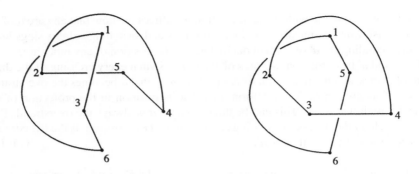

Figure 5.33   Two links in a placement of K_6.

the pairs $(136, 245)$ and $(156, 234)$, which have linking numbers 1 and 0 respectively (see Fig. 5.33). Hence, for this placement of K_6 we have $U = 1$.

By Lemma 5.3.3, any other placement of K_6 differs from the above by a sequence of ambient isotopies and crossing switches. It follows from Theorem 5.3.1 that the isotopies will not change the numbers $\mathrm{lk}(C_{2i}, C_{2i+1})$. The switching of a crossing, however, will in general change these parameters if the switched arcs belong to *distinct* cycles, and this case needs to be examined more closely.

It may be assumed without loss of generality that the switched crossing in question involves the edges 12 and 34. The only disjoint cycle pairs to which these arcs belong are $(125, 346)$ and $(126, 345)$. Hence, when the crossing is switched, only their two linking numbers are affected, and each is modified by ± 1. It follows that $U = \sum_{i=1}^{10} \mathrm{lk}(C_{2i}, C_{2i+1})$ is modified by $\pm 1 \pm 1$, which is either 0 or ± 2. Thus, the parity of U is unchanged by switchings, and hence it must always be odd.   Q.E.D.

Example 5.3.5 Figure 5.34 displays an indirect placement of K_6 in $\mathbb{R}^3$. First K_6 is drawn on a polygonal presentation of the torus, and then the reader is asked to imag-

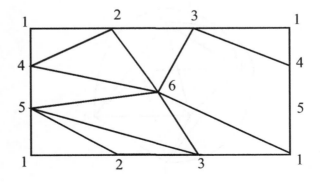

Figure 5.34

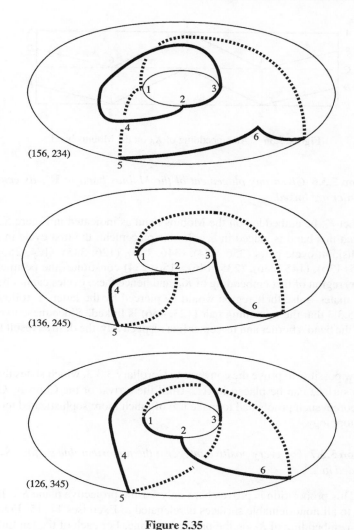

Figure 5.35

ine that this presentation is bent and its edges zipped so as to create an embedding of the torus in $\mathbb{R}^3$. The fact that the cycle 124 bounds a triangular region of the torus demonstrates that the cycle pair (124, 356) is unlinked. Similarly, the fact that each of the cycles 246, 236, 134, 456, 125, 235 bounds a region on the torus demonstrates that none of the cycle pairs (135, 246), (145, 236), (134, 256), (123, 456), (125, 346), and (146, 235) are linked. Figure 5.35 demonstrates that the pairs (156, 234) and (136, 245) are also unlinked, whereas the pair (126, 345) is linked.

Theorem 5.3.4 has two surprising applications to surfaces.

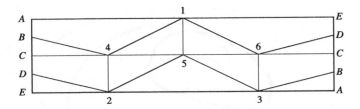

Figure 5.36   An embedding of K_6 on the Möbius band.

Proposition 5.3.6 *Given any placement of the Möbius band in* $\mathbb{R}^3$*, its center line and its border are linked.*

PROOF: Let K_6 be embedded on the Möbius band as indicated in Figure 5.36, and suppose that this band is placed in $\mathbb{R}^3$. In this placement, the first cycle in each of the nine disjoint cycle pairs (356, 124), (346, 125), (126, 345), (134, 256), (246, 135), (245, 136), (145, 236), (235, 146), (156, 234) constitutes the perimeter of a (triangular) region of this embedding of K_6, and hence these cycles cannot be linked with their mates—else their region would be pierced by the latter. It follows from Theorem 5.3.3 that the remaining pair (123, 456) is linked. Since these two cycles constitute the band's border and its center line respectively, the desired result follows.
                                                                                  Q.E.D.

It is now possible to prove the converse to Corollary 3.3.3, which states that every orientable surface can be placed in $\mathbb{R}^3$. Until the arrival of the Conway–Gordon–Sachs Theorem such proofs had to make use of much more sophisticated tools from algebraic topology.

Proposition 5.3.7 *For every positive integer n the nonorientable surface* $\tilde{S}_n$ *cannot be embedded in* $\mathbb{R}^3$*.*

PROOF: This proposition is proven here only for the projective plane $\tilde{S}_1$. The easy extension to all nonorientable surfaces is relegated to Exercises 14–15. Figure 5.37 exhibits an embedding of K_6 on the projective plane. For each of the ten links (123, 456), (356, 124), (346, 125), (126, 345), (256, 134), (135, 246), (245, 136), (145, 236), (146, 235), (234, 156), the first triangle is the perimeter of a region of this embedding, and hence all of these pairs constitute unlinks in any placement of the ambient projective plane in $\mathbb{R}^3$. It now follows from Theorem 5.3.4 that the projective plane cannot be embedded in $\mathbb{R}^3$, since any such embedding would result in a placement of K_6 in $\mathbb{R}^3$ that is free of linked cycles.   Q.E.D.

Exercises 5.3

1. For which values of n does the link $L_{1,n}$ of Figure 5.23 have two components? In those cases compute the linking numbers of all four of the orientations of $L_{1,n}$.

2. For which values of n does the link $L_{2,n}$ of Figure 5.23 have two components? In those cases compute the linking numbers of all four of the orientations of $L_{2,n}$.

3. For which values of n does the link $L_{3,n}$ of Figure 5.23 have two components? In those cases compute the linking numbers of all four of the orientations of $L_{3,n}$.

4. For which values of n does the link $L_{4,n}$ of Figure 5.23 have two components? In those cases compute the linking numbers of all four of the orientations of $L_{4,n}$.

5. Prove that for every positive integer k there exists a k-component link such that the linking number of any two of its components is 1.

6. Prove that for every sequence of integers $a_1, a_2, \ldots a_{k(k-1)/2}$ there exists a k-component link for which this sequence constitutes the sequence of all the linking numbers of its pairs of components.

7.* Prove that in every placement of $K_{1,3,3}$ in $\mathbb{R}^3$ there is a pair of cycles that constitute a nontrivial link.

8. For each of the placements of K_6 of Figure 5.38 list all of its linked pairs of cycles.

9. Figure 5.39 describes two embeddings of K_6 on the standard presentation of the torus. Place each of these tori in $\mathbb{R}^3$ by bending the rectangles and pasting the opposite edges. Find all the linked pairs of cycles in the resulting placements of K_6.

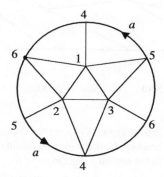

Figure 5.37   An embedding of K_6 on the projective plane.

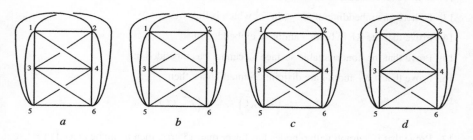

Figure 5.38   Some placements of K_6.

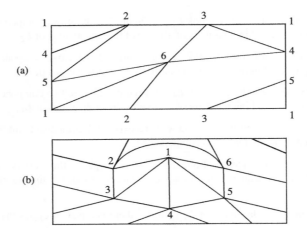

Figure 5.39 Toroidal embeddings of K_6.

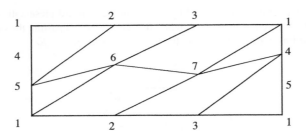

Figure 5.40 A toroidal embedding of $K_{1,3,3}$.

10. Figure 5.40 describes an embedding of $K_{1,3,3}$ on the standard presentation of the torus. Place this torus in $\mathbb{R}^3$ by bending the rectangle and pasting the opposite edges. Find all the linked pairs of cycles in the resulting placement of $K_{1,3,3}$.

11**. Prove that every placement of the graph $K_{4,4}$ has a linked pair of cycles.

12. Is there an embedding of $K_{1,2,3}$ in $\mathbb{R}^3$ which is link-free?

13. Is there an embedding of $K_{2,2,3}$ in $\mathbb{R}^3$ which is link-free?

14. Prove that the Klein bottle cannot be embedded in $\mathbb{R}^3$.

15. Prove that the nonorientable surface $\tilde{S}_n$ cannot be embedded in $\mathbb{R}^3$.

16. Prove that if $\overrightarrow{K_1}$ and $\overrightarrow{K_2}$ are disjoint oriented knots, then

$$\mathrm{lk}\left(\overrightarrow{K_1},\overrightarrow{K_2}\right) = -\mathrm{lk}\left(\overrightarrow{K_1},\overrightarrow{K_2}\right).$$

17. Prove that if a graph with 6 nodes has fewer than 15 arcs, then it can be placed in $\mathbb{R}^3$ so that none of its cycles are linked.

18. Prove that K_6 can be placed in $\mathbb{R}^3$ so that none of its cycles forms a nontrivial knot.

An unoriented knot both of whose orientations are isotopic is said to be reversible.

19. Use a $180°$ rotation to prove that the knot 3_b is reversible.

20. Prove that the knot 4 is reversible.

21. Prove that any orientation of the Whitehead link of Figure 5.24 is isotopic to its reverse.

22**. Find a knot that is not reversible.

23. Prove that every oriented link with two components and two crossings is isotopic to 0^2, 2_a^2, or 2_b^2.

5.4   The Jones Polynomial

In 1928 the topologist James W. Alexander (1888–1971) associated to each link a polynomial which was an isotopy invariant. This polynomial proved very useful in distinguishing between non-equivalent links. In 1986, as part of groundbreaking work that won him the prestigious Field Medal Award, Vaughan Jones developed a new polynomial which separated links much more efficiently. While the original description of the Jones polynomial was very complicated, subsequent work by Louis H. Kauffman resulted in a definition that is suitable for inclusion in textbooks.

Strictly speaking, the Jones polynomial is not a polynomial; it is a *Laurent polynomial*, as it includes powers of its variable with negative exponents. Nevertheless, this abuse of terminology is standard and will be followed here as well. The Jones polynomial of the link $\overrightarrow{L}$ is denoted by $V(\overrightarrow{L})$, and its independent variable is A. Not surprisingly, $V(\text{unknot}) = 1$. Several more polynomials are listed in Table 5.1, but this table requires a clarification, as it takes advantage of hindsight to simplify the notation. Notwithstanding the fact that $\overrightarrow{L}$ and $\overrightarrow{L^r}$ are not isotopic in general, it so happens (see Proposition 5.4.13 below) that $V(\overrightarrow{L^r}) = V(\overrightarrow{L})$ and hence, in the important special case where $\overrightarrow{L}$ is in fact an oriented knot $\overrightarrow{K}$, we may speak of the Jones polynomial of the unoriented knot K and define it as

$$V(K) = V\left(\overrightarrow{K}\right) = V\left(\overrightarrow{K^r}\right).$$

It will be demonstrated later that the Jones polynomial satisfies a relation which is more useful than the actual definition for the purposes of computation. Specifically, if $\overrightarrow{L}_+$, $\overrightarrow{L}_-$, and $\overrightarrow{L}_0$ are three oriented links with diagrams that differ only at one location, where the differences are depicted in Figure 5.41, then the Jones polynomials of $\overrightarrow{L}_+$, $\overrightarrow{L}_-$, and $\overrightarrow{L}_0$ are related by the equation

$$A^4V\left(\overrightarrow{L}_+\right) - A^{-4}V\left(\overrightarrow{L}_-\right) + \left(A^2 - A^{-2}\right)V\left(\overrightarrow{L}_0\right) = 0, \tag{3}$$

which is known as the *skein relation*.

This relation is now applied to the computation of some specific polynomials.

Example 5.4.1 $V(0^2)$: In Figure 5.42, both $\overrightarrow{L}_+$ and $\overrightarrow{L}_-$ are equivalent to the oriented

Table 5.1

K	$V(K)$
0	1
0^2	$-A^2 - A^{-2}$
2_a^2	$-A^{10} - A^2$
2_b^2	$-A^{-10} - A^{-2}$
3_a	$-A^{16} + A^{12} + A^4$
3_b	$-A^{-16} + A^{-12} + A^{-4}$
4	$A^8 - A^4 + 1 - A^{-4} + A^{-8}$
5_1	$-A^{28} + A^{24} - A^{20} + A^{16} + A^8$
5_2	$-A^{24} + A^{20} - A^{16} + 2A^{12} - A^8 + A^4$
6_1	$A^{16} - A^{12} + A^8 - 2A^4 + 2 - A^{-4} + A^{-8}$
6_2	$A^{20} - 2A^{16} + 2A^{12} - 2A^8 + 2A^4 - 1 + A^{-4}$
6_3	$-A^{12} + 2A^8 - 2A^4 + 3 - 2A^{-4} + 2A^{-8} - A^{-12}$
7_1	$-A^{40} + A^{36} - A^{32} + A^{28} - A^{24} + A^{20} + A^{12}$
7_2	$-A^{32} + A^{28} - A^{24} + 2A^{20} - 2A^{16} + 2A^{12} - A^8 + A^4$
7_3	$A^{-8} - A^{-12} + 2A^{-16} - 2A^{-20} + 3A^{-24} - 2A^{-28} + A^{-32} - A^{-36}$
7_4	$A^{-4} - 2A^{-8} + 3A^{-12} - 2A^{-16} + 3A^{-20} - 2A^{-24} + A^{-28} - A^{-32}$
7_5	$-A^{36} + 2A^{32} - 3A^{28} + 3A^{24} - 3A^{20} + 3A^{16} - A^{12} + A^8$
7_6	$-A^{24} + 2A^{20} - 3A^{16} + 4A^{12} - 3A^8 + 3A^4 - 2 + A^{-4}$
7_7	$-A^{12} + 3A^8 - 3A^4 + 4 - 4A^{-4} + 3A^{-8} - 2A^{-12} + A^{-16}$

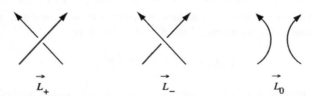

Figure 5.41   Related links.

unknot 0, which has polynomial 1; and $\overrightarrow{L}_0$ is the unlink 0^2. Hence,

$$A^4 \cdot 1 - A^{-4} \cdot 1 + \left(A^2 - A^{-2}\right) V\left(0^2\right) = 0,$$

or

$$V\left(0^2\right) = -\frac{A^4 - A^{-4}}{A^2 - A^{-2}} = -\left(A^2 + A^{-2}\right) = -A^2 - A^{-2}.$$

Example 5.4.2 $V(2_a^2)$: In Figure 5.43, it is $\overrightarrow{L}_+$ that is equivalent to the unlink 0^2 (and so, by Example 5.4.1, has polynomial $-A^2 - A^{-2}$), whereas $\overrightarrow{L}_0$ is now the unknot. It follows from the skein relation that

$$A^4 \left(-A^2 - A^{-2} \right) - A^{-4} V \left(2_a^2 \right) + \left(A^2 - A^{-2} \right) \cdot 1 = 0,$$

or

$$V \left(2_a^2 \right) = A^4 \left(-A^6 - A^2 + A^2 - A^{-2} \right) = -A^{10} - A^2.$$

Example 5.4.3 $V(2_b^2)$: In Figure 5.44, $\overrightarrow{L}_-$ is equivalent to the unlink 0^2, $\overrightarrow{L}_0$ is the unknot with polynomial 1, and $\overrightarrow{L}_+$ is the link 2_b^2. It follows from the skein relation that

$$A^4 V \left(2_b^2 \right) - A^{-4} \left(-A^2 - A^{-2} \right) + \left(A^2 - A^{-2} \right) \cdot 1 = 0,$$

or

$$V \left(2_b^2 \right) = A^{-4} \left(-A^{-2} - A^{-6} - A^2 + A^{-2} \right) = -A^{-10} - A^{-2}.$$

Since $V(2_a^2) \neq V(2_b^2)$, it follows that 2_a^2 and 2_b^2 are inequivalent links, an intuitively plausible observation. We now go on to show that the trefoil knots 3_a and 3_b are also inequivalent.

Example 5.4.4 $V(3_a)$: The polynomials $V(0)$ and $V(2_a^2)$ can be substituted into the skein relation of Figure 5.45 to obtain the Jones polynomial of the trefoil knot 3_a:

$$A^4 \cdot 1 - A^{-4} V \left(3_a \right) + \left(A^2 - A^{-2} \right) V \left(2_a^2 \right) = 0,$$

or

$$V(3_a) = A^4 \left[A^4 + \left(A^2 - A^{-2} \right) \left(-A^{10} - A^2 \right) \right]$$
$$= A^4 \left(A^4 - A^{12} - A^4 + A^8 + 1 \right) = A^4 + A^{12} - A^{16}.$$

Example 5.4.5 $V(3_b)$: The polynomials $V(0)$ and $V(2_b^2)$ can be substituted into the skein relation of Figure 5.46 to obtain the Jones polynomial of the trefoil knot 3_b:

$$A^4 V \left(3_b \right) - A^{-4} \cdot 1 + \left(A^2 - A^{-2} \right) V \left(2_b^2 \right) = 0,$$

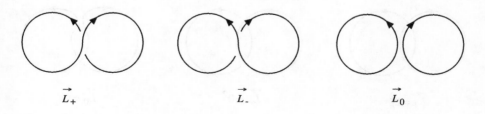

$$\overrightarrow{L}_+ \qquad\qquad \overrightarrow{L}_- \qquad\qquad \overrightarrow{L}_0$$

Figure 5.42   A derivation of $V(0^2)$.

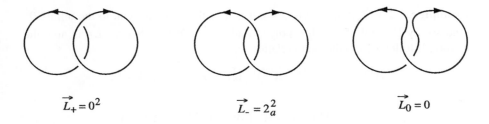

$$\overrightarrow{L}_+ = 0^2 \qquad\qquad \overrightarrow{L}_- = 2_a^2 \qquad\qquad \overrightarrow{L}_0 = 0$$

Figure 5.43   Deriving the Jones polynomial of 2_a^2.

or

$$V(3_b) = A^{-4} \left[A^{-4} + \left(A^2 - A^{-2}\right) \left(-A^{-10} - A^{-2}\right) \right]$$
$$= A^{-4} \left(A^{-4} + A^{-8} + 1 - A^{-12} - A^{-4}\right) = A^{-4} + A^{-12} - A^{-16}.$$

Since $V(3_a) \neq V(3_b)$ it follows that the trefoil knots 3_a and 3_b are not isotopic.

The above examples may be misleading. The skein relation does not necessarily provide an algorithm for computing the (as yet undefined) Jones polynomial. The reason this relation fails to yield an algorithm is that $\overrightarrow{L}_+$ and $\overrightarrow{L}_-$ have the same number of crossings and so there is no way of predicting which of the links $\overrightarrow{L}_+$ or $\overrightarrow{L}_-$ is "simpler." Another question left unanswered by the skein relation is that of invariance. Even in those cases where the skein relation can be used to compute the Jones polynomial, it is not yet clear that the derived polynomial is independent of the crossings at which the relation is applied.

The definition of the Jones polynomial calls for two auxiliary polynomials whose ratio turns out to be the Jones polynomial. Curiously, neither of these auxiliary polynomials is invariant in and of itself—only their product has that crucial property.

The first of these, the *bracket polynomial* $[D]$, is defined for unoriented diagrams and is based upon a close examination of the crossings of link diagrams. At each crossing there are four *angles* (or *corners*). Two types of angles can be distinguished: those that lie counterclockwise from the overstrand and those that lie clockwise from it (Fig. 5.47). The former are called *A-angles* and the latter *B-angles*. Borrowing

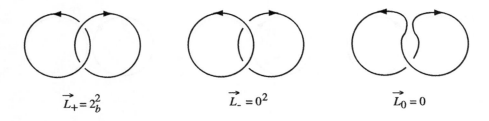

$$\overrightarrow{L}_+ = 2_b^2 \qquad\qquad \overrightarrow{L}_- = 0^2 \qquad\qquad \overrightarrow{L}_0 = 0$$

Figure 5.44

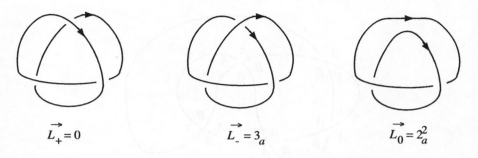

$$\overrightarrow{L_+} = 0 \qquad\qquad \overrightarrow{L_-} = 3_a \qquad\qquad \overrightarrow{L_0} = 2_a^2$$

Figure 5.45   Deriving the Jones polynomial of the trefoil knot 3_a.

terminology from the theory of electronic circuits, each crossing is thought of as a "switch" with two possible *settings* or *connections*. Each of these two settings is in actuality a modification of the diagram obtained by eliminating the crossing and connecting either the A-angles or the B-angles (Fig. 5.48). Of course, once a connection has been made, the diagram depicts a different link altogether. A *state* of a diagram D is any of the diagrams obtained by replacing all the crossings of D with appropriate connections (i.e., setting all of its switches). A diagram with two crossings therefore has four states (Fig. 5.49), a diagram with three crossings has eight states, and, in general, a diagram with c crossings has 2^c states. It is clear that each state of a diagram is itself a diagram of some unlink.

The number of component loops of the state S is denoted by $|S|$, the number of its A-connections by $\alpha_D(S)$, and the number of its B-connections by $\beta_D(S)$. The subscript D will be dropped from these parameters whenever possible. The *bracket polynomial* of an unoriented diagram D is a polynomial in the variables A, B, d:

$$[D] = \sum_S A^{\alpha(S)} B^{\beta(S)} d^{|S|-1}.$$

For example, the bracket polynomial of the four-state diagram of Figure 5.49 is $A^2 d + AB + ABd^2 + B^2 d$, whereas that of the trefoil diagram of Figure 5.50 is

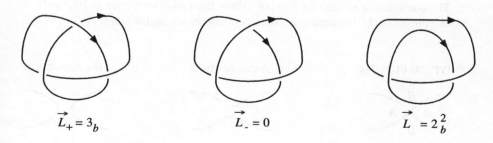

$$\overrightarrow{L_+} = 3_b \qquad\qquad \overrightarrow{L_-} = 0 \qquad\qquad \overrightarrow{L} = 2_b^2$$

Figure 5.46   Deriving the Jones polynomial of the trefoil knot 3_b.

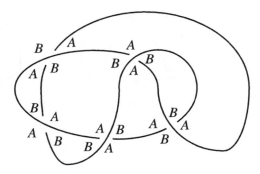

Figure 5.47   *A-angles and B-angles.*

$$[3_b] = A^3d + A^2B + A^2B + AB^2d + A^2B + AB^2d + AB^2d + B^3d^2$$
$$= 3A^2B + A^3d + 3AB^2d + B^3d^2.$$

The next lemma establishes some of the basic properties of the bracket polynomial. It can also be construed as an alternative, inductive definition of this polynomial.

Lemma 5.4.6 *The bracket polynomial satisfies the following equations:*

1. $[0] = 1$.

2. $[D] = A[D_A] + B[D_B]$, *where D_A and D_B are the diagrams obtained by replacing some crossing C by its A and B-connections, respectively.*

3. $[D \cup 0] = d[D]$.

PROOF: Part 1 follows immediately from the definition of the bracket polynomial, and part 3 follows from the fact that the diagram of $D \cup 0$ differs from that of D only in that it has one additional component (the crossings are identical in both). As for part 2, note that each state S of D_A is also a state of D in which C has been set to A. The contribution of such an S to $[D]$ differs from its contribution to $[D_A]$ only in having one more A. The analogous statement also holds for the states of D_B. As this

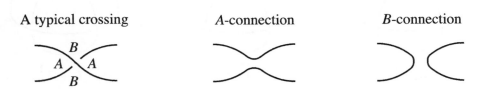

A typical crossing   **A-connection**   **B-connection**

Figure 5.48   The two settings of a crossing.

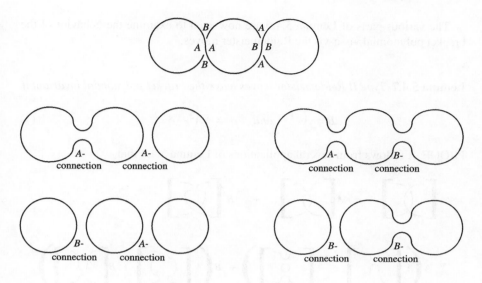

Figure 5.49   A diagram with four states.

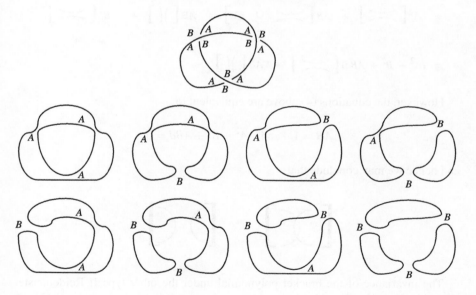

Figure 5.50   The eight states of a trefoil diagram.

accounts for all the states of D, it follows that

$$[D] = A[D_A] + B[D_B].$$ Q.E.D.

The various parts of Lemma 5.4.6 are now used to examine the behavior of the bracket polynomial vis-à-vis the Reidemeister moves.

Lemma 5.4.7 *Type II Reidemeister moves leave the bracket polynomial invariant if*

$$B = A^{-1} \quad and \quad d = -A^2 - A^{-2}. \tag{4}$$

PROOF: It follows from several applications of Lemma 5.4.6 that

$$\left[\,\rotatebox{90}{)(}\,\right] = A\left[\,\smallsmile \atop \smallfrown\,\right] + B\left[\,)(\,\right]$$

$$= A\left(A\left[\,\asymp\,\right] + B\left[\,\rotatebox{90}{)(}\,\right]\right) + B\left(A\left[\,)(\,\right] + B\left[\,\smallsmile\atop\smallfrown\,\right]\right)$$

$$= A^2\left[\,\asymp\,\right] + AB\left[\,\asymp \cup \circ\,\right] + AB\left[\,)(\,\right] + B^2\left[\,\asymp\,\right]$$

$$= (A^2 + B^2 + ABd)\left[\,\asymp\,\right] + AB\left[\,)(\,\right].$$

However, the equations (4) above are equivalent to

$$AB = 1 \quad and \quad A^2 + B^2 + ABd = 0,$$

and hence we may conclude that

$$\left[\,\rotatebox{90}{)(}\,\right] = \left[\,)(\,\right].$$

The invariance of the bracket polynomial under the other type II Reidemeister move is relegated to Exercise 16.   Q.E.D.

Lemma 5.4.8 *Type III Reidemeister moves leave the bracket polynomial invariant if* $B = A^{-1}$ *and* $d = -A^2 - A^{-2}$.

PROOF: It follows from several applications of Lemmas 5.4.6 and 5.4.7 that

$$\left[\text{✕} \right] = A \left[\smile\frown \right] + B \left[ )(\right]$$

$$= A \left[= \right] + B \left[ )(\right]$$

$$= A \left[\frown\smile \right] + B \left[ )(\right] = \left[\text{△} \right].$$

The proof of the invariance of the bracket polynomial for the other type III Reidemeister moves is relegated to Exercise 19.   Q.E.D.

The next lemma demonstrates that the bracket polynomial is *not* an isotopy invariant.

Lemma 5.4.9 *If $B = A^{-1}$ and $d = -A^2 - A^{-2}$ then*

$$\left[\rho \right] = -A^{-3} \left[ ) \right],$$

$$\left[\rho \right] = -A^3 \left[ ) \right].$$

PROOF: By Lemma 5.4.6,

$$\left[\rho \right] = A \left[\text{⌐} \right] + B \left[ ) \, \circ \right]$$

$$= A \left[ ) \right] + Bd \left[ ) \right] = (A + Bd) \left[ ) \right] = -A^{-3} \left[ ) \right]$$

and

$$\left[\rho \right] = A \left[ ) \, \circ \right] + B \left[\text{⌐} \right]$$

$$= Ad\left[\;\right)\;\right] + B\left[\;\right)\;\right] = (Ad + B)\left[\;\right)\;\right] = -A^3\left[\;\right)\;\right].$$

If D is a link diagram, then the *Kauffman bracket polynomial* $\langle D \rangle$ is the polynomial obtained from the bracket polynomial $[D]$ by replacing B with A^{-1} and d with $-A^2 - A^{-2}$. Since $\langle D \rangle$ contains negative powers of its variable A, it likewise is not properly speaking a polynomial, but a Laurent polynomial. Nevertheless, this inaccurate appellation is accepted by all. The above lemmas yield information about the new polynomial too.

Proposition 5.4.10 *The Kauffman bracket polynomial* $\langle D \rangle$ *is invariant under type II and III Reidemeister moves, and*

$$\left\langle\; \right\rangle = -A^{-3}\left\langle\; \right\rangle\;,$$

$$\left\langle\; \right\rangle = -A^{3}\left\langle\; \right\rangle\;.\qquad\qquad \square$$

For the second auxiliary polynomial required by the Jones polynomial it is necessary to define a numerical parameter $w(\vec{D})$, the *writhe* of $\vec{D}$, as the sum of all the values of its crossings, including the self-crossings of its components. Thus, since the crossings of the diagram of the Whitehead link of Figure 5.24 have values $-1, 1, -1, 1, -1, -1$, its writhe equals -2, whereas its linking number is 0. Since the reversal of every link diagram $\vec{D}$ does not change the value of any of its crossings, it follows that $w(\vec{D}) = w(\vec{D'})$. In particular, $w(\vec{D})$ is well defined whenever D is the diagram of an unoriented *knot*.

Vis-à-vis the Reidemeister moves, the writhe behaves very much like the Kauffman bracket polynomial.

Proposition 5.4.11 *Type II and III Reidemeister moves leave the writhe invariant and*

$$w\left(\; \right) = -1 + w\left(\; \right)\;,\qquad w\left(\; \right) = 1 + w\left(\; \right)\;,$$

$$w\left(\; \right) = 1 + w\left(\; \right)\;,\qquad w\left(\; \right) = -1 + w\left(\; \right)\;.$$

PROOF: See Exercise 17. $\square$

The *Jones polynomial* $V(\vec{L})$ is defined as

$$V\left(\vec{L}\right) = (-A)^{-3w\left(\vec{D}\right)}\langle D \rangle,$$

where $\vec{D}$ is any diagram of $\vec{L}$. For example, it was computed in Figure 5.49 that

$$\left[\;\bigcirc\!\!\bigcirc\!\!\bigcirc\;\right] \;=\; A^2 d + AB + ABd^2 + B^2 d\,.$$

It follows that

$$\left\langle\;\bigcirc\!\!\bigcirc\!\!\bigcirc\;\right\rangle$$

$$= A^2\left(-A^2 - A^{-2}\right) + AA^{-1} + AA^{-1}\left(-A^{-2} - A^2\right)^2 + A^{-2}\left(-A^2 - A^{-2}\right)$$

$$= -A^4 - 1 + 1 + A^{-4} + 2 + A^4 - 1 - A^{-4} = 1.$$

Since each orientation of this diagram has writhe $1 - 1 = 0$, it follows that its knot has the Jones polynomial

$$(-A)^{-3 \cdot 0} \cdot 1 = 1,$$

which is consistent with the fact that this is an unknot. Less trivially, it was demonstrated on the basis of Figure 5.50 that for that diagram D of 3_b,

$$[D] = 3A^2 B + A^3 d + 3AB^2 d + B^3 d^2,$$

so that

$$\langle D \rangle = 3A^2 A^{-1} + A^3\left(-A^2 - A^{-2}\right) + 3AA^{-2}\left(-A^2 - A^{-2}\right)$$

$$+ A^{-3}\left(-A^2 - A^{-2}\right)^2$$

$$= 3A - A^5 - A - 3A - 3A^{-3} + A + 2A^{-3} + A^{-7}$$

$$= -A^5 - A^{-3} + A^{-7}.$$

Since $w(D) = 3$, it follows that

$$V(3_b) = (-A)^{-3 \cdot 3}\left(-A^5 - A^{-3} + A^{-7}\right) = A^{-4} + A^{-12} - A^{-16},$$

which agrees with the previous derivation that was based on the skein relation.

This section's main theorem proves that the Jones polynomial of an oriented link does not depend on the specific diagram that is used for its computation.

Theorem 5.4.12 *The Jones polynomial is invariant under all Reidemeister moves and is therefore an isotopy invariant of oriented links.*

PROOF: By Lemmas 5.4.7 and 5.4.8 and Proposition 5.4.11, the Jones polynomial is invariant under type II and III Reidemeister moves. To prove its invariance under type I moves, suppose first that

$$\vec{D} = \quad\text{and}\quad \vec{D'} = \;\Big).$$

Figure 5.51 A crossing and its reversal.

It then follows from Lemma 5.4.9 and Proposition 5.4.10 that

$$V\left(\overrightarrow{D}\right) = (-A)^{-3w(\overrightarrow{D})}\langle D \rangle$$

$$= (-A)^{-3[-1+w(\overrightarrow{D'})]}\left(-A^{-3}\right)\langle D' \rangle$$

$$= (-A)^{-3w(\overrightarrow{D'})}\langle D' \rangle = V\left(\overrightarrow{D'}\right).$$

The invariance of the Jones polynomial under the other three oriented type I moves (listed in the statement of Proposition 5.4.11) is relegated to Exercise 18.   Q.E.D.

While oriented links and knots are of interest, it is their unoriented versions that are of primary interest. It is therefore fortunate that the Jones polynomial can be shown to be an invariant of unoriented knots (although not of unoriented links).

Proposition 5.4.13 *If* $\overrightarrow{L}$ *is any oriented link, then* $V(\overrightarrow{L}) = V(\overrightarrow{L^r})$.

PROOF: Let $\overrightarrow{D}$ be a diagram of the link $\overrightarrow{L}$. Since the bracket polynomial is defined for unoriented links, it follows that $[\overrightarrow{D}] = [\overrightarrow{D^r}]$ and so $\langle \overrightarrow{D} \rangle = \langle \overrightarrow{D^r} \rangle$. Moreover, the reversal of both the orientations of the two strings that constitute a crossing leaves the value of the crossing unchanged (see Figure 5.51), and hence it follows that $w(\overrightarrow{D}) = w(\overrightarrow{D^r})$. The conclusion of the proposition now follows immediately from the definition of the Jones polynomial.   Q.E.D.

It follows from this proposition that of the four links in Figure 5.25 the two in the middle have the same Jones polynomial, as have the other two. More importantly, the next proposition also holds.

Proposition 5.4.14 *If K is an unoriented knot, then both its orientations have the same Jones polynomial, so that* $V(K)$ *is well defined.*   □

The skein relation, so useful for the purpose of computations, can now be proven.

Proposition 5.4.15 (Skein relation). *If the oriented links* $\overrightarrow{L_+}, \overrightarrow{L_-}, \overrightarrow{L_0}$ *are related by Figure 5.41, then*

$$A^4 V\left(\overrightarrow{L_+}\right) - A^{-4} V\left(\overrightarrow{L_-}\right) + \left(A^2 - A^{-2}\right) V\left(\overrightarrow{L_0}\right) = 0.$$

PROOF: Recall that both the bracket and the Kauffman bracket polynomials are defined for *unoriented* links. Let D_+, D_-, and D_0 denote the diagrams of L_+, L_-, and L_0 respectively. It follows from Lemmas 5.4.6.2 and the fact that $B = A^{-1}$ that

$$\langle D_+ \rangle \;=\; \langle \times \rangle \;=\; A\,\langle\, \rangle\langle\, \rangle \;+\; A^{-1}\langle\,\smile\,\frown\,\rangle ,$$

$$\langle D_- \rangle \;=\; \langle \times \rangle \;=\; A\,\langle\,\smile\,\frown\,\rangle \;+\; A^{-1}\,\langle\,\rangle\langle\,\rangle ,$$

and consequently

$$A\langle D_+ \rangle - A^{-1}\langle D_- \rangle = \left(A^2 - A^{-2}\right)\langle D_0 \rangle.$$

Since $w(\overrightarrow{D_\pm}) = w(\overrightarrow{D_0}) \pm 1$ and $\langle D \rangle = (-A)^{3w(\overrightarrow{D})}V(\overrightarrow{L})$, this equation becomes

$$A(-A)^{3w(\overrightarrow{D_0})+3}V\left(\overrightarrow{L_+}\right) - A^{-1}(-A)^{3w(\overrightarrow{D_0})-3}V\left(\overrightarrow{L_-}\right)$$
$$= \left(A^2 - A^{-2}\right)(-A)^{3w(\overrightarrow{D_0})}V\left(\overrightarrow{L_0}\right),$$

or

$$-A^4 V\left(\overrightarrow{L_+}\right) + A^{-4}V\left(\overrightarrow{L_-}\right) = \left(A^2 - A^{-2}\right)V\left(\overrightarrow{L_0}\right). \qquad \text{Q.E.D.}$$

The reader is reminded that whereas the expression of the Jones polynomial in terms of the writhe and the Kauffman bracket polynomial is a useful device for establishing its theoretical properties, the skein relation is the better calculational tool. The following three propositions are also useful in this regard. Suppose $\overrightarrow{L_1}$ and $\overrightarrow{L_2}$ are two oriented links whose diagrams D_1 and D_2 are drawn in the same plane so that they can be separated by a single straight line. The *split union* $\overrightarrow{L_1} \cup \overrightarrow{L_2}$ is the link whose diagram is $D_1 \cup D_2$.

Proposition 5.4.16 $V(\overrightarrow{L_1} \cup \overrightarrow{L_2}) = (-A^2 - A^{-2})V(\overrightarrow{L_1})V(\overrightarrow{L_2})$.

PROOF: Let D_1 and D_2 be the split unoriented diagrams of $\overrightarrow{L_1}$ and $\overrightarrow{L_2}$, with c_1 and c_2 components respectively, and set $D = D_1 \cup D_2$. We first prove, by induction on the number of crossings in D, that

$$[D] = d\,[D_1]\,[D_2]. \tag{5}$$

If D has no crossings, then

$$[D] = d^{c_1+c_2-1} = dd^{c_1-1}d^{c_2-1} = d\,[D_1]\,[D_2].$$

Next assume that D has $k > 0$ crossings and that equation (5) holds whenever D has fewer than k crossings. It may be assumed without loss of generality that D_1 has a crossing C. Note that if we resort again to the notational device of Lemma 5.4.6, then

$$D_A = (D_1)_A \cup D_2 \quad \text{and} \quad D_B = (D_1)_B \cup D_2.$$

It follows from Lemma 5.46 and the induction hypothesis that

$$
\begin{aligned}
[D] &= A[D_A] + B[D_B] = A(d[(D_1)_A][D_2]) + B(d[(D_1)_B][D_2]) \\
&= (Ad[(D_1)_A] + Bd[(D_1)_B])[D_2] = d[D_1][D_2],
\end{aligned}
$$

and the induction is complete. It follows that

$$\langle D \rangle = \left(-A^2 - A^{-2}\right) \langle D_1 \rangle \langle D_2 \rangle.$$

The disjointness of the diagrams D_1 and D_2 implies that

$$w\left(\vec{D}\right) = w\left(\vec{D_1}\right) + w\left(\vec{D_2}\right)$$

and hence

$$
\begin{aligned}
V\left(\vec{L_1} \cup \vec{L_2}\right) &= (-A)^{-3w(\vec{D})} \langle D \rangle \\
&= (-A^2 - A^{-2})(-A)^{-3w(\vec{D_1})} \langle D_1 \rangle \\
&\quad \times (-A)^{-3w(\vec{D_2})} \langle D_2 \rangle \\
&= \left(-A^2 - A^{-2}\right) V\left(\vec{L_1}\right) V\left(\vec{L_2}\right). \qquad \text{Q.E.D.}
\end{aligned}
$$

The *sum* $\vec{L_1} + \vec{L_2}$ of two oriented links is defined by means of Figure 5.52. In order for this definition to work it is important that the two links be visually disjoint. This is formalized by the requirement that there exist a sphere Σ that contains one of the links in its interior and the other in its exterior. With this stipulation, the sum of two oriented *knots* is well defined in the sense that when the exact locations on the diagrams where the two summand knots are connected are allowed to vary, the results remain isotopic (see Exercise 12).

Proposition 5.4.17 $V(\vec{L_1} + \vec{L_2}) = V(\vec{L_1})V(\vec{L_2})$.

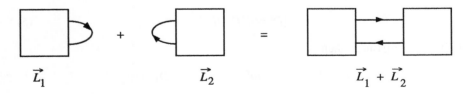

$$\vec{L_1} \qquad\qquad \vec{L_2} \qquad\qquad \vec{L_1} + \vec{L_2}$$

Figure 5.52 The sum of two oriented links.

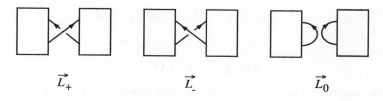

$$\vec{L}_+ \qquad\qquad \vec{L}_- \qquad\qquad \vec{L}_0$$

Figure 5.53

PROOF: In Figure 5.52 flip the link $\vec{L}_1$ across its horizontal axis, so that $\vec{L}_1 \cup \vec{L}_2$ is represented by the link $\vec{L}_0$ in Figure 5.53. Note that the links $\vec{L}_+$ and $\vec{L}_-$ of Figure 5.53 are both equivalent to $\vec{L}_1 + \vec{L}_2$. It follows from the skein relation (Proposition 5.4.15) that

$$A^4 V\left(\vec{L}_1 + \vec{L}_2\right) - A^{-4} V\left(\vec{L}_1 + \vec{L}_2\right)$$
$$+ \left(A^2 - A^{-2}\right)\left(-A^2 - A^{-2}\right) V\left(\vec{L}_1\right) V\left(\vec{L}_2\right) = 0$$

and hence

$$V\left(\vec{L}_1 + \vec{L}_2\right) = V\left(\vec{L}_1\right) V\left(\vec{L}_2\right). \qquad\qquad \text{Q.E.D.}$$

The *obverse* or *mirror image* of L (or $\vec{L}$) is the link L^m (or $\vec{L^m}$) obtained by assuming it to be in the half-space $z > 0$ and multiplying the z-coordinates of all of its points by -1. A link and its obverse have closely related diagrams. To be precise, suppose the link L (or $\vec{L}$) lies entirely between the planes $z = 0$ and $z = d > 0$. If D is the diagram obtained by projecting the link L (or $\vec{L}$) onto the $z = 0$ plane, and if D^m is the diagram of L^m ($\vec{L^m}$) obtained by a projection onto the plane $z = -d$, then D^m is obtained from D by simply switching all of its crossings. For example, the links 2_a^2 and 2_b^2 are obverses of each other, as are the trefoils 3_a and 3_b of Figure 5.5.

Proposition 5.4.18 *If $\vec{L}$ and $\vec{L^m}$ are obverse oriented links, then*

$$V\left(\vec{L^m}\right)(A) = V\left(\vec{L}\right)(A^{-1}).$$

PROOF: Let D be a diagram of L. For any state S of D let S^m be the state of D^m obtained by setting each crossing of D^m to A (or B) if and only if it is set to B (or A) in D. Note that this defines a one-to-one correspondence between the states of D and those of D^m, such that S and S^m are homeomorphic (not to say identical). Then

$$\alpha(S) = \beta\left(S^m\right), \qquad \beta(S) = \alpha\left(S^m\right), \qquad |S| = |S^m|.$$

It follows that

$$[D](A,B,d) = \sum_S A^{\alpha(S)} B^{\beta(S)} d^{|S|-1} = \sum_S A^{\beta(S^m)} B^{\alpha(S^m)} d^{|S^m|-1}$$

$$= \sum_S B^{\alpha(S)} A^{\beta(S)} d^{|S|-1} = [D^m](B,A,d)$$

and hence

$$\langle D \rangle (A) = \langle D^m \rangle (A^{-1}).$$

Since the switching of crossings of oriented knots reverses their values,

$$w\left(\overrightarrow{D^h}\right) = -w\left(\overrightarrow{D}\right).$$

It now follows from the definition of the Jones polynomial that

$$V\left(\overrightarrow{L^h}\right)(A) = (-A)^{-3w(\overrightarrow{D^h})} \langle D^m \rangle (A) = (-A)^{3w(\overrightarrow{D})} \langle D \rangle (A^{-1})$$

$$= \left(-A^{-1}\right)^{-3w(\overrightarrow{D})} \langle D \rangle (A^{-1})$$

$$= V\left(\overrightarrow{L}\right)(A^{-1}). \qquad\qquad \text{Q.E.D.}$$

A knot whose orientations cannot be expressed as the connected sums of nontrivial knots is said to be *prime*. The knots of Figure 5.7 are all prime, whereas those of Figures 5.56 and 5.57 are not. The *crossing number* of a knot K is the minimum number of crossings in any diagram of that knot. For each integer $n = 3, 4, \ldots, 16$ the following table lists the number of inequivalent prime knots that have crossing number n, with the stipulation that only one of each obverse pair of knots is counted:

Crossing number:	3	4	5	6	7	8	9	10	11
Number of prime knots:	1	1	2	3	7	21	49	165	552

Crossing number:	12	13	14	15	16
Number of prime knots:	2,176	9,988	46,972	253,293	1,388,705

Exercises 5.4

1. Compute the bracket polynomials of the links in Figure 5.54. Use this information to show that their Jones polynomials are:

$$a,d: \ -A^{22} + A^{18} - A^{14} - A^6;$$
$$b,c: \ -A^{-2} + A^{-6} - A^{-10} - A^{-18};$$
$$e,h: \ -A^{-6} - A^{-14} + A^{-18} - A^{-22};$$
$$f,g: \ -A^{18} - A^{10} + A^6 - A^2.$$

Explain why these results prove that the unoriented links 4_a^2 and 4_b^2 of Figure 5.8 are not equivalent.

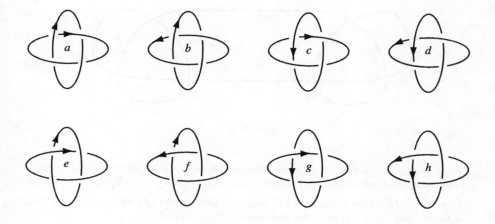

Figure 5.54   Some links.

2. Compute the Jones polynomial of the knot 4 of Figure 5.55 by applying the skein relation to the following label:

(a) *A*,   (b) *B*,   (c) *C*,   (d) *D*.

3. Compute the Jones polynomials of the following knots (you may use Exercise 1):

(a) 5_1,   (b) 5_2.

4. Compute the Jones polynomials of the following knots (you may use Exercise 1):

(a) 6_1,   (b) 6_2,   (c) 6_3.

5. Compute the Jones polynomials of the following knots:

(a) 7_1,   (b) 7_2,   (c) 7_3,   (d) 7_4,

(e) 7_5,   (f) 7_6,   (g) 7_7.

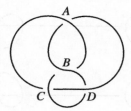

Figure 5.55

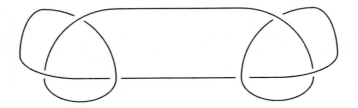

Figure 5.56

The Conway polynomial $\nabla_L(z)$ is a link invariant whose definition falls outside the bounds of this book. However, it is known to satisfy the skein relation $\nabla_{\overrightarrow{L_+}}(z) - \nabla_{\overrightarrow{L_-}}(z) = -z\nabla_{\overrightarrow{L_0}}(z)$, where $\overrightarrow{L_+}, \overrightarrow{L_-}, \overrightarrow{L_0}$ are as in Figure 5.41. Assuming that analogs of Propositions 5.4.13 and 5.4.14 hold and $\nabla_0(z) = 1$, prove the statements in Exercises 6–8.

6. The Conway polynomial of every splittable link is 0.

7. The Conway polynomials of 2_a^2 and 2_b^2 are $\pm z$.

8. The Conway polynomial of each trefoil knot is $z^2 + 1$.

9. Let $\overrightarrow{L}$ be an oriented two-component link neither of whose components has self crossings, and let $\overrightarrow{L}^*$ be an oriented link obtained from $\overrightarrow{L}$ by reversing the orientation of one of its components. Prove that there is an integer n such that $V(\overrightarrow{L}^*) = A^n V(\overrightarrow{L})$.

10. Compute the Jones polynomial of the knot of Figure 5.56.

11. Compute the Jones polynomial of the knot of Figure 5.57.

12. Prove that the sum of two oriented knots $\overrightarrow{K_1}$ and $\overrightarrow{K_2}$ is a well-defined oriented knot.

13. For $n \geq 3$ let K_n be the knot given by the n-crossing diagram of Figure 5.58. Show that K_3 and K_4 are equivalent to the knots 3_b and 4, and find a recursive formula for $V(K_n)$ for all values of $n \geq 3$.

14. Find the Jones polynomial of the n-component oriented link of Figure 5.59.

15. Show by means of an example that the sum of two oriented links need not be well defined.

16. Complete the proof of Lemma 5.4.7.

17. Prove Proposition 5.4.11.

18. Complete the proof of Theorem 5.4.12.

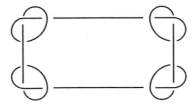

Figure 5.57   A knot.

19. Complete the proof of Lemma 5.4.8.

20. Find the Jones polynomials of all the orientations of link a of Figure 5.60.

21. Find the Jones polynomials of all the orientations of link b of Figure 5.60.

22. A knot K is said to be *achiral* or *amphicheiral* (no, this is not a spelling error) if K and K^m are equivalent. A link that is not achiral is chiral.

    (a) Prove that if the knot K is achiral, then $V(K)(A) = V(K)(A^{-1})$.

    (b) Prove that the trefoil knots are chiral.

    (c) Prove that the knot K_n of Figure 5.58 is chiral for $n \neq 4$.

5.5 The Jones Polynomial and Alternating Diagrams

Besides being a powerful tool for distinguishing between inequivalent links, the Jones polynomial has also been used to resolve some of the oldest questions of knot theory. A case in point is the proof of Tait's conjecture regarding the invariance of the number of crossings in alternating diagrams of knots.

The diagram of a link is said to be *alternating* if the traversal of any component of the link encounters the diagram's crossings alternatingly as overstrands and understrands. All of the diagrams in Figures 5.7 and 5.61 are alternating, whereas the diagrams of Figures 5.31 and 5.57 are not. In this section the Jones polynomial will be used to demonstrate that for a large class of easily recognized knots, the number of crossings in an alternating diagram is an isotopy invariant of its knot. The proof begins with a series of lemmas about alternating diagrams. In these the difference between a diagram and its associated projection (see Fig. 5.1) will be glossed over. This is more convenient than creating new notation to distinguish between the two concepts and should not lead to any confusion.

Lemma 5.5.1 *In an alternating diagram each region has all of its angles of one kind only.*

PROOF: It follows from the definition of alternating diagrams that every arc that appears on the perimeter of a region does so as an overstrand at one end and an understrand at its other end. Hence the perimeter of each region must look like one

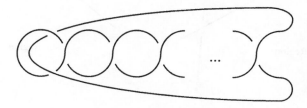

Figure 5.58   A generalization of the knots 3_b and 4.

Figure 5.59   An oriented link with *n* components.

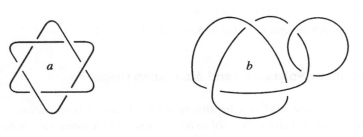

Figure 5.60

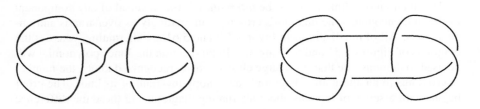

Figure 5.61   Two alternating diagrams of equivalent knots.

of the two displayed in Figure 5.62. It is clear that the region on the left contains only *A*-angles whereas that on the right contains only *B*-angles.   Q.E.D.

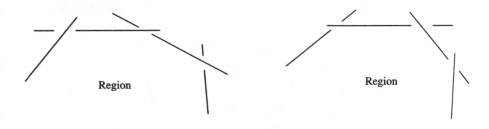

Figure 5.62   Perimeters of regions in alternating diagrams.

The number of crossings of a knot diagram D is denoted by $v(D)$. If a diagram is alternating, then the number of its regions that have only A-angles (A-*regions*) is denoted by $r_A(D)$. Similarly, the number of its B-regions is denoted by $r_B(D)$.

Lemma 5.5.2 *If D is an alternating knot diagram, then*

$$v(D) = r_A(D) + r_B(D) - 2.$$

PROOF: The diagram D is a plane graph whose p nodes are the $v(D)$ crossings of D. Since every node has degree 4, it follows from Proposition 2.1.1 that the number of its arcs is $q = 2p = 2v(D)$. Moreover, since D is connected as a graph, Euler's equation yields

$$2 = p - q + r = v(D) - 2v(D) + r_A(D) + r_B(D),$$

and the desired equation follows immediately. Q.E.D.

The state obtained from an alternating knot diagram D by setting all the crossings to A (or B)-connections is called the A-*state* (or B-*state*) and is denoted by $S_A(D)$ (or $S_B(D)$).

Lemma 5.5.3 *If D is an alternating knot diagram, then $|S_A(D)| = r_B(D)$ and $|S_B(D)| = r_A(D)$.*

PROOF: Since D is connected as a graph, it follows that all the $r_A(D)$ A-regions of D have been unified into a single region of $S_A(D)$ whereas all the $r_B(D)$ B-regions of D have been passed on intact to $S_A(D)$. Thus, $S_A(D)$ consists of $r_B(D)$ loops each of which bounds an erstwhile B-region of D. Hence $|S_A(D)| = r_B(D)$. The other equation follows from a similar argument. Q.E.D.

Lemma 5.5.4 *If the states S_1 and S_2 of a knot diagram D differ by the setting of a single crossing, then $|S_2| = |S_1| \pm 1$.*

PROOF: Let S be an arbitrary state of D, and let the four corners of the setting of some set crossing c be labeled as in Figure 5.63. Suppose both of the X points are contained in the same component loop a of S. It then follows that a must also contain both the Y points, and moreover, because the arc a on the right could not possibly be completed to a loop of S, these four points must be traversed by a in the cyclic order $\cdots X \cdots Y \cdots X \cdots Y \cdots$. Hence each of the loops of S either contains neither of the X and Y pairs or else alternates between them. Since each of the arcs of every set crossing is contained in some such loop, it follows that the transition from S_1 to S_2 and vice versa is depicted by one of the two halves of Figure 5.64.

In the left case $|S_2| = |S_1| + 1$, and on the right $|S_2| = |S_1| - 1$. Q.E.D.

For any state S of a diagram D of a knot K, let $\langle D|S \rangle$ be that state's contribution to $\langle D \rangle$, i.e.,

$$\langle D|S \rangle = A^{\alpha(S) - \beta(S)} \left(-A^2 - A^{-2} \right)^{|S|-1}.$$

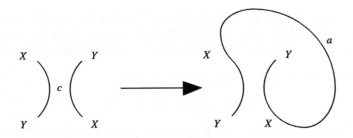

Figure 5.63   An impossible connection at a set crossing.

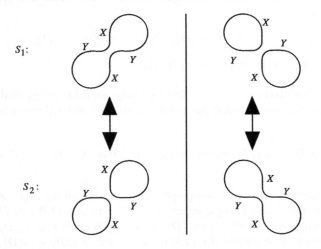

Figure 5.64   Switching a single crossing.

Lemma 5.5.5 *If S_2 is the state of D obtained from S_1 by flipping a single crossing from an A-setting to a B-setting, then*

$$\max \deg\langle D|S_1\rangle \geq \max \deg\langle D|S_2\rangle.$$

The inequality is strict if $|S_1| > |S_2|$.

PROOF: By definition,

$$
\begin{aligned}
\max \deg\langle D|S_1\rangle &= \alpha(S_1) - \beta(S_1) + 2|S_1| - 2 \\
&= (\alpha(S_1) - 1) - (\beta(S_1) + 1) + 2(|S_1| - |S_2|) + 2|S_2| \\
&= \alpha(S_2) - \beta(S_2) + 2|S_2| - 2 + 2(|S_1| - |S_2|) + 2 \\
&\geq \max \deg\langle D | S_2\rangle,
\end{aligned}
$$

the last inequality being justified by the fact that $|S_1| - |S_2| = \pm 1$. The asseverated strict inequality follows immediately.   Q.E.D.

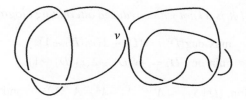

Figure 5.65   A knot diagram with an isthmus at v.

A crossing of a diagram is said to be an *isthmus* if some two of the angles at the crossing belong to the same region of the diagram (see Figure 5.65 and Exercise 4). A diagram is said to be *reduced* if none of its crossings is an isthmus.

Lemma 5.5.6 *Let K be a knot with a reduced diagram D. If the state S is obtained from the A-state S_A by altering the setting of a single crossing, then*

$$\max \deg\langle D \mid S_A \rangle > \max \deg\langle D \mid S \rangle.$$

PROOF: Let the switched crossing be c. By Proposition 5.5.4, $|S_A| - |S| = \pm 1$. If $|S_A| - |S| = -1$ then it follows from the left half of Figure 5.64 with $S_1 = S_A$ and $S_2 = S$ that the two arcs of S_A that replaced the crossing in D are contained in a single loop, say L, of S_A (Fig. 5.66). Since all of the A-regions of D have been coalesced into a single region of S_A, it follows that all the other loops of S_A lie in the interior of L and consequently this crossing of D is in fact an isthmus. This, however, contradicts the fact that K is reduced. It follows that $|S_A| - |S| = 1$ and hence, by the strict version of Lemma 5.5.5 (with $S_1 = S_A$ and $S_2 = S$), this lemma is proved.
Q.E.D.

Lemma 5.5.7 *Let K be a knot with a reduced diagram D. If the state S is obtained from the B-state S_B by altering the setting of a single crossing, then*

$$\min \deg\langle D \mid S_B \rangle < \min \deg\langle D \mid S \rangle.$$

PROOF: Exercise 5.   □

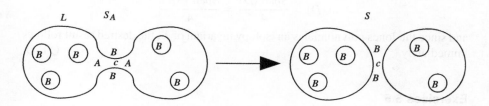

Figure 5.66   The only way to get more loops by changing an A-setting.

Proposition 5.5.8 *If K is a knot with a reduced alternating diagram D, then*

$$\max \deg\langle D\rangle = v(D) + 2\left(r_B(D) - 1\right),$$
$$\min \deg\langle D\rangle = -v(D) - 2\left(r_A(D) - 1\right).$$

PROOF: By definition, $\langle D \mid S_A\rangle = A^{v(D)-0}(-A^2 - A^{-2})^{|S_A|-1}$ and hence, by Lemma 5.5.3, $\max \deg\langle D \mid S_A\rangle = v(D) + 2(r_B(D) - 1)$. If S is any other state and $S_A = S_1, S_2, \ldots, S_k = S$ is a sequence of states such that for each $i = 2, 3, \ldots, k$ the state S_i is obtained from S_{i-1} by altering one crossing of D from an A-setting to a B-setting, then it follows from Lemmas 5.5.5 and 5.5.6 that

$$\max \deg\langle D \mid S_A\rangle > \max \deg\langle D \mid S_2\rangle \geq \max \deg\langle D \mid S_3\rangle \geq \cdots$$
$$\geq \max \deg\langle D \mid S_k\rangle = \max \deg\langle D \mid S\rangle.$$

Since $\langle D\rangle$ is the sum of all the polynomials $\langle D \mid S\rangle$ where S varies over all the states of D, it follows that

$$\max \deg\langle D \mid S\rangle = \max \deg\langle D \mid S_A\rangle = v(D) + 2\left(r_B(D) - 1\right).$$

The proof of the second equation is relegated to Exercise 6.   Q.E.D.

This section's main result was conjectured over a century ago by P. G. Tait. It was proved independently by Louis H. Kauffman, Kunio Murasugi, and Morwen B. Thistlethwaite in 1986. This proof utilizes the notion of the *span* of a polynomial, which is the difference between the smallest and largest exponents of its variable.

Theorem 5.5.9 *The number of crossings in the reduced alternating diagrams of a knot is an isotopy invariant of the knot.*

PROOF: It follows from Proposition 5.5.8 and Lemma 5.5.2 that for any reduced alternating diagram of D of K,

$$\text{Span } \langle D\rangle = [v(D) + 2\left(r_B(D) - 1\right)] - [-v(D) - 2\left(r_A(D) - 1\right)]$$
$$= 2v(D) + 2\left(r_A(D) + r_B(D)\right) - 4 = 4v(D).$$

Hence,

$$v(D) = \frac{\text{Span } \langle D\rangle}{4} = \frac{\text{Span } V(K)}{4}$$

and since the Jones polynomial is an isotopy invariant of K, the desired result follows immediately.   Q.E.D.

Exercises 5.5

1. (a) Prove that if D is a reduced alternating diagram of an achiral knot K, then $3w(K) = r_B(D) - r_A(D)$.

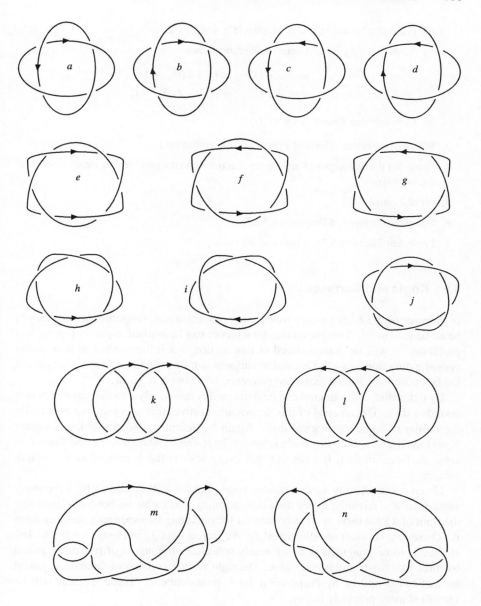

Figure 5.67   An assortment of oriented links.

(b) Prove that if D is a reduced alternating diagram of a knot K such that $|w(K)| \geq v/3$, then K is chiral.

2. The *dual* S^* of a state S of a knot diagram is that state whose settings all differ from those of S. Let D be a diagram of some knot K.

(a*) Prove that for any state S of D, $|S| + |S^*| \leq v(D) + 2$.

(b) Show that if S_A is the A-state of diagram D, then

$$\text{max deg } \langle K \rangle \leq v(D) + 2(|S_A| - 1),$$
$$\text{min deg } \langle K \rangle \geq -v(D) - 2(|S_A^*| - 1).$$

(c) Conclude that Span $V(K) \leq 4v(D)$.

3. Which of the oriented links of Figure 5.67 are equivalent?

4. Prove that the two angles of an isthmus that belong to the same region are necessarily of the same type.

5. Prove Lemma 5.5.7.

6. Complete the proof of Proposition 5.5.8.

7. Prove that Theorem 5.5.9 is also true for links.

5.6   Knots and Surfaces

In Proposition 5.3.6 knot theory was used to prove that nonorientable surfaces cannot be embedded in $\mathbb{R}^3$. This shows that knot theory can be applied to purely topological problems. It will be demonstrated in this section that information can flow in the opposite direction as well. Orientable surfaces can be used to settle questions about knots whose formulation makes no reference whatsoever to surfaces.

By definition, every unknot can be deformed continuously into the plane, where it bounds a disk. The reversal of this deformation proves that every unknot is actually the border of some topological disk. Again by definition, nontrivial knots cannot bound such disks, but, surprisingly, every knot can be shown to be the border of some surface. In fact, it turns out that every knot is the border of an *orientable* surface.

Given a knot K, any surface whose single border is K is said to be a *spanning surface* of K. Trivially, every disk is a spanning surface of its border. Given any diagram of a knot there is an easy method for obtaining two spanning surfaces from it. These two surfaces are illustrated for the trefoil knot 3_a in Figure 5.68. The left surface is to be imagined as being nearly spherical. It consists of two disks joined by three half twists and is orientable. The right surface consists of three disks joined by half twists and is, by Proposition 3.4.9, nonorientable. These surfaces will be identified more precisely below.

What makes these two spanning surfaces possible is the fact that the plane map formed by the projection of the trefoil knot is 2-colorable (see Fig. 5.69). However, all knot projections are graphs in which every node has degree 4, and so they all define Eulerian maps which, by Proposition 2.4.10, are 2-colorable. Hence this construction works for all knot diagrams.

Proposition 5.6.1 *Every knot has a spanning surface.*   □

A polygonal presentation of these surfaces is obtained by introducing a cut at each crossing (Fig. 5.70) of the diagram of the knot K. If v is the number of crossings of the diagram, then this yields a presentation with $2v$ nodes, v cut arcs, and $2v$ border arcs. The number of polygons equals the number of regions (of the plane map formed by the projection) that have the same color. Thus, the orientable surface of Figure 5.68 has characteristic $6 - 9 + 2 = -1$. Since it has exactly one border (K itself), the closure of the surface has characteristic $-1 + 1 = 0$ and genus 1. Thus this spanning surface is $S^{1,1}$. The nonorientable surface of Figure 5.68 has characteristic $6 - 9 + 3 = 0$, so that its closure has characteristic $0 + 1 = 1$ and (nonorientable) genus 1. This spanning surface is therefore $\tilde{S}^{1,1}$.

For reasons that will be become clear below, it would be useful to have a method that produces orientable spanning surfaces. Orientability cannot be guaranteed by the above construction, and the existence of such orientable spanning surfaces was first demonstrated by Lev S. Pontrjagin (1908–1988) and F. Frankl in 1930. Orientable spanning surfaces are called *Seifert surfaces* after Herbert Seifert, whose construction of 1934 is now described.

Given a diagram D of a knot K, let it be oriented. Replace each crossing of D by a pair of arcs that are both consistent with the orientation, but follow the "wrong" strings (see Fig. 5.71). This modifies D into a collection of disjoint oriented plane loops known as *Seifert circles* (see Fig. 5.72). Each such circle, together with its interior, constitutes a *Seifert disk*. Since the Seifert circles may be nested, it is necessary to think of their disks as being stacked, or rather, floating, one on top of the next (Fig. 5.73). Finally, at each modified crossing paste a pair of small antiparallel arcs, one on the border of each disk (the pasted arcs are labeled a in Figure 5.74). The pasting of the a's should be visualized by means of a twisted band connecting them. Of the two possible twists choose the one that gives you back the original value (± 1) of the crossing. If each of the Seifert disks is oriented by the direction its perimeter inherits from the original knot diagram, then, by construction, one of the antiparallel pasted arcs is consistent and the other is inconsistent with these orientations. Consequently, the presentation consisting of the Seifert disks with the identified antiparallel arcs is

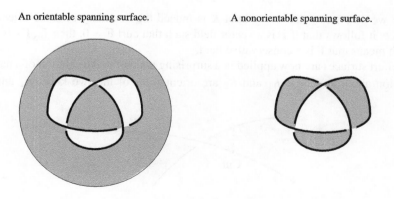

An orientable spanning surface.   A nonorientable spanning surface.

Figure 5.68   Two spanning surfaces for the trefoil knot.

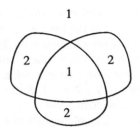

Figure 5.69   A 2-coloring of the regions of the projection of the trefoil knot.

coherently oriented. That the border of the underlying surface consists of the given knot K follows from the oppositeness of the directions of the identified arcs. Note that in the right portion of Figure 5.75, after the border is traversed along the upper arc from A to u, it reemerges on the lower arc at u and goes on to B, just as it does in the knot crossing on the left. Hence this presentation has a single border, which is in fact a copy of the given knot K.

Thus, the following theorem has been proven.

Theorem 5.6.2 *Every knot has a Seifert surface.*   □

The Seifert surface constructed above can be embedded in $\mathbb{R}^3$. To accomplish this, visualize the nested Seifert disks as floating above each, other and join them at each crossing by means of a half-twisted band instead of identifying anti parallel arcs (see Fig. 5.76, where the joining bands are triangulated).

The existence of Seifert surfaces has an important consequence for multivariable calculus and mathematical physics. Recall that if $\mathbf{F} = (F_1, F_2, F_3)$ is a continuously differentiable vector field, and S is an orientable bounded surface, then Stokes's Theorem states that

$$\int_{\partial S} \mathbf{F} \cdot d\mathbf{s} = \int\int_S (\text{curl } \mathbf{F}) \cdot d\mathbf{S}.$$

Since we now know that every knot K is indeed the boundary of some orientable surface, it follows that if $\mathbf{F}$ is a vector field such that curl $\mathbf{F} = 0$, then $\int_{\partial S} \mathbf{F} \cdot d\mathbf{s} = 0$, which means that $\mathbf{F}$ is a conservative field.

Seifert surfaces are now applied in a surprising manner to the solution of a natural question about knots. If $\overrightarrow{K_1}$ and $\overrightarrow{K_2}$ are orientations of the two knots K_1 and K_2

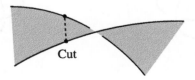

Figure 5.70   Cutting a crossing.

Figure 5.71 Modifying a crossing.

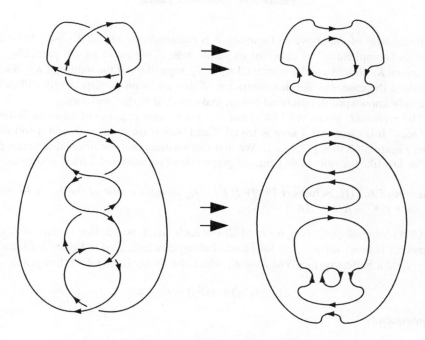

Figure 5.72 A modification of two knot diagrams.

respectively, then $K_1 + K_2$ denotes the unoriented knot that underlies $\overrightarrow{K_1} + \overrightarrow{K_2}$. Since each of K_1 and K_2 has two distinct orientations, the sum $K_1 + K_2$ is ambiguous, as it may denote any of four knots. These four can in fact be grouped into equivalent pairs, so that $K_1 + K_2$ has at most two possibilities (see Exercises 5, 6).

Figure 5.73 A side view of stacked Seifert disks.

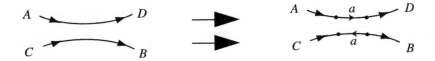

Figure 5.74   Arcs to be pasted.

Regardless of the above ambiguities, it is reasonable to expect $K_1 + K_2$ to be at least as "complicated" as either of its summands, if not more so. In particular, a nontrivial K_1 should yield a nontrivial $K_1 + K_2$, regardless of the nature of K_2. Such is indeed the case, but the demonstration of this assertion is surprisingly difficult. The only known proof, described below, makes use of Seifert surfaces.

The *orientable genus* $\gamma(K)$ of a knot K is the minimum genus of all of its Seifert surfaces. It is clear that a knot is trivial if and only if its genus is 0, and it follows from Figure 5.68 that $\gamma(3_a) = 1$. We first demonstrate that the orientable genus of knots, just like the orientable genus of graphs (see Proposition 4.3.6), is additive.

Theorem 5.6.3 (H. Schubert 1949) *If $K_1 + K_2$ denotes a sum of the knots K_1 and K_2, then $\gamma(K_1 + K_2) = \gamma(K_1) + \gamma(K_2)$.*

PROOF: Suppose $\gamma(K_i) = n_i$ for $i = 1, 2$. For each i let S_i be a Seifert surface of K_i of genus n_i. If these surfaces are then pasted along a as indicated in Figure 5.77, then they yield a Seifert surface S of $K_1 + K_2$ which, by Proposition 3.4.13, has genus

$$\gamma(S) = \gamma(S_1) + \gamma(S_2) = n_1 + n_2.$$

Consequently

$$\gamma(K_1 + K_2) \leq \gamma(S) = n_1 + n_2 = \gamma(K_1) + \gamma(K_2).$$

Conversely, suppose S is a Seifert surface of $K_1 + K_2$ such that $\gamma(S) = \gamma(K_1 + K_2)$. It is tempting to visualize S and $K_1 + K_2$ in the form displayed in the right side of Figure 5.77, in which case a simple cut along a would produce the spanning surfaces S_1 and S_2 on the left side, for which

$$\gamma(K_1 + K_2) = \gamma(S) = \gamma(S_1) + \gamma(S_2) \geq \gamma(K_1) + \gamma(K_2), \tag{6}$$

Figure 5.75   A knot on the border of a presentation.

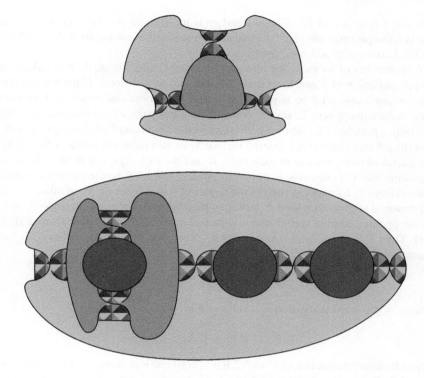

Figure 5.76   Two Seifert surfaces in $\mathbb{R}^3$.

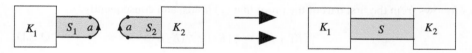

Figure 5.77   Creating a Seifert surface S for $K_1 + K_2$.

and the proof would be concluded. Unfortunately, the convolutions of S may render Figure 5.77 a case of wishful thinking. These convolutions may be such as to alternate between the vicinities of K_1 and K_2 several times. Still, even then it is possible to modify S into a Seifert surface of the form of Figure 5.77.

Let Σ be a sphere which cuts $K_1 + K_2$ in two points only and otherwise separates K_1 and K_2 in the sense that it contains one in its interior and the other in its exterior (Fig. 5.78). There is no telling where the Seifert surface S lies relative to this separating sphere Σ, and they will, in general, intersect many times. These intersections may consist of isolated points, open arcs, loops, or regions in Σ. However, it is plausible to assume that after some small perturbations of the Seifert surface S, its intersections with Σ will consist of the open arc a that joins two distinct points of $K_1 + K_2$ and a finite number of loops. This can be argued as follows: Let $\Sigma \cap S \cap (K_1 + K_2) = \{A, B\}$.

Then every open arc of $\Sigma \cap S$ begins and ends in different points of $\{A,B\}$. Since there is a unique open arc of $\Sigma \cap S$ emanating from each of the points A,B, it follows that $\Sigma \cap S$ contains exactly one open arc.

If the number of loops in the intersection is zero, then Figure 5.78 is indeed an accurate depiction of S and Σ, so that (1) holds and we are done. If the intersection does contain loops, let b be any such an intersection loop, one of whose two sides on the separating sphere Σ, say D, contains no other parts of the intersection (Fig. 5.79; only a portion of Σ is shown here). Excise the Seifert surface S along b, thereby converting S into either one bordered surface S_1 or two bordered surfaces S_1 and S_2, with a total of three borders in each case: K and the two copies of b. In either case we assume that S_1 contains K as one of its borders and let S' be the bordered surface obtained from S_1 by capping each of the copies of b that occur on S'. In other words, S' is trimmed slightly at a loop b' that runs parallel to b at a small distance from it, and the opening at b' is sealed with a disk that is parallel to the disk D on Σ (Fig. 5.79). Then S' is also a Seifert surface of K. Moreover, by either Corollary 3.3.2 or Proposition 3.3.7,

$$\gamma(S') = \gamma(S'^c) = \gamma(S_1^c) = \gamma(S_1) \le \gamma(S).$$

Because of the minimality of $\gamma(S)$ it follows that

$$\gamma(S') = \gamma(S) = \gamma(K_1 + K_2).$$

Note that the intersection of S' with Σ has at least one less component than does the original S. Repeat this procedure until a Seifert surface of K, of genus $\gamma(K_1 + K_2)$, is obtained, whose intersection with P consists of only a single open arc. It then follows from the first part of the proof that (1) holds, and consequently,

$$\gamma(K_1 + K_2) = \gamma(K_1) + \gamma(K_2). \hspace{2cm} \text{Q.E.D.}$$

Corollary 5.6.4 *If K is a non-trivial knot then so are all the knots of the form $K + K'$.*

PROOF: If K is a nontrivial knot, then $\gamma(K) > 0$. It follows from the above theorem that $\gamma(K + K') = \gamma(K) + \gamma(K') > 0$ and hence $K + K'$ is also nontrivial. \hspace{1cm} Q.E.D.

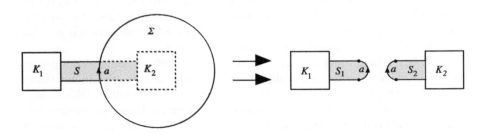

Figure 5.78   A case of wishful thinking.

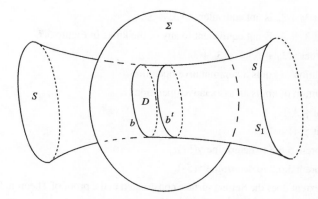

Figure 5.79   Creating a new Seifert surface.

The following theorem, whose proof is not included, yields the genus of most of the knots discussed in this book.

Theorem 5.6.5 *The genus of every knot with an alternating diagram equals that of its Seifert surface.*   □

Exercises 5.6

1. Prove that if a knot diagram has v crossings, then the construction of Theorem 5.6.2 yields a Seifert surface of genus $(v-s+1)/2$, where s is the number of Seifert circles of the diagram.

2. For the knot 4 of Figure 5.7:

   (a) Draw and identify the two spanning surfaces obtained by the method of Figure 5.68.

   (b) Draw and identify the Seifert surface obtained by the method of Figure 5.76.

   (c) What is the genus of this knot?

3. For the knot 5_1 of Figure 5.7:

   (a) Draw and identify the two spanning surfaces obtained by the method of Figure 5.68.

   (b) Draw and identify the Seifert surface obtained by the method of Figure 5.76.

   (c) What is the genus of this knot?

4. For the knot 6_1 of Figure 5.7:

   (a) Draw and identify the two spanning surfaces obtained by the method of Figure 5.68.

   (b) Draw and identify the Seifert surface obtained by the method of Figure 5.76.

   (c) What is the genus of this knot?

5. Suppose for each $i = 1, 2$, $\overleftarrow{K_i}$ and $\overrightarrow{K_i}$ are the two orientations of the knot K_i. Show that $\overrightarrow{K_1} + \overrightarrow{K_2}$ and $\overleftarrow{K_1} + \overleftarrow{K_2}$ have the same underlying knots.

6. Prove that $3_a + 3_b$ is not equivalent to $3_a + 3_a$.

7. Prove that $3_a + 3_b$ is not equivalent to any of the knots in Figure 5.7.

8. Is it true that $\gamma(3_a + 3_b) = \gamma(3_a + 3_a)$?

9. Is the addition of knots a commutative operation?

10. Is the addition of knots an associative operation?

11. Prove that $\gamma(3_a) = 1$.

12. Prove that $\gamma(4) = 1$.

13. Use Theorem 5.6.5 to determine $\gamma(5_1)$.

14. Use Theorem 5.6.5 to determine $\gamma(5_2)$.

15. To what extent does the Seifert surface constructed in the proof of Theorem 5.6.2 depend on the choice of an orientation of the knot?

16. Prove that for every positive integer n there exists a knot of orientable genus n.

17. For the knot of Figure 5.58:

    (a) Draw and identify the two spanning surfaces obtained by the method of Figure 5.68.

    (b) Draw and identify the Seifert surface obtained by the method of Figure 5.76.

18. Let $\tilde{\gamma}(K)$ denote the *minimum genus* of all the nonorientable spanning surfaces of the knot K. Show that for every knot K, $\tilde{\gamma}(K) \leq 2\gamma(K) + 1$.

19.* It is known that $\tilde{\gamma}(7_4) = 3$. Use this fact to show that $\tilde{\gamma}(7_4 + 7_4) \neq \tilde{\gamma}(7_4) + \tilde{\gamma}(7_4)$.

Chapter Review Exercises

1. Show that 6_3 is its own obverse.

2. Show that link a of Figure 5.80 has no p-labelings.

3. Show that link b of Figure 5.80 has a 3-labeling.

4. Find the Jones polynomials of all the orientations of link a of Figure 5.80.

5. Find the Jones polynomials of all the orientations of link b of Figure 5.80.

6. Are the following statements true or false? Justify your answers.

    (a) Every two knots are homeomorphic.

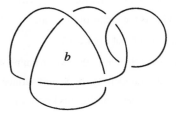

Figure 5.80

(b) Every two knots are isotopic.

(c) Every two links are homeomorphic.

(d) Every two 2-component links are homeomorphic.

(e) Every knot diagram has a p-labeling for some odd prime p.

(f) Every split diagram has a p-labeling for some odd prime p.

(g) Every split diagram has a p-labeling for every odd prime p.

(h) Every knot has a Jones polynomial.

(i) Every oriented knot has a Jones polynomial.

(j) Every knot diagram has a Jones polynomial.

(k) Every link has a Jones polynomial.

(l) Every oriented link has a Jones polynomial.

(m) Isotopic oriented knots have the same Jones polynomials.

(n) Isotopic oriented links have the same Jones polynomials.

(o) Homeomorphic knots have the same Jones polynomials.

(p) Isotopic knots have the same labeling equations.

(q) The unknot is not a knot.

CHAPTER 6

THE DIFFERENTIAL GEOMETRY OF SURFACES

Besides the plane and space, the ancient Greeks also studied the surfaces of the cone and the sphere. The investigation of other surfaces had to await the advent of calculus in the mid seventeenth century. The brothers Jacob (1654–1705) and John (1667–1748) Bernoulli initiated the study of geodesics on surfaces. Euler characterized these geodesics in terms of differential equations in 1728 and established the first results regarding the curvature of surfaces in 1760. In his *General Investigations of Curved Surfaces* of 1827 Gauss provided a unified framework for these two seemingly disparate aspects of surfaces. This monograph determined the direction of the development of differential geometry for the following two centuries, and the present chapter is an exposition of its main ideas.

6.1 Surfaces, Normals, and Tangent Planes

A *smooth parametrized surface* is a subset of $\mathbb{R}^3$ of the form

$$\{\mathbf{X}(u,v) = (x(u,v), y(u,v), z(u,v))\}, \qquad (u,v) \in D,$$

Introduction to Topology and Geometry, Second Edition.
By Saul Stahl and Catherine Stenson Copyright © 2013 John Wiley & Sons, Inc.

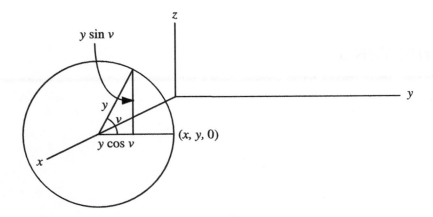

Figure 6.1   Parametrizing a surface of rotation.

where x, y, and z have all derivatives of order 3, D is some open subset of $\mathbb{R}^2$, and

$$\mathbf{X}_u \times \mathbf{X}_v \neq \mathbf{0} \qquad \text{for all } (u,v) \in D.$$

It is quite common to think of the function $\mathbf{X}(u,v)$ itself as the surface, and such will be the case here as well. The domain D, as well as the qualifications *smooth* and *parametrized*, will mostly be left implicit. The requirement that $\mathbf{X}_u \times \mathbf{X}_v \neq 0$ excludes such troublesome points as the tip of a cone.

Examples of such surfaces are

$$\mathbf{X}(u,v) = \begin{cases} (\sin u \cos v, \sin u \sin v, \cos u), & 0 < u < \pi, 0 < v < 2\pi, \\ (u^2, v^2, uv), & (u,v) \neq (0,0), \\ (r\cos u, r\sin u, v), & D = \mathbb{R}^2. \end{cases}$$

The first of these three surfaces is the sphere minus its north and south poles. The third one is an infinitely extended circular cylinder of radius r whose generating lines are parallel to the z-axis.

One important class of surfaces are the *surfaces of revolution*. We will parametrize those whose axis of rotation is the x-axis. Let $\mathbf{C}(u) = (x(u), y(u), 0)$ be a parametrized curve in the xy-plane, and suppose it is rotated about the x-axis. It is clear from Figure 6.1 that when the point $(x, y, 0)$ is rotated about the x-axis, it traces out the circle $(x, y\cos v, y\sin v)$, where v denotes the angle of rotation. Consequently, the curve $\mathbf{C}(u)$ yields the surface

$$\mathbf{X}(u,v) = (x(u), y(u)\cos v, y(u)\sin v).$$

The *meridians* of such a surface of revolution are its intersections with the half-planes that emanate from the axis of revolution. The *latitudes* are its intersections with the planes perpendicular to the axis of rotation. Both of these types of curves are displayed in Figures 6.2–6.4.

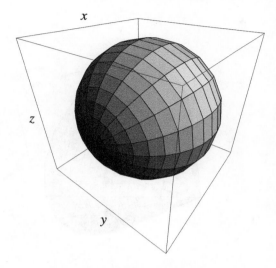

Figure 6.2   The sphere.

Example 6.1.1 The sphere of radius R (Fig. 6.2) is generated by revolving the semi-circle $\mathbf{C}(u) = (R\cos u, R\sin u, 0), 0 < u < \pi$, about the x-axis. Consequently it has the parametrization

$$\mathbf{X}(u,v) = (R\cos u, R\sin u\cos v, R\sin u\sin v), \qquad 0 < u < \pi, \, 0 < v < 2\pi.$$

Of course, this parametrization misses two points, but this will not cause us any difficulties in the subsequent discussion.

Example 6.1.2 When the ray $\mathbf{C}(u) = (1 - u, u, 0), 0 < u < \infty$, is rotated about the x-axis, it forms a right-angled cone (Fig. 6.3) with parametrization

$$\mathbf{X}(u,v) = (1 - u, u\cos v, u\sin v), \qquad 0 < u < \infty, \, v \in \mathbb{R}.$$

Example 6.1.3 To parametrize the torus we rotate a circle of radius r, centered at the point $(0, R, 0)$, about the x-axis (Fig. 6.4). As this circle can be given the parametrization $\mathbf{C}(u) = (r\cos u, R + r\sin u, 0)$, it follows that the torus is parametrized as

$$\mathbf{X}(u,v) = (r\cos u, (R + r\sin u)\cos v, (R + r\sin u)\sin v).$$

Let c be any constant. In the surface $\mathbf{X}(u,v)$, the parametrized curves $\mathbf{X}(u,c)$ and $\mathbf{X}(c,v)$ are, respectively, the *u-parameter* and *v-parameter* curves. They are also collectively known as the surface's *parametric curves*.

Example 6.1.4 For the sphere $\mathbf{X}(u,v) = (R\cos u, R\sin u\cos v, R\sin u\sin v)$ (see Example 6.1.1) the u-parameter curves are the meridians (see Fig. 6.2), and the v-parameter curves are the latitudes (relative to a north pole situated at $(R, 0, 0)$).

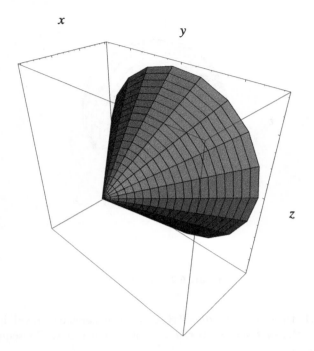

Figure 6.3   The right cone.

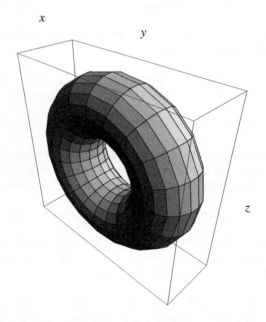

Figure 6.4   The torus.

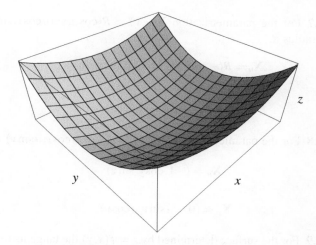

Figure 6.5   Parameter curves on $z = f(x, y)$.

The foregoing discussion of topological spaces, homeomorphisms, and isotopies is an informal working introduction that will serve for the purposes of this text. Experience indicates that this lack of precision will not hamper comprehension of the subsequent material. Rigorous definitions are provided in Chapter 10, which can be read out of sequence.

Example 6.1.5 The u-parameter curves of the cone $\mathbf{X}(u, v) = (1 - u, u\cos v, u\sin v)$ (see Example 6.1.2) are the generating straight lines. The v-parameter curves are the circular latitudes.

Example 6.1.6 The v-parameter curves of the torus of Example 6.1.3, parametrized by $\mathbf{X}(u, v) = (r\cos u, (R + r\sin u)\cos v, (R + r\sin u)\sin v)$, are the circles that are parallel to the yz-plane (see Fig. 6.4) and surround the hole in the torus. The u-parameter curves are circles that pass through that hole.

As these examples indicate, the u-parameter curves of the surface of revolution

$$\mathbf{X}(u, v) = (x(u), y(u)\cos u, y(u)\sin u)$$

are its meridians, whereas the v-parameter curves are its latitudes.

The graph of the function $z = f(x, y)$ can be parametrized by $\mathbf{X}(x, y) = (x, y, f(x, y))$. The x-parameter curves lie in the surface over the straight lines $y = c$ in the xy-plane (Fig. 6.5). The y-parameter curves bear an analogous relationship to the straight lines $x = c$.

Given a parametrization $\mathbf{X}(u, v)$, the *tangent vectors* to its parameter curves are

$$\frac{\partial \mathbf{X}}{\partial u} = \mathbf{X}_u \quad \text{and} \quad \frac{\partial \mathbf{X}}{\partial v} = \mathbf{X}_v.$$

Example 6.1.7 For the parametrization $\mathbf{X}(u,v) = R(\cos u, \sin u \cos v, \sin u \sin v)$ of the sphere of radius R,

$$\mathbf{X}_u = R(-\sin u, \cos u \cos v, \cos u \sin v)$$

and

$$\mathbf{X}_v = R(0, -\sin u \sin v, \sin u \cos v).$$

Example 6.1.8 For the parametrization $\mathbf{X}(u,v) = (1-u, u\cos v, u\sin v)$ of the right cone,

$$\mathbf{X}_u = (-1, \cos v, \sin v)$$

and

$$\mathbf{X}_v = (0, -u\sin v, u\cos v).$$

Example 6.1.9 For the surface determined by $z = f(x,y)$ the tangents to the parameter curves are

$$\mathbf{X}_x = (1,0,f_x) \quad \text{and} \quad \mathbf{X}_y = (0,1,f_y).$$

The tangent plane at a point $P = \mathbf{X}(u_0,v_0)$ on the surface contains the tangent vectors, at P, to all the curves on the surface that also pass through P. Since these tangent vectors are visualized at the origin, the tangent plane too should be visualized at the origin rather than at P. As both of the tangent vectors $\mathbf{X}_u = \mathbf{X}_u(u_0,v_0)$ and $\mathbf{X}_v = \mathbf{X}_v(u_0,v_0)$ are contained in the tangent plane, it follows that the vector $\mathbf{X}_u \times \mathbf{X}_v$ is orthogonal to the tangent plane. Consequently, the *unit normal* to the surface at P is defined as

$$\mathbf{n} = \mathbf{n}_P = \frac{\mathbf{X}_u \times \mathbf{X}_v}{|\mathbf{X}_u \times \mathbf{X}_v|}. \tag{1}$$

Equation (1) does not yield a well-defined normal. The normals associated with the parametrization $\mathbf{X}(u,v)$ and $\mathbf{X}(s,t)$ where $s = v$ and and $t = u$ have opposite directions. The normal's line of direction, however, is well defined (Exercise 3).

Example 6.1.10 For the parametrization $\mathbf{X}(u,v) = R(\cos u, \sin u \cos v, \sin u \sin v)$ of the sphere of radius R (see Example 6.1.7),

$$\mathbf{X}_u \times \mathbf{X}_v = R^2(\sin u \cos u \cos^2 v + \sin u \cos u \sin^2 v, \sin^2 u \cos v, \sin^2 u \sin v)$$
$$= R^2 \sin u(\cos u, \sin u \cos v, \sin u \sin v).$$

Since the vector $(\cos u, \sin u \cos v, \sin u \sin v)$ has unit length and $R^2 \sin u$ is positive, it follows that

$$\mathbf{n} = (\cos u, \sin u \cos v, \sin u \sin v) = \frac{1}{R}\mathbf{X}(u,v).$$

Note that in this case the normal at P has the same line of direction as the position vector $\overrightarrow{OP}$, as should indeed be the case, since the tangent plane to the sphere is perpendicular to the radius through the point of contact.

Example 6.1.11 For the parametrization $\mathbf{X}(u,v) = (1 - u, u\cos v, u\sin v)$ of the right cone, we have

$$\mathbf{X}_u \times \mathbf{X}_v = u(1, \cos v, \sin v),$$

and hence, since $u > 0$,

$$\mathbf{n} = \frac{\mathbf{X}_u \times \mathbf{X}_v}{|\mathbf{X}_u \times \mathbf{X}_v|} = \frac{1}{\sqrt{2}}(1, \cos v, \sin v).$$

In particular, note that in this case $\mathbf{n}$ is independent of u, implying that $\mathbf{n}$ is constant along the generating rays of the cone. This can easily be corroborated in Figure 6.3.

Example 6.1.12 If the surface $z = f(x,y)$ is parametrized as

$$\mathbf{X}(x,y) = (x, y, f(x,y)),$$

then

$$\mathbf{X}_x \times \mathbf{X}_y = (1,0,f_x) \times (0,1,f_y) = (-f_x, -f_y, 1)$$

and hence

$$\mathbf{n} = \frac{(-f_x, -f_y, 1)}{\sqrt{f_x^2 + f_y^2 + 1}}.$$

Exercises 6.1

1. Parametrize the surface generated by revolving a parametric curve in the yz-plane about the y-axis.

2. Parametrize the surface generated by revolving a parametric curve in the xz-plane about the z-axis.

3. Prove formally that if $u = f(\bar{u}, \bar{v}), v = g(\bar{u}, \bar{v})$ is a reparametrization of $\mathbf{X}(u,v)$, then the unit normals, as defined by (1) for both parametrizations, have the same line of direction.

4.* The straight line joining the points (a,b,c) and $(-a,-b,c)$, $a,b,c > 0$, is revolved about the x-axis. Prove that it describes a hyperboloid of revolution of one sheet.

Surfaces can also be defined implicitly as the graphs of equations of the form $W(x,y,z) = 0$, provided one of the independent variables can be expressed as a function of the other two. The precise conditions that guarantee such an expression are provided by the Implicit Function Theorem, which can be found in Lang (1987) and elsewhere. In the sequel such surfaces will occur only in the exercises, and the necessary existence and differentiability conditions will be taken for granted.

5. Explain why the normal to the surface defined implicitly by the equation $W(x,y,z) = 0$ has the same line of direction as (W_x, W_y, W_z).

6. The graph of the equation $f(x,y) = 0$ in the xy-plane is rotated about the x-axis. Show that the resulting surface of revolution has the equation $f(x, \sqrt{y^2 + z^2}) = 0$.

6.2 The Gaussian Curvature

We begin the investigation of surfaces with Gauss's definition of their curvature.

For any given surface S, the *Gauss map* g assigns to each point P of S that point $P^* = g(P)$ such that

$$\mathbf{n}_P = \overrightarrow{OP^*}.$$

Note that $g(S)$ is by definition a portion of the unit sphere. In general, whenever S is a surface, $g(S)$ will also be a surface, but there are some notable exceptions to this rule. If S is a plane, then $g(S)$ consists of a single point. If S is either a cylinder or a cone, then $g(S)$ consists of a circle. When $g(S)$ is a surface, it can be parametrized indirectly by any parametrization $\mathbf{X}(u,v)$ by means of Equation (1).

Lemma 6.2.1 *Let S be a surface with a point P on it. If $g(S)$ is also a surface, then the tangent plane to S at P and the tangent plane to $g(S)$ at $g(P)$ are identical.*

PROOF: Let $\mathbf{X}(u,v)$ be a parametrization of S with unit normal $\mathbf{n} = \mathbf{n}(u,v)$. Since $\mathbf{n} \cdot \mathbf{n} = 1$, it follows that

$$\mathbf{n} \cdot \mathbf{n}_u + \mathbf{n}_u \cdot \mathbf{n} = 0$$

and hence

$$\mathbf{n} \cdot \mathbf{n}_u = 0.$$

Since $\mathbf{n}$ is perpendicular to the tangent plane to S, which in turn contains the vectors $\mathbf{X}_u$ and $\mathbf{X}_v$, it follows that there exist scalars p_1 and p_2 such that

$$\mathbf{n}_u = p_1 \mathbf{X}_u + p_2 \mathbf{X}_v. \tag{2}$$

Similarly, there exist scalars q_1 and q_2 such that

$$\mathbf{n}_v = q_1 \mathbf{X}_u + q_2 \mathbf{X}_v.$$

It follows that

$$\mathbf{n}_u \times \mathbf{n}_v = (p_1 q_2 - p_2 q_1)\mathbf{X}_u \times \mathbf{X}_v, \tag{3}$$

which implies the equality of the tangent plane to the surface described by $\mathbf{n}$ with the tangent plane to the surface described by $\mathbf{X}$.   Q.E.D.

It follows from Equation (3) that the unit normals to S and $g(S)$ at the points P and $g(P)$, respectively, have the same line of direction but may have opposite directions. The Gauss map g is said to be *sense-preserving* if these two normals have the same direction, and *sense-reversing* otherwise. In the first case, a sequence of points A, B, C that describes a counterclockwise arc about P is mapped by g to the counterclockwise sequence $g(A), g(B), g(C)$ (Fig. 6.6a). In the second case such a counterclockwise sequence is mapped into a clockwise sequence (Fig. 6.6(b)).

In calculus, the curvature of a curve is defined in terms of the turning of its unit tangent vector. The curvature of a surface is similarly defined by means of the turning

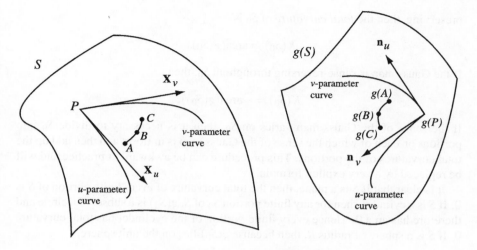

(a) A sense-preserving Gauss map.

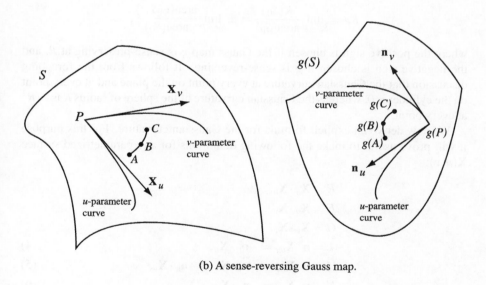

(b) A sense-reversing Gauss map.

Figure 6.6   Sense preservation and reversal.

of the unit normal. If S_0 is a finite portion of S on which the Gauss map is sense-preserving, then the *total curvature* of S_0 is

$$K(S_0) = \text{area}(g(S_0)).$$

If the Gauss map is sense-reversing throughout S_0, then

$$K(S_0) = -\text{area}(g(S_0)).$$

If the sense of the Gauss map varies on S_0, then it is necessary to divide S_0 into portions on each of which the sense of the Gauss map is uniform and then add up the total curvatures of the portions. This procedure can be awkward in practice and will be replaced by a very explicit formula.

It is clear that if S is a plane, then the total curvature of every finite portion of S is 0. If S is a cylinder, then for any finite portion S_0 of S, $g(S_0)$ is a subset of a circle and therefore has area 0. Hence every finite portion of the cyclinder has total curvature 0. If S is a sphere of radius R, then because $g(S_0)$ lies on the unit sphere,

$$K(S_0) = \text{area}(g(S_0)) = \frac{\text{area}(S_0)}{R^2}.$$

The *Gaussian curvature* of the surface S at the point P is defined as

$$K_P = \lim_{S_0 \to P} \frac{K(S_0)}{\text{area}(S_0)} = \pm \lim_{S_0 \to P} \frac{\text{area}(g(S_0))}{\text{area}(S_0)},$$

where the positive sign is chosen if the Gauss map g is sense-preserving at P, and the negative sign is chosen if g is sense-reversing. It follows from the foregoing discussion that the Gaussian curvature at every point on the plane and at every point on the cylinder is 0, whereas the Gaussian curvature of the sphere of radius R is $1/R^2$ at every point.

We now derive an explicit formula for the Gaussian curvature. For this purpose it will prove useful to make the following definitions for any parametrized surface $\mathbf{X}(u,v)$:

$$E = \mathbf{X}_u \cdot \mathbf{X}_u,$$
$$F = \mathbf{X}_u \cdot \mathbf{X}_v,$$
$$G = \mathbf{X}_v \cdot \mathbf{X}_v,$$
$$L = \mathbf{n} \cdot \mathbf{X}_{uu} = -\mathbf{n}_u \cdot \mathbf{X}_u, \tag{4}$$
$$M = \mathbf{n} \cdot \mathbf{X}_{uv} = -\mathbf{n}_v \cdot \mathbf{X}_u = -\mathbf{n}_u \cdot \mathbf{X}_v, \tag{5}$$
$$N = \mathbf{n} \cdot \mathbf{X}_{vv} = -\mathbf{n}_v \cdot \mathbf{X}_v. \tag{6}$$

The equality of the different formulations of $L, M,$ and N follows from the differentiation of the identities $\mathbf{n} \cdot \mathbf{X}_u = \mathbf{n} \cdot \mathbf{X}_v = 0$ (Exercise 5).

Example 6.2.2 For the surface of revolution parametrized as

$$\mathbf{X}(u,v) = (x(u), y(u)\cos v, y(u)\sin v)$$

we have

$$\mathbf{X}_u = (x', y'\cos v, y'\sin v),$$
$$\mathbf{X}_v = (0, -y\sin v, y\cos v),$$
$$\mathbf{X}_{uu} = (x'', y''\cos v, y''\sin v),$$
$$\mathbf{X}_{uv} = (0, -y'\sin v, y'\cos v),$$
$$\mathbf{X}_{vv} = (0, -y\cos v, -y\sin v),$$

where the prime denotes differentiation with respect to u. Hence,

$$E = x'^2 + y'^2, \qquad F = 0, \qquad G = y^2,$$
$$\mathbf{X}_u \times \mathbf{X}_v = y(y', -x'\cos v, -x'\sin v),$$
$$\mathbf{n} = \frac{(y', -x'\cos v, -x'\sin v)}{\sqrt{x'^2 + y'^2}},$$
$$L = \frac{x''y' - x'y''}{\sqrt{x'^2 + y'^2}}, \qquad M = 0, \qquad N = \frac{x'y}{\sqrt{x'^2 + y'^2}}.$$

Lemma 6.2.3 *The area of a portion of a parametrized surface* $\mathbf{X}(u,v)$ *defined over the parameter domain D is*

$$\iint_D |\mathbf{X}_u \times \mathbf{X}_v|\, du\, dv = \iint_D \sqrt{EG - F^2}\, du\, dv.$$

PROOF: The points $\mathbf{X}(u,v), \mathbf{X}(u+\Delta u, v), \mathbf{X}(u, v+\Delta v), \mathbf{X}(u+\Delta u, v+\Delta v)$ form a near-parallelogram whose area is approximately

$$|(\mathbf{X}(u+\Delta u, v) - \mathbf{X}(u,v)) \times (\mathbf{X}(u, v+\Delta v) - \mathbf{X}(u,v))|$$
$$\approx |\mathbf{X}_u \Delta u \times \mathbf{X}_v \Delta v| = |\mathbf{X}_u \times \mathbf{X}_v|\Delta u \Delta v,$$

and by Lagrange's identity (Exercises 9 and 10), this equals

$$\sqrt{(\mathbf{X}_u, \mathbf{X}_u)(\mathbf{X}_v, \mathbf{X}_v) - (\mathbf{X}_u, \mathbf{X}_v)(\mathbf{X}_v, \mathbf{X}_u)}\, \Delta u \Delta v = \sqrt{EG - F^2}\,\Delta u \Delta v.$$

The lemma now follows from standard integration procedures. Q.E.D.

Proposition 6.2.4 *For any parametrized surface* $\mathbf{X}(u,v)$,

$$\mathbf{n}_u \times \mathbf{n}_v = \frac{LN - M^2}{EG - F^2}\mathbf{X}_u \times \mathbf{X}_v. \qquad (7)$$

PROOF: As noted in Equation (2),

$$\mathbf{n}_u = p_1 \mathbf{X}_u + p_2 \mathbf{X}_v$$

When this equation is dotted with $\mathbf{X}_u$ and $\mathbf{X}_v$, we obtain the two equations

$$-L = -\mathbf{n} \cdot \mathbf{X}_{uu} = \mathbf{n}_u \cdot \mathbf{X}_u = p_1 E + p_2 F,$$

$$-M = -\mathbf{n} \cdot \mathbf{X}_{vu} = \mathbf{n}_u \cdot \mathbf{X}_v = p_1 F + p_2 G,$$

and Cramer's rule yields

$$p_1 = \frac{MF - LG}{EG - F^2} \quad \text{and} \quad p_2 = \frac{LF - ME}{EG - F^2}.$$

In a similar manner, the equation

$$\mathbf{n}_v = q_1 \mathbf{X}_u + q_2 \mathbf{X}_v$$

yields

$$q_1 = \frac{NF - MG}{EG - F^2} \quad \text{and} \quad q_2 = \frac{MF - NE}{EG - F^2}.$$

Hence

$$\begin{aligned}
\mathbf{n}_u \times \mathbf{n}_v &= (p_1 q_2 - p_2 q_1)\mathbf{X}_u \times \mathbf{X}_v \\
&= \frac{(MF - LG)(MF - NE) - (LF - ME)(NF - MG)}{(EG - F^2)^2}\mathbf{X}_u \times \mathbf{X}_v \\
&= \frac{M^2 F^2 + LNEG - LNF^2 - M^2 EG}{(EG - F^2)^2}\mathbf{X}_u \times \mathbf{X}_v \\
&= \frac{(LN - M^2)(EG - F^2)}{(EG - F^2)^2}\mathbf{X}_u \times \mathbf{X}_v \\
&= \frac{LN - M^2}{EG - F^2}\mathbf{X}_u \times \mathbf{X}_v.
\end{aligned} \qquad \Box \ (8)$$

Theorem 6.2.5 *For any parametrized surface* $\mathbf{X}(u,v)$,

$$K_P = \frac{LN - M^2}{EG - F^2}.$$

PROOF: It follows from the definition of the Gaussian curvature and the mean value theorem for double integrals (see Lang 1987) that

$$\begin{aligned}
K_P &= \pm \lim_{S_0 \to P} \frac{\text{area}(g(S_0))}{\text{area}(S_0)} = \pm \lim_{S_0 \to P} \frac{\int \int_D |\mathbf{n}_u \times \mathbf{n}_v|\, du\, dv}{\int \int_D |\mathbf{X}_u \times \mathbf{X}_v|\, du\, dv} \\
&= \pm \frac{|\mathbf{n}_u \times \mathbf{n}_v|}{|\mathbf{X}_u \times \mathbf{X}_v|} = \pm \left| \frac{LN - M^2}{EG - F^2} \right|.
\end{aligned} \qquad (9)$$

By definition, the Gaussian curvature is positive (negative) exactly when $\mathbf{n}_u \times \mathbf{n}_v$ and $\mathbf{X}_u \times \mathbf{X}_v$ have identical (opposite) directions, and by Equation (8) this happens exactly when the fraction

$$\frac{LN - M^2}{EG - F^2} \qquad (10)$$

is positive (negative). It therefore follows from Equations (7) and (9) that

$$K_P = \frac{LN - M^2}{EG - F^2}.$$

Q.E.D.

Example 6.2.6 If Example 6.2.2 is applied to the right-angled cone parametrized as $\mathbf{X}(u,v) = (1 - u, u\cos v, u\sin v)$, we have $x' = -1, y' = 1, x'' = y'' = 0$ and hence

$$E = 2, \qquad F = 0, \qquad G = u^2,$$
$$L = 0, \qquad M = 0, \qquad N = \frac{-u}{\sqrt{2}}.$$

Thus $K_P = 0$ at every point P of this cone.

Example 6.2.7 For the parametrized surface $\mathbf{X}(u,v) = (u + v, u - 2v, uv)$ we have

$$\mathbf{X}_u = (1, 1, v),$$
$$\mathbf{X}_v = (1, -2, u),$$
$$\mathbf{X}_u \times \mathbf{X}_v = (u + 2v, v - u, -3),$$
$$|\mathbf{X}_u \times \mathbf{X}_v| = \sqrt{(u + 2v)^2 + (v - u)^2 + 9} = \sqrt{9 + 2u^2 + 2uv + 5v^2},$$
$$\mathbf{n} = \frac{(u + 2v, v - u, -3)}{\sqrt{9 + 2u^2 + 2uv + 5v^2}},$$
$$\mathbf{X}_{uu} = (0, 0, 0),$$
$$\mathbf{X}_{uv} = (0, 0, 1),$$
$$\mathbf{X}_{vv} = (0, 0, 0),$$

$$E = 2 + v^2, \qquad F = -1 + uv, \qquad G = 5 + u^2,$$
$$L = 0, \qquad M = \frac{-3}{\sqrt{9 + 2u^2 + 2uv + 5v^2}}, \qquad N = 0,$$

and hence

$$K = \frac{-9}{(9 + 2u^2 + 2uv + 5v^2)[(2 + v^2)(5 + u^2) - (-1 + uv)^2]}$$
$$= \frac{-9}{(9 + 2u^2 + 5v^2 + 2uv)^2}.$$

Example 6.2.8 For the graph of $z = f(x,y)$, parametrized as $\mathbf{X}(x,y) = (x, y, f(x,y))$, we have (see Exercise 1).

$$K = \frac{f_{xx}f_{yy} - f_{xy}^2}{(1 + f_x^2 + f_y^2)^2}.$$

(11)

Example 6.2.9 The Gaussian curvature of the surface of revolution parametrized by $(x(u), y(u) \cos v, y(u) \sin v)$ is (see Exercise 2)

$$\frac{(x''y' - x'y'')x'}{(x'^2 + y'^2)^2 y}, \tag{12}$$

where a prime denotes differentiation with respect to u.

Example 6.2.10 Suppose c is a positive real number and $y = y(x)$ is a solution of the differential equation $y' = -y/\sqrt{c^2 - y^2}$. Then the Gaussian curvature of the surface of revolution $(x, y(x) \cos v, y(x) \sin v)$ is $-1/c^2$ (see Exercise 4). In other words, these surfaces have constant negative Gaussian curvatures. Because spheres have constant positive Gaussian curvatures, these "negative" analogs are known as *pseudospheres*.

In conclusion we note that the computationally awkward definition of the total curvature given at the beginning of this section can be replaced by the formula

$$K(S_0) = \int \int_S K dS = \int \int_D K \sqrt{EG - F^2} du dv = \int \int_D \frac{LN - M^2}{\sqrt{EG - F^2}} du dv.$$

Exercises 6.2

1. Prove that the Gaussian curvature of the surface $z = f(x, y)$ is

$$(f_{xx} f_{yy} - f_{xy}^2)/(1 + f_x^2 + f_y^2)^2.$$

2. Prove that the Gaussian curvature of the surface of revolution $(x(u), y(u) \cos v, y(u) \sin v)$ is

$$\frac{(x''y' - x'y'')x'}{(x'^2 + y'^2)^2 y},$$

   where a prime denotes differentiation with respect to u.

3. Compute E, F, G, L, M, N for the parametric form of the sphere of radius R given in Example 6.1.1, and use them to compute the total curvature of this sphere.

4. Suppose c is a positive real number and $y = y(x)$ is a solution of the (separable) differential equation $dy/dx = -y/\sqrt{c^2 - y^2}$. Prove that the Gaussian curvature of the surface of revolution $(x, y(x) \cos v, y(x) \sin v)$ is $-1/c^2$.

5. Prove Equations (4)–(6).

6. Prove that the surface of revolution $\mathbf{X}(u, v) = (x(u), y(u) \cos v, y(u) \sin v)$, $a \le u \le b$, has area

$$2\pi \int_a^b y \sqrt{x_u^2 + y_u^2} \, du.$$

   Prove that the surface area of the torus of Example 6.1.3 is $4\pi^2 Rr$.

7. Prove that the total curvature of the torus of Example 6.1.3 is 0.

8**. Prove that the Gaussian curvature of the surface defined by the equation $W(x, y, z) = 0$ satisfies the equation

$$(W_x^2 + W_y^2 + W_z^2)^2 K$$

$$= W_x^2 (W_{yy}W_{zz} - W_{yz}^2) + W_y^2 (W_{xx}W_{zz} - W_{xz}^2) + W_z^2 (W_{xx}W_{yy} - W_{xy}^2)$$
$$+ 2W_y W_z (W_{xz}W_{xy} - W_{xx}W_{yz}) + 2W_x W_z (W_{yz}W_{xy} - W_{yy}W_{xz})$$
$$+ 2W_x W_y (W_{yz}W_{xz} - W_{zz}W_{xy}).$$

9*. Prove that if $\mathbf{a}$ and $\mathbf{b}$ are two vectors in $\mathbb{R}^3$, then $|\mathbf{a} \times \mathbf{b}| = \sqrt{(\mathbf{a} \cdot \mathbf{a})(\mathbf{b} \cdot \mathbf{b}) - (\mathbf{a} \cdot \mathbf{b})^2}$.

10*. Prove Lagrange's identity: If $\mathbf{a}, \mathbf{b}, \mathbf{c}, \mathbf{d}$ are three-dimensional vectors, then

$$(\mathbf{a} \times \mathbf{b}, \mathbf{c} \times \mathbf{d}) = (\mathbf{a}, \mathbf{c})(\mathbf{b}, \mathbf{d}) - (\mathbf{a}, \mathbf{d})(\mathbf{b}, \mathbf{c}).$$

6.3 The First Fundamental Form

Surfaces contain curves, and these curves have lengths and form angles. The functions E, F, and G of Section 6.2 provide formulas for evaluating these lengths and angles.

Given a parametrized surface $\mathbf{X}(u, v)$, a parametrization

$$u = u(t) \quad \text{and} \quad v = v(t) \tag{13}$$

yields a curve $\mathbf{X}(u(t), v(t))$ on the surface.

Example 6.3.1 The parametrization $x = \cos t, y = \sin 3t$ yields the curve

$$\mathbf{X}(t) = (\cos t, \sin 3t, 2\cos^2 t - \sin^2 3t)$$

on the surface $z = 2x^2 - y^2$.

Example 6.3.2 The parametrization $u = 2 + 3t, v = 3 - 2t$ yields the curve

$$\mathbf{X}(t) = (-1 - 3t, (2 + 3t)\cos(3 - 2t), (2 + 3t)\sin(3 - 2t))$$

on the right cone parametrized by $\mathbf{X}(u, v) = (1 - u, u\cos v, u\sin v)$.

The length of such a curve on a surface can of course be easily evaluated by means of the formula

$$\int_a^b |\mathbf{X}'(t)| \, dt$$

that is taught within the framework of calculus. Nevertheless, for theoretical purposes it is useful to separate the contribution to this length of the parametrization $\mathbf{X}(u, v)$ on the one hand, and the parametrization $u = u(t), v = v(t)$ on the other hand. By the above equation, this arclength is in general given by the following integral (from this point on, a prime will always denote differentiation with respect to t, unless otherwise stated):

$$\int_a^b |\mathbf{X}'(u(t), v(t))| \, dt = \int_a^b |\mathbf{X}_u u' + \mathbf{X}_v v'| \, dt$$

$$= \int_a^b \sqrt{(\mathbf{X}_u \cdot \mathbf{X}_u)u'^2 + 2(\mathbf{X}_u \cdot \mathbf{X}_v)u'v' + (\mathbf{X}_v \cdot \mathbf{X}_v)v'^2} \, dt$$

$$= \int_a^b \sqrt{Eu'^2 + 2Fu'v' + Gv'^2} \, dt. \tag{14}$$

The integrand of (14) motivates the following definitions. A *quadratic form* is a function $Q(x, y) = Ax^2 + Bxy + Cy^2$ where A, B, C are independent of x and y. Given a parametrized surface $\mathbf{X}(u, v)$, its *first fundamental form* is the expression

$$ds^2 = E \, du^2 + 2F \, du \, dv + G \, dv^2. \tag{15}$$

The differentials contained in this form are to be understood as abbreviations of derivatives with respect to an implicit parameter t. In other words, Equation (15) is simply shorthand for the equation

$$\left(\frac{ds}{dt} \right)^2 = E \left(\frac{du}{dt} \right)^2 + 2F \frac{du}{dt} \frac{dv}{dt} + F \left(\frac{dv}{dt} \right)^2. \tag{16}$$

Example 6.3.3 The surface $\mathbf{X}(u, v) = (u + v, u - 2v, uv)$ of Example 6.2.7 was seen to have $E = 2 + v^2$, $F = -1 + uv$, $G = 5 + u^2$. Hence, its first fundamental form is

$$ds^2 = (2 + v^2) \, du^2 + 2(-1 + uv) \, du \, dv + (5 + u^2) \, dv^2.$$

It is clear from Equations (14)–(16) that in order to compute the length of a curve on a surface, all that is required is a knowledge of the surface's first fundamental form rather than its explicit parametrization.

Example 6.3.4 Suppose a surface is parametrized so that $E = 2v^2$, $F = v^2$, $G = v^2$. Then the length of the curve determined by $u = 2 - 3t$, $v = t^2$, $1 \leq t \leq 2$, is

$$\int_1^2 \sqrt{2t^4(-3)^2 + 2t^4(-3)(2t) + t^4(2t)^2} \, dt \approx 7.13117.$$

Example 6.3.5 The first fundamental form of the surface of revolution parametrized by $\mathbf{X}(u, v) = (x(u), y(u) \cos v, y(u) \sin v)$ is

$$ds^2 = \left(\left(\frac{dx}{du} \right)^2 + \left(\frac{dy}{du} \right)^2 \right) du^2 + y^2 \, dv^2.$$

(See Exercise 6.3.16.)

Example 6.3.6 The first fundamental form of the graph of $z = f(x, y)$, when parametrized as $\mathbf{X}(x, y) = (x, y, f(x, y))$, is

$$ds^2 = (1 + f_x^2) \, dx^2 + 2f_x f_y \, dx \, dy + (1 + f_y^2) \, dy^2.$$

(See Exercise 6.3.17.)

The first fundamental form can also be used to determine the angle at which two curves intersect on a surface. Let $\mathbf{X}(u,v)$ be a parametrization of a surface S. Suppose

$$u = u(t), \quad v = v(t), \qquad a \le t \le b,$$

and

$$u = \tilde{u}(\tau), \quad v = \tilde{v}(\tau), \qquad c \le \tau \le d,$$

are the parametrizations of two curves on S that intersect at the point

$$P = \mathbf{X}(u(t_0), v(t_0)) = \mathbf{X}(\tilde{u}(\tau_0), \tilde{v}(\tau_0)),$$

and let θ be the angle between them. Then, if the derivatives below are taken with respect to t and τ respectively (and evaluated at t_0 and τ_0),

$$
\begin{aligned}
\cos\theta &= \frac{\mathbf{X}'(u(t),v(t)) \cdot \mathbf{X}'(\tilde{u}(\tau),\tilde{v}(\tau))}{|\mathbf{X}'(u(t),v(t))| \cdot |\mathbf{X}'(\tilde{u}(\tau),\tilde{v}(\tau))|} \\
&= \frac{(\mathbf{X}_u u' + \mathbf{X}_v v') \cdot (\mathbf{X}_u \tilde{u}' + \mathbf{X}_v \tilde{v}')}{|\mathbf{X}_u u' + \mathbf{X}_v v'| \cdot |\mathbf{X}_u \tilde{u}' + \mathbf{X}_v \tilde{v}'|} \\
&= \frac{E u' \tilde{u}' + F(u'\tilde{v}' + \tilde{u}'v') + G v' \tilde{v}'}{\sqrt{E u'^2 + 2F u'v' + G v'^2}\sqrt{E \tilde{u}'^2 + 2F \tilde{u}'\tilde{v}' + G \tilde{v}'^2}}.
\end{aligned}
\tag{17}
$$

Example 6.3.7 Suppose the surface S is parametrized by $\mathbf{X}(u,v)$ so that

$$E = 3 + v^2, \qquad F = 1 + uv, \qquad G = 1 + u^2.$$

Find the angle at which the two curves parametrized on S by

$$u = t^2, \qquad v = t + 1 \quad \text{and} \quad \tilde{u} = 1 - \tau^2, \qquad \tilde{v} = 3\tau - 2$$

intersect.

The simultaneous equations

$$t^2 = 1 - \tau^2, \qquad t + 1 = 3\tau - 2$$

have the (nonunique) solution $t = 0$, $\tau = 1$ with corresponding values $u(0) = \tilde{u}(1) = 0$, $v(0) = \tilde{v}(1) = 1$, $u'(0) = 0$, $v'(0) = 1$, $\tilde{u}'(1) = -2$, $\tilde{v}'(1) = 3$. The given curves therefore intersect at the point $\mathbf{X}(0,1)$. At that point, $E = 4$, $F = 1$, and $G = 1$, and hence

$$
\begin{aligned}
\cos\theta &= \frac{4 \cdot 0 \cdot (-2) + 1((0 \cdot 3) + (-2) \cdot 1) + 1 \cdot 1 \cdot 3}{\sqrt{4 \cdot 0^2 + 2 \cdot 1 \cdot 0 \cdot 1 + 1 \cdot 1^2}\sqrt{4 \cdot (-2)^2 + 2 \cdot 1 \cdot (-2) \cdot 3 + 1 \cdot 3^2}} \\
&= \frac{1}{\sqrt{1}\sqrt{16 - 12 + 9}}.
\end{aligned}
$$

It follows that

$$\theta = \cos^{-1}\frac{1}{\sqrt{13}}.$$

It should be stressed that the above angle is understood to be determined by *oriented* arcs. The reversal of either of its sides replaces an angle by its supplementary angle.

A parametrization is said to be *orthogonal* if its *u*-parameter and *v*-parameter curves intersect at right angles.

Proposition 6.3.8 *A parametrization* $\mathbf{X}(u,v)$ *of a surface S is orthogonal if and only if F is identically* 0.

PROOF: See Exercise 7.   □

It is convenient to begin our discussion of geodesics by defining a *strict geodesic* on a surface S to be a curve g on S such that the arc of g between any two of its points is not longer than any other curve on S joining those endpoints. In other words, strict geodesics realize the minimum length for all the curves joining their endpoints on S. The strict geodesics of the plane are its straight line segments. In fact, whenever a surface contains a straight line segment, that segment constitutes a strict geodesic of the surface. Such is the case for cones and cylinders. On the other hand, a *geodesic* is a curve that is the union of a set of strict geodesics with the property that the intersection of any two is either empty or a strict geodesic itself. For example, it will be demonstrated towards the end of this section that the arcs of *great circles* of the sphere, that is, those circles that bisect the sphere, are its geodesics. On the other hand, the strict geodesics of the sphere are those arcs of the great circles that do not contain antipodal points.

It will be assumed that all of our surfaces are *geodesically complete* in the sense that given any two points on the surface, there is indeed a geodesic that joins them. Nevertheless, noncomplete surfaces abound. For example, the surface obtained by deleting the origin $(0,0)$ from the plane is not geodesically complete. No two points of the form (a,b) and $(-a,-b)$ can be joined by a geodesic.

We now set out to characterize the geodesics of surfaces in terms of the first fundamental form. For our purposes here it suffices to solve this problem in the context of *orthogonal* parametrizations of surfaces. The analogous characterization for arbitrary forms is the subject of Proposition 6.3.18 and Exercise 14. We begin with a series of technical lemmas that require a definition. Given any $\delta > 0$, the function $\varepsilon(x) : [a,b] \to \mathbb{R}$ is said to be a δ-*perturbation* on $[a,b]$ if it is three times differentiable, $\varepsilon(a) = \varepsilon(b) = 0$, and $\varepsilon(t) < \delta$ for all $t \in [a,b]$.

Lemma 6.3.9 *Let* δ *be a positive number, and let* $f(x)$ *be a continuous function on* $[a,b]$ *such that*

$$\int_a^b f(x)\varepsilon(x)\,dx = 0$$

whenever $\varepsilon(x)$ *is a* δ-*perturbation on* $[a,b]$. *Then* $f(x) = 0$ *for all x in* $[a,b]$.

PROOF: By contradiction. If the conclusion of the theorem is false, then there exists $x_0 \in (a,b)$ such that $f(x_0) \neq 0$. Suppose first that $f(x_0) > 0$. It follows from the continuity of f on $[a,b]$ that there exists a positive real number c such that

$$a < x_0 - c < x_0 + c < b$$

and

$$f(x) > 0 \qquad \text{for all } x \in (x_0 - c, x_0 + c).$$

Let M be the maximum value of the function

$$A(x) = \begin{cases} [(x - x_0)^2 - c^2]^4, & x \in (x_0 - c, x_0 + c), \\ 0 & \text{elsewhere,} \end{cases}$$

and set

$$\varepsilon(x) = \frac{\delta}{M} A(x).$$

It is easily verified (see Exercise 20) that $\varepsilon(x)$ is three times differentiable on $[a,b]$, $|\varepsilon(x)| < \delta$, and $\varepsilon(a) = \varepsilon(b) = 0$. Moreover, since

$$f(x)\varepsilon(x) \geq 0 \qquad \text{for all } x \in [a,b]$$

and

$$f(x)\varepsilon(x) > 0 \qquad \text{for all } x \in (x_0 - c, x_0 + c),$$

it follows that

$$\int_a^b f(x)\varepsilon(x)\,dx > 0,$$

contradicting the stated properties of $f(x)$. A similar contradiction can be derived if $f(x_0) < 0$. Hence $f(x)$ vanishes throughout $[a,b]$.   Q.E.D.

Lemma 6.3.10 *If $f(x)$ is a differentiable function on $[a,b]$ and $\varepsilon(x)$ is a perturbation on $[a,b]$, then*

$$\int_a^b f(x)\varepsilon'(x)\,dx = -\int_a^b f'(x)\varepsilon(x)\,dx.$$

PROOF: It follows from the method of integration by parts that

$$\int_a^b f(x)\varepsilon'(x)\,dx = f(b)\varepsilon(b) - f(a)\varepsilon(a) - \int_a^b f'(x)\varepsilon(x)\,dx$$

$$= -\int_a^b f'(x)\varepsilon(x)\,dx.$$ Q.E.D.

Let $\gamma(t)$ be a curve, parametrized by $u(t), v(t)$, on a surface S parametrized by $\mathbf{X}(u,v)$. The *inclination* of $\gamma(t)$ at any one of its points P is the acute (or right) angle θ measured from the branch of the u-parameter curve through P along which $u(t)$ increases, to $\gamma(t)$. The angle is *positive* if it lies on the side of the u-parameter curve that corresponds to increasing values of v, and *negative* in the opposite case. Thus, in Figure 6.7, the arrowheads indicate the direction of increasing u, and $v_1 < v_2 < v_3 < v_4$. This selection guarantees that θ and $\sin \theta$ have the same sign as $v'(t)$. If the

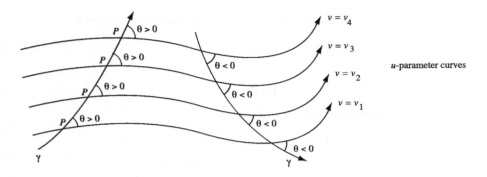

Figure 6.7   The angle of inclination.

surface and curve are fixed, then the angle of inclination is in general a function of t. The cosine of the angle of inclination is easily derived from Equation (17), and this also yields its sine (Exercise 18). For our purposes it suffices to obtain these trigonometric functions for orthognal parametrizations.

Lemma 6.3.11 *Let γ be a curve, parametrized by $u(t), v(t)$, on a surface S with orthogonal parametrization $\mathbf{X}(u,v)$. Then*

$$\cos\theta = \frac{\sqrt{E}u'}{\sqrt{Eu'^2 + Gv'^2}}, \qquad \sin\theta = \frac{\sqrt{G}v'}{\sqrt{Eu'^2 + Gv'^2}},$$

where θ is the inclination of γ.

PROOF: Exercise 15.   □

The last of the lemmas is a well-known fact that is widely used in mathematics and theoretical physics. Its proof can be found in many standard texts on analysis. It is also commonly paraphrased as a permission to differentiate under the integral sign.

Lemma 6.3.12 *If in the rectangle $a \leq x \leq b$, $c \leq y \leq d$ the function $f(x,y)$ is continuously differentiable with respect to x, then*

$$\frac{d}{dx}\int_c^d f(x,y)\,dy = \int_c^d f_x(x,y)\,dy.$$

The next theorem characterizes geodesics by means of a differential equation. We only prove the necessity of the equation, as the sufficiency requires tools that cannot be presented here.

Theorem 6.3.13 (Gauss's Equation) *On a surface S with the orthogonal parametrization* $\mathbf{X}(u,v)$*, let* $u = u(t)$*,* $v = v(t)$*,* $a \leq t \leq b$*, be the parametrization of a curve* γ*. The curve* γ *is a geodesic if and only if its angle of inclination* θ *satisfies the equation*

$$\theta' = \frac{E_v u' - G_u v'}{2\sqrt{EG}}. \tag{18}$$

PROOF: As only necessity is to be proven, it may be assumed that γ is a strict geodesic. Let δ be a positive real number such that for each $t \in [a,b]$, $(u(t) + \lambda\delta, v(t))$ is in the domain of $\mathbf{X}(u,v)$ for each $\lambda \in [-1,1]$. Let $\varepsilon(t)$ be a δ-perturbation function on $[a,b]$.

For each $\lambda \in [-1,1]$ let γ_λ be the curve on S parametrized by

$$u = \bar{u}(t) = u(t) + \lambda\varepsilon(t), \quad v = v(t), \qquad t \in [a,b].$$

Note that $\bar{u}' = u'(t) + \lambda\varepsilon'(t)$. We now set

$$L(\lambda) = \text{Length}(\gamma_\lambda) = \int_a^b \sqrt{E\bar{u}'^2 + Gv'^2}\, dt$$

and note that since $\gamma_0 = \gamma$ is a geodesic, the function $L(\lambda)$ assumes a minimum value at $\lambda = 0$. By Lemma 6.3.12

$$\frac{dL}{d\lambda} = \int_a^b \frac{\left(\dfrac{\partial E}{\partial u}\dfrac{\partial \bar{u}}{\partial \lambda} + \dfrac{\partial E}{\partial v}\dfrac{\partial v}{\partial \lambda}\right)\bar{u}'^2 + E\dfrac{\partial \bar{u}'^2}{\partial \lambda} + \left(\dfrac{\partial G}{\partial u}\dfrac{\partial \bar{u}}{\partial \lambda} + \dfrac{\partial G}{\partial v}\dfrac{\partial v}{\partial \lambda}\right)v'^2 + G\dfrac{\partial v'^2}{\partial \lambda}}{2\sqrt{E\bar{u}'^2 + Gv'^2}}\, dt$$

$$= \int_a^b \frac{(E_u \cdot \varepsilon + E_v \cdot 0)\bar{u}'^2 + E \cdot 2\bar{u}' \cdot \varepsilon' + (G_u \cdot \varepsilon + G_v \cdot 0)v'^2 + G \cdot 0}{2\sqrt{E\bar{u}'^2 + Gv'^2}}\, dt$$

It follows that at $\lambda = 0$ we have

$$0 = \int_a^b \frac{(E_u u'^2 + G_u v'^2)\varepsilon + 2Eu'\varepsilon'}{2\sqrt{Eu'^2 + Gv'^2}}\, dt,$$

which, by Lemma 6.3.10, equals

$$\int_a^b \left[\frac{E_u u'^2 + G_u v'^2}{2\sqrt{Eu'^2 + Gv'^2}} - \frac{d}{dt}\left(\frac{Eu'}{\sqrt{Eu'^2 + Gv'^2}}\right)\right] \varepsilon(t)\, dt$$

By Lemmas 6.3.9 and 6.3.11,

$$E_u u'^2 + G_u v'^2 = 2\sqrt{Eu'^2 + Gv'^2}\,\frac{d}{dt}\left(\frac{Eu'}{\sqrt{Eu'^2 + Gv'^2}}\right)$$

$$= 2\sqrt{Eu'^2 + Gv'^2}\,\frac{d}{dt}(\sqrt{E}\cos\theta)$$

$$= 2\sqrt{Eu'^2 + Gv'^2} \cdot \frac{1}{2\sqrt{E}} \cdot E'\cos\theta + 2\sqrt{Eu'^2 + Gv'^2}\sqrt{E}(-\sin\theta)\theta'$$

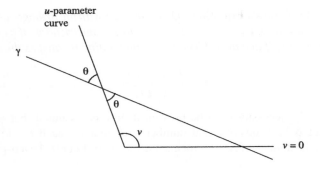

Figure 6.8   A geodesic in the plane.

$$= E'u' - 2\sqrt{EG}\theta'v' = (E_u u' + E_v v')u' - 2\sqrt{EG}\theta'v'.$$

It follows that

$$\theta' = \frac{1}{2v'\sqrt{EG}}(E_u u'^2 + E_v u'v' - E_u u'^2 - G_u v'^2)$$

$$= \frac{1}{2\sqrt{EG}}(E_v u' - G_u v'). \qquad\qquad \text{Q.E.D.}$$

Example 6.3.14 The parametrization $\mathbf{X}(u,v) = (u,v,0)$ of the plane has the first fundamental form $du^2 + dv^2$, which is clearly orthogonal. Here Gauss's Equation (18) takes the form $\theta' = 0$, meaning that every geodesic has constant inclination. This is clearly consistent with the fact that the geodesics of this surface are straight lines.

Example 6.3.15 The parametrization $\mathbf{X}(u,v) = u(\cos v, \sin v, 0)$ of the plane has the orthogonal first fundamental form $du^2 + u^2\,dv^2$. Hence, in this case, Gauss's Equation (18) reduces to

$$\theta' = -\frac{2uv'}{2u} = -v',$$

or

$$\theta + v = \text{constant}.$$

This is clearly borne out by Figure 6.8.

Proposition 6.3.16 *If a surface of revolution S is parametrized as*

$$\mathbf{X}(u,v) = (x(u), y(u)\cos v, y(u)\sin v),$$

then the meridians are geodesics of S.

PROOF: As noted in Example 6.3.5, the first fundamental form of this parametrization is

$$\left(\left(\frac{dx}{du}\right)^2 + \left(\frac{dy}{du}\right)^2\right)du^2 + y^2 dv^2.$$

The typical meridian has parametrization $u = t$, $v =$ constant, and inclination $\theta = 0$. Along this meridian, Gauss's Equation (18) takes the form

$$0' = \frac{0 \cdot 1 - 2y\dfrac{dy}{du} \cdot 0}{2\sqrt{EG}},$$

which is clearly valid.   Q.E.D.

Of course, it is now possible to conclude that the sphere's meridians are geodesics. Because of the symmetry of the sphere, this property is also possessed by all the arcs of its great circles.

Corollary 6.3.17 *The geodesics of the sphere are portions of its great circles.*   □

The statement and proof of Gauss's Equation (18) can be generalized to cover arbitrary parametrizations as well. The details are left to Exercise 14.

Proposition 6.3.18 *On a surface S with the parametrization* $\mathbf{X}(u,v)$, *let* $u = u(t)$, $v = v(t)$, $a \le t \le b$, *be the parametrization of a curve* γ. *Then* γ *is a geodesic if and only if its angle of inclination* θ *satisfies the equation*

$$\sqrt{EG - F^2}\,\theta'$$
$$= \frac{1}{2} \cdot \frac{F}{E} \cdot E_u \cdot u' + \frac{1}{2} \cdot \frac{F}{E} \cdot E_v \cdot v' + \frac{1}{2} \cdot E_v \cdot u' - F_u \cdot u' - \frac{1}{2} \cdot G_u \cdot v'. \qquad \square$$

This section concludes with a useful general observation regarding the first fundamental form. For any real numbers A, B, C, the quadratic form $f(x,y) = Ax^2 + 2Bxy + Cy^2$ is said to be *positive definite* provided that $f(x,y) > 0$ for all real x, y that are not both 0. This is tantamount to the inequalities $A, C > 0$ and $AC - B^2 > 0$ (Exercise 8).

Proposition 6.3.19 *If* $ds^2 = E\,du^2 + 2F\,du\,dv + G\,dv^2$ *is a first fundamental form, then the quadratic form* $Ex^2 + 2Fxy + Gy^2$ *is positive definite.*

PROOF: See Exercise 9.   □

Exercises 6.3

1. Find the length of the curve parametrized by the functions below on the surface $\mathbf{X}(u,v) = (u + v^2, u - v, u^2)$:

    (a) $u = t^2$, $v = t^3$, $1 \le t \le 2$.

    (b) $u = \sin t$, $v = t^2$, $1 \le t \le 2$.

    (c) $u = e^t$, $v = \ln t$, $1 \le t \le 2$.

2. Find the length of the curve parametrized by the functions below on the right helicoid $X(u, v) = (2v\cos u, 2v\sin u, 3u)$:

   (a) $u = t^2, v = t^3, 1 \le t \le 2$.

   (b) $u = \sin t, v = t^2, 1 \le t \le 2$.

   (c) $u = e^t, v = \ln t, 1 \le t \le 2$.

3. Find the length of the curve parametrized by the functions below on the surface $z = 1 + x^2 - y^3$:

   (a) $x = t^2, y = t^3, 1 \le t \le 2$.

   (b) $x = \sin t, y = t^2, 1 \le t \le 2$.

   (c) $x = e^t, y = \ln t, 1 \le t \le 2$.

4. Find the first fundamental form of the surface $X(u, v) = (u^2, uv, u - v)$. Use this form to find the angle formed on the surface, at the specified points, by the following pairs of curves:

   (a) $X(2, 1), u = t^2 - 2, v = t - 1$ and $u = -2\tau, v = \tau^2$.

   (b) $X(0, 3), u = 3 - t, v = t$ and $u = -3\tau - 3, v = 4 + \tau$.

5. Suppose a surface has $E = 4u^2 + 4v^2 + 1$, $F = 2u + uv - v + 2$, $G = u^2 - 2u + 6$. Find the angle formed on the surface, at the specified point, by the following pairs of curves:

   (a) $X(2, 1); u = t^2 - 2, v = t - 1$ and $u = -2\tau, v = \tau^2$.

   (b) $X(0, 3); u = 3 - t, v = t$ and $u = -3\tau - 3, v = 4 + \tau$.

6. Let $\omega (0 < \omega < \pi)$ be the angle between the u-parameter and the v-parameter curves of the surface parametrized by $X(u, v)$. Prove that $\cos \omega = F/\sqrt{EG}$.

7. Prove that the u- and v-parametric curves of the surface $X(u, v)$ are orthogonal to each other if and only if $F = 0$.

8. Prove that the quadratic form $f(x, y) = Ax^2 + 2Bxy + Cy^2$ is positive definite if and only if $A, C > 0$ and $AC - B^2 > 0$.

9. Prove Proposition 6.3.19.

10. Let $u = f(t), v = g(t)$ be a parametrization of a curve C that bisects the angle between the parametric curves on a surface with an orthgonal parametrization. Prove that $E f'^2 - G g'^2 = 0$.

11. Prove that on the surface $X(u, v) = (u\cos v, u\sin v, v + \ln\cos u)$, any two v-parameter curves cut equal segments from all u-parameter curves.

12. A parametrized surface has first fundamental form $du^2 + G dv^2$. Prove that its u-parameter curves are geodesics.

13. The *stationary* latitudes of a surface of revolution parametrized as

$$(x(u), y(u)\cos v, y(u)\sin v)$$

are those that correspond to values of u such that $dy/du = 0$. Prove that a latitude is a geodesic if and only if it is stationary.

14**. Prove Proposition 6.3.18.

15. Prove Proposition 6.3.11.

16. Verify the details of Example 6.3.5.

17. Verify the details of Example 6.3.6.

18. Show that for the angle of inclination θ

$$\cos\theta = \frac{Eu' + Fv'}{\sqrt{Eu'^2 + 2Fu'v' + Gv'^2}\sqrt{E}},$$

$$\sin\theta = \frac{v'\sqrt{EG - F^2}}{\sqrt{Eu'^2 + 2Fu'v' + Gv'^2}\sqrt{E}}.$$

19. Let k and r be constant positive real numbers. Prove that on the cylinder $\mathbf{X}(u,v) = (u, r\cos v, r\sin v)$, the helix parametrized by $u(t) = kt$, $v(t) = t$ is a geodesic.

20. Prove that the function $\varepsilon(x)$ defined in the proof of Lemma 6.3.9 is three times differentiable.

6.4 Normal Curvatures

Euler had already investigated the curvature of surfaces before Gauss. His approach expressed this two-dimensional curvature at a point as an assembly of the curvatures of certain of its cross sections. We describe Euler's view in detail and then relate it to the Gaussian curvature.

Fix a point P on a surface and anchor all the vectors below at P. A *direction* $\mathbf{d}$ at P is a vector in the tangent plane to S at P. The *normal section* $\gamma_{\mathbf{d}}$ is the intersection of S with the plane that contains both $\mathbf{n}_P$ and $\mathbf{d}$.

Let s be an arclength parameter of this normal section $\gamma_{\mathbf{d}}$ with unit tangent $\mathbf{T}$, normal $\mathbf{N}$, and curvature κ. Since $\mathbf{T}$ is contained in the cutting plane, so is

$$\mathbf{N} = \frac{1}{\kappa}\frac{d\mathbf{T}}{ds}.$$

The surface normal $\mathbf{n}_P$ is also perpendicular to $\mathbf{T}$ and, by the definition of the cutting plane, it is parallel to this plane. It follows that

$$\mathbf{N} = \pm\mathbf{n}_P.$$

Given a direction of the surface S at P, with normal section $\gamma_{\mathbf{d}}$, the *normal curvature* is the scalar

$$\kappa_{\mathbf{d}} = \begin{cases} \kappa & \text{if } \mathbf{N} = \mathbf{n}_P, \\ -\kappa & \text{if } \mathbf{N} = -\mathbf{n}_P. \end{cases}$$

Thus, the normal curvature is positive if $\gamma_{\mathbf{d}}$ curves towards $\mathbf{n}_P$ and negative if $\gamma_{\mathbf{d}}$ curves away from $\mathbf{n}_P$. Note that necessarily

$$\frac{d\mathbf{T}}{ds} = \kappa\mathbf{N} = \kappa_{\mathbf{d}}\mathbf{n}.$$

Lemma 6.4.1 *If the normal section γ_d of the surface of $X(u,v)$ is parametrized by $u = u(t)$, $v = v(t)$, then*

$$\kappa_d = \frac{Lu'^2 + 2Mu'v' + Nv'^2}{Eu'^2 + 2Fu'v' + Gv'^2}.$$

PROOF: The section γ_d has the expression

$$\gamma_d(t) = X(u(t), v(t)).$$

Let s be the arclength parameter of this section. Since $n \cdot T = 0$ along γ_d, it follows that

$$\kappa_d = n \cdot \frac{dT}{ds} = -\frac{dn}{ds} \cdot T = -\frac{dn}{ds} \cdot \frac{d\gamma_d}{ds} = \frac{-n' \cdot \gamma_d'}{(ds/dt)^2}. \tag{19}$$

However,

$$\gamma_d' = X_u u' + X_v v',$$

and hence

$$\gamma_d'' = X_{uu} u'^2 + 2X_{uv} u'v' + X_{vv} v'^2 + X_u u'' + X_v v''.$$

Thus, since $n \cdot \gamma_d' = 0$ along γ_d,

$$-n' \cdot \gamma_d' = n \cdot \gamma_d'' = Lu'^2 + 2Mu'v' + Nv'^2 + 0 \cdot u'' + 0 \cdot v'',$$

so that, by Equations (16) and (19),

$$\kappa_d = \frac{Lu'^2 + 2Mu'v' + Nv'^2}{Eu'^2 + 2Fu'v' + Gv'^2}. \qquad \text{Q.E.D.}$$

It should be noted that the formula of Lemma 6.4.1, explicit as it may seem, is in general rather difficult to apply. Parametrizations of arbitrary normal sections are not readily available. However, for surfaces defined by means of an equation of the form $z = f(x, y)$ the normal curvatures are easily described.

Example 6.4.2 Let S be the graph of $z = x^2 - y^2 + 10$ parametrized as $X(x, y) = (x, y, x^2 - y^2 + 10)$. Here

$$X_x = (1, 0, 2x), \qquad X_y = (0, 1, -2y),$$

$$E = 1 + 4x^2, \qquad F = -4xy, \qquad G = 1 + 4y^2,$$

$$n = \frac{(-2x, 2y, 1)}{\sqrt{1 + 4x^2 + 4y^2}},$$

$$X_{xx} = (0, 0, 2), \qquad X_{xy} = (0, 0, 0), \qquad X_{yy} = (0, 0, -2),$$

$$L = \frac{2}{\sqrt{1 + 4x^2 + 4y^2}}, \qquad M = 0, \qquad N = \frac{-2}{\sqrt{1 + 4x^2 + 4y^2}}.$$

It follows that at the point $P(0, 0, 10)$ the tangent plane is the xy-plane, and the typical direction vector $d(\alpha)$ is $(\cos\alpha, \sin\alpha, 0)$, where α is the angle of inclination of the cutting plane to the xz-plane (Fig. 6.9).

The normal section is parametrized by $x = t \cos \alpha$, $y = t \sin \alpha$. At P the surface has

$$E = 1, \qquad F = 0, \qquad G = 1, \qquad L = 2, \qquad F = 0, \qquad N = -2,$$

and hence

$$\kappa_\alpha = \kappa_{\mathbf{d}(\alpha)} = 2\frac{x'^2 - y'^2}{x'^2 + y'^2} = 2(\cos^2 \alpha - \sin^2 \alpha) = 2\cos 2\alpha.$$

This description of the dependence of the normal curvature on the direction $\mathbf{d}(\alpha)$ is clearly consistent with the graph of $z = x^2 - y^2 + 10$ depicted in Figure 6.10.

Example 6.4.3 At the point $(1, 2, 7)$ on the surface of the previous example we have

$$E = 5, \quad F = -8, \quad G = 17, \quad L = 2/\sqrt{21}, \quad M = 0, \quad N = -2/\sqrt{21}$$

and hence

$$\kappa_{\mathbf{d}} = \frac{2}{\sqrt{21}} \cdot \frac{x'^2 - y'^2}{5x'^2 - 16x'y' + 17y'^2}.$$

The simple description of sectional curvatures obtained in Example 6.4.2 can be generalized to arbitrary surfaces. First note that given any point P on any surface S, the surface can be so rotated and translated that $P = (0, 0, 0)$ and S is the graph of some equation $z = f(x, y)$. It may be necessary to trim S down a bit to obtain such a description, but some portion of S containing P always can be so placed and described. A further rotation will guarantee that the tangent plane to S at P is the xy-plane. All this is of course tantamount to the equations

$$f(0, 0) = f_x(0, 0) = f_y(0, 0) = 0.$$

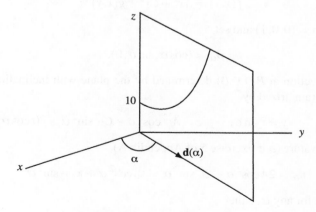

Figure 6.9   A normal section.

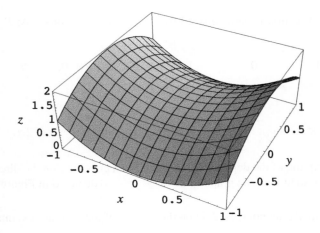

Figure 6.10   A surface.

It follows from Taylor's Theorem, and the assumption that all of our functions have all of their derivatives of order 3, that there exists a function $g(x,y)$ such that

$$f(x,y) = Ax^2 + 2Bxy + Cy^2 + g(x,y), \tag{20}$$

where

$$2A = f_{xx}(0,0), \quad 2B = f_{xy}(0,0), \quad 2C = f_{yy}(0,0)$$

and

$$0 = g(0,0) = g_x(0,0) = g_y(0,0) = g_{xx}(0,0) = g_{xy}(0,0) = g_{yy}(0,0).$$

Finally, a suitable rotation of the xy-plane (see Exercise 3) can be used to eliminate the xy term in Equation (20), so that

$$f(x,y) = Ax^2 + Cy^2 + g(x,y). \tag{21}$$

Now choose $\mathbf{n} = (0,0,1)$ and set

$$\mathbf{d}_\alpha = (\cos\alpha, \sin\alpha, 0).$$

The normal section at $P(0,0,0)$ determined by the plane with inclination α to the xz-plane is parametrized by

$$x = t\cos\alpha, \qquad y = t\sin\alpha, \qquad z = At^2\cos^2\alpha + Ct^2\sin^2\alpha + g(t\cos\alpha, t\sin\alpha),$$

and it has curvature (see Exercise 5 of Appendix A)

$$\kappa_\alpha = 2A\cos^2\alpha + 2C\sin^2\alpha = \kappa_0\cos^2\alpha + \kappa_{\pi/2}\sin^2\alpha.$$

It is clear that for any α, either

$$\kappa_0 \le \kappa_\alpha \le \kappa_{\pi/2} \quad \text{or} \quad \kappa_{\pi/2} \le \kappa_\alpha \le \kappa_0.$$

The scalars κ_0 and $\kappa_{\pi/2}$ are called the *principal curvatures* of S at P, because they constitute the extremal values of all the sectional curvatures; every other sectional curvature is an appropriately weighted average of κ_0 and $\kappa_{\pi/2}$. The foregoing discussion is now summarized as a theorem.

Theorem 6.4.4 *Let P be a point on the surface S. Then there exist two orthogonal directions $\mathbf{d}_1$ and $\mathbf{d}_2$ such that if the direction $\mathbf{d}$ has inclination α to $\mathbf{d}_1$, then*

$$\kappa_{\mathbf{d}} = \kappa_{\mathbf{d}_1} \cos^2 \alpha + \kappa_{\mathbf{d}_2} \sin^2 \alpha. \qquad \square$$

We next relate the principal curvatures to the Gaussian curvature.

Proposition 6.4.5 *If the surface S has Gaussian curvature K and principal curvatures $\kappa_{\mathbf{d}_1}$ and $\kappa_{\mathbf{d}_2}$ at the point P, then*

$$K = \kappa_{\mathbf{d}_1} \kappa_{\mathbf{d}_2}.$$

PROOF: Suppose the surface S is placed as it was in the proof of Theorem 6.4.4, so that $P = (0,0,0)$. Then it follows from Example 6.2.8 that

$$K = \frac{f_{xx}f_{yy} - f_{xy}^2}{\sqrt{1 + f_x^2 + f_y^2}} = \frac{f_{xx}f_{yy} - 0}{\sqrt{1+0+0}} = f_{xx}f_{yy} = 4AC = \kappa_{\mathbf{d}_1} \kappa_{\mathbf{d}_2}. \qquad \text{Q.E.D.}$$

The remainder of this section is devoted to the explicit computation of principal curvatures from arbitrary parametrizations of surfaces—not just those associated with the graphs of $z = f(x,y)$. This is a digression that can be skipped without any adverse effects on the readers' ability to comprehend the subsequent material.

Let S be a fixed surface parametrized as $\mathbf{X}(u,v)$, and let P be a fixed point on S. The *normal curvature* of any curve $\mathbf{C}$ on S, parametrized by $u = u(t), v = v(t)$, at P, is defined as

$$\kappa_n(\mathbf{C}) = \frac{Lu'^2 + 2Mu'v' + Nv'^2}{Eu'^2 + 2Fu'v' + Gv'^2}.$$

Here it is understood that u' and v', as well as the other derivatives appearing below, are evaluated at P. Since

$$\mathbf{C}' = \mathbf{X}_u u' + \mathbf{X}_v v',$$

it follows from the definition that for any two curves $\mathbf{C}$ and $\mathbf{D}$ on S and real number r,

$$\kappa_n(\mathbf{C}) = \kappa_n(\mathbf{D}) \quad \text{whenever } \mathbf{C}' = r\mathbf{D}'.$$

Lemma 6.4.6 *If $\mathbf{C}$ is any curve on S through P, then*

$$\kappa_{\mathbf{C}'} = \kappa_n(\mathbf{C}).$$

PROOF: By the preceding discussion,

$$\kappa_n(\mathbf{C}) = \kappa_n(\gamma_{\mathbf{C}'}).$$

By the definition of the normal curvature and Lemma 6.4.1,

$$\kappa_{C'} = \frac{Lu'^2 + 2Mu'v' + Nv'^2}{Eu'^2 + 2Fu'v' + Gv'^2} = \kappa_n(\gamma_{C'}),$$

where $u = u(t), v = v(t)$ is any parametrization of $\gamma_{C'}$. The lemma now follows immediately.   Q.E.D.

Lemma 6.4.6 implies that every normal curvature is a sectional curvature and vice versa. Consequently, in searching for the extremal values of

$$\frac{Lu'^2 + 2Mu'v' + Nv'^2}{Eu'^2 + 2Fu'v' + Gv'^2}$$

it may be assumed that (u', v') varies over all $\mathbb{R}^2 - \{(0,0)\}$, since such is clearly the case when this fraction is regarded as a normal curvature. We are now ready to find the principal curvatures.

Proposition 6.4.7 *If the surface S is parametrized by* $\mathbf{X}(u, v)$, *then its principal curvatures at any point are the solutions, for* λ, *of the equation*

$$(EG - F^2)\lambda^2 - (EN + GL - 2FM)\lambda + (LN - M^2) = 0.$$

PROOF: By the preceding discussion it suffices to find the extremal values of the function

$$h(x,y) = \frac{Lx^2 + 2Mxy + Ny^2}{Ex^2 + 2Fxy + Gy^2}$$

where (x, y) varies over all pairs of reals except $(0,0)$. Since $h(rx, ry) = h(x, y)$ for every nonzero r, it follows that the extremal values of κ_d are also the extremal values of

$$Lx^2 + 2Mxy + Ny^2$$

subject to the constraint

$$Ex^2 + 2Fxy + Gy^2 = 1.$$

The Lagrange multiplier method calls for the location of the extremal values of the function

$$Lx^2 + 2Mxy + Ny^2 - \lambda(Ex^2 + 2Fxy + Gy^2 - 1)$$
$$= (L - \lambda E)x^2 + 2(M - \lambda F)xy + (N - \lambda G)y^2 + \lambda.$$

For these extremal values, partial differentiation with respect to x and y yields

$$(L - \lambda E)x + (M - \lambda F)y = 0, \tag{22}$$
$$(M - \lambda F)x + (N - \lambda G)y = 0. \tag{23}$$

When these equations are multiplied by x and y respectively, and then added, we obtain

$$(L - \lambda E)x^2 + 2(M - \lambda F)xy + (N - \lambda G)y^2 = 0,$$

or

$$\lambda = \frac{Lx^2 + 2Mxy + Ny^2}{Ex^2 + 2Fxy + Gy^2},$$

implying that λ is indeed a normal curvature. Moreover, since Equations (22), (23) are consistent, the determinant of their coefficients is 0, and so

$$(L - \lambda E)(N - \lambda G) - (M - \lambda F)^2 = 0,$$

from which the desired equation is easily derived. Q.E.D.

Example 6.4.8 The right circular cylinder of radius R centered about the x-axis has the parametrization $\mathbf{X}(u,v) = (u, R\cos v, R\sin v)$. For this surface

$$\mathbf{X}_u = (1,0,0), \qquad \mathbf{X}_v = (0, -R\sin v, R\cos v),$$
$$E = 1, \qquad F = 0, \qquad G = R^2,$$
$$\mathbf{X}_u \times \mathbf{X}_v = (0, -R\cos v, -R\sin v), \qquad \mathbf{n} = (0, -\cos v, -\sin v),$$
$$\mathbf{X}_{uu} = (0,0,0), \qquad \mathbf{X}_{uv} = (0,0,0), \qquad \mathbf{X}_{vv} = (0, -R\cos v, -R\sin v),$$
$$L = 0, \qquad M = 0, \qquad N = R.$$

It follows that the principal curvatures are the solutions of the quadratic

$$0 = (1 \cdot R^2 - 0^2)\lambda^2 - (1 \cdot R + R^2 \cdot 0 - 2 \cdot 0 \cdot 0)\lambda + (0 \cdot R - 0^2) = R^2\lambda^2 - R\lambda,$$

that is, 0 and $1/R$.

Example 6.4.9 Determine the principal curvatures of the surface parametrized by $\mathbf{X}(u,v) = (u + v, u - 2v, uv)$ at the point $\mathbf{X}(2,1)$.

By Example 6.2.7 we have at $\mathbf{X}(2,1)$

$$E = 3, \qquad F = 1, \qquad G = 9, \qquad L = 0, \qquad M = \frac{-3}{\sqrt{26}}, \qquad N = 0,$$

and the principal curvatures are the solutions of

$$26\lambda^2 - \frac{6}{\sqrt{26}}\lambda - \frac{9}{26} = 0,$$

that is,

$$\frac{3 \pm 9\sqrt{3}}{26\sqrt{26}}.$$

Exercises 6.4

1. Find the expression for $\kappa_{d(\alpha)}$ for the surface $z = f(x, y)$.

2. Show that the transformation $x = \bar{x}\cos\alpha - \bar{y}\sin\alpha$, $y = \bar{x}\sin\alpha + \bar{y}\cos\alpha$ is a rotation of the plane by demonstrating that it

   (a) has exactly one fixed point (provided α is not an integer multiple of $360°$);

   (b) preserves the distance between points.

3. Show that when the rotation of Exercise 2 is applied to the expression $Ax^2 + Bxy + Cy^2$, the result has the form $A'\bar{x}^2 + C'\bar{y}^2$ provided that $\tan 2\alpha = B/(A - C)$.

4. Compute the principal curvatures of the surface $\mathbf{X}(u, v) = (u, v, uv)$ at the point $\mathbf{X}(1, 2)$.

5. Compute the principal curvatures of the surface $\mathbf{X}(u, v) = (u, u + v, u + v^2)$ at the point $\mathbf{X}(1, 2)$.

6. Prove that the principal curvatures of the general surface of revolution $\mathbf{X}(u, v) = (x(u), y(u)\cos v, y(u)\sin v)$ are

$$\frac{x''y' - x'y''}{(x'^2 + y'^2)^{3/2}} \quad \text{and} \quad \frac{x'}{y(x'^2 + y'^2)^{1/2}}.$$

(Note: Here a prime denotes differentiation with respect to u.) Relate these curvatures to those of the surface's parametric curves.

7. The torus can be parametrized as

$$\mathbf{X}(u, v) = (r\cos u, (R + r\sin u)\cos v, (R + r\sin u)\sin v).$$

Use Exercise 6 to derive the principal curvatures of this torus and explain why one of them is constant.

8. A point on a surface at which all the normal curvatures are equal is called an *umbilic*. Prove that a point P is an umbilic if and only if E, F, G, are proportional to L, M, N at P.

9. Find the umbilics on the surface of $z = 1/xy$.

10. Prove that the sum of the normal curvatures at a point in two directions is constant provided the two directions are perpendicular to each other.

6.5   The Geodesic Polar Parametrization

Imagine that the typical surface is made of a fabric that is flexible but not stretchable. It may then be embedded in $\mathbb{R}^3$ in many different ways. Some of its properties may depend on the actual embedding, and others may not. For example, if a sheet of paper is laid out flat, then its sectional curvatures are all 0. If, on the other hand, the same sheet is rolled up into a right circular cylinder of radius R, then, as computed in Example 6.4.8, its principal curvatures become 0 and $1/R$. Geodesics, on the other hand, do not depend on the embedding. A geodesic on the flat sheet of paper is bent into a geodesic on the cylinder when the sheet is rolled up. More generally, lengths of curves are also not affected by the bending of their ambient surfaces.

Aspects of surfaces that are not affected by bending are said to be *intrinsic*, whereas those that may be altered by bending are *extrinsic*. These concepts can be formalized in terms of the first fundamental form. A property, or an aspect, of surfaces is intrinsic if it is expressible in terms of the coefficients E, F, G of the first fundamental form. It is clear that lengths of curves and their angles of intersection are intrinsic. The principal and sectional curvatures, on the other hand, are extrinsic, as they depend on the way the surface is embedded in space. The bending of a surface clearly has the potential for changing all of its sectional curvatures.

On the face of it, the Gaussian curvature would also seem to be extrinsic, as its definition employs the normal vector, which clearly varies with the embedding. Moreover, the formula derived in Theorem 6.2.5 involves the parameters L, M, N, which are not part of the first fundamental form. Nevertheless, we are about to demonstrate that the Gaussian curvature is in fact intrinsic. This curvature does not vary even when the surface is bent out of shape (as long as the surface is not stretched)—a very surprising and important fact.

The following proposition was tacitly assumed by Gauss, and we shall do so as well.

Proposition 6.5.1 *Given a point P on a surface S, there is a loop L on S, surrounding P, such that every point Q on L can be joined by a unique geodesic γ_Q to P. Moreover,*

1. *if Q and Q' are distinct points on L, then γ_Q and $\gamma_{Q'}$ intersect only at P;*

2. *every point on the side of L that contains P is contained in some γ_Q.* □

This system of geodesics that radiates from P can be used to parametrize the vicinity of P in S. For this purpose one of the geodesics is designated as γ_0. The arbitrary geodesic γ through P is then designated as γ_v, where v is the oriented angle from γ_0 to γ_v. Every point Q of this vicinity of P can then be parametrized as $\mathbf{X}(u, v)$, where γ_v is the geodesic from P to Q, and u is the length of this geodesic (from P to Q). This parametrization is said to be *geodesic polar*.

Example 6.5.2 A geodesic polar parametrization of the plane $z = 0$ at the point $(a, b, 0)$ is given by $\mathbf{X}(u, v) = (a + u \cos v, b + u \sin v, 0)$. Here,

$$\mathbf{X}_u = (\cos v, \sin v, 0), \qquad \mathbf{X}_v = (-u \sin v, u \cos v, 0),$$
$$E = 1, \qquad F = 0, \qquad G = u^2.$$

Example 6.5.3 A geodesic polar parametrization of the sphere of radius R, centered at $P(0, 0, R)$ (see Fig. 6.11), is

$$\mathbf{X}(u, v) = R\left(\sin\frac{u}{R}\cos v, \sin\frac{u}{R}\sin v, \cos\frac{u}{R}\right).$$

Here,

$$\mathbf{X}_u = \left(\cos\frac{u}{R}\cos v, \cos\frac{u}{R}\sin v, -\sin\frac{u}{R}\right),$$

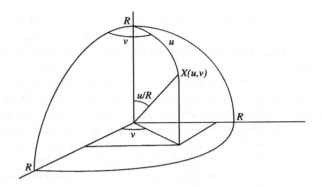

Figure 6.11   A geodesic polar parametrization of the sphere.

$$\mathbf{X}_v = R\left(-\sin\frac{u}{R}\sin v, \sin\frac{u}{R}\cos v, 0\right),$$

$$E = 1, \qquad F = 0, \qquad G = R^2\sin^2\frac{u}{R}.$$

As these examples indicate, geodesic polar parametrizations have a particularly simple first fundamental form. Given such a parametrization, the v-parameter curves are called *geodesic circles*.

Proposition 6.5.4 *If* $\mathbf{X}(u,v)$ *is a geodesic polar parametrization, then* $E = 1$, *and* $F = 0$.

PROOF: By its definition, u is the arclength parameter of the geodesic parametrized by $u = t$ and $v = $ constant. It follows that

$$E = \mathbf{X}_u \cdot \mathbf{X}_u = (|\mathbf{X}_u|)^2 = 1^2 = 1.$$

We offer Gauss's informal argument to prove the vanishing of F. As noted in Proposition 6.3.8, it is necessary to demonstrate that the u-parameter curves intersect the v-parameter curves, that is, the geodesic circles, orthogonally. This is a natural generalization of the fact that in Euclidean circles the tangent line is perpendicular to the radius through the point of contact. Suppose such is not the case. Then there is a u-parameter curve PB (see Fig. 6.12) that intersects the geodesic circle $u = d$ at the point B in an angle of $90° - \omega < 90°$. Let B' be another point on this geodesic circle that is close enough to B so that $\angle PB'B$ is obtuse where PB' is a geodesic which, of necessity, has the same length d as the geodesic PB. Let C be a point on the geodesic PB such that the geodesic CB' is orthogonal to the geodesic circle d, so that

$$\angle BB'C = 90°.$$

Since, in Gauss's own words, "the infinitesimally small $\triangle BB'C$ may be regarded as plane" (that is, Euclidean), it follows that

$$\angle BCB' = \omega \quad \text{and} \quad CB' = BC\cos\omega.$$

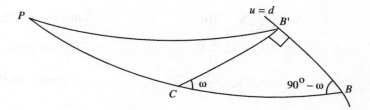

Figure 6.12   An impossible configuration.

Hence,

$$PC + CB' = PC + BC\cos\omega = PB - BC + BC\cos\omega$$
$$= PB - BC(1 - \cos\omega) = PB' - BC(1 - \cos\omega) < PB',$$

contradicting the fact that PB' is a geodesic. Hence, the u-parameter curves intersect the v-parameter curves orthogonally.   Q.E.D.

We now go on to derive the value of the Gaussian curvature for a geodesic polar parametrization.

Lemma 6.5.5 *Let* $\mathbf{X}(u,v)$ *be a geodesic polar parametrization of the surface S. Then*

$$\mathbf{X}_u \cdot \mathbf{X}_{uu} = 0, \qquad\qquad \mathbf{X}_u \cdot \mathbf{X}_{uv} = 0, \qquad\qquad (24)$$

$$\mathbf{X}_v \cdot \mathbf{X}_{uu} = -\mathbf{X}_u \cdot \mathbf{X}_{uv} = 0, \qquad \mathbf{X}_v \cdot \mathbf{X}_{uv} = -\mathbf{X}_u \cdot \mathbf{X}_{vv}, \qquad (25)$$

$$\mathbf{X}_v \cdot \mathbf{X}_{uv} = \frac{1}{2}G_u, \qquad\qquad \mathbf{X}_v \cdot \mathbf{X}_{vv} = \frac{1}{2}G_v. \qquad\qquad (26)$$

PROOF: Equations (24)–(26) follow from the respective differentiations of $\mathbf{X}_u \cdot \mathbf{X}_u = 1$, $\mathbf{X}_u \cdot \mathbf{X}_v = 0$, $\mathbf{X}_v \cdot \mathbf{X}_v = G$ with respect to both u and v.   Q.E.D.

Theorem 6.5.6 *If* $\mathbf{X}(u,v)$ *is a geodesic polar parametrization of the surface S, then*

$$K = -\left(\frac{G_u}{2G}\right)_u - \left(\frac{G_u}{2G}\right)^2. \qquad (27)$$

PROOF: Since the three vectors $\mathbf{X}_u, \mathbf{X}_v, \mathbf{n}$ are pairwise orthogonal, they form a basis of $\mathbb{R}^3$. Moreover, because

$$\mathbf{X}_u \cdot \mathbf{X}_u = 1, \qquad \mathbf{X}_v \cdot \mathbf{X}_v = G, \qquad \mathbf{n} \cdot \mathbf{n} = 1,$$

it follows that if $\mathbf{W}$ is any vector, then

$$\mathbf{W} = (\mathbf{W} \cdot \mathbf{X}_u)\mathbf{X}_u + \frac{\mathbf{W} \cdot \mathbf{X}_v}{G}\mathbf{X}_v, + (\mathbf{W} \cdot \mathbf{n})\mathbf{n}.$$

Then

$$\mathbf{X}_{uu} = L\mathbf{n} \qquad\qquad\qquad \text{follows from (24), (25),}$$

$$\mathbf{X}_{uv} = \frac{G_u}{2G}\mathbf{X}_v + M\mathbf{n} \qquad\qquad \text{follows from (25), (26),}$$

$$\mathbf{n}_u = -L\mathbf{X}_u - \frac{M}{G}\mathbf{X}_v \qquad\qquad \text{follows from (4), (5),}$$

$$\mathbf{n}_v = -M\mathbf{X}_u - \frac{N}{G}\mathbf{X}_v \qquad\qquad \text{follows from (5), (6).}$$

Yet another differentiation yields

$$\mathbf{X}_{uuv} = L_v\mathbf{n} + L\mathbf{n}_v = -LM\mathbf{X}_u - \frac{LN}{G}\mathbf{X}_v + L_v\mathbf{n}, \tag{28}$$

$$\mathbf{X}_{uvu} = \left(\frac{G_u}{2G}\right)_u \mathbf{X}_v + \frac{G_u}{2G}\mathbf{X}_{vu} + M_u\mathbf{n} + M\mathbf{n}_u$$

$$= (\cdot)\mathbf{X}_u + \left[\left(\frac{G_u}{2G}\right)_u + \left(\frac{G_u}{2G}\right)^2 + M\left(-\frac{M}{G}\right)\right]\mathbf{X}_v + (\cdot)\mathbf{n}. \tag{29}$$

However, $\mathbf{X}_{uuv} = \mathbf{X}_{uvu}$, and hence it follows from the equality of the coefficients of $\mathbf{X}_v$ in (28) and (29) that

$$-\frac{LN}{G} = \left(\frac{G_u}{2G}\right)_u + \left(\frac{G_u}{2G}\right)^2 - \frac{M^2}{G}.$$

Thus,

$$K = \frac{LN - M^2}{EG - F^2} = \frac{LN - M^2}{G} = -\left(\frac{G_u}{2G}\right)_u - \left(\frac{G_u}{2G}\right)^2. \qquad \text{Q.E.D.}$$

Example 6.5.7 The geodesic polar parametrization $\mathbf{X}(u,v) = (u\cos v, u\sin v, 0)$ of the plane has the first fundamental form $du^2 + u^2\,dv^2$ with $G = u^2$ (see Example 6.5.2). It follows that $G_u = 2u$ and hence

$$-\left(\frac{G_u}{2G}\right)_u - \left(\frac{G_u}{2G}\right)^2 = -\left(\frac{2u}{2u^2}\right)_u - \left(\frac{2u}{2u^2}\right)^2$$

$$= -\left(\frac{1}{u}\right)_u - \left(\frac{1}{u}\right)^2 = \frac{1}{u^2} - \frac{1}{u^2} = 0 = K.$$

Example 6.5.8 It follows from Example 6.5.3 that the geodesic polar parametrization of the sphere of radius R, centered at $P = (0, 0, R)$, has the first fundamental form

$$du^2 + R^2\sin^2\frac{u}{R}\,dv^2.$$

It follows that

$$\frac{G_u}{2G} = \frac{R^2 2\sin\dfrac{u}{R}\cos\dfrac{u}{R}\cdot\dfrac{1}{R}}{2R^2\sin^2\dfrac{u}{R}} = \frac{1}{R}\cot\frac{u}{R}$$

and hence

$$\left(-\frac{G_u}{2G}\right)_u - \left(\frac{G_u}{2G}\right)^2 = -\frac{1}{R}\frac{1}{R}\left(-\csc^2\frac{u}{R}\right) - \frac{1}{R^2}\cot^2\frac{u}{R} = \frac{1}{R^2} = K.$$

Theorem 6.5.6 expresses the Gaussian curvature in terms of the coefficients of a special first fundamental form. It follows that the Gaussian curvature is an intrinsic property of the surface.

Theorem 6.5.9 *The Gaussian curvature is an intrinsic property of surfaces.*   ☐

The following two propositions can be proved by the same method that was used to prove Theorem 6.5.6. The details are relegated to Exercises 1 and 4.

Proposition 6.5.10 *A surface with first fundamental form $E\,du^2 + G\,dv^2$ has Gaussian curvature*

$$-\frac{1}{2\sqrt{EG}}\left(\left(\frac{E_v}{\sqrt{EG}}\right)_v + \left(\frac{G_u}{\sqrt{EG}}\right)_u\right).$$ ☐

Proposition 6.5.11 *The curvature K of a surface with first fundamental form $E\,du^2 + 2F\,du\,dv + G\,dv^2$ satisfies the equation*

$$\begin{aligned}
4(EG - F^2)^2 K &= E\left[E_vG_v - 2F_uG_v + (G_u)^2\right] \\
&\quad + F\left[E_uG_v - E_vG_u - 2E_vF_v + 4F_uF_v - 2F_uG_u\right] \\
&\quad + G\left[E_uG_u - 2E_uF_v + (E_v)^2\right] \\
&\quad - 2(EG - F^2)\left[E_{vv} - 2F_{uv} + G_{uu}\right].
\end{aligned}$$ ☐

A parametrization $\mathbf{X}(u,v)$ of a surface is said to be *u-geodesic* provided it has $E = 1$ and $F = 0$—in other words, its first fundamental form is

$$du^2 + G\,dv^2.$$

It is clear that every geodesic polar parametrization is also *u*-geodesic. The converse, however, is not true. The parametrization $\mathbf{X}(u,v) = (u,v,0)$ has the first fundamental form $du^2 + dv^2$ and is therefore *u*-geodesic. As its *u*-parameter curves are all parallel straight lines, this parametrization is not geodesic polar.

Exercises 6.5

1*. Prove Proposition 6.5.10.

2. Compute the Gaussian curvature of a surface with first fundamental form $du^2/v^2 + dv^2/v^2$.

3. (a) Verify Proposition 6.5.11 for $\mathbf{X}(u,v) = (u,v,uv)$.

(b) Verify Proposition 6.5.11 for $\mathbf{X}(u, v) = (u, v, f(u, v))$.

(c) Verify Proposition 6.5.11 for surfaces of revolution.

4**. Prove Proposition 6.5.11.

5. Show that Proposition 6.5.10 implies Theorem 6.5.6.

6. Suppose u is an arclength parameter of a curve $(x(u), y(u))$ used to generate a surface of revolution. Prove that the resulting parametrization $(x(u), y(u) \cos v, y(u) \sin v)$ is u-geodesic.

7. Prove that the u-parameter curves of a u-geodesic parametrization are indeed geodesics.

8. Use arguments of the same informal nature as the proof of Propositiion 6.5.4 to prove the following.

(a) Given a curve γ and a point P on a surface S such that $P \notin \gamma$, prove that a shortest geodesic of all those that join P to γ is orthogonal to γ.

(b) Let $\gamma(t), a < t < b$, be a curve on the surface S. For each such t, let β_t be the geodesic of S that contains $\gamma(t)$ and is orthogonal to γ. For each real number s, let $\beta_t(s)$ be the point of β_t such that the length of β_t from $\gamma(t) = \beta_t(0)$ to $\beta_t(s)$ is s (we are assuming that this is done so that $\beta_t(s)$ is a continuous function of t). Prove that the parametrization $\mathbf{X}(s, t) = \beta_t(s)$ is s-geodesic.

6.6   Polyhedral Surfaces I

A *polyhedral surface* is the union of a finite connected system of plane polygons of which every two intersect, if at all, in either an edge or a vertex. Moreover, the intersection of three distinct polygons is either null or a vertex. The surfaces of cubes, boxes, pyramids, and prisms are all polyhedral. Figures 6.2–6.5, while meant to represent smooth surfaces, are in fact polyhedral surfaces each of whose polygons is a quadrilateral.

Surprisingly, it is possible to define an analog of the Gaussian curvature for the polyhedral surfaces in a natural manner. This easily computed variant will be demonstrated to also be intrinsic, thus providing us with an informal infinitesimal-based supporting argument for the rather difficult, and surprising, Theorem 6.5.9 (that K is intrinsic).

We begin by replacing the notion of a tangent plane with that of a *supporting plane*. Every plane in $\mathbb{R}^3$ separates it into two sides, and it is understood that the plane is actually contained in each of the sides. A plane is said to support the subset S of $\mathbb{R}^3$ if they intersect, but S is completely contained in one side of the plane. A polyhedral surface is said to be *convex* if it has a supporting plane at each of its points.

If P is any one of the points in the interior of a polygonal face of the convex polyhedral surface S, then P is contained in only one supporting plane. Moreover, all the points in the interior of the same face of S are contained in the same supporting plane. In other words, the unit normal $\mathbf{n}_P$ (pointing away from S) is well defined at each interior point P and is constant over the interior of a face. As was the case for smooth surfaces, this normal is anchored at the origin.

If P is a point on an edge e of a convex polyhedron, the supporting plane is no longer necessarily unique and can be envisioned as rotating, about e as axis, from one of the faces abutting on e to the other. The unit normal to this rotating plane then describes an arc $g(e)$ on the unit sphere. Moreover, since $g(e)$ has radius 1, it must be an arc of a great circle, i.e., a geodesic of this sphere.

These arcs require a closer look. Suppose the vertex v of the polyhedral surface S is surrounded by the polygons $\Pi_1, \ldots, \Pi_{i-1}, \Pi_i, \Pi_{i+1}, \ldots, \Pi_d$ (see Fig. 6.13), and that for each $i = 1, 2, \ldots, d$, the faces Π_i and Π_{i+1} (indices modulo d) share the edge vu_i. Then, while the supporting plane along vu_i rotates from the plane of Π_i to that of Π_{i+1}, its unit normal begins as $\mathbf{n}_i$ and ends up as $\mathbf{n}_{i+1}$; in the process, this unit normal has traced the arc, or rather the spherical geodesic, $a_i = C_i C_{i+1}$. Moreover, if the supporting plane is then successively rotated onto the planes of $\Pi_{i+2}, \Pi_{i+3}, \ldots, \Pi_d, \Pi_1, \ldots, \Pi_{i-1}$, using the successive edges $vu_{i+1}, vu_{i+2}, \ldots, vu_{i-1}$ as axes, then the unit normal will trace out the spherical polygon $\wp_v = C_1 C_2 \cdots C_d$. The convexity of the polyhedral surface in the vicinity of v guarantees the *convexity* of $\wp_v$ on the unit sphere in the sense that $\wp_v$ lies in its entirety in one of the sides of each of the great circles that contain the sides of $\wp_v$. Equivalently, a spherical polygon is convex if each of its angles is less than π.

Finally, suppose a supporting plane is placed at v and is allowed to wobble freely as long as it actually remains supporting, that is, as long as it does not penetrate into the surface's interior. The terminal point of the unit vector then roams freely over the interior of the previous geodesic polygon $\wp_v$.

In view of these considerations it is reasonable to define the *Gaussian curvature* K_P of a point P on a convex polyhedral surface to be *zero* unless that point is a vertex, in which case

$$K_P = \text{area}(\wp_P).$$

This brings us to the issue of evaluating the areas of spherical (geodesic) polygons. Surprisingly, this question admits of an answer that is much simpler than that of its planar analog.

We begin by examining two-sided polygons on the sphere. Since any two great circles on a sphere must intersect in diametrically opposite points, it follows that the two vertices of such a 2-gon must be antipodal. The angles at the 2-gon's vertices are therefore equal, and hence, if this common angle is α, it is possible to refer to this 2-gon unambiguously as lune(α). Because any two lunes of the same angle on the sphere of radius R are necessarily congruent, it follows that the area of a lune is proportional to its angle. The constant of proportionality is determined by observing that

$$\text{area}(\text{lune}(2\pi)) = (\text{surface area of sphere of radius } R) = 4\pi R^2.$$

It follows that

$$\text{area}(\text{lune}(\alpha)) = 2\alpha R^2,$$

where the angle α is measured in radians.

The formula for the area of a spherical triangle was first derived by Albert Girard (1595–1632). The wonderful proof presented here is due to Euler.

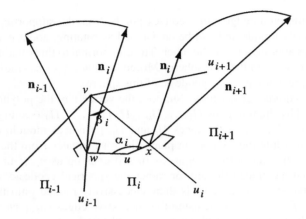

Polyhedral surface

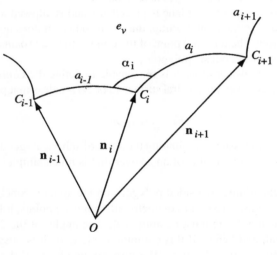

Unit sphere

Figure 6.13   Unit normals near a vertex.

Proposition 6.6.1 *On a sphere of radius R, the area of a geodesic triangle with angles* α, β, γ *is*

$$(\alpha + \beta + \gamma - \pi)R^2.$$

PROOF: Let the spherical triangle in question have vertices A, B, C, with respective antipodes A', B', C' (Fig. 6.14). Draw the great circles that underlie the geodesics AB,

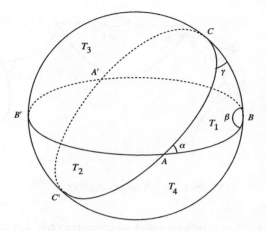

Figure 6.14   A spherical triangle.

BC, CA. These circles divide the surface of the sphere into eight geodesic triangles, of which $\triangle ABC$, $\triangle AB'C'$, $\triangle AB'C$, $\triangle ABC'$ have areas T_1, T_2, T_3, T_4 respectively. The areas of the lunes of α, β, and γ are denoted by L_α, L_β, and L_γ respectively.

As $\triangle A'BC$ and $\triangle AB'C'$ are antipodal, they have the same areas. Hence,

$$T_1 + T_2 = L_\alpha.$$

Also,

$$T_1 + T_3 = L_\beta$$

and

$$T_1 + T_4 = L_\gamma.$$

It follows that

$$
\begin{aligned}
2T_1 &= L_\alpha + L_\beta + L_\gamma - (T_1 + T_2 + T_3 + T_4) \\
&= 2\alpha R^2 + 2\beta R^2 + 2\gamma R^2 - 2\pi R^2,
\end{aligned}
$$

and the desired formula now follows immediately.   Q.E.D.

This proposition is now easily extended to arbitrary convex spherical polygons.

Theorem 6.6.2 *The area of a convex spherical geodesic polygon with consecutive angles* $\alpha_1, \alpha_2, \ldots, \alpha_n$ $(n \geq 2)$ *is*

$$\left(\sum_{i=1}^{n} \alpha_i - (n-2)\pi \right) R^2.$$

PROOF: By induction on n. Note that the theorem has already been established for $n = 2, 3$. Fix n, and assume that the theorem is valid for all convex spherical polygons

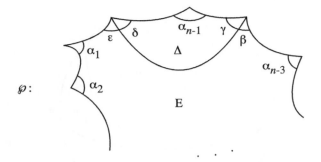

Figure 6.15   A spherical polygon.

with less than n sides. Let $\wp$ be an n-sided spherical polygon. Let Δ be the spherical triangle formed by joining the vertices of the angles α_{n-2} and α_n (Fig. 6.15). Since the polygon is assumed to be convex, this diagonal passes through the interior of the polygon. The remainder of the polygon $\wp$ is an $(n-1)$-gon denoted by E. It now follows from the induction hypothesis that

$$\text{area}(\wp) = \text{area}(\Delta) + \text{area}(E)$$

$$= (\alpha_{n-1} + \gamma + \delta - \pi)R^2 + \left(\sum_{i=1}^{n-3} \alpha_i + \beta + \varepsilon - (n-3)\pi \right) R^2$$

$$= \left(\sum_{i=1}^{n-3} \alpha_i + \beta + \gamma + \alpha_{n-1} + \delta + \varepsilon - (n-2)\pi \right) R^2$$

$$= \left(\sum_{i=1}^{n} \alpha_i - (n-2)\pi \right) R^2. \qquad\qquad \text{Q.E.D.}$$

Let us now return to Figure 6.13. On the unit sphere the angle α_i is the angle between the planes $OC_{i-1}C_i$ and OC_iC_{i+1}. Pictured on the polyhedral surface, this is the dihedral angle, at u, formed by the planes perpendicular to vu_{i-1} and vu_i at w and x respectively. Note that the plane quadrilateral $vwux$ has right angles at w and x. Since the sum of its angles is 2π it follows that

$$\alpha_i = \pi - \beta_i, \qquad i = 1, 2, \ldots, n,$$

and consequently,

$$K_v = \text{area}(\wp_v) = \sum_{i=1}^{n} \alpha_i - (n-2)\pi$$

$$= \sum_{i=1}^{n}(\pi - \beta_i) - (n-2)\pi = 2\pi - \sum_{i=1}^{n} \beta_i. \qquad (30)$$

The reader is reminded that our discussion so far has been restricted to convex polyhedra. The reason for this is that in the nonconvex case the behavior of the point $g(P)$ on the unit sphere is much more complicated and is in fact not well understood. In particular, the analog of $\wp_v$ fails to be a convex polygon and may intersect itself many times. Nevertheless, Equation (30) constitutes sufficient motivation for us to extend the notion of Gaussian curvature to nonconvex vertices as well. Thus, the definition of the *Gaussian curvature* at any point P on any polyhedral surface S is

$$K_P = 2\pi - \sum_{i=1}^{n} \beta_i \quad \text{or} \quad 0$$

according as P is a vertex or not, where $\beta_1, \beta_2, \ldots, \beta_n$ are the angles that surround the vertex P.

The polyhedral analog of the bending of smooth surfaces is the *flexing*, in which, while the shapes of the individual faces remain unchanged, they are allowed a limited amount of rotation about their edges. Of course, a polyhedral surface in which each vertex is surrounded by three polygons allows no such flexing. However, vertices with four or more faces do, in general, allow for flexing. Moreover, since flexing leaves all angles unchanged, it follows that flexing does not change the Gaussian curvature at any any point on the polyhedral surface. Thus, the polyhedral analog of Theorem 6.5.9 is immediate.

Exercise 6.6

1. Show that the number s equals the sum of the angles of a spherical triangle if and only if $\pi < s < 3\pi$.

6.7   Gauss's Total Curvature Theorem

We now go on to state and prove this chapter's main theorem. First, however, two technical lemmas are needed.

It will prove convenient to reformulate Equation (27) for the Gaussian curvature and Gauss's Equation (18) in terms of $\sqrt{G}$.

Lemma 6.7.1 *If a surface has a geodesic polar parametrization, then*

$$K = -\frac{1}{\sqrt{G}} \frac{\partial^2 \sqrt{G}}{\partial u^2},$$

and Gauss's Equation (18) *can be written as*

$$\theta' = -\frac{\partial \sqrt{G}}{\partial u} v'.$$

PROOF: See Exercise 8. $\qquad\qquad\qquad\qquad\qquad\qquad\qquad\qquad\qquad\qquad\qquad$ □

The proof of the second lemma is also incomplete. However, the shortcuts it employs were used by Gauss himself and are quite plausible. (See Exercise 7.)

Lemma 6.7.2 *If a surface has a geodesic polar parametrization, then*

$$G = 0 \quad and \quad \frac{\partial \sqrt{G}}{\partial u} = 1 \qquad at \ u = 0.$$

PROOF: Note that for all v, $\mathbf{X}(0, v) = P$, where P is the center of the polar parametrization. It follows that $\mathbf{X}_v(0, v) = \mathbf{0}$ and hence $G(0, v) = 0$.

Figure 6.16 portrays two geodesic rays, parametrized respectively by $u = t$, $v = v_1$ and $u = t$, $v = v_2$, intersected by the geodesic circle γ parametrized as $u = h$, $v = t$. The length of arc AB is

$$\text{arc}(AB) = \int_{v_1}^{v_2} \sqrt{u'^2 + Gv'^2}\, dt = \int_{v_1}^{v_2} \sqrt{G}\, dv = \sqrt{G(h, \tilde{v})} \cdot \alpha$$

for some $\tilde{v}, v_1 \leq \tilde{v} \leq v_2$. Hence

$$\lim_{v_2 \to v_1} \frac{\text{arc}(AB)}{\alpha} = \sqrt{G(h, v_1)}.$$

On the other hand, at $u = 0$, because $G(0, v_1) = 0$,

$$\frac{\partial \sqrt{G}}{\partial u}(0, v_1) = \lim_{h \to 0} \frac{\sqrt{G(h, v_1)} - \sqrt{G(0, v_1)}}{h} = \lim_{h \to 0} \frac{\sqrt{G(h, v_1)}}{h}$$

$$= \lim_{h \to 0} \lim_{v_2 \to v_1} \frac{\text{arc}(AB)}{h\alpha},$$

which double limit Gauss assumed to be 1. Since v_1 is arbitrary, the proof of the lemma is complete.

Gauss's assumption is tantamount to saying that the lengths of small arcs of small geodesic circles behave much like the arcs of Euclidean circles.   Q.E.D.

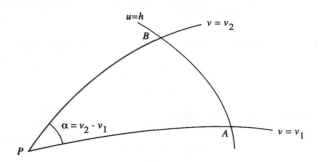

Figure 6.16   A small geodesic angle.

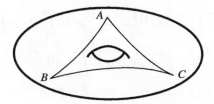

Figure 6.17   This is not a geodesic triangle.

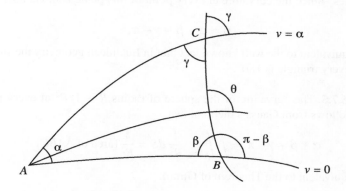

Figure 6.18   A geodesic triangle.

A *geodesic triangle* is a topological disk on a surface whose border consists of three open geodesic arcs. The requirement that the triangle be a disk is of the essence, and the union of the arcs AB, BC, AC on the torus in Figure 6.17 does not constitute the border of any triangle.

Theorem 6.7.3 (Gauss) *If $\triangle ABC$ is a geodesic triangle with respective interior angles α, β, γ on a surface S, then its total curvature is*

$$\int\int_{\triangle ABC} K\,dS = \alpha + \beta + \gamma - \pi.$$

PROOF: Suppose the surface S has the geodesic polar parametrization $\mathbf{X}(u,v)$ centered at A such that the geodesics AB and AC are the u-parameter curves $v = 0$ and $v = \alpha$ respectively (Fig. 6.18). Suppose further that the geodesic BC is parametrized by $u = f(v), v = v$. Then

$$\int\int_{\triangle ABC} K\,dS = \int_0^\alpha \int_0^{f(v)} K|\mathbf{X}_u \times \mathbf{X}_v|\,du\,dv = \int_0^\alpha \int_0^{f(v)} K\sqrt{EG - F^2}\,du\,dv$$

$$= \int_0^\alpha \int_0^{f(v)} -\frac{1}{\sqrt{G}}\frac{\partial^2 \sqrt{G}}{\partial u^2}\sqrt{G}\,du\,dv = \int_0^\alpha \left[-\frac{\partial \sqrt{G}}{\partial u}\right]_0^{f(v)}\,dv$$

$$= \int_0^\alpha \left[1 - \frac{\partial \sqrt{G(f(v), v)}}{\partial u} \right] dv = \int_0^\alpha dv + \int_0^\alpha - \frac{\partial \sqrt{G}}{\partial u} dv$$

$$= \int_0^\alpha dv + \int_0^\alpha \theta' dv = \alpha + \int_{\pi - \beta}^\gamma d\theta$$

$$= \alpha + (\gamma - (\pi - \beta)) = \alpha + \beta + \gamma - \pi. \qquad\qquad \text{Q.E.D.}$$

Example 6.7.4 In the Euclidean plane a geodesic triangle is one whose sides are straight lines. Since the curvature at every point of this plane is 0, Gauss's theorem reduces to

$$0 = \alpha + \beta + \gamma - \pi,$$

which is equivalent to the well known fact that in Euclidean geometry the sum of the angles of every triangle is $180°$.

Example 6.7.5 The curvature of the sphere of radius R is $1/R^2$ at every point. It therefore follows from Gauss's theorem that

$$\alpha + \beta + \gamma - \pi = \int\int_{\Delta ABC} \frac{1}{R^2} \, dS = \frac{1}{R^2} (\text{area of } \Delta ABC) \qquad (31)$$

which is equivalent to the Theorem of Girard.

Example 6.7.6 It follows from Example 6.2.10 that in the pseudosphere with $c = 1$, the Gaussian curvature is -1 at every point. It follows from Gauss's theorem that for the geodesic ΔABC

$$\alpha + \beta + \gamma - \pi = \int\int_{\Delta ABC} (-1) \, dS = -(\text{area of } \Delta ABC)$$

and hence the area of such a triangle on the pseudosphere is

$$\pi - \alpha - \beta - \gamma. \qquad (32)$$

Note that this is the area that would be obtained from Equation (31) if we pretended that the pseudosphere is a sphere of imaginary radius i. Unlikely as it may sound, this quasimathematical observation did play a role in the evolution of non-Euclidean geometry.

A *geodesic polygon* is a topological disk on a surface whose border consists of geodesic arcs. We assume the plausible, though not easily proven, fact that every such geodesic polygon, with at least four sides, can be decomposed into two geodesic polygons with fewer sides by means of a suitably chosen diagonal.

Corollary 6.7.7 *Let Π denote a geodesic polygon with interior angles $\alpha_1, \alpha_2, \ldots, \alpha_n$ on a surface S. Then*

$$\int\int_\Pi K \, dS = \sum_{i=1}^n \alpha_i - (n-2)\pi.$$

PROOF: By induction on n. For $n = 3$, this is Gauss's Theorem. The case $n = 2$ can be reduced to that of $n = 3$ by cutting the digon into two triangles by means of a geodesic. The induction step is nearly identical with the proof of Theorem 6.6.2 and is relegated to Exercise 1.   Q.E.D.

Theorem 6.7.8 *If S is a closed orientable surface, then*

$$\int\int_S K \, dS = 2\pi\chi(S).$$

PROOF: Cut S into the geodesic polygons $\Pi_1, \Pi_2, \ldots, \Pi_r$, and suppose each Π_i has n_i sides. These polygons constitute a presentation of S with, say, p nodes and q arcs. Then, if $\sum_i \alpha$ denotes the sum of the interior angles of Π_i, we have

$$\int\int_S K \, dS = \sum_{i=1}^{r} \left[\sum_i \alpha - (n_i - 2)\pi \right]$$

$$= \sum_{i=1}^{r} \sum_i \alpha - \pi \sum_{i=1}^{r} n_i + \sum_{i=1}^{r} 2\pi = 2\pi p - \pi 2q + 2\pi r = 2\pi\chi(S). \quad \text{Q.E.D.}$$

Since the sphere has characteristic 2, the Gauss–Bonnet Theorem implies that it, as well as any of its homeomorphs, has total curvature 4π. This fact, of course, follows immdiately from the definition of the total curvature when the surface has a reasonable convex shape. It is much less obvious if the surface is highly convoluted. The torus, having characteristic 0, has, according to the Gauss–Bonnet Theorem, total curvature also 0. This is much less easily verified directly (see Exercise 6.2.7).

The reader might object to the cutting procedure used to prove Theorem 6.7.8 on the grounds that it is not obvious that geodesics, which are so hard to locate with precision, can be used to create a polygonal presentation. This concern can be easily allayed. First cut the surface along any curves. Then replace each such curve with a succession of small geodesic arcs that approximate the original curves in much the same way that plane polygons can be used to approximate plane curves.

Exercises 6.7

1. Complete the proof of Corollary 6.7.7.

2**. Prove that every plane polygon has an internal diagonal.

3. Prove that on a surface of negative curvature two geodesics cannot enclose a topological disk.

4. Prove that the total curvature of the portion of a surface of revolution that is bounded by two stationary latitudes and two u-parameter curves is 0.

A geodesic is said to be closed *provided it is a (smooth) loop.*

5. Prove that the total curvature of a disk that is bounded by a closed geodesic is 2π. Conclude that on a surface of negative curvature a closed geodesic cannot bound a disk.

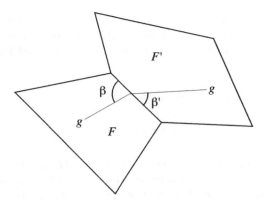

Figure 6.19   The incidence angles of a geodesic.

6.* Prove that the surface of revolution $\mathbf{X}(u,v) = (\sinh u, \cosh u \cos v, \cosh u \sin v)$ has exactly one closed geodesic.

7. Prove Lemma 6.7.1.

8.* Criticize Gauss's proof of Lemma 6.7.2.

6.8 Polyhedral Surfaces II

Just like smooth surfaces, polyhedral surfaces also have *strict geodesics*, which are defined by the minimality of their lengths, as well as mere *geodesics*. It is the purpose of this section to state and prove polyhedral analogs of Gauss's Theorem 6.7.3 and Corollary 6.7.7. We begin with a close examination of the polyhedral geodesics.

Let a polyhedral surface S be given, and let g be a geodesic on S. The portion of a geodesic that is restricted to a single polygonal face must clearly be a straight line. However, when we try to extend such a straight line into neighboring faces, things get more complicated. Suppose first that the geodesic g joins two points in two faces F and F' of S that abut along the common edge e (Fig. 6.19). The angles from e to g (β and β') are the geodesic's *angles of incidence*.

Lemma 6.8.1 *On a polyhedral surface the two angles of incidence that a geodesic makes with any edge are equal.*

PROOF: Observe that when face F' of Figure 6.19 is rotated about e as a hinge, g remains a geodesic and the angles of incidence are unchanged. This is because such a rotation changes no distances on $F \cup F'$. Suppose F' is rotated into the plane of F. Since g is a geodesic, it is now a straight line, and consequently the angles of incidence are now vertically opposite angles and therefore equal. It follows that the angles of incidence must have been equal to begin with.   Q.E.D.

We next examine the passage of a geodesic g through a vertex v of a polyhedral surface. Suppose first that v has positive Gaussian curvature so that the sum of the facial angles at v is less than 2π. It is then possible to cut the surface S near v along one of the two rays of g determined by v and flatten it out, without distortion, so that it spreads out as indicated in Figure 6.20. This, however, contradicts the minimality property of g, as the distance along the dotted line is clearly shorter than the length of the corresponding (broken) portion of g. It follows that no geodesic can pass through a vertex with positive Gaussian curvature.

Finally, at a vertex with negative curvature, a geodesic entering a vertex can be extended in at least two ways (Exercise 1).

For these reasons it will henceforth be assumed that a geodesic polygon on a polyhedral surface contains none of the surface's vertices. For such polygons a straightforward analog of Gauss's Theorem 6.7.3 and Corollary 6.7.7 holds. A vertex of a polyhedral surface is said to be *internal* to a topological disk on the surface provided it is in the disk but not on its border.

Theorem 6.8.2 *Let Π denote a geodesic n-gon with interior angles $\alpha_1, \alpha_2, \ldots, \alpha_n$ and internal vertices $v_1, v_2, \ldots, v_d$ on a polyhedral surface S. Then the total curvature of Π is*

$$\sum_{i=1}^{d} K_{v_i} = \sum_{i=1}^{n} \alpha_i - (n-2)\pi.$$

PROOF: The given n-gon Π encloses angles of three types (some of these are displayed in Figure 6.21):

1. interior angles $\alpha_1, \alpha_2, \ldots, \alpha_n$ whose vertices are also the vertices of Π;

2. incidence angles $\beta_1, \beta_2, \ldots, \beta_{2m}$ whose vertices are formed by the intersection of the perimeter of Π with the edges of the faces of S;

3. internal angles $\gamma_1, \gamma_2, \ldots, \gamma_w$ whose vertices lie in the interior of Π.

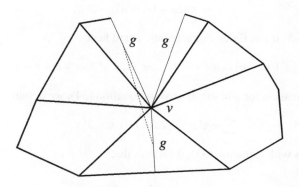

Figure 6.20   A flattening of the surface near V.

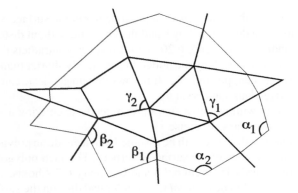

Figure 6.21   A geodesic polygon on a polyhedral surface.

Let $\sum$ angles denote the sum of all the angles of all three types. Then

$$
\begin{aligned}
\sum \text{angles} &= \sum_{i=1}^{n} \alpha_i + \sum_{i=1}^{2m} \beta_i + \sum_{i=1}^{w} \gamma_i \\
&= \sum_{i=1}^{n} \alpha_i + m\pi + \sum_{i=1}^{d} (2\pi - K_{v_i}) \\
&= \sum_{i=1}^{n} \alpha_i + m\pi + 2d\pi - \sum_{i=1}^{d} K_{v_i}.
\end{aligned}
\tag{33}
$$

On the other hand, if there are r polygons interior to Π and they have $n_1, n_2, \ldots, n_r$ sides respectively, and if q denotes the number of polygon sides that lie either on the perimeter or in the interior of Π, then

$$
\begin{aligned}
\sum \text{angles} &= \sum_{i=1}^{r} (n_i - 2)\pi = \pi \sum_{i=1}^{r} n_i - 2r\pi \\
&= \pi(2q - (m+n)) - 2r\pi.
\end{aligned}
\tag{34}
$$

Since Π is a disk, it has Euler characteristic 1 and hence

$$
(d + m + n) - q + r = 1, \quad \text{or} \quad q = d + m + n + r - 1.
$$

When this expression for q is substituted in Equation (34), we obtain

$$
\sum \text{angles} = (2d + m + n - 2)\pi.
$$

Comparing this with Equation (33), it follows that

$$
(n - 2)\pi = \sum_{i=1}^{n} \alpha_i - \sum_{i=1}^{d} K_{v_i},
$$

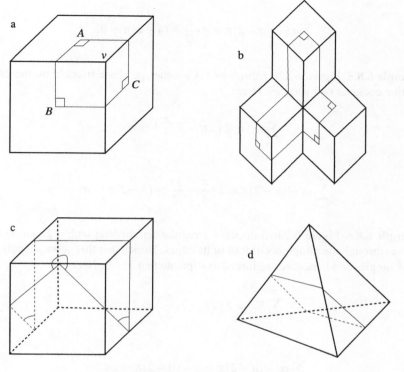

Figure 6.22 Some geodesic polygons.

which is equivalent to the desired equation. Q.E.D.

Example 6.8.3 Figure 6.22(a) displays a geodesic $\triangle ABC$ that surrounds the single vertex v on a cube. Here

$$\sum_{i=1}^{d} K_{v_i} = 2\pi - 3 \cdot \frac{\pi}{2} = \frac{\pi}{2}$$

and

$$\sum_{i=1}^{n} \alpha_i - (n-2)\pi = 3 \cdot \frac{\pi}{2} - (3-2)\pi = \frac{\pi}{2}.$$

Example 6.8.4 Figure 6.22(b) consists of four stacked cubes of which one is not visible. The geodesic polygon is in fact a geodesic quadrilateral each of whose angles is a right angle. It surrounds three of the vertices of the stacked cubes. Here

$$\sum_{i=1}^{3} K_{v_i} = \left(2\pi - 6 \cdot \frac{\pi}{2}\right) + 2\left(2\pi - 3 \cdot \frac{\pi}{2}\right) = -\pi + \pi = 0$$

and

$$\sum_{i=1}^{n} \alpha_i - (n-2)\pi = 4 \cdot \frac{\pi}{2} - (4-2)\pi = 0.$$

Example 6.8.5 Figure 6.22(c) displays yet another geodesic triangle on the cube, one that encloses two vertices. Here

$$\sum_{i=1}^{d} K_{v_i} = 2\left(2\pi - 3 \cdot \frac{\pi}{2}\right) = \pi$$

and

$$\sum_{i=1}^{n} \alpha_i - (n-2)\pi = 2 \cdot \frac{\pi}{4} + \frac{3\pi}{2} - (3-2)\pi = \pi.$$

Example 6.8.6 Figure 6.22(d) displays a regular tetrahedron with a geodesic that courses through the midpoints of four of its edges. Notice that this geodesic polygon has no angles and hence we are forced to stipulate that $n = 0$. Here

$$\sum_{i=1}^{2} K_{v_i} = 2\left(2\pi - 3 \cdot \frac{\pi}{3}\right) = 2\pi$$

and

$$\sum_{i=1}^{n} \alpha_i - (n-2)\pi = 0 - (0-2)\pi = 2\pi.$$

There is a polyhedral analog of the Gauss–Bonnet Theorem that was first stated and proved by George Polya (1887–1985). For the surfaces of convex polyhedra this had already been recognized by René Descartes (1596–1650).

Theorem 6.8.7 *The total Gaussian curvature of a polyhedral surface S is $2\pi\chi(S)$.*

PROOF: See Exercise 2.   □

Exercises 6.8

1. Prove that on a polyhedral surface, a geodesic entering a vertex with negative curvature can be extended in at least two ways.

2. Prove Theorem 6.8.7.

   Exercises 3–4 refer to the surface of the cube with vertices (a,b,c) where $a,b,c \in \{0,1\}$.

3. For any x, y between 0 and 1 find the distance between $(0,0,0)$ and $(x,y,1)$.

4. For any x, y between 0 and 1 find the distance between $(0.5, 0.5, 0)$ and $(x,y,1)$.

Chapter Review Exercises

1. Are the following statements true ot false? Justify your answer.

   (a) The normal to a surface at a point is unique.

   (b) The parametrization $\mathbf{X}(u,v) = (u\cos v, u\sin v, 1-u)$, $u,v \in \mathbb{R}$, defines a surface.

   (c) Let $f(x,y)$ be a three times differentiable function on $\mathbb{R}^2$. Then $(u+v, u-v, f(u,v))$ is the parametrization of a surface.

   (d) The Gaussian curvature of every surface of revolution is constant.

   (e) There is a surface whose Gaussian curvature is negative everywhere.

   (f) The Gaussian curvature of the torus is negative for some points, zero for others, and positive for others.

   (g) There is a surface whose first fundamental form is $u^2\,du^2 + uv\,du\,dv + v^2\,dv^2$.

   (h) Every strict geodesic is a geodesic.

   (i) For every real number r there exists a surface whose Gaussian curvature at some point equals r.

   (j) For any surface S, $\{\kappa_P \mid P \in S\}$ is a (possibly infinite) interval.

   (k) For any point P on a surface S, $\{\kappa_\mathbf{d} \mid \mathbf{d}$ is a direction at $\mathbf{P}\}$ is a closed interval.

   (l) A geodesic polar parametrization is necessarily orthogonal.

   (m) The parameter curves of an orthogonal parametrization are geodesics.

   (n) The principal curvatures at a point on a surface are an intrinsic property.

   (o) The sum of the principal curvatures at a point on a surface is an intrinsic property.

   (p) The product of the principal curvatures at a point on a surface is an intrinsic property.

   (q) Every polyhedral surface is a surface.

   (r) The Euler characteristic of a surface is an intrinsic property.

CHAPTER 7

RIEMANN GEOMETRIES

Many of the important theorems and concepts of Chapter 6 turned out to be intrinsic. This means that if a surface is viewed as a flexible but inextensible membrane, something like a piece of cloth, then these properties are independent of the particular shape the surface assumes and only express the interplay of the lengths of curves on the surface. Some time around the middle of the nineteenth century, Riemann came to the realization, already implicit in Gauss's work, that if these properties are independent of the surface then they could be studied without reference to the surface. This transcendence was accomplished by replacing the notion of the length of a curve on a surface with an application of the first fundamental form directly to those curves in the parameter space that underlie the curve on the surface.

Informally, a *Riemann metric* is a first fundamental form without an accompanying surface. Formally, a Riemann metric $\mathcal{M}$ is an ordered triple

$$\mathcal{M} = (E, F, G) = (E(u, v), F(u, v), G(u, v))$$

of real-valued, three times differentiable functions of some domain $D \subset \mathbb{R}^2$ such that

$$E, G, EG - F^2 > 0.$$

Introduction to Topology and Geometry, Second Edition.
By Saul Stahl and Catherine Stenson Copyright © 2013 John Wiley & Sons, Inc.

The pair $(\mathcal{M}, D)$ constitutes a *Riemann geometry*. The metric is now used to define lengths, angles, areas, and isometries in D. In this exposition the domain D will frequently be left implicit.

Given a parametrized curve

$$\gamma(t) = (u(t), v(t)), \qquad a \le t \le b, \tag{1}$$

in the domain D of a Riemann metric $\mathcal{M} = (E, F, G)$, the *length* of γ relative to $\mathcal{M}$ is

$$L_{\mathcal{M}}(\gamma) = \int_a^b \sqrt{Eu'^2 + 2Fu'v' + Gv'^2}\, dt. \tag{2}$$

It is customary to allow the term *Riemann metric* to also refer to the expressions

$$Eu'^2 + 2Fu'v' + Gv'^2 \tag{3}$$

and

$$ds^2 = E\, du^2 + 2F\, du\, dv + G\, dv^2. \tag{4}$$

Example 7.1 The constant metric $\mathcal{E} = (1, 0, 1)$ is known as the *Euclidean metric*

$$\mathcal{E} = u'^2 + v'^2.$$

The length of the curve γ of (2) relative to $\mathcal{E}$ is

$$L_{\mathcal{E}}(\gamma) = \int_a^b \sqrt{u'^2 + v'^2}\, dt$$

which is, of course, its Euclidean length. The letter $\mathbb{E}$ will denote the *Euclidean geometry* $(\mathcal{E}, \mathbb{R}^2)$.

Example 7.2 The *half-plane metric* is $\mathcal{H} = (1/v^2, 0, 1/v^2)$ or

$$\mathcal{H} = \frac{du^2 + dv^2}{v^2} \tag{5}$$

at every point (u, v) of the upper half-plane $v > 0$. The length of the curve γ of (2) is

$$L_{\mathcal{H}}(\gamma) = \int_a^b \frac{\sqrt{u'^2 + v'^2}}{v}\, dt$$

For instance, if γ_1 is the vertical line segment joining (c, a) to (c, b) (Fig. 7.1), then it has the parametrization $u = c$ and $v = t$, $a \le t \le b$. Its length relative to the half-plane metric $\mathcal{H}$ is

$$L_{\mathcal{H}}(\gamma_1) = \int_a^b \frac{\sqrt{0^2 + 1^2}}{t}\, dt = \ln \frac{b}{a}. \tag{6}$$

On the other hand, if γ_2 is the circular arc $u = d + r\cos t$, $v = r\sin t$, $\alpha \le t \le \beta$, then its length relative to the half-plane metric is

$$L_{\mathcal{H}}(\gamma_2) = \int_\alpha^\beta \frac{\sqrt{r^2 \sin^2 t + r^2 \cos^2 t}}{r\sin t} = \int_\alpha^\beta \csc t\, dt$$

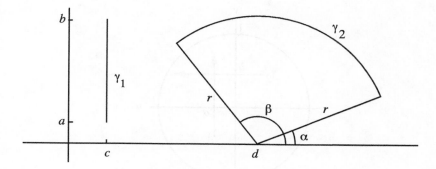

Figure 7.1   Arcs in the upper half-plane.

$$= \ln \frac{\csc \beta - \cot \beta}{\csc \alpha - \cot \alpha}. \tag{7}$$

The Riemann geometry that the metric $\mathscr{H}$ defines on the upper half-plane is called the *half-plane geometry* and is denoted by $\mathbb{H}$.

Example 7.3 The *Poincaré metric*

$$\mathscr{P} = \frac{4(du^2 + dv^2)}{(1 - u^2 - v^2)^2} \tag{8}$$

and the *Beltrami metric*

$$\mathscr{B} = \frac{(1 - v^2)\,du^2 + 2uv\,du\,dv + (1 - u^2)\,dv^2}{(1 - u^2 - v^2)^2} \tag{9}$$

have the interior of the unit disk as their common domain. If γ is the line segment joining $(u,v) = (t\cos\theta, t\sin\theta)$, $0 \le t \le r < 1$, then its length relative to these two metrics is, respectively,

$$L_{\mathscr{P}}(\gamma) = \int_0^r \frac{2\sqrt{\cos^2\theta + \sin^2\theta}}{1 - t^2\cos^2\theta - t^2\sin^2\theta}\,dt = \int_0^r \frac{2}{1 - t^2}\,dt = \ln\frac{1+r}{1-r},$$

and, with the details relegated to Exercise 1,

$$L_{\mathscr{B}}(\gamma) = \frac{1}{2}\ln\frac{1+r}{1-r}.$$

The corresponding Riemann geometries are denoted by $\mathbb{P}$ and $\mathbb{B}$.

In a Riemann geometry $(\mathscr{M}, D)$, the *distance* between any two points is the length of the shortest curve, relative to $\mathscr{M}$, that joins those two points and lies completely in D. As was the case for surfaces (Section 6.3), it is assumed that this shortest distance is realized by some such curve. This notion of distance turns every Riemann geometry into a metric space in the sense of Chapter 10.

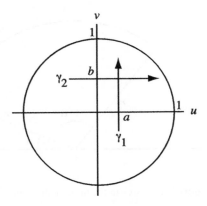

Figure 7.2   Oriented segments inside the unit circle.

For the definition of the measures of angles in a Riemann geometry, we once again take our cue from the first fundamental form as a surface. Suppose

$$u = u(t), \quad v = v(t), \qquad a \leq t \leq b,$$

and

$$u = \tilde{u}(\tau), \quad v = \tilde{v}(\tau), \qquad c \leq \tau \leq d,$$

are two curves in D that intersect at the point

$$(u(t_0), v(t_0)) = (\tilde{u}(\tau_0), \tilde{v}(\tau_0)).$$

Then, if the derivatives below are taken relative to t and τ respectively (and evaluated at t_0 and τ_0), the measure (with respect to the metric of (2)–(4)) of the angle determined by these two curves is

$$\cos^{-1}\left(\frac{E u' \tilde{u}' + F(u' \tilde{v}' + \tilde{u}' v') + G v' \tilde{v}'}{\sqrt{E u'^2 + 2F u' v' + G v'^2}\sqrt{E \tilde{u}'^2 + 2F \tilde{u}' \tilde{v}' + G \tilde{v}'^2}} \right). \qquad (10)$$

Example 7.4 Let γ_1 and γ_2 be the oriented line segments of Figure 7.2. Then,

$$\text{for } \gamma_1: \quad u = a, \quad v = t, \quad u' = 0, \quad v' = 1;$$
$$\text{for } \gamma_2: \quad \tilde{u} = \tau, \quad \tilde{v} = b, \quad \tilde{u}' = 1, \quad \tilde{v}' = 0.$$

At the point of intersection (a, b), with $c = (1 - a^2 - b^2)^{-2}$, we have

$$\mathscr{E} = du^2 + dv^2,$$
$$\mathscr{P} = 4c\,du^2 + 4c\,dv^2,$$

and

$$\mathscr{B} = (1 - b^2)c\,du^2 + abc\,du\,dv + (1 - a^2)c\,dv^2.$$

If $\theta_{\mathcal{M}}$ denotes the angle between these two oriented arcs relative to the metric $\mathcal{M}$, then

$$\theta_{\mathcal{E}} = \cos^{-1}\left(\frac{1 \cdot 0 \cdot 1 + 0(0 \cdot 0 + 1 \cdot 1) + 1 \cdot 1 \cdot 0}{\sqrt{1 \cdot 0^2 + 2 \cdot 0 \cdot 0 \cdot 1 + 1 \cdot 1^2}\sqrt{1 \cdot 1^2 + 2 \cdot 0 \cdot 1 \cdot 0 + \cdot 1 \cdot 0^2}}\right)$$

$$= \cos^{-1}(0) = \frac{\pi}{2},$$

$$\theta_{\mathcal{P}} = \cos^{-1}\left(\frac{4c \cdot 0 \cdot 1 + 0(0 \cdot 0 + 1 \cdot 1) + 4c \cdot 1 \cdot 0}{\sqrt{4c \cdot 0^2 + 2 \cdot 0 \cdot 0 \cdot 1 + 4c \cdot 1^2}\sqrt{4c \cdot 1^2 + 2 \cdot 0 \cdot 1 \cdot 0 + 4c \cdot 0^2}}\right)$$

$$= \frac{\pi}{2},$$

whereas

$$\theta_{\mathcal{B}} = \cos^{-1}\left(\left((1-b^2)c \cdot 0 \cdot 1 + abc(0 \cdot 0 + 1 \cdot 1) + (1-a^2)c \cdot 1 \cdot 0\right)\right.$$

$$\left/\left(\sqrt{(1-b^2)c \cdot 0^2 + 2abc \cdot 0 \cdot 1 + (1-a^2)c \cdot 1^2}\right.\right.$$

$$\left.\left. \times \sqrt{(1-b^2)c \cdot 1^2 + 2abc \cdot 1 \cdot 0 + (1-a^2)c \cdot 0^2}\right)\right)$$

$$= \cos^{-1}\left(\frac{ab}{\sqrt{1-a^2}\sqrt{1-b^2}}\right).$$

It is noteworthy that in Example 7.4, the given Euclidean right angles remain right angles relative to the Poincaré metric, whereas they are not right angles relative to the Beltrami metric. In general, a metric $\mathcal{M}$ is said to be *conformal* if the angles of the underlying domain D have the same measure relative to $\mathcal{M}$ as they do relative to the Euclidean metric $\mathcal{E}$. The following important proposition is easily proved (Exercise 17). Its converse is also true, but is of less use to us.

Proposition 7.5 *If $f(u,v)$ is everywhere positive and three times differentiable, then*

$$f(u,v)(du^2 + dv^2)$$

is a conformal metric.   □

The half-plane geometry $\mathbb{H}$ is the main subject of the next chapter. For that reason, examples regarding this geometry will often be stated as propositions.

Proposition 7.6 *The half-plane metric $\mathcal{H}$ is conformal.*

PROOF: This follows immediately from the preceding proposition.   Q.E.D.

A *geodesic* of the metric $\mathcal{M}$ is a curve in the domain D whose length between any two of its points, relative to $\mathcal{M}$, realizes their distance relative to $\mathcal{M}$. This definition is not easily applied, even in the easiest case where $\mathcal{M}$ is the Euclidean metric $\mathcal{E}$. Fortunately, the proof of Gauss's Theorem 6.3.13 is intrinsic, meaning that it only depends on the first fundamental form of the surface rather than the explicit parametrization of the surface. (The reader is urged to reexamine this proof while keeping this new context in mind.) It follows that Gauss's Equation (6.18) can be used to find geodesics of Riemann geometries as well. Theorem 6.3.13 is now restated in this new context. In a Riemann geometry with metric $\mathcal{M}$, the *angle of inclination* θ of the curve γ is the angle it makes, relative to $\mathcal{M}$, with the Euclidean straight line $v = c$.

Theorem 7.7 *Let $u = u(t)$, $v = v(t)$ be a curve γ in the domain D of the metric*

$$\mathcal{M} = E\,du^2 + G\,dv^2.$$

The curve γ is a geodesic if and only if its angle of inclination θ satisfies the equation

$$\theta' = \frac{E_v u' - G_u v'}{2\sqrt{EG}}. \tag{11}$$

$\square$

Example 7.8 For the Euclidean metric $\mathcal{E} = du^2 + dv^2$, Gauss's Equation (11) becomes

$$\theta' = 0.$$

Hence, the geodesics of this metric are the Euclidean straight lines.

Proposition 7.9 *The geodesics of the half-plane geometry $\mathbb{H}$ are of two kinds:*

(a) *Semicircles centered on the u-axis.*

(b) *Rays perpendicular to the u-axis.*

PROOF: By Proposition 7.6 the half-plane metric $\mathcal{H}$ is conformal, and so we may treat the angle of inclination θ as a Euclidean angle. Since $E = G = v^{-2}$, Gauss's equation takes the form

$$\theta' = \frac{-\dfrac{2}{v^3}u'}{2\dfrac{1}{v^2}} = -\frac{u'}{v}.$$

Parametrize the semicircle of Figure 7.3 as $u = c + r\cos t$, $v = r\sin t$, and note that $\theta = t - \pi/2$. We then obtain

$$\theta' = 1 = -\frac{-r\sin t}{r\sin t} = -\frac{u'}{v}.$$

Next, parametrize the vertical ray of Figure 7.3 as $u = d, v = t$, and note that $\theta = \pi/2$. We then obtain

$$\theta' = 0 = -\frac{u'}{v}.$$

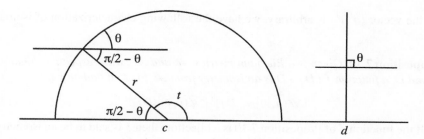

Figure 7.3   Geodesics in the upper half-plane.

Note that given any point P in $\mathbb{H}$ and any direction at P, one of the above geodesics will pass through P in that direction. Since Equation (11) is an ordinary second-order differential equation, it follows from the uniqueness of the solutions of such equations that we have described all the geodesics of $\mathscr{H}$.   Q.E.D.

In the most general context, an *isometry* is a transformation that preserves lengths. This concept is now interpreted in the more restricted context of a Riemann geometry. Let $\mathscr{M} = E\,du^2 + 2F\,du\,dv + G\,dv^2$ and $\tilde{\mathscr{M}} = \tilde{E}\,du^2 + 2\tilde{F}\,du\,dv + \tilde{G}\,dv^2$ be two metrics with domains D and $\tilde{D}$ respectively. An isometry from $\mathscr{M}$ to $\tilde{\mathscr{M}}$ is a function $f : D \to \tilde{D}$ such that for every curve γ parametrized as $u = u(t)$, $v = v(t)$, $a \leq t \leq b$, in D,

$$L_{\mathscr{M}}(\gamma) = L_{\tilde{\mathscr{M}}}(f(\gamma)). \tag{12}$$

If $(\tilde{u}, \tilde{v}) = f(u, v)$, then it follows from the definition of lengths (2) and the Fundamental Theorem of Calculus that this condition is tantamount to

$$Eu'^2 + 2Fu'v' + Gv'^2 = \tilde{E}\tilde{u}'^2 + 2\tilde{F}\tilde{u}'\tilde{v}' + \tilde{G}\tilde{v}'^2. \tag{13}$$

It is convenient to restate Equation (13) in the language of matrices. Set

$$[\mathscr{M}] = \begin{pmatrix} E & F \\ F & G \end{pmatrix}, \qquad [Df] = \begin{pmatrix} \dfrac{\partial \tilde{u}}{\partial u} & \dfrac{\partial \tilde{u}}{\partial v} \\[2mm] \dfrac{\partial \tilde{v}}{\partial u} & \dfrac{\partial \tilde{v}}{\partial v} \end{pmatrix}.$$

Since

$$\begin{pmatrix} \tilde{u}' \\ \tilde{v}' \end{pmatrix} = \begin{pmatrix} \dfrac{\partial \tilde{u}}{\partial u} & \dfrac{\partial \tilde{u}}{\partial v} \\[2mm] \dfrac{\partial \tilde{v}}{\partial u} & \dfrac{\partial \tilde{v}}{\partial v} \end{pmatrix} \begin{pmatrix} u' \\ v' \end{pmatrix},$$

it follows that Equation (13) can be rewritten as

$$\begin{pmatrix} u' \\ v' \end{pmatrix}^{\mathrm{T}} [\mathscr{M}] \begin{pmatrix} u' \\ v' \end{pmatrix} = \begin{pmatrix} u' \\ v' \end{pmatrix}^{\mathrm{T}} [Df]^{\mathrm{T}} [\tilde{M}] [Df] \begin{pmatrix} u' \\ v' \end{pmatrix}.$$

As the vector (u', v') is arbitrary, we have the following characterization of isometries.

Proposition 7.10 *Given two Riemann metrics $\mathcal{M}$ and $\tilde{\mathcal{M}}$, with respective domains D and $\tilde{D}$, a function $f : D \to \tilde{D}$ is an isometry from $\mathcal{M}$ to $\tilde{\mathcal{M}}$ if and only if*

$$[\mathcal{M}(u,v)] = [Df]^T[\tilde{\mathcal{M}}(\tilde{u},\tilde{v})][Df]. \qquad \square \ (14)$$

If the function f of Proposition 7.10 is a bijection, then f is said to be an isometry from $(\mathcal{M}, D)$ to $(\tilde{\mathcal{M}}, \tilde{D})$. We shall mostly be concerned with situations where $\mathcal{M} = \tilde{\mathcal{M}}$, in which case f is said to be an *isometry of $\mathcal{M}$*.

Example 7.11 Let $\mathcal{E} = \tilde{\mathcal{E}}$ be the Euclidean metric, and let α be a fixed angle. Set $f(u, v) = (\tilde{u}, \tilde{v})$, where

$$\tilde{u} = u\cos\alpha - v\sin\alpha, \qquad \tilde{v} = u\sin\alpha + v\cos\alpha.$$

It follows from elementary linear algebra that f is a bijection. Moreover,

$$[Df]^T[\tilde{\mathcal{E}}][Df] = \begin{pmatrix} \cos\alpha & \sin\alpha \\ -\sin\alpha & \cos\alpha \end{pmatrix} \begin{pmatrix} 1 & 0 \\ 0 & 1 \end{pmatrix} \begin{pmatrix} \cos\alpha & -\sin\alpha \\ \sin\alpha & \cos\alpha \end{pmatrix}$$

$$= \begin{pmatrix} 1 & 0 \\ 0 & 1 \end{pmatrix} = [\mathcal{E}].$$

Hence the function f is indeed an isometry of the Euclidean plane. It is, in fact, a counterclockwise rotation by the angle α about the origin.

Example 7.12 We show that the function

$$(\tilde{u},\tilde{v}) = f(u,v) = \left(\frac{2u}{u^2 + (v-1)^2}, \frac{1 - u^2 - v^2}{u^2 + (v-1)^2} \right)$$

is an isometry from $\mathscr{P}$ to $\mathscr{H}$. For

$$[Df] = \frac{2}{[u^2 + (v-1)^2]^2} \begin{pmatrix} (v-1)^2 - u^2 & -2u(v-1) \\ 2u(v-1) & (v-1)^2 - u^2 \end{pmatrix},$$

$$[\mathscr{P}(u,v)] = \frac{4}{(1 - u^2 - v^2)^2} \begin{pmatrix} 1 & 0 \\ 0 & 1 \end{pmatrix},$$

$$[\mathscr{H}(\tilde{u},\tilde{v})] = \frac{1}{\tilde{v}^2} \begin{pmatrix} 1 & 0 \\ 0 & 1 \end{pmatrix} = \left[\frac{u^2 + (v-1)^2}{1 - u^2 - v^2} \right]^2 \begin{pmatrix} 1 & 0 \\ 0 & 1 \end{pmatrix},$$

and the required Equation (14) is tantamount to the easily verified matrix multiplication

$$\begin{pmatrix} (v-1)^2 - u^2 & 2u(v-1) \\ -2u(v-1)^2 & (v-1)^2 - u^2 \end{pmatrix} \begin{pmatrix} (v-1)^2 - u^2 & -2u(v-1)^2 \\ 2u(v-1) & (v-1)^2 - u^2 \end{pmatrix}$$

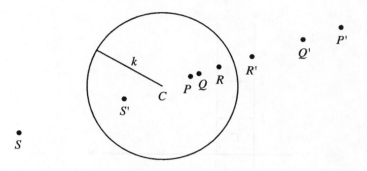

Figure 7.4   An inversion.

$$= [u^2 + (v-1)^2]^2 \begin{pmatrix} 1 & 0 \\ 0 & 1 \end{pmatrix}.$$

We now go on to describe some of the isometries of $\mathbb{H}$. Given a point C of the Euclidean plane $\mathbb{R}^2$, and a real number k, the *inversion* $I = I_{C,k}$ is a transformation of the plane such that if $P' = I_{C,k}(P)$ then (see Fig. 7.4)

a. P' is a point on the ray from C to P;

b. $CP \cdot CP' = k^2$.

It is clear that $I_{C,k}$ interchanges the exterior and interior of the circle of radius k centered at C, while it leaves the points on that circle fixed. Strictly speaking, $I_{C,k}$ fails to be defined at C. This is a minor nuisance that will be dealt with in Section 8.4. Note also that in its domain, $I_{C,k}^2 = \text{Id}$.

Proposition 7.13 *The restriction f of the inversion $I_{(0,0),k}$ to the half-plane $v > 0$ is an isometry of the half-plane geometry $\mathbb{H}$.*

PROOF: If we set $w = u^2 + v^2$, then

$$(\tilde{u}, \tilde{v}) = f(u,v) = I_{(0,0),k}(u,v) = \frac{k^2}{w}(u,v).$$

It follows that

$$[Df] = \frac{k^2}{w^2} \begin{pmatrix} v^2 - u^2 & -2uv \\ -2uv & u^2 - v^2 \end{pmatrix}$$

and

$$[\mathscr{H}(u,v)] = \frac{1}{v^2} \begin{pmatrix} 1 & 0 \\ 0 & 1 \end{pmatrix}, \qquad [\mathscr{H}(\tilde{u}, \tilde{v})] = \frac{w^2}{k^4 v^2} \begin{pmatrix} 1 & 0 \\ 0 & 1 \end{pmatrix}.$$

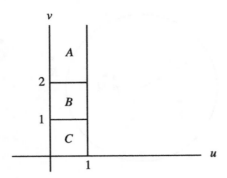

Figure 7.5   Three regions in the upper half-plane.

The required inequality

$$[\mathscr{H}(u,v)] = [Df]^T [\mathscr{H}(\tilde{u},\tilde{v})][Df]$$

is now easily verified. Moreover, it is clear that f is indeed a bijection of the upper half-plane.   Q.E.D.

Since isometries must transform shortest curves into shortest curves, we have the following proposition.

Proposition 7.14 *Isometries transform geodesics into geodesics.*   □

For the definition of area we take our cue from Lemma 6.2.3 and define the *area* of the subset R of the domain D relative to the metric $\mathscr{M}$ of Equations (2)–(4) to be

$$a_{\mathscr{M}}(R) = \int\int_R \sqrt{EG - F^2}\, du\, dv. \tag{15}$$

For example, for the half-plane metric $\mathscr{H}$, Equation (15) becomes

$$\int\int_R \frac{du\, dv}{v^2}.$$

Example 7.15 Compute the half-plane areas of the regions A, B, and C in Figure 7.5. Here,

$$\text{area of } A = \int_2^\infty \int_0^1 \frac{du\, dv}{v^2} = \int_2^\infty \frac{dv}{v^2} = \frac{1}{2},$$

$$\text{area of } B = \int_1^2 \int_0^1 \frac{du\, dv}{v^2} = \int_1^2 \frac{dv}{v^2} = \frac{1}{2},$$

and

$$\text{area of } C = \int_0^1 \int_0^1 \frac{du\, dv}{v^2} = \int_0^1 \frac{dv}{v^2} = \infty.$$

The following proposition is very plausible. Nevertheless, it does require proof. The demonstration relies heavily on the properties of the determinant function.

Proposition 7.16 *The area function in a Riemann geometry is invariant under its isometries.*

PROOF: Let $(\tilde{u}, \tilde{v}) = f(u, v)$ be an isometry of the Riemann metric $\mathcal{M} = E\,du^2 + 2F\,du\,dv + G\,dv^2$. If $R \subset D$, then the area of $f(R)$ relative to $\mathcal{M}$ is

$$\iint_{f(R)} \sqrt{E(\tilde{u}, \tilde{v})G(\tilde{u}, \tilde{v}) - [F(\tilde{u}, \tilde{v})]^2}\, d\tilde{u}\, d\tilde{v}$$

$$= \iint_{f(R)} \sqrt{\det[M(\tilde{u}, \tilde{v})]}\, d\tilde{u}\, d\tilde{v}$$

$$= \iint_{R} \sqrt{\det[Df]^{-1} \det[\mathcal{M}(u, v)] \det[Df]^{-1}}\, \frac{\partial(\tilde{u}, \tilde{v})}{\partial(u, v)}\, du\, dv.$$

However,

$$\frac{\partial(\tilde{u}, \tilde{v})}{\partial(u, v)} = |\det[Df]|,$$

and hence the above area equals

$$\iint_{R} \sqrt{\det[\mathcal{M}(u, v)]}\, du\, dv = \iint_{R} \sqrt{E(u, v)G(u, v) - [F(u, v)]^2}\, du\, dv,$$

which is the area of R relative to the metric $\mathcal{M}$.   Q.E.D.

Since the Gaussian curvature was demonstrated in Theorem 6.5.9 to be intrinsic, it makes sense to speak of the curvature of a metric. Taking our cue from Proposition 6.5.11, we define the curvature K of the metric $\mathcal{M} = E\,du^2 + 2F\,du\,dv + G\,dv^2$ by means of the equation

$$4(EG - F^2)^2 K_{\mathcal{M}} = E\left[\frac{\partial E}{\partial v}\frac{\partial G}{\partial v} - 2\frac{\partial F}{\partial u}\frac{\partial G}{\partial v} + \left(\frac{\partial G}{\partial u}\right)^2\right]$$

$$+ F\left[\frac{\partial E}{\partial u}\frac{\partial G}{\partial v} - \frac{\partial E}{\partial v}\frac{\partial G}{\partial u} - 2\frac{\partial E}{\partial v}\frac{\partial F}{\partial v} + 4\frac{\partial F}{\partial u}\frac{\partial F}{\partial v} - 2\frac{\partial F}{\partial u}\frac{\partial G}{\partial u}\right]$$

$$+ G\left[\frac{\partial E}{\partial u}\frac{\partial G}{\partial u} - 2\frac{\partial E}{\partial u}\frac{\partial F}{\partial v} + \left(\frac{\partial E}{\partial v}\right)^2\right]$$

$$- 2(EG - F^2)\left[\frac{\partial^2 E}{\partial v^2} - 2\frac{\partial^2 F}{\partial u \partial v} + \frac{\partial^2 G}{\partial u^2}\right]. \tag{16}$$

If $F = 0$, this expression reduces to

$$K_{\mathcal{M}} = -\frac{1}{2\sqrt{EG}}\left[\left(\frac{E_v}{\sqrt{EG}}\right)_v + \left(\frac{G_u}{\sqrt{EG}}\right)_u\right]. \tag{17}$$

The intrinsic proof of Gauss's Theorem 6.7.3 constitutes a proof of the following theorem.

Theorem 7.17 *Let $\triangle ABC$ be a geodesic triangle of a Riemann geometry $(\mathcal{M}, D)$. If its interior angles are α, β, γ, then*

$$\int\int_{\triangle ABC} K_{\mathcal{M}} \sqrt{EG - F^2}\, du\, dv = \alpha + \beta + \gamma - \pi. \qquad \Box\ (18)$$

The similarity between Equations (15) and (18) implies that in a geometry with constant curvature, the total curvature of a region is proportional to its area. Theorem 7.17 can therefore yield very elegant area formulas.

Example 7.18 For the half-plane metric $\mathcal{H}$, Equation (17) yields

$$K_{\mathcal{H}} = -1$$

and hence, for any geodesic $\triangle ABC$,

$$\int\int_{\triangle ABC} (-1) \cdot \frac{1}{v^2}\, du\, dv = \alpha + \beta + \gamma - \pi,$$

or

$$a_{\mathcal{H}}(\triangle ABC) = \pi - \alpha - \beta - \gamma.$$

The following observations make it clear that the curvature constitutes a useful tool for distinguishing between nonisometric geometries.

Corollary 7.19 *The curvature K of a Riemann metric is invariant under its isometries.*

PROOF: Theorem 7.17 together with the invariance of geodesics and angles implies the invariance of the total curvature of a triangle. The remainder of the proof is relegated to Exercise 21. Q.E.D.

While Theorem 7.7, which characterizes geodesics when $F = 0$, is sufficient for the purposes of this text, some of the exercises deal with geodesics in more general settings. Those require the following version of Proposition 6.3.18.

Proposition 7.20 *Let $u = u(t)$, $v = v(t)$, $a \le t \le b$, be the parametrization of a curve γ in the domain D of the metric*

$$\mathcal{M} = E\, du^2 + 2F\, du\, dv + G\, dv^2.$$

Then γ is a geodesic if and only if its angle of inclination θ satisfies the equation

$$\sqrt{EG - F^2}\, \theta' = \frac{1}{2} \cdot \frac{F}{E} \cdot E_u \cdot u' + \frac{1}{2} \cdot \frac{F}{E} \cdot E_v \cdot v'$$

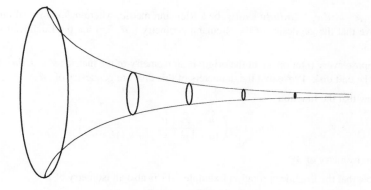

Figure 7.6   A pseudosphere.

$$+ \frac{1}{2} \cdot E_v \cdot u' - F_u \cdot u' - \frac{1}{2} \cdot G_u \cdot v'. \qquad \square$$

In conclusion we address a natural question that may have already occurred to the careful readers: *Is every Riemann metric the first fundamental form of some surface?* The answer is negative. For example, it was proved by David Hilbert that a metric with constant negative curvature, such as $\mathscr{H}$ and $\mathscr{P}$, is not the first fundamental form of any parametrization of a surface which is *complete* in the sense that every geodesic either is a closed curve or else can be extended indefinitely in both directions. The qualification "complete" is of the essence here, since the incomplete surface of Figure 7.6, known as the *pseudosphere* (see Exercise 6.2.4), does have constant negative curvature $-1/c^2$ and, when $c = 1$, can be parametrized so as to have $\mathscr{H}$ as its first fundamental form. This surface is incomplete because its meridianal geodesics cannot be extended past the rim on its left end. Metrics for which even incomplete corresponding surfaces fail to exist are also known.

Exercises 7

1. For each of the metrics below, compute the length of the Euclidean line segment that joins the origin to the point $r(\cos\theta, \sin\theta)$, $0 < r < 1$. What is the limit of this length as r approaches 1?

   (a) $\mathcal{M} = \dfrac{du^2 + dv^2}{1 - u^2 - v^2}$.

   (b) $\mathcal{M} = \dfrac{(1 - v^2)\,du^2 + 2uv\,du\,dv + (1 - u^2)\,dv^2}{1 - u^2 - v^2}$.

   (c) $\mathcal{B} = \dfrac{(1 - v^2)\,du^2 + 2uv\,du\,dv + (1 - u^2)\,dv^2}{(1 - u^2 - v^2)^2}$.

2. Prove that the diameters of the unit disk are geodesics of the metric of Exercise 1(a).

3**. Prove that the diameters of the unit disk are geodesics of the metrics of Exercise 1(b),(c).

4. Let $\mathcal{M} = a\,du^2 + 2b\,du\,dv + c\,dv^2$ be a Riemann metric, where a, b, c are real numbers. Prove that the geodesics of the Riemann geometry $(\mathcal{M}, \mathbb{R}^2)$ are the Euclidean straight lines.

5. Suppose every rotation about the origin is an isometry of the metric $\mathcal{M} = E\,du^2 + G\,dv^2$ on the unit disk. Prove that the diameters of this disk are geodesics of $\mathcal{M}$.

6. Show that the function

$$f(u, v) = \left(\frac{u^2 + v^2 + u}{(u+1)^2 + v^2}, \frac{v}{(u+1)^2 + v^2} \right)$$

is an isometry of $\mathcal{H}$.

7. Show that the Euclidean rotation Example 7.11 is also an isometry of $\mathcal{P}$.

8. Show that the Euclidean rotation of Example 7.11 is also an isometry of $\mathcal{B}$.

9. For each $0 < a < 1$ let Q_a be the Euclidean circle centered at the origin with radius a. Show that Q_a is also a circle relative to $\mathcal{P}$. Evaluate the radius, circumference, and area of Q_a relative to the Riemann metric $\mathcal{P}$.

10. Repeat Exercise 9 relative to the Beltrami metric $\mathcal{B}$.

11. Repeat Exercise 9 relative to the metric of Exercise 1(b).

12*. Show that the curvature of the Riemann metric of Exercise 1(b) is 1.

13. Compute the curvature of the Riemann metric of Exercise 1(a).

14. A Riemann metric has negative curvature throughout its domain. Prove that every pair of geodesics intersects in at most one point.

15. Show that the Riemann metric

$$\frac{du^2 + dv^2}{[1 + \frac{c}{4}(u^2 + v^2)]^2}$$

has curvature c throughout its domain.

16. Compute the curvature of the Riemann metric

$$v^c(du^2 + dv^2)$$

for every real c.

17. Prove Proposition 7.5.

18. Prove formally that the function $(\bar{u}, \bar{v}) = (u + a, v)$ is a half-plane isometry.

19. Prove formally that if m is the straight line $u = a$, then the reflection ρ_m is a half-plane isometry.

20. Prove formally that if $a > 0$, then the function $(\bar{u}, \bar{v}) = (au, av)$ is a half-plane isometry.

21. Complete the proof of Corollary 7.19.

If $\mathbb{M} = (\mathcal{M}, D)$ and a is a positive real number, then $a\mathbb{M} = (a\mathcal{M}, D)$.

22. Let a be a positive real number. Prove that the Riemann geometries $\mathbb{M}$ and $a\mathbb{M}$ have the same geodesics and the same isometries.

23. Compare the following in $\mathbb{M}$ and $a\mathbb{M}$.

(a) Lengths of curves.

(b) Measures of angles.

(c) Areas of regions.

(d) Curvatures.

24. For which a and b are $a\mathbb{H}$ and $b\mathbb{H}$ isometric?

25. For which a and b are $a\mathbb{H}$ and $b\mathbb{P}$ isometric?

26. For which a and b are $a\mathbb{P}$ and $\mathbb{P}_b = (\mathscr{P}_b, D_b)$ isometric, if

$$\mathscr{P}_b = \frac{4(du^2 + dv^2)}{(b^2 - u^2 - v^2)^2}$$

and D_b denotes the interior of the disk of radius b centered at the origin?

27*. Prove that the Riemann geometry with metric $\mathscr{M} = du^2 + e^{2u}\,dv^2$ and domain $D = \mathbb{R}^2$ is isometric to $\mathbb{H}$.

28**. For any real number q, let $\mathbb{B}_q$ be the Riemann geometry with metric

$$\mathscr{B}_q = \frac{(1 + qv^2)\,du^2 - 2quv\,du\,dv + (1 + qu^2)\,dv^2}{(1 + qu^2 + qv^2)^2}$$

and domain

$$D_q = \{(u, v) \mid 1 + qu^2 + qv^2 > 0\}.$$

Prove:

(a) The curvature of $\mathbb{B}_q$ is q.

(b) The geodesics of $\mathbb{B}_q$ are straight lines.

29. Determine the length of the plane curve $(t, t^2), 0.1 \le t \le 0.7$, relative to the following metrics: (a) $\mathscr{E}$, (b) $\mathscr{P}$, (c) $\mathscr{B}$, (d) $\mathscr{H}$, (e) $(1 + uv)^2\,du^2 + v^2\,dv^2$.

30. Prove that the transformation

$$(\tilde{u}, \tilde{v}) = f(u, v) = \left(\frac{2u}{u^2 + v^2 + 1}, \frac{2v}{u^2 + v^2 + 1} \right)$$

is an isometry from $\mathscr{P}$ to $\mathscr{B}$.

31. Prove that the transformation

$$(\tilde{u}, \tilde{v}) = f(u, v) = \left(\frac{2u^2 + 5u + 2 + 2v^2}{(u + 2)^2 + v^2}, \frac{3v}{(u + 2)^2 + v^2} \right)$$

is an isometry of $\mathscr{P}$.

32. In the Poincaré geometry $\mathscr{P}$ find the $\mathscr{P}$-circumference and the $\mathscr{P}$-area of the circle with $\mathscr{P}$-radius r.

33. In the half-plane geometry $\mathscr{H}$ find the $\mathscr{H}$-circumference and the $\mathscr{H}$-area of the circle with $\mathscr{H}$-radius r.

Chapter Review Exercises

Are the following statements true or false? Justify your answers.

1. The first fundamental form of every surface is a Riemann metric.

2. Every Riemann metric is the first fundamental form of a surface.

3. The half-plane and the Poincaré metrics are conformal.

4. Every two metrics that have the same domain are conformal.

5. The half-plane and the Beltrami metrics are isometric.

6. Every two metrics that have the same domain are isometric.

7. Let α be the angle formed by the positive x and y axes. If M_1 and M_2 are metrics on the unit disk, then α has the same measure relative to M_1 and M_2.

8. Let α be the angle formed by the positive x and y axes. Then α has the same measure relative to $\mathbb{P}$ and $\mathbb{B}$.

9. Let α be the angle formed by the positive x and y axes. Then α has the same measure relative to the metrics $du^2 + dv^2$ and $du^2 + 2dv^2$.

10. If R is the region of the plane that lies entirely in the half-plane $y > 1$, then the half-plane area of R is smaller than its Euclidean area.

11. The composition of two isometries of a geometry is an isometry of the same geometry.

12. The composition of two inversions is an inversion.

13. The transformation $(x, y) \rightarrow (2x, 2y)$ is an isometry of $\mathbb{E}$.

14. The transformation $(x, y) \rightarrow (2x, 2y)$ is an isometry of $\mathbb{H}$.

15. The transformation $(x, y) \rightarrow (2x, 2y)$ is an isometry of $\mathbb{P}$.

16. The transformation $(x, y) \rightarrow (x + 2, y)$ is an isometry of $\mathbb{E}$.

17. The transformation $(x, y) \rightarrow (x + 2, y)$ is an isometry of $\mathbb{H}$.

18. The transformation $(x, y) \rightarrow (x + 2, y)$ is an isometry of $\mathbb{P}$.

CHAPTER 8

HYPERBOLIC GEOMETRY

All high school students are subjected to a course in Euclidean geometry, and all of them take it for granted that this geometry actually describes the space in which they live. The validity of this belief, of profound interest to mathematicians, philosophers, and physicists, came under attack in the nineteenth century and was disproved in the twentieth. In this chapter we propose to display one of the alternatives to Euclidean geometry and to explain both its history and its relevance.

8.1 Neutral Geometry

Around 300 B.C. Euclid codified what has since come to be known as Euclidean geometry. His treatise, known as the *Elements*, consists of thirteen books that develop plane and solid geometry, as well as some number theory, from a small collection of axioms. This style of studying geometry is known as *synthetic*, as opposed to the *analytic* style that was employed in Chapters 6 and 7. Euclid's system is not error-free and has since been supplanted by others, of which the most notable is that published by David Hilbert (1862–1943) in 1899. It is nonetheless unique in its clarity and the easy access it provides to neutral geometry (to be defined below). For these rea-

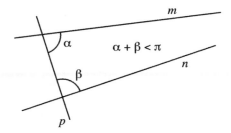

Figure 8.1   Euclid's Postulate 5.

sons the author decided to use a recapitulation of the first 28 propositions of Book I of the *Elements* as an introduction to both neutral and hyperbolic geometries. Our presentation here constitutes both a paraphrase and an abridgement of that book.

Euclid's geometrical axioms consist of five postulates and five common notions.

Postulates:

1. *Given any two points there exists exactly one straight line joining them.*

2. *Every straight line extends indefinitely in both directions.*

3. *Given a straight line segment and one of its endpoints, there exists a circle that has that point and segment as its respective center and radius.*

4. *All right angles are equal.*

5. *If two straight lines m and n are intersected by a third straight line p so that the sum of the interior angles on one side of p is less than π, then m and n, if produced indefinitely, intersect (see Fig. 8.1).*

Euclid defined a *right angle* as one half of a straight angle and used it as his unit for measuring angles. We shall, albeit anachronistically, denote the size of the straight angle by π and that of the right angle by $\pi/2$. Postulate 5 is this chapter's hub and will be reexamined repeatedly below.

The common notions we are about to list should be understood as properties shared by any sense of measurement (or size): length, area, volume, and angular measure, in particular. Euclid tacitly assumed that every geometric object has such an aspect of size, and when he said that two things are equal, he meant that they have equal sizes.

Common Notions:

1. *Figures which are equal to the same figure are also equal to one another.*

2. *If equals be added to equals, the wholes are equal.*

3. *If equals be subtracted from equals, the remainders are equal.*

4. *Congruent figures are equal.*

5. *The whole is greater than the part.*

From these meager beginnings Euclid developed what has become known as Euclidean geometry. Unfortunately, in this process, he did implicitly assume other postulates that he failed to acknowledge. Most important among these are the following two:

A. *Figures such as triangles and circles have well defined interiors and exteriors. Any line joining a point of the interior to a point of the exterior must intersect the separating figure.*

B. *Given any straight line segments AB and CD, there is an isometry f of the Euclidean plane such that $f(A) = C$ and $f(AB)$ falls along CD.*

Postulate A is a special case of the Jordan Curve Theorem 2.4.1. Geometries whose isometries possess the property displayed in Postulate B are said to be *transitive*.

Euclidean geometry is defined as the set of all the propositions that can be deduced from Euclid's (augmented) postulates and common notions. This is equivalent to the definition given in Chapter 7.

Euclid's Postulate 5 was deemed from early on to be qualitatively different from the other axioms. It seemed to possess more content than the others, and the suspicion arose that perhaps some clever arguments could be found to deduce it from them. This led naturally to an interest in that collection of Euclidean propositions that could be proved without the use of Postulate 5. This set of propositions is today called *neutral geometry* or *absolute geometry*. It is clear that neutral geometry is a subset of Euclidean geometry and that these two geometries are equal if and only if Postulate 5 can be proved on the basis of the other axioms.

It is a matter of fact that Euclid organized his development so that Postulate 5 is not required for the proofs of the first 28 propositions. Was this intentional on his part or was this a happy accident? The answer to this question will probably never be known. Still, these first 28 propositions form a part of the corpus of neutral geometry and, what's more important, also a part of hyperbolic geometry (to be defined below). These propositions are now described.

Of the first fifteen propositions in Book I we mention only the following ten, with which the reader is in all likelihood quite familiar. The appended number in parentheses refers to the numbering of this proposition in Heath's translation of the *Elements*.

Side-angle-side (SAS) congruence. (4)

Two sides of a triangle are equal if and only if their opposite angles are also equal. (5, 6)

Side-side-side (SSS) congruence. (8)

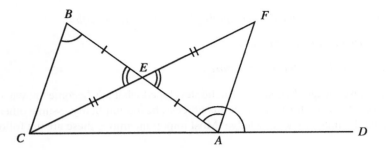

Figure 8.2   The neutral exterior angle theorem.

It is possible to bisect a given angle. (9)

It is possible to bisect a given line segment. (10)

It is possible to raise a perpendicular to a given straight line from a given point on it. (11)

It is possible to draw a perpendicular to a given straight line from a given point on it. (12)

The sum of two supplementary angles is π. (13)

Vertically opposite angles are equal. (15)

Euclid's sixteenth proposition (our Theorem 8.1.1) may seem like a weak version of the well-known Euclidean proposition that the exterior angle of a triangle equals the sum of the interior and opposite angles. It is, nonetheless, a crucial proposition of neutral geometry, and the best result that can be proved in that context.

Theorem 8.1.1 (Exterior Angle) *If one side of a triangle is extended, then the exterior angle it forms is greater than either of the interior and opposite angles.* (16)

PROOF: It suffices to prove that in Figure 8.2, $\angle ABC < \angle BAD$. For this purpose we bisect side AB at E and extend CE to F so that $CE = EF$. Note that $\triangle EBC \cong \triangle EAF$ by SAS. It follows that

$$\angle EBC = \angle EAF < \angle BAD. \qquad \text{Q.E.D.}$$

We continue with our summary of Euclid's first 28 propositions.

The sum of two angles of a triangle is less than π. (17)

One side of a triangle is greater than another if and only if the angle opposite to the first is greater than the angle opposite to the second. (18, 19)

The sum of two sides of a triangle exceeds the third side (triangle inequality). (20)

On a given straight line and at a given point on it, it is possible to construct an angle equal to a given angle. (23)

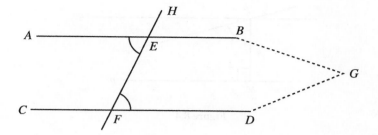

Figure 8.3   A proof of parallelism.

Angle-side-angle (ASA) and angle-angle-side (AAS) congruence. (26)

Euclid's Propositions 27 and 28 (Proposition 8.1.2 below) are, of course, well-known propositions of Euclidean geometry. Their equally well-known converses, however, constitute Proposition 29, whose proof does—unavoidably, it turns out—rely on Postulate 5. Consequently, while Propositions 27 and 28 belong to neutral geometry, their converses do not. For this reason we have chosen to include their proofs here. Two straight lines are said to be *parallel* if they lie in the same plane and do not intersect.

Proposition 8.1.2 *Suppose that one of the following equations holds in Figure 8.3:*

1. $\angle AEF = \angle EFD$;

2. $\angle BEH = \angle DFE$;

3. $\angle BEF + \angle DFE = \pi$.

Then AB and CD are parallel. (27, 28)

PROOF: Part 1 is proved by contradiction. Suppose that $\angle AEF = \angle EFD$ and that AB and CD do intersect at, say, G. Then, relative to $\triangle EFG$, $\angle AEF$ is an exterior angle whereas $\angle DFE$ is an interior and opposite angle. By the Exterior Angle Theorem 8.1.1, $\angle AEF > \angle DFE$, which contradicts our supposition.

As both the hypotheses of parts 2 and 3 imply that of part 1, the theorem is now proved in its entirety.   Q.E.D.

Proposition 8.1.2 can be used to prove the existence of parallel lines. The need to prove this fact is demonstrated by the nonexistence of parallel geodesics in the geometry of the sphere.

Theorem 8.1.3 *Given a straight line p and a point P not on it, there exists a straight line q that contains P and is parallel to p.*

PROOF: By Euclid's Propositions 12 and 11 it is possible to construct, through P, straight lines m and q such that $m \perp p$ and $q \perp m$. It follows from Euclid's Proposition 28 that $q \parallel p$.   Q.E.D.

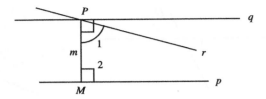

Figure 8.4

In today's high school geometry curriculum it is customary to replace Postulate 5 by the logically equivalent *Playfair's postulate*:

Given a straight line p and a point P not on it, there exists a unique straight line q that contains P and is parallel to p.

Of course, the existence of the line q of Playfair's postulate was established in Proposition 8.1.3. It is its uniqueness that will be shown to be equivalent to Postulate 5. The well-known Euclidean fact that the sum of the angles of every triangle is π also turns out to have the same logical status.

Theorem 8.1.4 *The following statements are equivalent in the context of neutral geometry:*

1. *Postulate 5.*

2. *Playfair's postulate.*

3. *The sum of the angles of every triangle is π.*

PROOF: 1 $\Rightarrow$ 2: Assume the validity of Postulate 5. Let p be a given line, and P a point not on it (Fig. 8.4). Let $m = PM \perp p$ and $q \parallel p$ be as constructed in Proposition 8.1.3. Suppose that r is another straight line, distinct from q and containing P. Since it may be assumed without loss of generality that $\angle 1 < \pi/2$, it follows that

$$\angle 1 + \angle 2 < \pi$$

and hence, by Postulate 5, r, if produced indefinitely, must intersect p. This proves the uniqueness of the parallel q.

2 $\Rightarrow$ 3: Assume the validity of Playfair's postulate, and let $\triangle ABC$ be given (Fig. 8.5). Let m and n be straight lines through P such that $\angle 2 = \beta$ and $\angle 3 = \gamma$. It follows from Euclid's Proposition 28 that $m \parallel BC \parallel n$ and hence, by Playfair's postulate, $m = n$. Consequently,

$$\alpha + \beta + \gamma = \alpha + \angle 2 + \angle 3 = \pi.$$

3 $\Rightarrow$ 1: Assume that the sum of the angles of every triangle is π. Let the transversal p intersect the straight lines $m = AM$ and $n = BC$ at A and B, respectively, with $\angle BAM + \angle ABC < \pi$ (Fig. 8.6). Set

$$\alpha = \angle BAM, \quad \beta = \angle ABC, \quad C_1 = C,$$

and, for $n = 2,3,4,\ldots$, define C_n recursively by means of the equation

$$C_n C_{n-1} = AC_{n-1}.$$

Since $\Delta C_{n-1} C_n A$ is isosceles, it follows from the assumption that the sum of the angles of every triangle is π that if $\gamma_n = \angle AC_n B$ then

$$\gamma_{n-1} = 2\gamma_n, \quad n = 2,3,4,\ldots,$$

and hence

$$\gamma_n = \frac{\gamma_1}{2^{n-1}}, \quad n = 1,2,3,\ldots.$$

However, if we set $\alpha_n = \angle BAC_n$, then, by our hypothesis,

$$\alpha_n + \beta + \gamma_n = \pi,$$

and hence,

$$\lim_{n \to \infty} \alpha_n = \pi - \beta - \lim_{n \to \infty} \gamma_n = \pi - \beta - 0 > \alpha.$$

It follows that for some $n_0, \alpha_{n_0} > \alpha$. Consequently, the ray AM falls *inside* ΔABC_{n_0}, and so it must intersect the line n. Q.E.D.

In Euclidean geometry, the sum of the angles of every triangle is π. We shall see later why the following weaker statement is the best that can be obtained for neutral geometry.

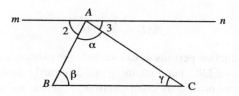

Figure 8.5

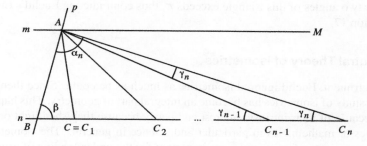

Figure 8.6

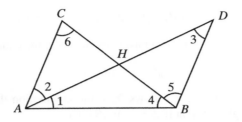

Figure 8.7

Theorem 8.1.5 *The sum of the angles of a neutral triangle is at most π.*

PROOF: We begin by showing that given any triangle, there exists a triangle with an arbitrarily small angle that nevertheless has the same angle sum as the given triangle. To prove this, let $\triangle ABC$ be an arbitrary triangle in which it may be supposed that $AB \geq AC$ (Fig. 8.7). Let H be the midpoint of BC, and extend AH to D so that $AH = HD$. Since $\triangle AHC \cong \triangle DHB$ (by SAS), it follows that

$$\angle 1 + \angle 4 + \angle 6 + \angle 2 = \angle 1 + \angle 4 + \angle 5 + \angle 3,$$

so that $\triangle ABD$ and $\triangle ABC$ have the same angle sums. On the other hand,

$$DB = AC \leq AB$$

and hence

$$\angle 1 \leq \angle 3 = \angle 2.$$

It follows that

$$\angle BAD \leq \frac{1}{2}\angle BAC.$$

A straightforward induction permits us to conclude the existence of a triangle having the same angle sum as $\triangle ABC$, but containing one arbitrarily small angle.

The theorem is now proved by contradiction. Suppose we are given a triangle with angle sum $s > \pi$. Then, by the first part of the proof, there exists a triangle with angle sum s, one of whose angles is smaller than $s - \pi$. It follows that the sum of the other two angles of this triangle exceeds π, thus contradicting Euclid's (neutral) Proposition 17.   Q.E.D.

The Neutral Theory of Isometries

To his detriment, Euclid ignored isometries as much as he could. Since then, however, the study of isometries has become an integral part of geometry. This happened partly because of their intrinsic interest and partly because they shed light on other disciplines of mathematics in particular and science in general. The isometries of Euclidean geometry, for example, are important for the understanding of crystalography, whereas the isometries of half-plane geometry form an indispensible part of

Andrew Wiles's recent proof of Fermat's Last Theorem. *Neutral isometries* are the isometries of neutral geometry.

Proposition 8.1.6 *Neutral isometries transform straight lines into straight lines.*

PROOF: Let f be an isometry of neutral geometry, let P, Q, and R be any three points, and set $A' = f(A), B' = f(B), C' = f(C)$. Since

$$AB + BC = AC \quad \text{if and only if} \quad A'B' + B'C' = A'C',$$

it follows from Euclid's Proposition 20 (the famed *triangle inequality*) that the points A, B, C are collinear if and only if A', B', C' are collinear. The proposition now follows immediately.   Q.E.D.

The proof of the following lemma is relegated to Exercise 16.

Lemma 8.1.7 *Let P and Q be two distinct points, and let f and g be neutral isometries. If f and g agree on P and Q, then they agree at every point on the straight line PQ.*   □

The next theorem demonstrates that every neutral isometry is determined by its action on any three noncollinear points.

Theorem 8.1.8 *Let P, Q, and R be three noncollinear points, and let f and g be neutral isometries. If f and g agree on P, Q, and R, then they agree everywhere.*

PROOF: It follows from Lemma 8.1.7 that f and g agree everywhere on $PQ \cup QR \cup PR$. If X is any point of the plane, let m be any straight line containing X that intersects $PQ \cup QR \cup PR$ in at least two distinct points, say Y and Z. Since f and g agree at Y and Z, they must also agree at X.   Q.E.D.

Corollary 8.1.9 *If a neutral isometry fixes three noncollinear points, then it is the identity.*   □

For any straight line m, the *reflection* ρ_m is a transformation of the plane into itself such that if $P' = \rho_m(P)$, then m is the perpendicular bisector of $\mathbb{P}'$. Not only are reflections neutral isometries (Exercise 17), but they are also the natural building blocks of *all* neutral isometries.

Theorem 8.1.10 *Suppose $\triangle ABC \cong \triangle A'B'C'$. Then there exists a sequence of no more than three reflections whose composition carries A, B, C onto A', B', C' respectively.*

PROOF: The theorem is trivially true if $A, B, C = A', B', C'$, respectively.

Suppose first that $A = A'$ and $B = B'$ but $C \neq C'$. Then, because $CA = C'A$ and $CB = C'B$, the reflection ρ_{AB} accomplishes the required task by itself.

Suppose next that only $A = A'$, and let D be the midpoint of BB'. Because $AB = AB'$, it follows that the reflection ρ_{AD} carries A, B onto A', B' respectively, and C onto some point, say C^*. Since $\triangle A'B'C^* \cong \triangle ABC \cong \triangle A'B'C'$, it follows from the previous paragraph that either the identity or the reflection $\rho_{A'B'}$ will finish the task of carrying A, B, C onto A', B', C'.

Finally, in the general case, let m be the perpendicular bisector of AA'. It is clear that $\rho_m(A) = A'$. Set $\rho_m(B) = B''$ and $\rho_m(C) = C''$. Since $\triangle A'B''C'' \cong \triangle ABC \cong \triangle A'B'C'$ it follows from the previous paragraph that there is a sequence of at most two reflections whose composition will finish the task of carrying A, B, C onto A', B', C'. Q.E.D.

Theorem 8.1.11 *Every neutral isometry is the composition of no more than three reflections.*

PROOF: Let f be a neutral isometry. Let $\triangle ABC$ be any triangle, and set $A' = f(A), B' = f(B), C' = f(C)$. By Theorem 8.1.10 there exists a sequence of at most three reflections whose composition carries A, B, C onto A', B', C' respectively. By Theorem 8.1.8, f equals this composition.   Q.E.D.

Given any point A and oriented angle α, the *rotation* $R_{A,\alpha}$ is a transformation of the plane into itself such that if $P' = R_{A,\alpha}(P)$, then

$$AP = AP' \quad \text{and} \quad \angle PAP' = \alpha.$$

Note that the identity transformation is a rotation by angle 0. Rotations are isometries, and the following proposition helps make Theorem 8.1.11 more concrete.

Proposition 8.1.12 *Let m and n be two straight lines that interesect in the point A so that the counterclockwise angle from m to n is α. Then*

$$\rho_n \circ \rho_m = R_{A,2\alpha}.$$

PROOF: Let P be a point such that $P' = \rho_m(P)$ is inside the angle α from m to n (Fig. 8.8), and let $P'' = \rho_n(P')$. It follows from the congruences

$$\triangle AMP \cong \triangle AMP', \qquad \triangle ANP' \cong \triangle ANP''$$

that

$$AP = AP' = AP''$$

and

$$\angle PAP'' = \angle PAP' + \angle P'AP'' = 2\angle MAP' + 2\angle P'AN = 2\alpha.$$

Hence

$$\rho_n \circ \rho_m(P) = R_{A,2\alpha}(P).$$

The validity of this equation for arbitrary P now follows from Theorem 8.1.8. Q.E.D.

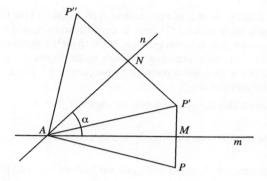

Figure 8.8

Exercises 8.1

All the following exercises are to be proved in the context of neutral geometry.

1. The sum of the angles of every quadrilateral is at most 2π.

   A *rectangle* is a quadrilateral each of whose angles is $\pi/2$.

2. The opposite sides of a rectangle are parallel.

   Exercises 3–7 constitute a proof of the equivalence of Postulate 5 to the existence of a *single* triangle with angle sum π.

3. If a rectangle exists, then there also exist rectangles with arbitrarily large sides.

4. If a rectangle exists, then there exists a rectangle with two adjacent sides having prescribed lengths.

5. If a rectangle exists, then every right triangle has angle sum π.

6. If a rectangle exists, then every triangle has angle sum π.

7. If there exists one triangle with angle sum π, then there exists a rectangle.

8. If one triangle has angle sum π, then every triangle has angle sum π.

9. Criticize the following "neutral proof" of Playfair's Postulate, offered by Proclus (410–485): "I say that if any straight line cuts one of two parallels, it will cut the other also. For let AB, CD be parallel and let EFG cut AB [at F, with G between AB and CD]; I say that it will cut CD also. For, since BF, FG are two straight lines from one point F, they have, when produced indefinitely, a distance greater than any magnitude, so that it will be greater than the interval between the parallels. Whenever, therefore, they are at a distance from one another greater than the distance between the parallels, FG will cut CD."

10. Criticize the following "neutral proof" of the fact that the sum of the interior angles of a triangle is π, which was offered by B. F. Thibault in 1809. Let ABC be a given triangle, and let d be a line segment that lies on the straight line AB with its center at A. Slide d along AB until its center falls on B, and then rotate it through the exterior of the triangle, about B as a pivot, until it falls along side BC. Next slide d along BC until its center

reaches C, and rotate it about C as a pivot through the exterior of the triangle until it falls along CA. Finally, slide d along CA until its center reaches A, and rotate it about A as pivot so that it comes into its initial position. If the triangle's interior angles are α, β, γ, then the segment d has been rotated successively by the angles $\pi - \beta, \pi - \gamma$, and $\pi - \alpha$ before it is returned to its original position. Consequently $\pi - \beta + \pi - \gamma + \pi - \alpha = 2\pi$, from which it follows that $\alpha + \beta + \gamma = \pi$.

11. Let ABC be a clockwise triangle with interior angles α, β, γ. Prove that $R_{C,2\gamma} \circ R_{B,2\beta} \circ R_{A,2\alpha} = \mathrm{Id}$.

12. Prove that if the point P is on the straight line p, then $\rho_p \circ R_{P,\alpha} \circ \rho_p = R_{P,-\alpha}$ for every angle α.

13. Which of Postulates 1–5, A, and B and Common Notions 1–5 hold in spherical geometry? Is spherical geometry neutral?

14. Comment on the spherical analog of Euclid's Proposition 16.

15. Prove that every neutral triangle has an internal altitude.

16. Prove Proposition 8.1.7.

17. Prove that every reflection is a neutral isometry.

The point P is said to be a fixed point *of the transformation f if $f(P) = P$.*

18. Prove that every neutral isometry with more than one fixed point is either the identity or a reflection.

19. Prove that every neutral isometry with exactly one fixed point is a rotation.

Exercises 20–26 refer to the Euclidean plane and constitute a classification of its isometries.

Given any two points A and B of the Euclidean plane, the translation τ_{AB} *is a transformation such that if $P' = \tau_{AB}(P)$, then $PP' = AB$ and either PP' is parallel to AB or else all four points are collinear. (By definition, $\tau_{AA} = \mathrm{Id}.$)*

20. Prove that the composition of two translations is a translation.

21. Prove that the composition of two reflections with parallel axes is a translation.

22. Prove that the composition of two rotations is either a rotation or a translation.

23. Prove that the composition of a translation and a rotation which is not the identity is a rotation.

Given two points A and B on a straight line m of the Euclidean plane, the composition $\tau_{AB} \circ \rho_m$ is called a glide reflection *and is denoted by γ_{AB} whenever $A \neq B$. Note that every reflection is a glide reflection.*

24. Prove that the composition of any translation with any reflection is a glide reflection.

25. Prove that the composition of any rotation with any reflection is a glide reflection.

26. Prove that every isometry of the Euclidean plane is a translation, a rotation, or a glide reflection.

The distance *from a point P to a line m is the length of the shortest line segment that joins P to m.*

27. Prove that in neutral geometry the distance from a point P to a straight line m not containing P is realized by the unique line segment from P that is perpendicular to m.

8.2 The Upper Half-plane

In 1840 Ferdinand Minding (1806–1885) demonstrated the existence of surfaces of constant negative curvature (see Example 6.2.10) and investigated their intrinsic geometry in great detail. A few years later Joseph Liouville (1809–1882) observed that every surface of constant curvature -1 could be reparametrized so that its first fundamental form assumed the very simple form of the half-plane metric

$$\mathscr{H} = \frac{du^2 + dv^2}{v^2}. \tag{1}$$

In the 1880s Poincaré published several papers on the half-plane geometry—work that he considered to be his greatest contribution to mathematics. In the ensuing pages we shall describe the half-plane geometry in great detail. It will be demonstrated that this geometry is every bit as interesting and as rich as Euclidean geometry. Moreover, this geometry will be shown to constitute the answer to one of the oldest difficult problems of mathematics, namely, whether non-Euclidean neutral geometries exist. Finally, we shall describe, albeit informally, the important role that this geometry plays in state-of-the art mathematical research. First, however, we present a different, both more naive and more accessible, motivation for the metric in (1).

Along the lines employed in Edwin A. Abbott's classic *Flatland: A Romance of Many Dimensions*, we ask our readers to imagine that the upper half-plane is inhabited by two-dimensional creatures. Moreover, let us assume that unbeknownst to these inhabitants, the u-axis is infinitely cold and, their ignorance of this fact notwithstanding, it produces on them the effect of shrinking their sizes by a factor that is proportional to their distance from this cold u-axis. In other words,

$$\text{half-plane length} = \frac{\text{Euclidean length}}{v}.$$

Since the metric

$$\mathscr{E} = du^2 + dv^2$$

defines Euclidean geometry, it follows that the infinitely cold u-axis induces on the upper half-plane the geometry defined by the half-plane metric

$$\mathscr{H} = \frac{du^2 + dv^2}{v^2}.$$

The geodesics of the geometry $\mathbb{H}$ were determined in Proposition 7.1.9. The two kinds of geodesics described there are, respectively, *bowed geodesics* and *straight geodesics*. As will be demonstrated below, this distinction is external only – the half-plane people could detect no difference between these two varieties.

The *half-plane distance* between two points is the half-plane length of the geodesic that joins them. If they are joined by a straight geodesic, then their half-plane distance is given by Equation (7.7). Otherwise, it can be computed by means of Equation (7.6).

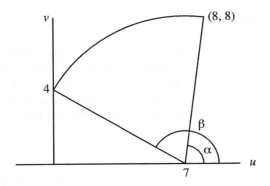

Figure 8.9

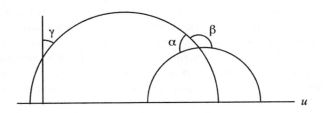

Figure 8.10   Three types of angles in $\mathbb{H}$.

Example 8.2.1 Compute the half-plane distance between the points $(0,4)$ and $(8,8)$.

The unique point on the u-axis that is equidistant from $(0,4)$ and $(8,8)$ is $(7,0)$ (see Fig. 8.9). It is clear from the figure that $\csc\alpha = \sqrt{65}/8, \cot\alpha = 1/8, \csc\beta = \sqrt{65}/4, \cot\beta = -7/4$. It follows from Equation (7.7) that the half-plane distance in question is

$$\ln \frac{\dfrac{\sqrt{65}}{4} - \left(-\dfrac{7}{4}\right)}{\dfrac{\sqrt{65}}{8} - \dfrac{1}{8}} = \ln \frac{2(\sqrt{65}+7)}{\sqrt{65}-1} \approx 1.451.$$

It was noted in Chapter 7 that the half-plane metric $\mathscr{H}$ is conformal. This means that given two intersecting curves in $\mathbb{H}$, the half-plane and Euclidean measures of the angles they form agree. Here we are mainly interested in the measures of the angles formed by half-plane geodesics. The computation of the measures of these angles is greatly facilitated by the fact that the tangent to a Euclidean circle is perpendicular to the radius through the point of contact. From our point of view there are three possible dispositions of the pairs of geodesic rays that constitute the sides of a half-plane angle:

(a) Both geodesics are bowed, and both rays lie on one side of the angle's vertex (α in Fig. 8.10).

(b) Both geodesics are bowed, and the angle's rays lie on both sides of its vertex (β in Fig. 8.10).

(c) One geodesic is straight and the other is bowed (γ in Fig. 8.10).

A *half-plane triangle* has three geodesics as its sides.

Example 8.2.2 Determine the measures of the angles of the half-plane $\triangle ABC$ where $A = (0,1), B = (4,1)$, and $C = (8,1)$ (see Fig. 8.11).

It is clear that the Euclidean centers of the geodesics AB, AC, and BC are $P(2,0)$, $Q(4,0)$, and $R(6,0)$, respectively. Because of the perpendicularity of the radii PA and QA to the respective geodesics AB and AC, it follows from the Euclidean law of cosines that

$$\angle BCA = \angle BAC = \angle PAQ = \cos^{-1}\left(\frac{5+17-4}{2\sqrt{5}\sqrt{17}}\right)$$
$$\approx 0.219 \text{ radians} \approx 12.5°.$$

Because of the different disposition of the sides of $\angle ABC$ relative to its vertex, this angle is supplementary to $\angle PBR$ rather than equal to it. To be precise,

$$\angle ABC = \pi - \cos^{-1}\left(\frac{5+5-16}{2\sqrt{5}\sqrt{5}}\right) \approx 0.927 \text{ radians} \approx 53.1°.$$

Example 8.2.3 Determine the measures of the half-plane angles of $\triangle ABC$ where $A = (0,16), B = (0,4)$, and $C = (8,4)$.

The geodesics AC and BC are centered at $P(-11,0)$ and $Q(4,0)$ respectively (see Fig. 8.12). Hence,

$$\angle BAC = \frac{\pi}{2} - \angle OAP = \angle OPA = \tan^{-1}\left(\frac{16}{11}\right) \approx 0.969 \text{ radians} \approx 55.5°,$$

$$\angle ABC = \frac{\pi}{2} - \angle OBQ = \angle OQB = \tan^{-1}\frac{4}{4} = \frac{\pi}{4} = 45°,$$

and

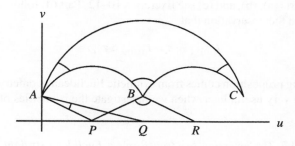

Figure 8.11   A triangle in $\mathbb{H}$.

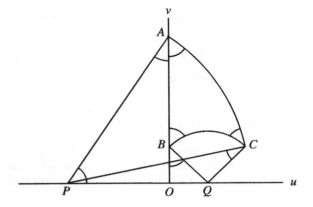

Figure 8.12   A triangle in $\mathbb{H}$.

$$\angle ACB = \angle PCQ = \cos^{-1}\left(\frac{377 + 32 - 225}{2\sqrt{377}\sqrt{32}}\right) \approx 0.577 \approx 33.1°.$$

It is noteworthy that in both of the half-plane triangles of Examples 8.2.2 and 8.2.3 the sum of the angles is considerably less than 180°: 78.1° in one case and 133.6° in the other.

We next describe four types of isometries of the metric $\mathscr{H}$ that suffice for our purposes here. The complete description of the isometries of $\mathscr{H}$ appears in Section 8.4. The domain of these transformations is understood to be the upper half-plane, and all four are easily seen to be bijections.

Proposition 8.2.4 *The following transformations are isometries of* $\mathbb{H}$*:*

(a) *Horizontal translations* $\tau_a(u,v) = (u+a,v)$ *where a is a real number.*

(b) *Reflections* ρ_m *where m is perpendicular to the u-axis.*

(c) *Dilations* $D_a(u,v) = (au,av)$ *where a is a positive constant.*

(d) *Inversions* $I_{C,k}$ *where C is on the u-axis.*

PROOF: For parts (a), (b), and (c) see Exercises 10–12. Part (d) follows from Proposition 7.1.13 and the observation that

$$I_{(c,0),k} = \tau_c \circ I_{(0,0),k} \circ \tau_{-c}. \qquad \text{Q.E.D.}$$

The following proposition comes from synthetic Euclidean geometry. It will, nevertheless, prove very useful later when we investigate the isometries of $\mathbb{H}$ in greater detail.

Proposition 8.2.5 *The inversion* $I_{C,k}$ *transforms a Euclidean straight line not containing C to a circle that does contain C, and vice versa.*

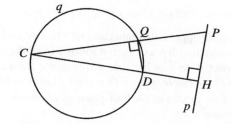

Figure 8.13   An inversion.

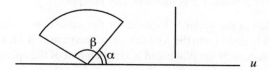

Figure 8.14   Two finite geodesics in the half-plane.

PROOF: In Figure 8.13, let CD be a diameter of the Euclidean circle q that is perpendicular to the Euclidean straight line p at H. Since

$$\angle CQD = \frac{\pi}{2} = \angle CHP,$$

it follows that $\triangle CQD$ and $\triangle CHP$ are similar, so that

$$\frac{CD}{CQ} = \frac{CP}{CH}, \quad \text{or} \quad CD \cdot CH = CP \cdot CQ.$$

Since the lengths of CD and CH are arbitrary, they may be chosen so that $k^2 = CD \cdot CH$, from which it follows that

$$I_{C,k}(P) = Q \quad \text{and so} \quad I_{C,k}(p) = q$$

and vice versa.   Q.E.D.

We now reexamine the postulates that define Euclidean geometry in the context of $\mathbb{H}$.

Postulate 1 holds because two points (a,b) and (a,c) are joined by a unique straight geodesic, whereas two points (a,b) and (a',c), $a \neq a'$, are joined by a unique bowed geodesic.

Postulate 2 also holds provided the finite geodesics of Figure 8.14 can be extended to arbitrary half-plane lengths in both directions. Such is indeed the case, because, for example,

$$\lim_{\alpha \to 0} \ln \frac{\csc \beta - \cot \beta}{\csc \alpha - \cot \alpha} = \lim_{\alpha \to 0} \ln[(\csc \beta - \cot \beta)(\csc \alpha + \cot \alpha)] = \infty.$$

Similar arguments dispose of all the other cases.

Let C be any point of the upper half-plane, and let r be any positive real number. If g is any geodesic ray emanating from C, then it follows from the validity of Postulate 2 that there exists a point P on g at half-plane distance r from C. The locus of all such points P is a *half-plane circle*. Hence *Postulate 3* holds in half-plane geometry. The Euclidean shape of the circles of the half-plane is, of course, of interest, and will be disclosed in Proposition 8.4.8.

That *Postulate 4* holds follows from the conformality of the Poincaré metric and the validity of this postulate in the Euclidean plane.

The discussion of *Postulate 5* is momentarily postponed.

Postulate A holds in the upper half-plane for the same reason that it holds in the Euclidean plane. It follows from the Jordan Curve Theorem (2.4.1), which is inherent in the topological nature of the plane and is independent of the specific metric used.

Given our imperfect understanding of the isometries of $\mathbb{H}$, the validity of *Postulate B* in this context requires careful proof.

Proposition 8.2.6 *Given any two geodesic rays g and h of $\mathbb{H}$, emanating from the points P and Q, respectively, there is a sequence of isometries of $\mathbb{H}$ whose composition carries P and g onto Q and h, respectively.*

PROOF: We first prove the theorem for entire geodesics g and h rather than rays.

It may be supposed, without loss of generality, that $P = (0,1)$ and g is the positive v-axis. We proceed by considering the several possibilities for (Q, h).

Case 1: If $Q = (0,b)$ and h is the v-axis, then $I_{(0,0),\sqrt{b}}$ does the job.

Case 2: If $Q = (a,b)$ and h is the straight geodesic $u = a$, then the horizontal translation $\tau_{-a}(u,v) = (u - a, v)$ reduces this situation to the previous one.

Case 3: If $Q = (a,b)$ and h is a bowed geodesic through Q, let C and D be the endpoints of h on the u-axis. By Proposition 8.2.5, the inversion $I_{C,CD}$ then transforms the pair (Q, h) into a pair (Q', h') where h' is the straight geodesic above D. By Case 2 we are done.

Finally, to show that the theorem also holds when g and h are rays, we observe that the inversion $I_{(0,0),1}$ interchanges the two half-plane rays that $P(0,1)$ determines on the positive v-axis.   Q.E.D.

So far we have seen that Postulates 1–4, A, and B all hold in $\mathbb{H}$. Its geometry is therefore neutral. Examples 8.2.2 and 8.2.3, as well as Figures 8.15 and 8.16, demonstrate that this geometry is not Euclidean. The sums of the angles of both of the triangles in Examples 8.2.2 and 8.2.3 are less than π. The distinct geodesics p, q, r in Figure 8.15 all contain the point M and are all parallel to m, since they do not intersect it. Finally, in Figure 8.16, $\alpha + \beta < \pi$, and yet the geodesics m and n do not intersect, thus providing a counterexample to Euclid's Postulate 5.

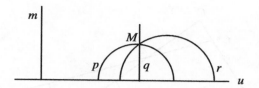

Figure 8.15   A counterexample to Playfair's postulate.

Hyperbolic geometry is formally defined as any neutral geometry in which Euclid's Postulate 5 does not hold. The geometry of the upper half-plane is hyperbolic, and other examples will be described later on. As hyperbolic geometry is a species of neutral geometry, it follows that Propositions 1–28 of Book I of Euclid's *Elements*, as well as the isometry Propositions 8.1.6–8.1.12, are all valid in $\mathbb{H}$ when correctly interpreted. It is strongly recommended that the reader take some time out at this point to reread these propositions with the understanding that they apply to half-plane figures.

The area of the geodesic triangle in $\mathbb{H}$ was already derived in Example 7.1.18. This important formula is restated here for the sake of completeness. An elementary derivation is outlined in Exercises 12–13.

Theorem 8.2.7 *The half-plane area of a half-plane triangle with angles α, β, γ is*

$$\pi - \alpha - \beta - \gamma. \qquad\qquad \Box\,(2)$$

Theorem 8.2.7 has many surprising consequences. To begin with, it implies that no half-plane triangle can have area greater than π. This stands in marked contrast to the situation in the Euclidean plane where, regardless of the units employed, triangles can have arbitrarily large areas. However, this restriction on the areas of triangles in the half-plane does resemble the situation in spherical geometry, where the area of a triangle is bounded by the area of the hemisphere.

Actually, the angles of the half-plane triangle determine more than just its area. They also determine its sides. Since the proof we offer is synthetic rather than analytic, the accompanying illustration is to be interpreted as a neutral figure, rather than a half-plane one.

Proposition 8.2.8 (AAA Congruence) *If the respective angles of two half-plane triangles are equal, then the triangles are congruent.*

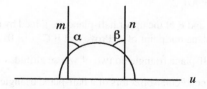

Figure 8.16   A counterexample to Postulate 5.

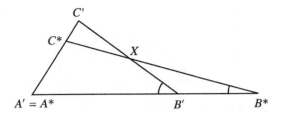

Figure 8.17   A proof of AAA congruence.

PROOF: Suppose $\triangle ABC$ and $\triangle A'B'C'$ are the triangles in question. Because the ASA congruence theorem holds in both neutral and hyperbolic geometry, it suffices to prove that $AB = A'B'$. Suppose they are unequal; then it may be assumed without loss of generality that $AB > A'B'$.

Let f be an isometry of the neutral plane such that $A^* = f(A) = A'$, the straight line $A^*B^* = f(A)f(B)$ falls along $A'B'$, and $A^*C^* = f(A)f(C)$ falls along $A'C'$. Because the triangles $A'B'C'$ and $A^*B^*C^*$ have equal areas, neither can contain the other, and hence it may be assumed that they overlap as in the neutral Figure 8.17. In that case however, $\angle A^*B^*C^*$ is interior and $\angle A'B'C'$ is exterior relative to $\triangle XB^*B'$, which contradicts the hypothesized equality. It follows that $AB = A'B'$ and hence $\triangle ABC \cong \triangle A'B'C'$. 　　　　　　　　　　　　　　　　　　　　　　　　Q.E.D.

Exercises 8.2

1. Find the lengths of the sides, the angles, and the area of the half-plane triangle with vertices $A(1,2), B(3,2), C(7,2)$.

2. Find the lengths of the sides, the angles, and the area of the half-plane triangle with vertices $A(0,1), B(0,2), C(2,1)$.

3. Show that if the points $P(x_1, y_1), Q(x_2, y_2)$ are contained in a bowed geodesic with Euclidean center $(c,0)$ and radius r, then the hyperbolic distance between P and Q is

$$\left| \ln \frac{(x_1 - c - r)y_2}{y_1(x_2 - c - r)} \right|.$$

4. The midpoints D, E, F of a half-plane equilateral triangle ABC are joined to form a new triangle. Prove that $\triangle DEF$ is also equilateral and compare its angles to those of $\triangle ABC$. Also prove that $DE < AB/2$.

5*. Suppose the points A and B of the upper half-plane are joined by a bowed geodesic g. Let $M(c,d)$ be the half-plane midpoint of g. Prove that if $C = (c,0)$, then $\angle BCM = \angle ACM$.

6*. Does there exist a half-plane triangle no two of whose altitudes intersect?

7. Suppose that two perpendicular bisectors of a half-plane triangle intersect in a point P of the upper half-plane. Prove that the perpendicular bisector to the third side also passes through P.

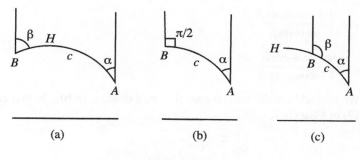

Figure 8.18

8*. Suppose that two perpendicular bisectors of a half-plane triangle intersect in a point P on the u-axis. Prove that the perpendicular bisector to the third side also passes through P.

9*. Find a half-plane triangle no two of whose perpendicular bisectors intersect.

10. Prove formally that the horizontal translation τ_a is a half-plane isometry.

11. Prove formally that if m is a straight geodesic of the half-plane, then the reflection ρ_m is an isometry.

12. Prove formally that the dilation $D_a, a > 0$, is a half-plane isometry.

13*. Use a direct integration, like that of Example 7.1.15, to prove that the half-plane area of the regions depicted in Figure 8.18 is $\pi - \alpha - \beta$.

14*. Use the previous exercise to prove Theorem 8.2.7.

Given a geodesic g, a curve h is said to be equidistant *from g provided all the points of h have the same distance from g.*

15*. Prove that if g is the straight geodesic of the half-plane above the point $(a, 0)$, then its equidistant curves are the Euclidean rays that emanate from $(a, 0)$.

16*. Prove that if g is the bowed geodesic that joins the points $(a, 0)$ and $(b, 0)$, then its equidistant curves are arcs of circles that join the same two points.

8.3   The Half-Plane Theorem of Pythagoras

In view of the fundamental importance of the Theorem of Pythagoras in Euclidean geometry, it is natural to look for a half-plane analog. By this is meant an equation that relates the sides of a right half-plane triangle. Above and beyond its own intrinsic interest, this theorem sheds light on the relation of half-plane geometry to both Euclidean and spherical geometries.

Lemma 8.3.1 *Let c denote the half-plane length of the bowed geodesic segment AB in Figure 8.18. Then,*

1. $\sinh c = \dfrac{\cos \alpha + \cos \beta}{\sin \alpha \sin \beta}$;

2. $\cosh c = \dfrac{1 + \cos \alpha \cos \beta}{\sin \alpha \sin \beta}$;

3. $\tanh c = \dfrac{\cos \alpha + \cos \beta}{1 + \cos \alpha \cos \beta}$.

PROOF: We first address the special case $\beta = \pi/2$ (Figure 18(b)). In that case, by Equation (7) of Chapter 7,

$$c = \ln \frac{\csc \dfrac{\pi}{2} - \cot \dfrac{\pi}{2}}{\csc \alpha - \cot \alpha} = \ln \frac{\sin \alpha}{1 - \cos \alpha}.$$

It follows that

$$
\begin{aligned}
2 \sinh c = e^c - e^{-c} &= \frac{\sin \alpha}{1 - \cos \alpha} - \frac{1 - \cos \alpha}{\sin \alpha} \\
&= \frac{\sin^2 \alpha - 1 + 2 \cos \alpha - \cos^2 \alpha}{\sin \alpha (1 - \cos \alpha)} \\
&= \frac{2 \cos \alpha (1 - \cos \alpha)}{\sin \alpha (1 - \cos \alpha)} = 2 \cot \alpha.
\end{aligned}
$$

Thus,

$$\sinh c = \cot \alpha \qquad \text{when } \beta = \frac{\pi}{2}.$$

Since

$$\cosh^2 c - \sinh^2 c = 1 \quad \text{and} \quad \csc^2 \alpha - \cot^2 \alpha = 1,$$

we have

$$\cosh c = \csc \alpha \quad \text{and} \quad \tanh c = \cos \alpha \qquad \text{when } \beta = \frac{\pi}{2}.$$

If $\beta \neq \pi/2$ then the high point H on the geodesic containing AB is either inside AB (Fig. 8.18(a)) or outside it (Fig. 8.18(c)). In the first case

$$\sinh c = \sinh(AH + HB) = \sinh(AH)\cosh(HB) + \cosh(AH)\sinh(HB)$$

$$= \cot \alpha \csc \beta + \csc \alpha \cot \beta = \frac{\cos \alpha + \cos \beta}{\sin \alpha \sin \beta}.$$

The remainder of the proof is relegated to Exercise 16.   Q.E.D.

We are now ready to state and prove the half-plane analog of the Theorem of Pythagoras.

Theorem 8.3.2 *Let ABC be a half-plane triangle with a right angle at C. If a, b, c are the half-plane lengths of the sides opposite the vertices A, B, C respectively, then*

$$\cosh c = \cosh a \cosh b.$$

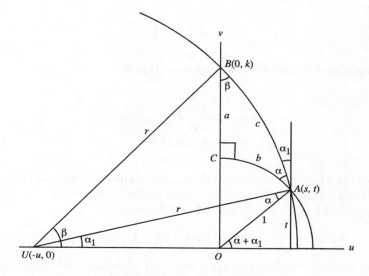

Figure 8.19   The Theorem of Pythagoras in $\mathbb{H}$.

PROOF: It follows from Proposition 8.2.6 that $\triangle ABC$ can be positioned as in Figure 8.19 with r as the radius and $U(-u,0)$ as the center of the bowed geodesic AB and $C = (1,0)$. By Equation (6) of Chapter 7, $a = \ln k$, so that

$$\cosh a = \frac{e^a + e^{-a}}{2} = \frac{k + \dfrac{1}{k}}{2} = \frac{k^2 + 1}{2k}.$$

By Lemma 8.3.1,

$$\cosh b = \csc(\alpha + \alpha_1) = \frac{1}{t}.$$

Again by Lemma 8.3.1,

$$\cosh c = \frac{1 + \cos \alpha_1 \cos(\pi - \beta)}{\sin \alpha_1 \sin(\pi - \beta)}$$

$$= \frac{1 - \dfrac{u + s}{r} \dfrac{u}{r}}{\dfrac{t}{r} \dfrac{k}{r}} = \frac{r^2 - u^2 - us}{kt}$$

$$= \frac{k^2 - us}{kt}. \tag{3}$$

However, by the Euclidean Theorem of Pythagoras,

$$s^2 + t^2 = 1 \quad \text{and} \quad (s + u)^2 + t^2 = u^2 + k^2 = r^2,$$

so that

$$2us + u^2 = r^2 - 1,$$

or

$$us = \frac{r^2 - u^2 - 1}{2} = \frac{k^2 - 1}{2}. \tag{4}$$

The substitution of Equation (4) into Equation (3) yields

$$\cosh c = \frac{k^2 - \dfrac{k^2 - 1}{2}}{kt} = \frac{k^2 + 1}{2kt}$$

$$= \cosh a \cosh b. \qquad\qquad \text{Q.E.D.}$$

It should be of interest to compare the numerical aspects of the half-plane Theorem of Pythagoras with those of its Euclidean analog. Accordingly, if a half-plane right triangle has legs $a = b = 1$, then it has hypotenuse

$$c = \cosh^{-1}(\cosh^2(1)) \approx 1.51.$$

This is not very different from the hypotenuse $c \approx 1.41$ in the Euclidean right triangle with legs $a = b = 1$. On the other hand, if a half-plane right triangle has legs $a = b = 10$, then it has hypotenuse

$$c = \cosh^{-1}(\cosh^2(10)) \approx 19.31,$$

which is about as far off from the Euclidean hypotenuse of 14.15 as it could be (because of the triangle inequality, this hypotenuse couldn't be greater than 20).

We now argue formally that, as these examples indicate, for small triangles the Euclidean and half-plane geometries of the right triangle are approximately the same.

Recall that

$$e^x = 1 + x + \frac{x^2}{2!} + \frac{x^3}{3!} + \cdots.$$

Hence, if terms of degree 4 or higher are ignored, we obtain

$$\cosh x = \frac{e^x + e^{-x}}{2} = 1 + \frac{x^2}{2}.$$

Consequently, if we persist in ignoring terms of degree 4 or more, the half-plane Theorem of Pythagoras becomes

$$1 + \frac{c^2}{2} = \left(1 + \frac{a^2}{2}\right)\left(1 + \frac{b^2}{2}\right)$$

which simplifies to

$$c^2 = a^2 + b^2.$$

These observations bring out the possibility that our own universe might actually be hyperbolic rather than Euclidean. It is conceivable that the geometry of our universe is hyperbolic rather than Euclidean, but that that part of the universe that we can observe is so small that our instruments cannot detect the difference between the

Euclidean and hyperbolic geometries. The situation is entirely analogous to that of people who live on the surface of a sphere but believe their world to be flat.

We would like to answer the question of how small a portion of half-plane geometry our observable universe must be restricted to so as to allow for such a confusion to occur. First, however, it is necessary to address the issue of units. If $X = (0,1)$ and $Y = (0,e)$, then their distance in $\mathbb{H}$ is

$$\ln \frac{e}{1} = 1.$$

This half-plane distance is independent of the units used in the Cartesian coordinate system that was employed to define $\mathbb{H}$. For example, if the original units were meters and we decide to use millimeters instead, then the coordinates of X and Y become $(0,1000)$ and $(0,1000e)$ and their half-plane distance now becomes

$$\ln \frac{1000e}{1000} = 1.$$

Thus, the half-plane segment XY can be chosen as the natural unit for lengths in $\mathcal{H}$. This is also known as the *absolute unit* of the half-plane.

Now, if we occupied such a small portion of $\mathbb{H}$ that the fourth powers of our measurements, expressed in terms of this absolute unit, were physically undetectable, then we would also fail to detect any practical difference between half-plane and Euclidean geometries.

It was noted above that the relationship between the hyperbolic and the spherical expressions for the area of a triangle caused Lambert to comment that it looks as though the former is the geometry of a sphere with an imaginary radius. It will now be demonstrated that the same relationship holds between the half-plane and spherical Theorems of Pythagoras. This third version of the Theorem of Pythagoras is obtained from the spherical law of cosines.

Proposition 8.3.3 (The Spherical Law of Cosines) *Let ABC be a triangle on a sphere of radius R. If α, β, γ are the angles at A, B, C, respectively, and a, b, c are the lengths of the opposite sides (Fig. 8.20), then*

$$\cos \gamma = \frac{\cos \dfrac{c}{R} - \cos \dfrac{a}{R} \cos \dfrac{b}{R}}{\sin \dfrac{a}{R} \sin \dfrac{b}{R}}. \tag{5}$$

PROOF: Assume first that the sphere has radius 1. If the sphere's center is O, let

$$\mathbf{a} = \overrightarrow{OA}, \quad \mathbf{b} = \overrightarrow{OB}, \quad \mathbf{c} = \overrightarrow{OC}.$$

It follows from the definition of the dot product of vectors that

$$\cos \gamma = \frac{\mathbf{a} \times \mathbf{c}}{|\mathbf{a} \times \mathbf{c}|} \cdot \frac{\mathbf{b} \times \mathbf{c}}{|\mathbf{b} \times \mathbf{c}|}.$$

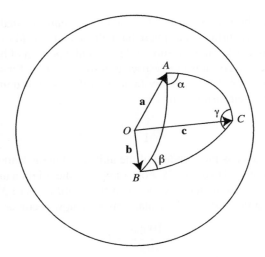

Figure 8.20   A spherical triangle.

However, by the identity of Lagrange,

$$(\mathbf{a} \times \mathbf{c}) \cdot (\mathbf{b} \times \mathbf{c}) = (\mathbf{a} \cdot \mathbf{b})(\mathbf{c} \cdot \mathbf{c}) - (\mathbf{a} \cdot \mathbf{c})(\mathbf{b} \cdot \mathbf{c})$$
$$= (\mathbf{a} \cdot \mathbf{b}) - (\mathbf{a} \cdot \mathbf{c})(\mathbf{b} \cdot \mathbf{c}).$$

Moreover, by the definition of the cross product,

$$|\mathbf{a} \times \mathbf{c}| = 2 \, \text{area}(\Delta OAC) = \sin b,$$

and similarly

$$|\mathbf{b} \times \mathbf{c}| = \sin a.$$

When these values are substituted into the above expression for $\cos \gamma$, we obtain

$$\cos \gamma = \frac{\cos c - \cos a \cos b}{\sin a \sin b}, \tag{6}$$

which is the desired equation in the case $R = 1$.

In the case where R is arbitrary, project the given sphere S to a concentric sphere S' of unit radius. The given triangle then projects to a triangle with the same respective angles as the original triangle, but whose sides have lengths $a/R, b/R, c/R$ respectively. When these values are substituted into Equation (6), we obtain the general Equation (5). Q.E.D.

Corollary 8.3.4 (The Spherical Theorem of Pythagoras) *On a sphere of radius R, let ABC be a triangle with a right angle at C. If a, b, c are the lengths of the sides opposite the vertices A, B, C respectively, then*

$$\cos \frac{c}{R} = \cos \frac{a}{R} \cos \frac{b}{R}. \tag{7}$$

PROOF: This follows immediately upon the substitution $\gamma = \pi/2$ in Equation (5).

Q.E.D.

Let i denote the imaginary number $\sqrt{-1}$. Note that the replacement of x with x/i in the Taylor series expansion of $\cos x$ yields

$$\cos \frac{x}{i} = 1 - \frac{x^2}{2!i^2} + \frac{x^4}{4!i^4} - \frac{x^6}{6!i^6} + \frac{x^8}{8!i^8} - + \cdots$$

$$= 1 + \frac{x^2}{2!} + \frac{x^4}{4!} + \frac{x^6}{6!} + \frac{x^8}{8!} + \cdots = \cosh x.$$

Thus, the substitution of $R = i$ converts the spherical Theorem of Pythagoras to the half-plane Theorem of Pythagoras! This observation supports Johann Lambert's (1729–1777) quip that (neutral) non-Euclidean geometry is the geometry of the surface of a sphere with an imaginary radius. There are other instances of this analogy between half-plane and spherical geometry. In fact, for each hyperbolic geometry there is a positive real number r such that the following equations hold for every triangle with sides a, b, c and respective opposite angles α, β, γ:

$$\cos \alpha = \frac{\cosh \dfrac{b}{r} \cosh \dfrac{c}{r} - \cosh \dfrac{a}{r}}{\sinh \dfrac{b}{r} \sinh \dfrac{c}{r}}, \tag{8}$$

$$\cosh \frac{a}{r} = \frac{\cos \alpha + \cos \beta \cos \gamma}{\sin \beta \sin \gamma}, \tag{9}$$

$$\frac{\sinh \dfrac{a}{r}}{\sin \alpha} = \frac{\sinh \dfrac{b}{r}}{\sin \beta} = \frac{\sinh \dfrac{c}{r}}{\sin \gamma}. \tag{10}$$

On the other hand, it is well known that the following equations hold for every such triangle on a sphere of radius R:

$$\cos \alpha = \frac{\cos \dfrac{a}{R} - \cos \dfrac{b}{R} \cos \dfrac{c}{R}}{\sin \dfrac{b}{R} \sin \dfrac{c}{R}}, \tag{11}$$

$$\cos \frac{a}{R} = \frac{\cos \alpha + \cos \beta \cos \gamma}{\sin \beta \sin \gamma}, \tag{12}$$

$$\frac{\sin \dfrac{a}{R}}{\sin \alpha} = \frac{\sin \dfrac{b}{R}}{\sin \beta} = \frac{\sin \dfrac{c}{R}}{\sin \gamma}. \tag{13}$$

In each case the hyperbolic equation can be obtained from the corresponding spherical equation by replacing R with ri (Exercise 17).

Nikolai I. Lobachevsky (1793–1856) and János Bolyai (1802–1860), the coinventors of hyperbolic geometry, were aware of the respective similarities between

Equations (8)–(10) and Equations (10)–(12). These analogies helped them overcome their nagging doubts regarding the consistency of the strange logical edifices they had constructed. For good reasons, they couldn't help but be concerned about the possibility of a contradiction arising further on down the line, when someone else derived more propositions of hyperbolic geometry. The aforementioned similarity gave them a heuristic argument for "proving" the relative impossibity of such a contradiction. For a contradiction resulting from the hyperbolic trigonometric formulas would also have a counterpart in the context of the spherical ones, whose consistency is taken for granted by all mathematicians. We say *relative* because even if this argument were rigorous, which it is not, it would at best have proven that hyperbolic geometry is at least as consistent as spherical geometry. Absolute proofs of consistency of such complicated logical systems as Euclidean, hyperbolic, and spherical geometries have been demonstrated to be impossible.

It should be mentioned that work done by Arthur Cayley (1821–1895) in the 1850s provides a satisfactory explanation for the strange role played by the imaginary i in relating the formulas of hyperbolic geometry to spherical geometry.

It might seem reasonable to assume that hyperbolic geometry derives its name from the role played in it by the hyperbolic functions. In reality, however, that is not the case. Hyperbolic trigonometric functions received that name because the hyperbola bears the same relation to the hyperbolic trigonometric functions as the circle does to the classical trigonometric functions. Hyperbolic geometry, on the other hand, owes its name to the fact that it opens up at infinity much faster than Euclidean geometry, much as the hyperbola $x^2 - y^2 = 1$ opens up faster than the parabola $y = x^2$. For example, the circle of radius R has half-plane circumference $2\pi \sinh(R)$ (see Exercise 8.4.9) which is, in general, much larger than the Euclidean circumference of $2\pi R$.

This is the place to explain the important role played by the half-plane $\mathbb{H}$ within the general framework of mathematical research. In the 1820s Niels H. Abel (1802–1829) and Carl G. J. Jacobi (1804–1851) solved an old and difficult problem by providing a context for the investigation of *elliptic integrals*. These are integrals of the type

$$I(x) = \int R\left(\sqrt{x^3 + ax^2 + bx + c}\right) dx \tag{14}$$

where R is a rational function. The elliptic integrals have important applications to physics and are the subjects of entire volumes. As part of their solutions both Abel and Jacobi took the following steps:

1. The focus was shifted to the *inverses* of the elliptic integrals, known as *elliptic functions*.

2. The variable x in the elliptic integral was allowed to assume complex values.

The advantage of the first step can be illustrated by the integral

$$\int \frac{dx}{\sqrt{1 - x^2}}, \tag{15}$$

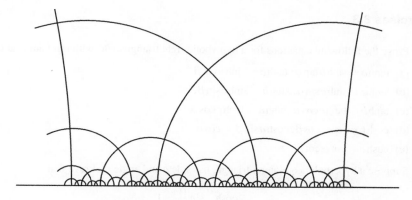

Figure 8.21   A pentagonal tiling of the half-plane.

whose inverse is the well known function $\sin x$. This integral can also be used to explain the significance of the second step. A fundamental property of the sine function is its periodicity. When considered as a function of a complex variable, elliptic functions (the inverses of the integrals of (14)) turn out to possesses similar periodicities in two directions. To be precise, for each elliptic function $f(z)$ there exist two nonzero complex numbers w_1 and w_2 such that w_1/w_2 is not real and

$$f(z) = f(z + mw_1 + nw_2), \qquad m, n \in \mathbb{Z}. \tag{16}$$

The set of points $z + mw_1 + nw_2$, $m, n \in \mathbb{Z}$, forms a grid in the plane whose cells are all congruent polygons. This is also commonly known as a *tiling* of the plane by this common polygon. The periodicity of (16), as well as both the geometric and the algebraic properties of the associated tiling, plays a crucial role in the study of elliptic integrals.

It turns out that a clever substitution converts integrals of the type

$$\int R \left(\sqrt{x^4 + ax^3 + bx^2 + cx + d} \right) dx$$

to the elliptic integrals of (14). However, when the degree of the radicand polynomial is raised further, or when the order of the radical itself is raised above 2, the tilings of (3) are no longer sufficient, and others must be called upon. These can be found in the half-plane.

It follows from the fact that the sum of the angles of the Euclidean triangle is $180°$ that a Euclidean regular pentagon's interior angle must be $108°$. Consequently, such pentagons cannot tile the Euclidean plane. The half-plane, on the other hand, does allow for regular pentagons with other angles. For example, Figure 8.21 displays a tiling of the half-plane with right-angled regular pentagons.

The algebraic structures that underlie tilings of the half-plane are known as the *Fuchsian groups*. They were classified around 1880, simultaneously and independently, by Poincaré and Klein. They also form the backdrop against which higher-order versions of the integrals of (14) are investigated.

Exercises 8.3

1*. Prove the following equations for a hyperbolic right triangle ABC with right angle at C:

   (a) $\tanh a = \sinh b \tan \alpha$, $\tanh b = \sinh a \tan \beta$.

   (b) $\sinh a = \sinh c \sin \alpha$, $\sinh b = \sinh c \sin \beta$.

   (c) $\tanh b = \tanh c \cos \alpha$, $\tanh a = \tanh c \cos \beta$.

   (d) $\cosh b \sin \alpha = \cos \beta$, $\cosh a \sin \beta = \cos \alpha$.

   (e) $\cosh c = \cot \alpha \cot \beta$.

2*. Suppose a hyperbolic equilateral triangle has side a and angle α. Prove that

$$2 \cosh \frac{a}{2} \sin \frac{\alpha}{2} = 1.$$

   Investigate what happens to α as $a \to 0$ or $a \to \infty$.

3. Let h_x denote the length of the hypotenuse of a half-plane right triangle with legs of length x each. Let e_x denote the length of the hypotenuse of a Euclidean right triangle with legs of length x each. Use a graphing calculator to compare h_x and e_x for all $0 < x < \infty$.

4. Explain why the inhabitants of a half-plane universe might, in principle, be able to verify that they occupy a hyperbolic, rather than Euclidean, universe.

5. Assuming the inevitability of some measurement errors, explain why the inhabitants of a Euclidean universe could never be absolutely sure that their universe is Euclidean rather than hyperbolic.

6*. Let ABC be a triangle in $\mathbb{H}$ with interior angles α, β, γ opposite sides of respective lengths a, b, c. Prove the following equations:

   (a) $\cos \alpha = \dfrac{\cosh b \cosh c - \cosh a}{\sinh b \sinh c}$,

   (b) $\cosh a = \dfrac{\cos \beta \cos \gamma + \cos \alpha}{\sin \beta \sin \gamma}$,

   (c) $\dfrac{\sinh a}{\sin \alpha} = \dfrac{\sinh b}{\sin \beta} = \dfrac{\sinh c}{\sin \gamma}$.

7*. Suppose a half-plane triangle has area $\pi/2$ radians. How large can its sides be? How small can they be?

8**. Prove that the length of the altitude to the hypothenuse of a half-plane right triangle cannot exceed $\ln(1 + \sqrt{2})$.

9**. Prove that the circle inscribed in a half-plane equilateral triangle has diameter at most $\ln 3$.

10. Show that the half-plane length of the geodesic joining the midpoints of two legs of a half-plane isosceles right triangle is less than half the length of its hypotenuse.

11**. Let S be the area of a half-plane right triangle with legs a and b. Prove that

$$\sin S = \frac{\sinh a \sinh b}{1 + \cosh a \cosh b}.$$

12*. Show that the length of the geodesic joining the midpoints of a half-plane equilateral triangle is less than $2\ln[(1+\sqrt{5})/2]$.

In any geometry, if A,B,C are points on a geodesic, the ratio AB/BC is considered to be positive or negative according as B is between A and C or not.

13*. Let ABC be a half-plane triangle, and let P,Q,R be points (distinct from A,B,C) on the geodesics BC,AC,AB respectively. Prove that the points P,Q,R all lie on the same geodesic if and only if

$$\frac{\sinh(AR)}{\sinh(RB)} = \frac{\sinh(BP)}{\sinh(PC)} = \frac{\sinh(CQ)}{\sinh(QA)} = -1.$$

In any geometry, a cevian *is a geodesic segment that joins the vertex of a geodesic triangle to a point on the opposite side.*

14*. Let ABC be a half-plane triangle, and let P,Q,R be points (distinct from A,B,C) on the geodesics BC,AC,AB respectively. Prove that the cevians AP,BQ,CR all pass through the same point if and only if

$$\frac{\sinh(AR)}{\sinh(RB)} = \frac{\sinh(BP)}{\sinh(PC)} = \frac{\sinh(CQ)}{\sinh(QA)} = 1.$$

15*. Prove that the medians of a half-plane triangle all pass through the same point.

16. Complete the proof of Lemma 8.3.1.

17. Show that the substitution $R = ri$ converts Equations (11)–(13) into Equations (8)–(10) respectively.

18*. Show that if $r > 0$, then Equations (8)–(10) hold in the upper half-plane relative to the metric

$$\mathscr{H}_r = \frac{r^2(du^2 + dv^2)}{v^2}.$$

8.4 Half-Plane Isometries

The theory of complex numbers supplies us with a very succinct and useful description of the isometries of the half-plane geometry. Let $\mathbb{C}$ denote the set of complex numbers. For any 2×2 nonsingular matrix A with complex entries a, b, c, d, we define the function $A_\mu : \mathbb{C} \to \mathbb{C}$ as

$$A_\mu(z) = \begin{pmatrix} a & b \\ c & d \end{pmatrix}_\mu (z) = \frac{az+b}{cz+d}.$$

These functions are called *Möbius transformations*. Note that while these transformations are described in terms of matrices, their operation is *not* linear, and their arguments are *not* vectors. Moreover, unlike matrices, if $\lambda \neq 0$, then

$$\begin{pmatrix} a & b \\ c & d \end{pmatrix}_\mu = \begin{pmatrix} \lambda a & \lambda b \\ \lambda c & \lambda d \end{pmatrix}_\mu. \tag{17}$$

Nevertheless, the Möbius transformations do partake of some of the properties of matrices. Witness the following proposition:

Proposition 8.4.1 *If A_μ and B_μ are two Möbius transformations, then*

$$A_\mu \circ B_\mu = (AB)_\mu, \qquad (A_\mu)^{-1} = (A^{-1})_\mu.$$

PROOF: Suppose

$$A = \begin{pmatrix} a_{11} & a_{12} \\ a_{21} & a_{22} \end{pmatrix}, \qquad B = \begin{pmatrix} b_{11} & b_{12} \\ b_{21} & b_{22} \end{pmatrix}.$$

Then

$$A_\mu \circ B_\mu(z) = A_\mu(B_\mu(z)) = A_\mu \left(\frac{b_{11}z + b_{12}}{b_{21}z + b_{22}} \right)$$

$$= \frac{a_{11} \dfrac{b_{11}z + b_{12}}{b_{21}z + b_{22}} + a_{12}}{a_{21} \dfrac{b_{11}z + b_{12}}{b_{21}z + b_{22}} + a_{22}} = \frac{(a_{11}b_{11} + a_{12}b_{21})z + (a_{11}b_{12} + a_{12}b_{22})}{(a_{21}b_{11} + a_{22}b_{21})z + (a_{21}b_{12} + a_{22}b_{22})}$$

$$= (AB)_\mu(z).$$

The second equation follows easily from the first one.   Q.E.D.

The Möbius transformation of (17) is undefined at $z = -d/c$ and is injective everywhere else on the complex plane $\mathbb{C}$ (Exercise 21). It is customary to smooth out this slight imperfection by adding to $\mathbb{C}$ an ideal point, denoted by ∞, with the understanding that

$$\tau \left(-\frac{d}{c} \right) = \infty \quad \text{and} \quad \tau(\infty) = \frac{a}{c}.$$

The augmented complex plain is denoted by $\mathbb{C}^*$ and is called the *Riemann sphere*. Because of the equation

$$\begin{pmatrix} a & b \\ c & d \end{pmatrix}_\mu \begin{pmatrix} d & -b \\ -c & a \end{pmatrix}_\mu = \begin{pmatrix} 1 & 0 \\ 0 & 1 \end{pmatrix}_\mu = \text{Id},$$

it follows that each (augmented) Möbius transformation is a bijection of the Riemann sphere $\mathbb{C}^*$. A similar pairing of ∞ with C turns the inversion $I_{C,k}$ into a bijection of $\mathbb{C}^*$. It is also convenient to define the *circles* of $\mathbb{C}^*$ as the ordinary circles of $\mathbb{C}$ together with the augmented straight lines $\lambda^* = \lambda \cup \{\infty\}$ where λ is an ordinary straight line in $\mathbb{C}$.

Let $\rho(z) = -\bar{z}$ denote the reflection in the y-axis. The easy proof of the following technical lemma is relegated to Exercise 6.

Lemma 8.4.2 *For any transformation* $\begin{pmatrix} a & b \\ c & d \end{pmatrix}_{\mu}$,

$$\begin{pmatrix} a & b \\ c & d \end{pmatrix}_{\mu} \circ \rho = \rho \circ \begin{pmatrix} a & -b \\ -c & d \end{pmatrix}_{\mu}. \qquad \square$$

The four types of half-plane isometries described in Proposition 8.2.4 can all be expressed as Möbius transformations. To wit,

$$\tau_a = \begin{pmatrix} 1 & a \\ 0 & 1 \end{pmatrix}_{\mu},$$

$$D_a = \begin{pmatrix} a & 0 \\ 0 & 1 \end{pmatrix}_{\mu},$$

$$\rho_{u=a} = \begin{pmatrix} 1 & 2a \\ 0 & 1 \end{pmatrix}_{\mu} \circ \rho,$$

and

$$I_{(a.0),k} = \tau_a \circ I_{(0,0),k} \circ \tau_{-a} = \begin{pmatrix} a & a^2 - k^2 \\ 1 & a \end{pmatrix}_{\mu} \circ \rho.$$

The set of Möbius transformations A_{μ} with *real* entries and for which $\det A > 0$ is called the *real special linear group (of dimension 2)* and is denoted by $\mathrm{SL}(2, \mathbb{R})$.

The following theorem provides an algebraic description of all the half-plane isometries. A geometric description of the action of these isometries will be given in the concluding pages of this chapter.

Theorem 8.4.3 *The isometries of the half-plane geometry consist of all the transformations of the complex plane of the formats:*

(a) A_{μ},

(b) $A_{\mu} \circ \rho$,

where A varies over all the 2×2 matrices with real entries and positive determinant.

PROOF: We first show that every transformation of the given formats is indeed an isometry of $\mathbb{H}$. Observe that if $c \neq 0$, then

$$\rho_{u=(a-d)/(2c)} \circ I_{(-d/c,0),\sqrt{ad-bc}/c}$$

$$= \begin{pmatrix} 1 & \frac{a-d}{c} \\ 0 & 1 \end{pmatrix}_{\mu} \circ \rho \circ \begin{pmatrix} -\frac{d}{c} & \frac{d^2}{c^2} - \frac{ad-bc}{c^2} \\ 1 & -\frac{d}{c} \end{pmatrix}_{\mu} \circ \rho$$

$$
= \begin{pmatrix} 1 & \frac{a-d}{c} \\ 0 & 1 \end{pmatrix}_\mu \circ \begin{pmatrix} -\frac{d}{c} & -\frac{d^2}{c^2} + \frac{ad-bc}{c^2} \\ -1 & -\frac{d}{c} \end{pmatrix}_\mu \circ \rho^2
$$

$$
= \begin{pmatrix} a & b \\ c & d \end{pmatrix}_\mu
$$

and

$$
\tau_{(a-d)/c} \circ I_{(d/c,0),\sqrt{ad-bc}/c}
$$

$$
= \begin{pmatrix} 1 & \frac{a-d}{c} \\ 0 & 1 \end{pmatrix}_\mu \circ \begin{pmatrix} \frac{d}{c} & \frac{d^2}{c^2} - \frac{ad-bc}{c^2} \\ 1 & \frac{d}{c} \end{pmatrix}_\mu \circ \rho
$$

$$
= \begin{pmatrix} a & b \\ c & d \end{pmatrix}_\mu \circ \rho.
$$

On the other hand, if $c = 0$, then for any $d \neq 0$

$$
\begin{pmatrix} a & b \\ c & d \end{pmatrix}_\mu (z) = \frac{a}{d}z + \frac{b}{d},
$$

which is the composition of a dilation with a translation, both of which are known to be isometries. Thus, all transformations of formats (a) and (b) are half-plane isometries.

Conversely, let f be a half-plane isometry. Since the half-plane geometry is neutral, it follows that f is the composition of some half-plane reflections. However, these reflections are transformations of the types $I_{(a,0),k}$ or $\rho_{u=a}$, which are known to have format (b). Moreover, it follows from Proposition 8.4.1 and Lemma 8.4.2 that the composition of any two transformations of types (a) and/or (b) also have formats (a) and/or (b). Hence f has one of the required formats.   Q.E.D.

We turn next to the examination of the effect that Möbius transformations have on the straight lines and circles of the complex plane. For this purpose a very useful invariant needs to be introduced. The *cross ratio* (z_1, z_2, z_3, z_4) of four points of the complex plane is the value of

$$
\frac{\dfrac{z_1 - z_3}{z_1 - z_2}}{\dfrac{z_4 - z_3}{z_4 - z_2}}. \tag{18}
$$

A set of points is said to be *cocyclic* if they all lie on one circle or on one (Euclidean) straight line. In other words, they all belong to the same circle of the Riemann sphere $\mathbb{C}^*$. The cross ratio provides a simple criterion for cocyclicity.

Proposition 8.4.4 *Four distinct points are cocyclic if and only if their cross ratio is a real number.*

PROOF: Suppose the points z_1, z_2, z_3, z_4 lie, in that order, on either a straight line or a circle. By a well-known proposition of Euclidean geometry,

$$\angle z_2 z_1 z_3 = \angle z_2 z_4 z_3.$$

By the argument principle for complex numbers these angles also equal

$$\arg\left(\frac{z_3 - z_1}{z_2 - z_1}\right) = \arg\left(\frac{z_3 - z_4}{z_2 - z_4}\right).$$

Thus, the complex numbers

$$\frac{z_3 - z_1}{z_2 - z_1} \quad \text{and} \quad \frac{z_3 - z_4}{z_2 - z_4}$$

have the same arguments, and hence their ratio, which is the cross ratio (18), is real.

If the order of the points is different, then the validity of the proposition follows from Exercise 18. The proof of the converse is relegated to Exercise 22.   Q.E.D.

Proposition 8.4.5 *Möbius transformations preserve the cross ratio.*

SKETCH OF PROOF: Let

$$f = \begin{pmatrix} a & b \\ c & d \end{pmatrix}_\mu$$

be a Möbius transformation. It is then necessary to prove that for any distinct complex numbers z_1, z_2, z_3, z_4 we have

$$\frac{\dfrac{f(z_1) - f(z_3)}{f(z_1) - f(z_2)}}{\dfrac{f(z_4) - f(z_3)}{f(z_4) - f(z_2)}} = \frac{\dfrac{z_1 - z_3}{z_1 - z_2}}{\dfrac{z_4 - z_3}{z_4 - z_2}}.$$

The details are left to Exercise 23.   Q.E.D.

The next corollary follows from Propositions 8.4.4 and 8.4.5.

Corollary 8.4.6 *Möbius transformations map circles of $\mathbb{C}^*$ onto circles of $\mathbb{C}^*$.*

The following corollary was foreshadowed by Example 7.1.12.

Corollary 8.4.7 *The geometries $\mathbb{H}$ and $\mathbb{P}$ are isometric.*

PROOF: For the Möbius transformation

$$T = \begin{pmatrix} i & -1 \\ -1 & i \end{pmatrix}_\mu, \qquad T^{-1} = \begin{pmatrix} i & 1 \\ 1 & i \end{pmatrix}_\mu$$

we have $T(-i) = 0$, $T(1) = 1$, $T(-1) = -1$, and $T(0) = i$. It follows from Corollary 8.4.6 that T transforms the unit circle onto the u-axis and the unit disk onto the upper half-plane. Moreover, since

$$T(u+iv) = \frac{2u + i(1 - u^2 - v^2)}{u^2 + (v-1)^2},$$

it follows from Example 7.1.12 that T is an isometry from $\mathbb{P}$ to $\mathbb{H}$.   Q.E.D.

It was mentioned in Section 8.2 that half-plane circles obviously exist. We are now ready to disclose the guise they assume to our Euclidean eyes.

Proposition 8.4.8 *The circles of the half-plane geometry $\mathbb{H}$ are also Euclidean circles.*

PROOF: Let T be the Möbius transformation of Corollary 8.4.7, which is an isometry from $\mathbb{P}$ to $\mathbb{H}$. If A varies over all the Möbius isometries of $\mathbb{H}$, then $T^{-1} \circ A \circ T$ varies over all the Möbius isometries of $\mathbb{P}$. Since the Möbius isometries of $\mathbb{H}$ act transitively, the same holds for the Möbius isometries of $\mathbb{P}$.

Let p be an arbitrary circle of $\mathbb{P}$. Then there is a Möbius isometry f of $\mathbb{P}$ such that $f(p)$ is centered at $(0,0)$. By Example 7.1.9, $f(p)$ is also a Euclidean circle. It follows from Corollary 8.4.6 that p too is a Euclidean circle. Thus, all the $\mathbb{P}$ circles are also Euclidean circles. Since the Möbius transformation T transforms circles of $\mathbb{P}$ into circles of $\mathbb{H}$, it follows that the circles of $\mathbb{H}$ are also Euclidean circles. Q.E.D.

The isometries of the Euclidean plane are easily visualized. While those of the half-plane were given an algebraic characterization in Theorem 8.4.3, a geometric description of the way they move the points of $\mathbb{H}$ is still lacking. We now describe the action of those isometries of $\mathbb{H}$ that are also Möbius transformations.

A *track* of an isometry f is either a Euclidean circle or a straight line t such that

$$f(t) = t.$$

It is clear that the circles centered at C are tracks of all the Euclidean rotations $R_{C,\alpha}$. The tracks of the Euclidean translations are the straight lines that are parallel to the direction of the translation. We now turn to the isometries of the half-plane.

The tracks of the horizontal translation τ_a are the Euclidean straight lines $v = c$ (Fig. 8.22). An arrowhead is added to indicate the direction in which the translation is acting. Note that these tracks are *not* half-plane geodesics.

The tracks of the dilation D_a ($0 < a \neq 1$) are the Euclidean rays that emanate from $O = (0,0)$ into the upper half-plane (Fig. 8.23).

Only one of these tracks, the positive v-axis, is a geodesic. However, all the other tracks are this geodesic's equidistant curves (see Exercise 8.2.15).

We employ a fixed point analysis to describe the tracks of the general Möbius isometry. In other words, the tracks of the transformation

$$A_\mu = \begin{pmatrix} a & b \\ c & d \end{pmatrix}_\mu$$

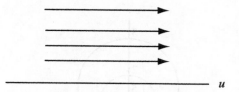

Figure 8.22   The tracks of a horizontal translation.

are located by solving the equation

$$\frac{az+b}{cz+d} = z. \tag{19}$$

If $c = 0$ and $a \neq d$, then this equation has the unique solution

$$z = \frac{b}{d-a}.$$

In this case

$$A_\mu(z) = \frac{a}{d}z + \frac{b}{d} = \tau_{b/(d-a)} \circ D_{a/d} \circ \tau_{-b/(d-a)}(z)$$

and hence its tracks are all the Euclidean rays that emanate from the fixed point $(b/(d-a), 0)$ into the half-plane. Once again, the tracks consist of a geodesic and its equidistant curves.

If $c = 0$ and $a = d$, then A_μ is the horizontal translation $\tau_{b/d}$, whose tracks have already been discussed.

If $c \neq 0$, then Equation (19) is a bona fide quadratic with real coefficients. If its roots are both imaginary, they must be complex conjugates, and hence exactly one of them, say C, lies in the upper half-plane. It follows from Exercise 8.1.19 that the restriction of A_μ to the half-plane is a half-plane rotation centered at C. The tracks of A_μ are therefore half-plane circles, which, by Proposition 8.4.8, are also Euclidean circles. Figure 8.24 displays the tracks of

$$A_\mu(z) = \frac{z-1}{z+1}.$$

In this case there is no point in placing a direction-indicating arrowhead in the diagram, because a counterclockwise rotation by the angle α can also be viewed as a clockwise rotation by the angle $2\pi - \alpha$.

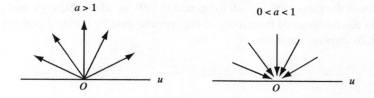

Figure 8.23   The tracks of the dilation D_a.

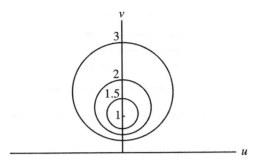

Figure 8.24   The tracks of a half-plane rotation.

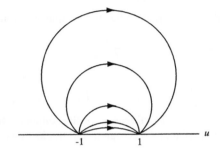

Figure 8.25   The tracks of a half-plane isometry.

If the roots of Equation (19) are real and distinct, say e and f, let g be the bowed geodesic joining $E(e,0)$ and $F(f,0)$. Since E and F are fixed points of the transformation A_μ and since isometries transform geodesics into geodesics, it follows that g is a track of A_μ. Hence, g's equidistant curves, the circular arcs that join E and F (see Exercise 8.2.16), are also tracks. Figure 8.25 displays the tracks of

$$A_\mu(z) = \frac{2z+1}{z+2}.$$

The direction of the arrowhead was determined by observing that in this case

$$A_\mu(i) = \frac{2i+1}{i+2} \cdot \frac{2-i}{2-i} = \frac{4}{5} + \frac{3}{5}i.$$

Finally, if the roots of the quadratic equation (19) are identical, say e and e, then the tracks can be obtained from those of the previous case by letting f converge to e. Figure 8.26 displays the tracks of

$$A_\mu(z) = \frac{2z}{z+2}.$$

These tracks are called *horocycles* (Exercise 3), and they play an important role in the synthetic development of hyperbolic geometry.

Exercises 8.4

1. Prove that the geodesics of $\mathbb{P}$ are arcs of circles of $\mathbb{C}^*$ that are orthogonal to the unit circle. (Hint: Go ahead and assume the fact that Möbius transformations, like all analytic functions of a complex variable, preserve angles.)

2**. Prove that the isometries of $\mathbb{P}$ are the transformations of the forms

$$(a) \quad \begin{pmatrix} a & \bar{c} \\ c & \bar{a} \end{pmatrix}_\mu \qquad \text{or} \qquad (b) \quad \begin{pmatrix} a & \bar{c} \\ c & \bar{a} \end{pmatrix}_\mu \circ \rho,$$

where a and c are any two complex numbers such that $|a| > |c|$.

3*. Prove that any two horocycles of $\mathbb{H}$ that are tangent to the same point on the u-axis, are at a constant distance from each other.

4. Prove that there is no Riemann metric $\mathcal{M}$ on the upper half-plane that is isometric to $\mathbb{H}$ and whose geodesics are the restrictions of Euclidean straight lines.

5. Prove that there is no conformal Riemann geometry that is isometric to $\mathbb{H}$ and whose geodesics are restrictions of Euclidean straight lines.

6. Prove Lemma 8.4.2.

7*. Suppose a circle in the upper half-plane has Euclidean center (a, b) and radius r, whereas it has half-plane center (A, B) and radius R. Prove that

$$A = a, \qquad B = \sqrt{b^2 - r^2}, \qquad R = \frac{1}{2} \ln \frac{b+r}{b-r}$$

and

$$a = A, \qquad b = B \cosh(R), \qquad r = B \sinh(R).$$

8*. Prove that if $k > r$, then the half-plane circumference of the circle with Euclidean center (a, b) and radius r is $2\pi r / \sqrt{b^2 - r^2}$.

9*. Prove that in either $\mathbb{H}$ or $\mathbb{P}$, a circle with radius R has circumference $2\pi \sinh(R)$.

10*. Prove that in either $\mathbb{H}$ or $\mathbb{P}$, a circle with radius R has area $2\pi(\cosh(R) - 1)$.

11. Does every half-plane triangle have an inscribed circle?

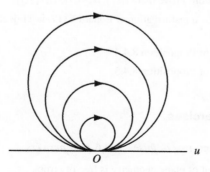

Figure 8.26 The tracks of a half-plane isometry.

12. Does every half-plane triangle have a circumscribed circle?

13. Draw track diagrams for the following half-plane isometries:

   (a) $5z$

   (b) $\dfrac{z}{5}$,

   (c) $z - 3$,

   (d) $5z - 3$,

   (e) $\dfrac{z}{2z+3}$,

   (f) $\dfrac{-2z}{z-1}$,

   (g) $\dfrac{13z+12}{-3z+1}$,

   (h) $\dfrac{2z+1}{-z+3}$.

14*. Let $f(z) = (az+b)/(cz+d)$ be a half-plane isometry. Prove that this isometry is a $180°$ rotation if and only if $a+d = 0$.

15*. Prove that the transformation $f(z) = [a(-\bar{z})+b]/[c(-\bar{z})+d]$ is a reflection of $\mathbb{H}$ if and only if $a = d$.

16*. Prove that the half-plane isometry $f(z) = \frac{z+1}{-z+1}$ is a $90°$ rotation.

17*. Prove that in $\mathbb{H}$

$$R_{i,2\alpha} = \begin{pmatrix} \cos\alpha & \sin\alpha \\ -\sin\alpha & \cos\alpha \end{pmatrix}_{\mu}.$$

18*. Prove that if $\lambda = (z_1, z_2, z_3, z_4)$ and k, l, m, n constitute a permutation of $1, 2, 3, 4$, then (z_k, z_l, z_m, z_n) has one of the values

$$\lambda, \quad \frac{1}{\lambda}, \quad 1-\lambda, \quad \frac{1}{1-\lambda}, \quad \frac{\lambda-1}{\lambda}, \quad \frac{\lambda}{\lambda-1}.$$

19. Prove that if M is any real Möbius transformation and z is any complex number, then the cross ratio $(z, M(z), M^2(z), M^3(z))$ is real.

20. Let $I(z)$ be an inversion. Prove that $(I(z_1), I(z_2), I(z_3), I(z_4)) = \overline{(z_1, z_2, z_3, z_4)}$.

21*. Prove that the Möbius transformation of Equation (17) is a bijection of $\mathbb{C} - \{-d/c\}$ onto $\mathbb{C} - \{a/c\}$.

22. Complete the proof of Proposition 8.4.4.

23. Complete the proof of Proposition 8.4.5.

Chapter Review Exercises

Are the following statements true or false? Justify your answers.

1. Euclid's development of plane geometry is free of errors.

2. Euclid's definition of area is axiomatic.

3. Every proposition of neutral geometry is also a proposition of Euclidean geometry.

4. Every proposition of neutral geometry is also a proposition of hyperbolic geometry.

5. Every proposition of Euclidean geometry is also a proposition of hyperbolic geometry.

6. Every proposition of hyperbolic geometry is also a proposition of neutral geometry.

7. Some proposition of hyperbolic geometry is also a proposition of Euclidean geometry.

8. Some proposition of Euclidean geometry is also a proposition of hyperbolic geometry.

9. Parallel lines exist in hyperbolic geometry.

10. Parallel straight lines exist in neutral geometry.

11. Parallel straight lines exist in Euclidean geometry.

12. Parallel straight lines exist in spherical geometry.

13. Möbius transformations transform Euclidean circles into Euclidean circles.

14. Euclidean circles are also circles of the half-plane geometry.

15. Circles of the half-plane geometry are also Euclidean circles.

16. If an isometry of $\mathbb{P}$ fixes the points $(0,0)$, $(.5,0)$, and $(.5,.5)$, then it must be the identity.

17. Hyperbolic geometry and spherical geometry are isometric.

Are the following statements true or false in the context of Euclidean geometry? Justify your answers.

18. The composition of two translations is a translation.

19. The composition of two rotations is a rotation.

20. The composition of two reflections is a reflection.

21. The composition of two glide reflections is never a glide reflection.

Are the following statements true or false in the context of hyperbolic geometry? Justify your answers.

22. The composition of two translations is a translation.

23. The composition of two rotations is a rotation.

24. The composition of two reflections is a reflection.

25. The composition of two glide reflections is never a glide reflection.

CHAPTER 9

THE FUNDAMENTAL GROUP

The fundamental group was first defined in 1895 by Poincaré for the purpose of describing the behavior of multivalued functions which arise naturally as the solutions of certain differential equations. He also recognized immediately this group's potential for classifying topological spaces. Our treatment of this topic is restricted to the utility of this algebraic structure in the classification of both surfaces and three-dimensional manifolds. Some attention is also given to the combinatorial structure of these manifolds. The chapter concludes with a description of the Poincaré Conjecture, which is widely recognized as the most important open question in topology.

9.1 Definitions and the Punctured Plane

The tale of the fundamental group begins with Green's Theorem. This well-known proposition states that for any continuously differentiable functions $P(x,y)$ and $Q(x,y)$, and for any region R in their domain whose counterclockwise boundary γ is piecewise smooth,

$$\int_{\gamma} P\,dx + Q\,dy = \int\int_{R} (Q_x - P_y)\,dx\,dy.$$

Introduction to Topology and Geometry, Second Edition.
By Saul Stahl and Catherine Stenson Copyright © 2013 John Wiley & Sons, Inc.

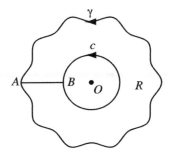

Figure 9.1   The simplification of an integral.

This theorem comes in very handy in the evaluation of some path integrals. Consider, by way of an example, the path integral

$$\int_\gamma (y^2 - x^2)\,dx + 2xy\,dy,$$

where γ is an arbitrary (piecewise smooth) loop in the plane. If R denotes the finite region bordered by γ, then

$$\int_\gamma (y^2 - x^2)\,dx + 2xy\,dy = \int\int_R (2y - 2y)\,dx\,dy = 0.$$

What made this easy evaluation possible was the fact that $Q_x - P_y = 0$ throughout R. While this observation cannot evaluate all path integrals, it and its modifications do apply in sufficiently many significant contexts to merit a deeper investigation.

As another example, consider the path integral

$$\int_\gamma \frac{x}{x^2 + y^2}\,dx + \frac{y}{x^2 + y^2}\,dy. \tag{1}$$

Note that the integrands $P = x/(x^2 + y^2)$ and $Q = y/(x^2 + y^2)$ are undefined only at $(0,0)$, and that

$$Q_x - P_y = \frac{-2xy}{(x^2 + y^2)^2} - \frac{-2xy}{(x^2 + y^2)^2} = 0$$

throughout their common domain. It follows that whenever γ is a loop that does not surround the origin, the integral of (1) is 0. Moreover, Green's Theorem can also be applied even when the origin does fall in γ's interior, though with a different conclusion. Let c be any other counterclockwise loop surrounding the origin (see Fig. 9.1) which does not intersect γ. Because $Q_x - P_y$ vanishes throughout the annular region R, whose counterclockwise boundary can be described as $(AB)c^{-1}(AB)^{-1}\gamma$, it follows that

$$0 = \int_{AB} P\,dx + Q\,dy + \int_{c^{-1}} P\,dx + Q\,dy + \int_{(AB)^{-1}} P\,dx + Q\,dy$$

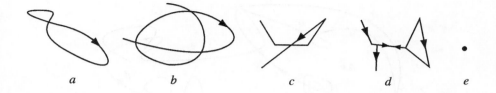

Figure 9.2   Some paths.

$$+ \int_\gamma P\,dx + Q\,dy$$

$$= \int_{AB} P\,dx + Q\,dy - \int_c P\,dx + Q\,dy - \int_{AB} P\,dx + Q\,dy$$

$$+ \int_\gamma P\,dx + Q\,dy$$

$$= \int_\gamma P\,dx + Q\,dy - \int_c P\,dx + Q\,dy$$

and hence,

$$\int_\gamma P\,dx + Q\,dy = \int_c P\,dx + Q\,dy.$$

In other words, the value of the integral of (1) is independent of the specific loop γ as long as the loop surrounds the origin in the counterclockwise sense. Alternatively, as long as the process avoids the singularity at the origin, planar deformations of the path γ will not change the value of the integral. (As the actual evaluation of this integral is of no concern to us, it is relegated to Exercise 1.) This clearly generalizes to any integrand of the form $P\,dx + Q\,dy$ whose components satisfy the condition

$$Q_x - P_y = 0$$

on some domain D, so long as the deformation process remains restricted to D.

For the purposes of integration it is necessary to allow for paths that are not arcs. Thus, a *path* of the topological space S is redefined in this chapter as a continuous image of the unit interval $[0, 1]$ rather than a homeomorph of it. Each of the five sets a–e of Figure 9.2 is a path. In particular, so is the set e, which consists of a single point. Paths, too, are assumed to have directions; the initial point of the path a is denoted by a_0, and its terminal point by a_1. Similarly, a *circuit* is a path with identical initial and terminal points, and the circuit is said to be *based* at that point. The trivial circuit that consists of the point P alone is denoted by 1_P. Only paths a and e of Figure 9.2 are circuits.

Suppose the paths a and b share the same initial and the same terminal points; they are said to be *homotopic* in S if there is a continuous deformation that transforms a onto b through S without moving the common endpoints at all. (A more formal definition appears in Section 10.2.) The homotopy of a and b is denoted by $a \simeq b$,

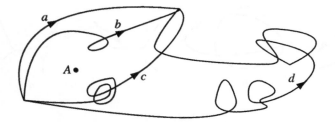

Figure 9.3   Homotopic and nonhomotopic loops.

and it is in fact an equivalence relation in the sense that it has the following properties (Exercise 2):

1. For any arc $a, a \simeq a$.

2. For any arcs a and b, $a \simeq b$ if and only if $b \simeq a$.

3. For any arcs a, b, and c, if $a \simeq b$ and $b \simeq c$, then $a \simeq c$.

The homotopy relation $\simeq$ therefore divides the set of arcs of a topological space S into equivalence classes, called *homotopy classes*. All the members of one homotopy class necessarily have the same initial and the same terminal points, but the converse is not true. The homotopy class of a will be denoted by $[a]$. In the space $\mathbb{R}^2 - \{A\}$ of Figure 9.3, $[a] = [b] \neq [c] = [d]$.

Given two paths a and b such that $a_1 = b_0$, their *product* $a \cdot b$ is simply their juxtaposition (see Fig. 9.4). Just as was the case in Chapter 3, the inverse a^{-1} of the path a is its reversal, and $(a^{-1})^{-1} = a$. These operations are consistent with the notion of homotopy (Exercise 3):

1. If $a \simeq b$ and $a' \simeq b'$ then $a \cdot a' \simeq b \cdot b'$.

2. If $a \simeq b$ then $a^{-1} \simeq b^{-1}$.

3. $a \cdot a^{-1} \simeq 1_{a_0}, a^{-1} \cdot a \simeq 1_{a_1}$.

4. $1_{a_0} \cdot a \cong a \cong a \cdot 1_{a_1}$.

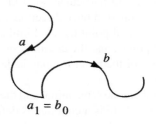

Figure 9.4   The product $a \cdot b$.

Note that if a is a circuit, then we do distinguish between a and $a^2 = a \cdot a$ and think of the latter as wrapping itself twice around the former. Similarly, for any positive integer n, $a^n = a \cdot a \cdot a \cdots \cdots a$ is wrapped n times around a. Products and inverses of arcs can be used to define *products* and *inverses* of homotopy classes (Exercise 4):

1. $[a][b] = [a \cdot b]$;

2. $[a]^{-1} = [a^{-1}]$.

Given a topological space S with a point X in it, the set of homotopy classes of all the circuits of S that are based at X is denoted by $\pi(S,X)$. It is clear that if X is any point in the plane $\mathbb{R}^2$, then $\pi(\mathbb{R}^2,X)$ consists of the single class $[1_X]$. On the other hand, if $O = (0,0), X = (1,0)$, and, for every positive integer n, the circuit a_n both begins and ends at X and winds n times around O in the counterclockwise sense, then a_m and a_n are homotopic if and only if $m = n$. Hence, $\pi(\mathbb{R}^2 - \{O\},X)$ has an infinite number of distinct homotopy classes.

If we set $1 = [1_X]$, then it follows from properties 1 and 2 of the product that, for any $[a], [b], [c] \in \pi(S,X)$,

1. $[a][b]$ is a well-defined element of $\pi(S,X)$;

2. $[a]1 = 1[a] = [a]$;

3. $([a][b])[c] = [a]([b][c])$;

4. $[a][a]^{-1} = [a]^{-1}[a] = 1$.

In other words, we have the following fact.

Theorem 9.1.1 *The product operation turns $\pi(S,X)$ into a group, called the* fundamental group. $\qquad\qquad\square$

A circuit a based at the point X is said to be *null homotopic* if it is homotopic to the constant circuit 1_X. A space S is said to be *simply connected* if every one of its circuits is null homotopic. Both the plane and the 2-sphere S_0 are simply connected, whereas the torus is not. By definition, a space S is simply connected if and only if its fundamental group $\pi(S,X)$ is trivial for every point X of S. The proofs of the following examples are immediate.

Example 9.1.2 If X is any point of the plane $\mathbb{R}^2$, then $\pi(\mathbb{R}^2,X) \cong \langle 1 \rangle$.

Example 9.1.3 If X is any point of the sphere S_0, then $\pi(S_0,X) \cong \langle 1 \rangle$.

Example 9.1.4 If A, X, Y are any two distinct points of the sphere S_0, then $\pi(S_0 - \{A\},X) \cong \langle 1 \rangle$.

We next argue that the aforementioned group $\pi(\mathbb{R}^2 - \{O\},X)$ is an infinite cyclic group generated by a single element α. First we extend the above definition of a_n to negative n by stipulating that for such n, a_n winds around the origin $|n|$ times in the

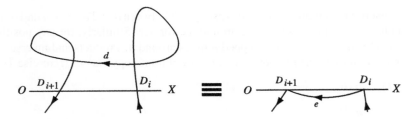

Figure 9.5   A homotopy.

clockwise sense. Also, $a_0 = 1_X$. It is clear that $a_m \simeq a_n$ if and only if $m = n$. Let c be an arbitrary circuit based at X. It will be shown that c is homotopic to some a_n. We take it for granted that c is homotopic to a circuit d that crosses the infinite ray from O through X in a finite number of points. Let $D_1, D_2, \ldots, D_k$ be these crossings listed as they occur on d. Assume OX is the positive x-axis of a Cartesian coordinate system centered at O. A crossing D_k is said to be positive if at that crossing the curve switches from below to above OX. The crossing is negative otherwise. It is clear that if $k = 0$ then $d \simeq 1_X$. If all the crossings are positive, then $d \simeq a_k$, and if all of them are negative, then $d \simeq a_{-k}$. If there is a positive crossing D_i such that D_{i+1} (addition modulo k) is negative (Fig. 9.5), let e be the circuit obtained from d by replacing the portion of d from D_i to D_{i+1} with a shallow arc that joins these points just below the x-axis. Then d is homotopic to the circuit e, where e has $k - 2$ crossings. It follows by mathematical induction that every circuit c based at X is homotopic to some a_n. In other words, the elements $[a_n]$ exhaust the group $\pi(\mathbb{R}^2 - \{O\}, X)$.

Finally, it is clear that for all n

$$[a_n] = [a]^n,$$

and hence $\pi(S, X)$ is the infinite cyclic group generated by $[a]$.

It is easy to see that the above discussion requires very little modification to prove the following proposition.

Proposition 9.1.5 *If A and X are two distinct points of the plane, then $\pi(\mathbb{R}^2 - \{A\}, X)$ is an infinite cyclic group.*   □

Raising the ante, let X, C, D be three distinct points of the plane, and let c and d be two counterclockwise loops based at X that surround C and D respectively (see Fig. 9.6). It is clear that c and d are not homotopic to each other and hence the fundamental group $\pi(\mathbb{R}^2 - \{C, D\}, X)$ contains the two distinct classes $[c]$ and $[d]$. The homotopy displayed in this diagram (with successive stages $e_1, e_2, \ldots$), demonstrates that the loop e that surrounds both C and D is actually homotopic to $c \cdot d$, so that

$$[e] = [c][d].$$

When a similar argument is applied to the loop f of Figure 9.7, we conclude that

$$[f] = [d]^3 [c]^{-3}.$$

In fact, every element of $\pi(\mathbb{R}^2 - \{C,D\}, X)$ is expressible in the form

$$[c]^{m_1}[d]^{n_1}[c]^{m_2}[d]^{n_2}\cdots[c]^{m_k}[d]^{n_k}$$

for some integers $m_1, n_1, m_2, n_2, \ldots, m_k, n_k$, and no nontrivial relation binds $[c]$ and $[d]$. In other words, this fundamental group is freely generated by two elements. These observations generalize to the following theorem.

Theorem 9.1.6 *Let P be a set of k distinct points in the plane, and X a point not in P. Then the fundamental group $\pi(\mathbb{R}^2 - \{P\}, X)$ is freely generated by k elements.*

We conclude this section by pointing out that homeomorphic topological spaces have isomorphic fundamental groups.

Proposition 9.1.7 *If $f : S \to T$ is a homeomorphism with $X \in S$ and $Y = f(X) \in T$, then the function*

$$f_*([c]) = [f(c)]$$

is an isomorphism of $\pi(S,X)$ and $\pi(T,Y)$.

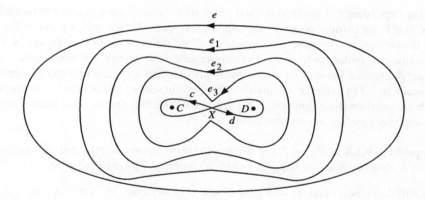

Figure 9.6   A homotopy.

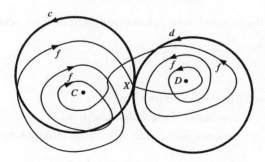

Figure 9.7   A homotopy.

PROOF: It is clear that if $c \simeq c'$ then $f(c) \simeq f(c')$, and hence the function f_* is well defined.

It follows from Exercises 3 and 4 that if c and c' are circuits of S at X, then

$$f_*([c][c']) = f_*([c \cdot c']) = [f(c \cdot c')]$$
$$= [f(c) \cdot f(c')] = [f(c)][f(c')] = f_*([c])f_*([c']),$$

and hence f_* is a homomorphism.

For any circuit d of T at Y, $c = f^{-1}(d)$ is a circuit of S at X such that

$$f_*([c]) = [f(c)] = [f(f^{-1}(d))] = [d].$$

It follows that f^* is surjective.

Finally, let c, c' be circuits of S at X such that $f_*([c]) = f_*([c'])$. It follows that $f(c) \simeq f(c')$,

$$c = f^{-1}(f(c)) \simeq f^{-1}(f(c')) = c'$$

and hence $[c] = [c']$, implying that f_* is injective. Thus, f_* is a homomorphism which is both surjective and injective. In other words, f_* is an isomorphism.   Q.E.D.

The dependence of the fundamental group on the base point needs to be clarified. If X and Y are distinct points of the topological space S, then $\pi(S, X)$ and $\pi(S, Y)$ are distinct groups because the circuits at X are different from the circuits at Y. Nevertheless, under fairly general circumstances these groups are isomorphic. A topological space S is *arcwise connected* if any two distinct points of S can be joined by an arc in S. For example, the metric space that consists of all the rational numbers, with the usual distance function, is not arcwise connected. On the other hand, every arc and every surface are arcwise connected.

Proposition 9.1.8 *If X and Y are two points of the arcwise connected topological space S, then the fundamental groups $\pi(S, X)$ and $\pi(S, Y)$ are isomorphic.*

PROOF: Let c be a fixed arc of S from X to Y. For any class $[d]$ of $\pi(S, X)$ set

$$f_*([d]) = [c^{-1} \cdot d \cdot c].$$

The function f_* is the required isomorphism (see Exercise 5 for details).   Q.E.D.

Exercises 9.1

1. Evaluate the integral of (1) over the path γ in Figure 9.1.

2. Prove formally that homotopy is an equivalence relation.

3. If $a(s), b(s) : I \to S$ are functions such that $a(1) = b(0)$, then we formally define

$$(a \cdot b)(s) = \begin{cases} a(2s), & 0 \le s \le 1/2, \\ a(2s-1), & 1/2 \le s \le 1, \end{cases}$$

and

$$a^{-1}(s) = a(1 - s).$$

Prove that

(a) if $a \simeq b$ and $a' \simeq b'$ then $a \cdot a' \simeq b \cdot b'$;

(b) if $a \simeq b$ then $a^{-1} \simeq b^{-1}$;

(c) $a \cdot a^{-1} \simeq 1_{a(0)}, a^{-1} \cdot a \simeq 1_{a(1)}$;

(d) $1_{a(0)} \cdot a \simeq a \simeq a \cdot 1_{a(1)}$.

(This exercise assumes that the reader is acquainted with the formal definition of homotopy given at the end of Section 10.2.)

4. Use Exercise 3 to explain why the expressions $[a][b]$ and $[a]^{-1}$ are well defined.

5. Complete the proof of Proposition 9.1.5.

6. Let S be the 2-sphere, and let X be any point of S. Describe $\pi(S, X)$.

7. Let S be the topological space obtained by deleting one point from the sphere S_0, and let X be any point of S. Describe $\pi(S, X)$.

8. Let S be the topological space obtained by deleting n distinct points from the sphere S_0, and let X be any point of S. Describe $\pi(S, X)$.

9. Let S be a circle and X a point of S. Describe $\pi(S, X)$.

10. Let S be the complement of a circle in the plane $\mathbb{R}^2$, and let X be a point of S. Describe $\pi(S, X)$.

11. Let S be the complement of a point in the space $\mathbb{R}^3$, and let X be a point of S. Describe $\pi(S, X)$.

12. Generalize Exercise 11.

13. Let S be the complement of an infinite straight line in $\mathbb{R}^3$ and let X be a point of S. Describe $\pi(S, X)$. (Hint: Attach to the given straight line a half-plane which is to be used in the same way that the positive x-axis was used in the proof of Proposition 9.1.5.)

14. Let S denote the space obtained by deleting a single point from a loop. Describe $\pi(S, X)$.

15. Let S be the complement of a circle in $\mathbb{R}^3$, and let X be a point of S. Describe $\pi(S, X)$.

16. A *tree* is a connected graph that contains no cycles. Describe the fundamental group of a tree.

17. A graph that contains a unique cycle is said to be *unicyclic*. Describe the fundamental group of a connected unicyclic graph.

18. The *bouquet* B_n is a graph that consists of a single node and n loops. Describe the fundamental group of B_n.

9.2 Surfaces

Considerations in complex analysis also led mathematicians to the study of path integrals of functions whose domains are the Riemann surfaces of Section 3.5. For the purpose of the computation of their fundamental groups, rather than visualizing these

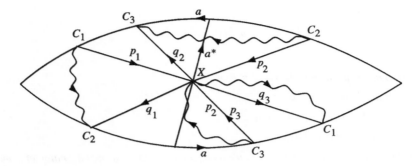

Figure 9.8   Part of a circuit on the projective plane.

surfaces as stacked sheets, it is more convenient to use the 1-polygon presentations of Sections 3.1 and 3.2.

We begin by computing the fundamental group of the sphere S_0. Here, it is unnecessary even to use presentations. Since the sphere is simply connected, it follows that for any base point X the group $\pi(S_0, X)$ is trivial.

Next we turn to the projective plane $\tilde{S}_1$. Let X be any point in the interior of the polygon in the canonical presentation of $\tilde{S}_1$ (Fig. 9.8). Let a^* be a loop based at X that crosses the perimeter aa of the polygon exactly once, and let c be an arbitrary circuit at X, denoted by a wavy line, that crosses the perimeter at, say, the points C_1, C_2, C_3. Note that the segment of c between C_i and C_{i+1} is homotopic to the composite arc $p_i \cdot q_i$, where p_i is a path that goes directly from C_i to X, and q_i is a path that goes directly from X to C_{i+1} (addition modulo 3). It follows that

$$c \simeq q_3 \cdot p_1 \cdot q_1 \cdot p_2 \cdot q_2 \cdot p_3.$$

However,

$$q_3 \cdot p_1 \simeq (a^*)^{-1}, \qquad q_1 \cdot p_2 \simeq (a^*)^{-1}, \qquad q_2 \cdot p_3 \simeq a^*,$$

and hence

$$c \simeq (a^*)^{-1} \cdot (a^*)^{-1} \cdot (a^*) \simeq (a^*)^{-1}.$$

A similar argument (which will also be generalized in Theorem 9.2.1) allows us to conclude that for any circuit d based at X there exists an integer n such that

$$d \simeq (a^*)^n.$$

We conclude the computation of $\pi(\tilde{S}_1)$ by showing that $[a^*] = [(a^*)^{-1}]$. The necessary homotopy is displayed in Figure 9.9, where the successive perturbations of a^* are denoted by $b_1, b_2, \ldots, b_7$ and it is clear that the last is homotopic to $(a^*)^{-1}$. It follows that the fundamental group of the projective plane has only two elements and is therefore isomorphic to $\mathbb{Z}_2$.

The elements of this computation are now applied to arbitrary 1-polygon presentations of surfaces. Let Π be such a polygon with a base point X in its interior, and

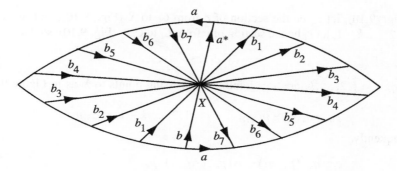

Figure 9.9 A homotopy on the projective plane.

let S be the associated surface. For each arc a on the perimeter of Π, let a^*, the *dual* of a, be a loop based at X that crosses a exactly once. It is necessary to give each a^* an orientation, but it is immaterial which one is selected. These choices are fixed throughout the computation of the fundamental group. It is clear that while there are many choices for a^*, the homotopy class $[a^*]$ is well defined.

Let ν be a node of a presentation of a surface S. Each of the two possible orientations of the vicinity of ν induces a cyclic ordering of the (undirected) arcs that are incident to ν, called a *star* of ν. Thus, each such node ν has two stars. For example, the node u of Example 3.2.2 has the two stars (d,a,f,g,h) and (h,g,f,a,d). A small directed loop surrounding a node ν is said to be *consistent* with a star provided its direction agrees with the cyclic order of the star.

Let c be an arbitrary curve. If this curve crosses the arc a at some point, then its *crossing index* at that point is 1 or -1 according as the direction of the crossing agrees or disagrees with that of a^*. For example, in Figure 9.8, the crossing indices of the wavy curve c at C_1, C_2, C_3 are -1, -1, and 1, respectively.

Theorem 9.2.1 *Let Π be a 1-polygon presentation of the surface S with arcs $a_1, a_2,$ $\dots, a_n$, nodes $A_1, A_2, \dots, A_k$, and a base point X. Then the fundamental group $\pi(S, X)$ has the presentation*

$$\langle \alpha_1, \alpha_2, \dots, \alpha_n \; ; \; W_1, W_2, \dots, W_k \rangle$$

where, for each $i = 1, 2, \dots, k$,

$$W_i = \alpha_{i_1}^{\varepsilon_1} \alpha_{i_2}^{\varepsilon_2} \cdots,$$

where each $(a_{i_1}, a_{i_2}, \dots)$ is a star of A_i, and the exponents $\varepsilon_1, \varepsilon_2, \dots$ are the corresponding crossing indices of a small consistent loop surrounding A_i.

PROOF: Let c be a circuit of S based at X. Since we are only concerned with the homotopy class of c, it may be assumed that c misses all of the presentation's nodes. If c does not cross any of the sides of Π, then it is null-homotopic and $[c] = 1$. Otherwise, let $C_1, C_2, \dots, C_k$ be the ordered list of c's crossings of the respective arcs $a_{j_1}, a_{j_2}, \dots, a_{j_k}$ on the perimeter of Π. Let c_0 be the section of c from X to

C_1 (Fig. 9.10), let c_k be the section of c from C_k to X (Fig. 9.10), and, for each $i = 1, 2, \ldots, k - 1$, let c_i be the section of c from C_i to C_{i+1} (Fig. 9.10), so that

$$c = c_0 \cdot c_1 \cdot c_2 \cdot \cdots \cdot c_k.$$

If p_i, q_i, $i = 1, 2, \ldots, k$, are arcs joining the crossing points to X as in Figure 9.10, then

$$c_0 \simeq q_k, c_i \simeq p_i \cdot q_i, \qquad i = 1, 2, \ldots, k - 1, \ c_k \simeq p_k.$$

Consequently,

$$c \simeq q_k \cdot (p_1 \cdot q_1) \cdot \cdots \cdot (p_{k-1} \cdot q_{k-1}) \cdot p_k$$
$$\simeq (q_k \cdot p_1) \cdot (q_1 \cdot p_2) \cdot \cdots \cdot (q_{k-2} \cdot p_{k-1}) \cdot (q_{k-1} \cdot p_k). \tag{2}$$

For each i let the crossing index of c at C_i be denoted by ε_i. Note that, as indicated by Figure 9.10, for each $i = 1, 2, \ldots, k$ we have $q_{i-1} \cdot p_i \simeq (a_{j_i}^*)^{\varepsilon_i}$ (subtraction modulo k). Hence

$$[c] = [a_{j_1}^*]^{\varepsilon_1} [a_{j_2}^*]^{\varepsilon_2} \cdots [a_{j_k}^*]^{\varepsilon_k}. \tag{3}$$

Thus, every element of the fundamental group $\pi(S, X)$ can be expressed in terms of the $[a_i^*]$'s.

The expressions provided by Equation (3) may not be unique, as homotopic curves may have altogether different crossing patterns. Nevertheless, it will shortly be shown that when this procedure is applied to homotopic curves, while it may produce different words in the $[a_i^*]$'s, it does yield equal elements of the group $\pi(S, X)$. First, however, it is necessary to derive the relations that bind the generators $[a_i^*]$. For this purpose it is necessary to describe how homotopies affect the set of crossings and how these changes, in turn, affect the associated expressions provided by Equation (3). If a homotopy of c does not sweep across any of the nodes of the presentation then it can be broken down into a sequence of homotopies of the type displayed in Figure 9.11. Note that in the transition from c to c', the factorizations associated with these circuits by Equation (2) differ only in the loss, or gain, of the product

$$q_{i-1} \cdot p_i \cdot q_i \cdot p_{i+1}.$$

Hence the expressions associated with $[c]$ and $[c']$ by Equation (3) differ only in the loss of one of the subwords

$$[a_{j_i}^*][a_{j_i}^*]^{-1} \quad \text{or} \quad [a_{j_i}^*]^{-1}[a_{j_i}^*].$$

In either case we get $[c] = [c']$.

It remains to examine the effect of a homotopy that sweeps across some node, say A. Here (see Fig. 9.12), we temporarily abandon the subscript notation, as it becomes too cumbersome. Instead we suppose that the loop c, which crosses the arcs p, q, r at the points P, Q, R, respectively, is homotopic to the loop c', which crosses s and t at S and T, respectively, and agrees with c off the diagram. This common portion off the diagram contributes some words, say w_1 and w_2, in the $[a_i^*]$'s such that

$$[c] = w_1 [p^*]^{\varepsilon_p} [q^*]^{\varepsilon_q} [r^*]^{\varepsilon_r} w_2 \quad \text{and} \quad [c'] = w_1 [t^*]^{-\varepsilon_t} [s^*]^{-\varepsilon_s} w_2,$$

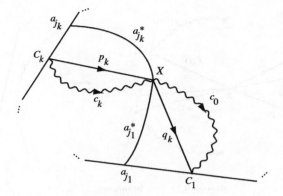

Initial and terminal sections of c

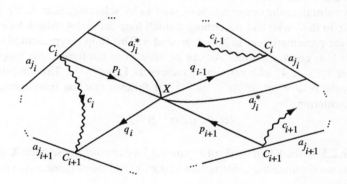

Intermediate sections of c

Figure 9.10   Finding an expression for a homotopy class.

where $\varepsilon_p, \ldots, \varepsilon_t$ are the crossing indices of a small loop d that surrounds the node A. It is clear that these two words differ exactly by the relation

$$[p^*]^{\varepsilon_p}[q^*]^{\varepsilon_q}[r^*]^{\varepsilon_r}[s^*]^{\varepsilon_s}[t^*]^{\varepsilon_t} = 1.$$

This accounts for the relators described in the statement of the theorem.   Q.E.D.

As it is customary to present surfaces by means of presentations alone, it will be necessary to describe the relators without recourse to diagrams like Figure 9.12. This is easily accomplished if we recall that the angles marked 1, 2, 3, 4, 5 in Figure 9.12 are in fact the same angles that were used in Section 3.1 to identify the nodes of the presentation.

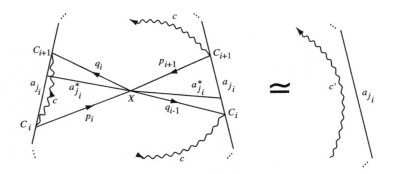

Figure 9.11   A homotopy.

In all of the following examples, the crossing C_i will be denoted by "i" alone.

Example 9.2.2 In the torus of Figure 9.13, we choose a base point X and duals a^*, b^*. Next we simulate the construction of star(A) by selecting, near A, a point 1 on some arc (a, in this case) and describing a small loop around A, which loop crosses the arcs of the presentation at 1, 2, 3, 4, around A. This loop is represented in Figure 9.13 by the four quarter circles. As can be seen from the figure, this loop crosses a, b, a, b at 1, 2, 3, 4, with respective indices 1, 1, -1, -1. This loop therefore contributes the relator $[a^*][b^*][a^*]^{-1}[b^*]^{-1}$. It follows that the fundamental group has the presentation

$$\langle \alpha, \beta; \alpha\beta\alpha^{-1}\beta^{-1} \rangle.$$

Example 9.2.3 In the Klein bottle of Figure 9.13 we choose a base point X and duals a^*, b^*. Next we simulate the construction of star(A) by selecting, near A, a point 1 on

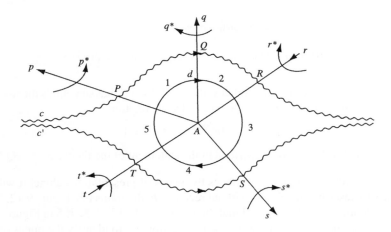

Figure 9.12   The creation of a relator.

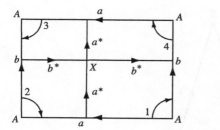 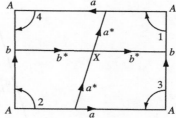

Figure 9.13   The torus and the Klein bottle.

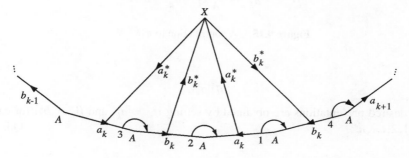

Figure 9.14   A contribution to $\pi(S^n, X)$.

some arc (b, in this case) and describing a small loop around A, which loop crosses the arcs of the presentation at 1, 2, 3, 4, around A. This loop is represented in Figure 9.13 by the four quarter circles. As can be seen from the figure, this loop crosses a, b, a, b at 1, 2, 3, 4, with respective indices -1, 1, -1, -1. This loop therefore contributes the relator $[a^*]^{-1}[b^*][a^*]^{-1}[b^*]^{-1}$. It follows that the fundamental group has the presentation

$$\langle \alpha, \beta; \ \alpha\beta\alpha^{-1}\beta \rangle.$$

We can now derive the fundamental groups of all the closed surfaces.

Theorem 9.2.4   *The fundamental group of the sphere S_0 is $\langle 1 \rangle$, and for each positive integer n,*

$$\pi(S_n, X) = \langle \alpha_1, \beta_1, \alpha_2, \beta_2, \ldots, \alpha_n, \beta_n; \alpha_1\beta_1\alpha_1^{-1}\beta_1^{-1}\alpha_2\beta_2\alpha_2^{-1}\beta_2^{-1}\cdots\alpha_n\beta_n\alpha_n^{-1}\beta_n^{-1} \rangle$$

and

$$\pi(\tilde{S}_n, X) = \langle \alpha_1 \alpha_2, \ldots, \alpha_n; \alpha_1^2\alpha_2^2\cdots\alpha_n^2 \rangle.$$

PROOF: As noted before, the fundamental group of the sphere S_0 is trivial because the sphere is simply connected. If we use the canonical presentations of Figures 3.38 and 3.39 for the other surfaces, and assign duals as in Figures 9.14 and 9.15, then the respective contributions of the depicted crossings to the unique relators are

$$[a_k^*][b_k^*][a_k^*]^{-1}[b_k^*]^{-1}$$

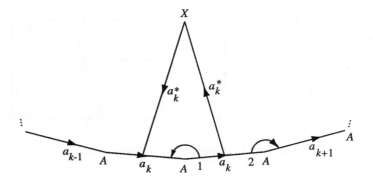

Figure 9.15   A contribution to $\pi(\tilde{S}^n, X)$.

and

$$[a_k^*][a_k^*].$$

The desired presentations are obtained by setting $\alpha_k = [a_k^*]$ and $\beta_k = [b_k^*]$ for each $k = 1, 2, \ldots, n$.   Q.E.D.

Exercises 9.2

1. Identify the surface with the presentation $abca^{-1}b^{-1}c^{-1}$, and use Theorem 9.2.1 to show that its fundamental group has the presentation $\langle \alpha, \beta, \gamma; \alpha\beta\gamma, \alpha\gamma\beta \rangle$. Explain why this group is isomorphic to the free abelian group of rank 2.

2. Identify the surface with the presentation $abcda^{-1}b^{-1}c^{-1}d$, and use Theorem 9.2.1 to show that its fundamental group has the presentation $\langle \alpha, \beta, \gamma, \delta; \alpha\beta\gamma\delta\gamma\beta\alpha\delta^{-1} \rangle$. Prove that the abelianization of this group has rank at most 3.

3. Identify the surface with the presentation $abd^{-1}f^{-1}e^{-1}b^{-1}cd^{-1}gacegf$, and use Theorem 9.2.1 to show that its fundamental group has the presentation

$$\langle \alpha, \beta, \gamma, \delta, \varepsilon, \phi, \kappa; \alpha^{-1}\beta\gamma, \beta^{-1}\delta\gamma\varepsilon^{-1}, \delta^{-1}\phi\alpha\kappa\phi^{-1}\varepsilon\kappa^{-1} \rangle.$$

4. Use Theorem 9.2.1 to find a presentation for the fundamental group of the surfaces defined by the following 1-polygon presentations:

   (a) $abcda^{-1}b^{-1}c^{-1}d^{-1}$;

   (b) $abcdabcd$;

   (c) $abba^{-1}ccdee^{-1}ffd$.

9.3   3-Manifolds

3-Manifolds are the three-dimensional analogs of closed surfaces. Now surfaces, although intrinsically two-dimensional, require at least three dimensions for a faithful

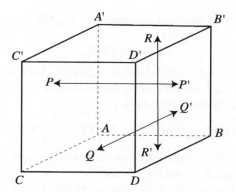

Figure 9.16   A presentation of the 3-torus.

display, and some, the nonorientable ones, actually require four. This difficulty was overcome by the employment of presentations, which allowed us to display these surfaces as plane polygons with a scheme for gluing segments of the perimeter in pairs. A similar device is used to represent 3-manifolds in space.

A *polyhedron* is the three-dimensional analog of a polygon. It is a homeomorph of the unit ball, whose periphery is subdivided by a network of arcs into faces that are themselves homeomorphic to polygons. In order to simplify the visualization problems inherent in this topic, it will be assumed that the perimeter of each face is a loop, i.e., it has no self intersections. In fact, this subject matter is so rich that most of our examples will be drawn from the well-known cube, octahedron, and dodecahedron.

A 3-*pseudomanifold* is the topological space formed from a finite collection of polyhedra by pairing their faces and gluing (technically, *identifying*) corresponding points in paired faces. The original collection of unglued polyhedra constitutes a *presentation* of the pseudomanifold. Many of the forthcoming examples are drawn from the series of articles in which Poincaré created the discipline of algebraic topology.

Example 9.3.1 The 3-torus $\mathscr{S}_1$ is an analog of the torus. It is obtained by gluing opposite faces in accordance with the pattern displayed in Figure 9.16.

More formally, if AC, AB, and AA' are selected as the respective positive x, y, and z axes of a Cartesian coordinate system, then the gluing process identifies

$$
\begin{array}{lll}
(x,y,0) & \text{with} & (x,y,1), \\
(x,0,z) & \text{with} & (x,1,z), \\
(0,y,z) & \text{with} & (1,y,z)
\end{array}
$$

on the surface of the given cube, where $0 \leq x, y, z \leq 1$.

These identifications will also be referred to as *equivalences*. It will prove convenient to establish a notational device for describing the identifications that define the pseudomanifold. If the two identified polygons are $ABC\ldots$ and $PQR\ldots$, we shall write

$$ABC\ldots \equiv PQR\ldots$$

to denote the fact that in this identification, $A,B,C,\ldots$ have been glued to $P,Q,R,$ $\ldots$, respectively, and the rest of the polygons follow suit in a linear fashion. Thus, the identifications that define the 3-torus are

$$ABDC \equiv A'B'D'C', \qquad ACC'A' \equiv BDD'B', \qquad DCC'D' \equiv BAA'B'.$$

Note that the restriction of such an equivalence is also an equivalence. Thus, on the above 3-torus, $ABD \equiv A'B'D'$ and $CAA' \equiv DBB'$.

As was the case for presentations of surfaces, a *node* is the result of the identification of vertices of the original polyhedra, and an *arc* is the result of the identification of their edges. Here, however, the identification of edges may lead to unexpected results. For example, the identifications

$$ABB'A' \equiv D'C'CD, \qquad ACC'A' \equiv DBB'D'$$

result in

$$BB' \equiv C'C \equiv B'B.$$

Thus, the edge BB' is equivalent to its reverse $B'B$. Such arcs are said to be *singular*. The number of nodes of a pseudosurface is α_0, and the number of *nonsingular* arcs is α_1.

In addition, a *pane* is the result of the identification of two faces, and the number of panes is α_2. The number of polyhedra that underlie the pseudomanifold is α_3. For the 3-torus, $\alpha_0 = 1$, $\alpha_1 = 3$, $\alpha_2 = 3$, $\alpha_3 = 1$.

Example 9.3.2 Let $\mathscr{S}_2$ be the 3-pseudomanifold defined on the cube of Figure 9.16 by the identifications

$$ABDC \equiv B'D'C'A', \qquad ABB'A' \equiv DD'C'C, \qquad ACC'A' \equiv DD'B'B.$$

Then $\mathscr{S}_2$ has two nodes

$$\{A \equiv B' \equiv C' \equiv D\}, \qquad \{A' \equiv B \equiv C \equiv D'\},$$

two arcs

$$\{AB \equiv B'D' \equiv C'C \equiv B'A' \equiv AC \equiv DD'\},$$
$$\{AA' \equiv DC \equiv C'A' \equiv B'B \equiv C'D' \equiv DB\},$$

and three panes.

Example 9.3.3 Let $\mathscr{S}_3$ be the 3-pseudomanifold defined on the cube of Figure 9.16 by the identifications

$$ABDC \equiv B'D'C'A', \qquad ABB'A' \equiv C'CDD', \qquad ACC'A' \equiv DD'B'B.$$

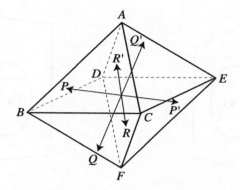

Figure 9.17   A presentation of the projective space.

Then $\mathscr{S}_3$ has two nodes

$$\{A \equiv B' \equiv C' \equiv D\}, \qquad \{A' \equiv B \equiv C \equiv D'\},$$

four arcs

$$\{AB \equiv B'D' \equiv C'C\},$$
$$\{AA' \equiv C'D' \equiv DB\},$$
$$\{AC \equiv B'A' \equiv DD'\},$$
$$\{BB' \equiv A'C' \equiv CD\},$$

and three panes.

Example 9.3.4 Let $\mathscr{S}_4$ be the 3-pseudomanifold defined on the cube of Figure 9.16 by the identifications

$$ABDC \equiv B'D'C'A', \qquad ABB'A' \equiv CDD'C', \qquad ACC'A' \equiv BDD'B'.$$

Then $\mathscr{S}_4$ has one node, three arcs

$$\{AB \equiv B'D' \equiv A'C' \equiv CD\},$$
$$\{AA' \equiv CC' \equiv DD' \equiv BB'\},$$
$$\{AC \equiv B'A' \equiv D'C' \equiv BD\},$$

and three panes.

Example 9.3.5 The projective space $\mathscr{S}_5$ is the analog of the projective plane. It can be obtained by gluing opposite faces according to the pattern displayed in Figure 9.17. More formally, if we imagine a coordinate system whose axes are the straight lines AF, BE, CD, then the gluing process identifies the point (x, y, z) with $(-x, -y, -z)$ on the surface of the octahedron. This presentation has $\alpha_0 = 3$, $\alpha_1 = 6$,

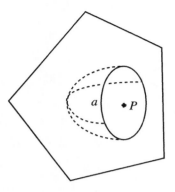
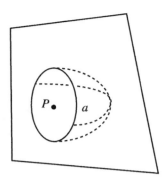

Figure 9.18   Two hemiballs.

$\alpha_2 = 4$. Alternatively, the projective space can be defined by identifying diamterically opposide points on the surface of the cube, in which case $\alpha_0 = 4$, $\alpha_1 = 6$, $\alpha_2 = 3$.

Note that with the exception of the pseudomanifold $\mathscr{S}_2$ of Example 9.3.2, the alternating sum

$$\alpha_0 - \alpha_1 + \alpha_2 - \alpha_3 \tag{4}$$

has the value 0 in the above examples. We now set out to explain the difference between the exceptional $\mathscr{S}_2$ and the other pseudomanifolds.

One of the traits of closed surfaces is that every node has a vicinity that is homeomorphic to the disk. It is therefore reasonable to expect their three-dimensional analogs to possess the similar property that every node should have a vicinity that is homeomorphic to a ball. 3-Pseudomanifolds that possess this property are called *3-manifolds*. It turns out that with the sole exception of $\mathscr{S}_2$ all of the above examples are in fact manifolds, and this difference accounts for the behavior of the alternating sum of (4), known as the *Euler characteristic* of the presentation.

We now examine the vicinities of all the points of a 3-pseudomanifold. These points fall into four classes: the interiors of the polyhedra, the interiors of the faces, the interiors of the arcs (or edges), and the nodes.

It is clear that every point of a 3-pseudomanifold $\mathscr{S}$ that lies in the interior of one of its polyhedra is indeed surrounded by a ball inside $\mathscr{S}$. The same is true, though less obvious, for any point P that lies in the interior of one of the panes of $\mathscr{S}$. At each of the identified faces there is a "hemiball" (see Figure 9.18) that "surrounds P to the best of its ability." Together, these two hemiballs do indeed properly surround P, but do they necessarily form a ball? To show that such is indeed the case, it suffices to demonstrate that their bounding hemispheres join up to form a 2-sphere S_0. However, regardless of how the two faces that contain P are identified, these hemispheres form a presentation consisting of two disks, whose perimeters are both labeled a. As this presentation is orientable and has Euler characteristic $1 - 1 + 2 = 2$, it does indeed yield a sphere.

Next, suppose that the point P lies in the interior of some of the edges of the presentation's polyhedra (see Figure 9.19).

We may suppose that these edges are labeled so that

$$A_i B_i \equiv A_{i+1} B_{i+1}, \qquad i = 1, 2, \ldots, n-1,$$

and that the vicinity of P in $\mathscr{S}$ consists of the union of the depicted wedges. Note that no claim is being made as to whether a_i appears to the left or to the right of $A_i B_i$. Moreover, neither occurrence of a_1 is assigned a direction, because we don't know whether

$$A_n B_n \equiv A_1 B_1 \qquad \text{(the arc is nonsingular)}$$

or

$$A_n B_n \equiv B_1 A_1 \qquad \text{(the arc is singular)}.$$

These wedges assemble into a ball if and only if their lune-like curved surfaces assemble into a topological sphere—in other words, if and only if the digons of the presentation of Figure 9.20 define a sphere. It is convenient to draw these digons so that their labeled sides appear in monotone increasing order, and it is clear that the required "flipping" changes neither the topological nature of the individual digons

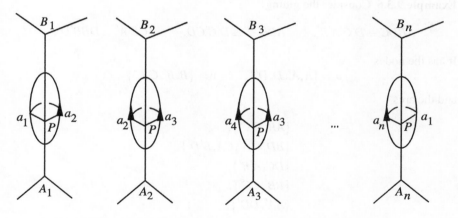

Figure 9.19 The vicinity of the midpoint of a singular arc.

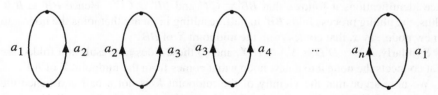

Figure 9.20 The link of the midpoint of a singular arc.

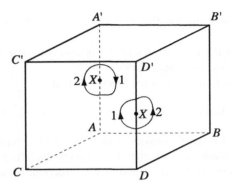

Figure 9.21   A vicinity at a singular arc.

nor the overall presentation. The surface of this presentation, of course, depends on the directions of the two occurrences of a_1. If both occurrences have the same direction relative to the page, that is, if the arc is nonsingular, this presentation is orientable and defines a topological sphere. On the other hand, if the two occurrences of a_1 have opposite directions, that is, if the arc is singular, then the presentation of Figure 9.20 defines a projective plane.

Example 9.3.6   Consider the gluing

$$ABDC \equiv D'C'A'B', \qquad ABB'A' \equiv D'C'CD, \qquad ACC'A' \equiv DBB'D'.$$

It has the nodes

$$u = \{A, A', D, D'\}, \qquad v = \{B, B', C, C'\}$$

and the arcs

$$\{AB, D'C'\},$$
$$\{BD, C'A', CA, B'D'\},$$
$$\{DC, A'B'\},$$
$$\{BB', CC'\},$$
$$\{AA', DD'\}.$$

The last two of these arcs, however, are singular. From the second and third defining face identifications it follows that $BB' \equiv C'C$ and $BB' \equiv CC'$. Hence $BB' \equiv B'B$. Thus, the gluing process folds BB' in half, resulting in an arc that joins the node v to a new node, say x, that comes from the midpoint X of BB'.

Similarly, $AA' \equiv D'D \equiv A'A \equiv DD'$, and so these sides too result in a folded arc that connects the node u to a new node y that comes from the midpoint Y of AA'.

We now argue that the vicinity of the midpoint X is not a ball and hence the pseudomanifold is not a manifold. This vicinity is pictured in Figure 9.21 as two quarter balls, and its boundary consists of two quarter spheres. These two quarter spheres constitute the 2-polygon presentation $\{12, 12^{-1}\}$ of the link of X. It follows that this link is in fact the projective plane $\tilde{S}^1$. Thus, the vicinity of X is not a ball.

The vicinity of a node on a surface's presentation is easily visualized (see Fig. 3.21 for an example). The vicinities of nodes on pseudomanifolds are more complicated and may not even be realizable in $\mathbb{R}^3$. For this reason they will mostly be portrayed in terms of their portions in the underlying polyhedra. An exceptionally simple example is provided by the vicinity of the single node of the 3-torus $\mathscr{S}_1$ of Example 9.3.1, which consists of the eight solid angles displayed in Figure 9.22. The eight curved triangles in the cube are assembled, by the gluing process, into a sphere that surrounds the unique node on the 3-torus. In this particular case, this assemblage can be visualized by translating all the solid angles of the cube to the vertex A, where they form the eight octants of the coordinate system mentioned in Example 9.3.1 (Fig. 9.23). We stress again that diagrams such as Figure 9.23 are very rarely feasible and that the assemblage must in general be left to the imagination and other techniques. What makes this figure possible in this case is the easily described nature of the identifications that define the 3-torus in Figure 9.16. Every face is identified with its parallel face by a straightforward translation whose direction is perpendicular to both. This facilitates a construction of the vicinity on the manifold by means of mere translations.

Let us turn to $\mathscr{S}_2$ of Example 9.3.2. It has the two nodes

$$u = \{A, B', C', D\}, \qquad v = \{A', B, C, D'\}.$$

The vicinity of u on $\mathscr{S}_2$ therefore consists of (deformed) copies of the solid angles at A, B', C', D with the curved triangles assembling to form the outside (or boundary) surface of the vicinity (Fig. 9.24). This boundary surface is in general called the *link* of u and is denoted by lk u. The reason these triangles form the presentation of a surface is that the equivalences that define $\mathscr{S}_2$ also identify the sides of these triangles in pairs. In Figure 9.24 identified sides bear identical labels, and they are the consequences of the following list of restrictions of the identifications that were used in Example 9.3.2 to define $\mathscr{S}_2$:

$$a: \quad ABD \equiv B'D'C',$$

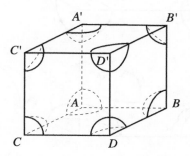

Figure 9.22   The vicinity of a node on a polyhedron.

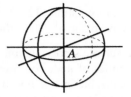

Figure 9.23   The vicinity of the node on the 3-torus.

$$
\begin{aligned}
b &: \quad BDC \equiv D'C'A', \\
c &: \quad DCA \equiv C'A'B', \\
d &: \quad CAB \equiv A'B'D', \\
e &: \quad ABB' \equiv DD'C', \\
f &: \quad BB'A' \equiv D'C'C, \\
g &: \quad B'A'A \equiv C'CD, \\
h &: \quad A'AB \equiv CDD', \\
i &: \quad ACC' \equiv DD'B', \\
j &: \quad CC'A' \equiv D'B'B, \\
k &: \quad C'A'A \equiv B'BD, \\
l &: \quad A'AC \equiv BDD'.
\end{aligned}
$$

The eight triangular sails illustrated in Figure 9.24 constitute two oriented presentations:

$$
\Pi = \{cig, ea^{-1}i^{-1}, ak^{-1}e^{-1}, kg^{-1}c^{-1}\}
$$

with nodes

$$
x = \{g, i^{-1}, a, e, i, c^{-1}\}, \qquad y = \{g^{-1}, c, k, e^{-1}, a^{-1}, k^{-1}\},
$$

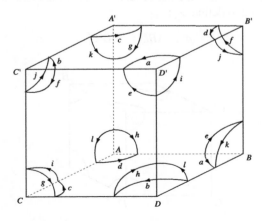

Figure 9.24   Vicinities of nodes on $\mathscr{S}_2$.

and

$$\Pi' = \{bhl^{-1}, ldh^{-1}, fjb^{-1}, d^{-1}f^{-1}j^{-1}\}$$

with nodes

$$x' = \{b, l, h, b^{-1}, j^{-1}, f\}, \qquad y' = \{l^{-1}, h^{-1}, d^{-1}, j, f^{-1}, d\}.$$

Each of these presentations defines an orientable surface with Euler characteristic

$$2 - 6 + 4 = 0$$

and is therefore a torus. Since the boundaries of the vicinities of u and v are not spheres, it follows that these vicinities are not balls, so that $\mathscr{S}_2$ is not a manifold.

We are fortunate in that there is a simple and effective criterion for identifying 3-manifolds. The following lemma is plausible and will therefore not be proved.

Lemma 9.3.7 *A 3-pseudomanifold is a 3-manifold if and only if the boundary of the vicinity of each node is a 2-sphere.*

The plausability and validity of this lemma notwithstanding, its n-dimensional analog is false for $n \geq 5$ (see Thurston 1997, p. 121).

Theorem 9.3.8 (H. Poincaré, W. P. Thurston) *If $\mathscr{S}$ is a 3-pseudomanifold, then $\chi(\mathscr{S}) \geq 0$ and equality holds if and only if $\mathscr{S}$ is a 3-manifold.*

PROOF: Let $\mathscr{S}$ be a 3-pseudomanifold with presentation Π, α_0 nodes, α_1 nonsingular arcs, α_2 panes, and α_3 polyhedra. If Π has a singular edge, subdivide it, and each equivalent edge, into two edges by placing a vertex in each edge's middle. This adds 1 to both α_0 and α_1 and so does not change the value of the Euler characteristic. Consequently, it may be assumed, without loss of generality, that Π has only nonsingular arcs.

Let a vicinity be selected for each node u of $\mathscr{S}$ with boundary surface lk u. The intersection of lk u with the solid angles at the vertices of the polyhedra of Π defines a presentation Π_u of lk u with p_u, q_u, r_u nodes, arcs, and faces, respectively. Since the nodes of Π_u are determined by the directed arcs of $\mathscr{S}$ emanating from u, it follows that

$$\sum_u p_u = 2\alpha_1.$$

The arcs of Π_u correspond to the angles on the panes of $\mathscr{S}$. Hence $\sum_u q_u$ equals one-half the number of angles on the polyhedra of Π, which, in turn, equals the number of edges (not arcs) of Π. We record this as

$$\sum_u q_u = e(\Pi).$$

The polygons of Π_u, that is, the sails that surround u, correspond to the solid angles at u, and hence

$$\sum_u r_u = v(\Pi),$$

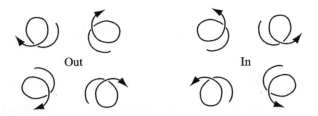

Out   In

Figure 9.25   Curls.

where $v(\Pi)$ denotes the number of vertices (not nodes) in Π.

Let $f(\Pi)$ denote the total number of faces on the polyhedra of Π. Since each polyhedron has Euler characteristic 2, it follows that

$$v(\Pi) - e(\Pi) + f(\Pi) = 2\alpha_3.$$

Recall that, by Corollary 3.2.13, for each u,

$$\chi(\mathrm{lk}\, u) = p_u - q_u + r_u \leq 2,$$

with equality holding if and only if $\mathrm{lk}\, u$ is a sphere. Hence

$$\sum_u \chi(\mathrm{lk}\, u) = \sum_u p_u - \sum_u q_u + \sum_u r_u = 2\alpha_1 - e(\Pi) + v(\Pi)$$

$$= 2\alpha_1 - f(\Pi) + 2\alpha_3 = 2\alpha_1 - 2\alpha_2 + 2\alpha_3 = 2\alpha_0 - 2\chi(\mathscr{S}),$$

and so

$$\chi(\mathscr{S}) = \alpha_0 - \frac{1}{2}\sum_u \chi(\mathrm{lk}\, u) = \sum_u \left(1 - \frac{1}{2}\chi(\mathrm{lk}\, u)\right) \geq 0. \tag{5}$$

Moreover, by the lemma, $\mathscr{S}$ is a 3-manifold if and only if each $\mathrm{lk}\, u$ is a sphere, that is, if and only if equality holds in (5).   Q.E.D.

It follows from Theorem 9.3.8 that $\mathscr{S}_1$, $\mathscr{S}_3$, $\mathscr{S}_4$, and $\mathscr{S}_5$ are all 3-manifolds.

As noted in Theorem 9.3.8, the Euler characteristic of every 3-manifold is 0, and so there is no hope that the surfaces' classification by means of Euler characteristic and orientability character can be carried over to 3-manifolds. Nevertheless, the orientability character of a 3-manifold is of interest.

Three-dimensional orientations are defined in terms of screw motions, which are of two types: *in* and *out* (Fig. 9.25). The former describes the pressure used to drive a screw in, the latter to extract it. The arcs used to denote these motions are *curls*. A presentation of a 3-pseudomanifold $\mathscr{S}$ is *orientable* provided that a curl moving about in $\mathscr{S}$ (and through the various panes) will always return to its point of departure as a curl of the same nature. Note that if we assume the equivalence $ABCD \equiv A'B'C'D'$ in the box of Figure 9.26, the in-curl exiting on the left returns as an out-curl on the right. To prove this we argue as follows. Project the (directed) in-curl on the left onto the directed circle c_1 on the face $ABCD$. This circle is equivalent to the directed circle c_2, which is the projection of the curl as it reappears on the right side. The sense of c_2 indicates that the curl must reemerge as an out-curl. On the other hand, Figure 9.27, which portrays two faces of the same polyhedron Π, shows how a curl leaving Π through the face $ABCD$ reenters Π through the equivalent face $A'B'C'D'$ in its original state.

It is clear that if the curl's tour is restricted to the interior of a polyhedron then a consistent return is inevitable. Things can only go wrong if one or more panes are crossed during the trip. As it turns out, Figures 9.26 and 9.27 describe all the relevant possibilities. To show this it is necessary to define a term.

An *orientation* of a polyhedron in $\mathbb{R}^3$ assigns to it one of the two screw motions. An orientation of a polyhedron Π *induces* an orientation of each of its faces F as follows: project onto F the curl assigned to Π as it is leaving through F (see the left portions of Figs. 9.26–9.27). The sense of this circle is the induced orientation of F.

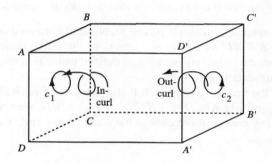

Figure 9.26   Nonorientability.

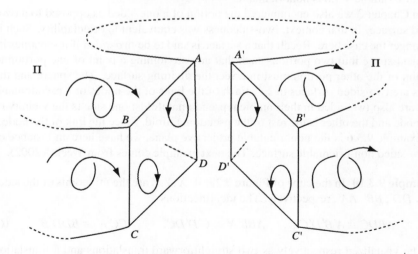

Figure 9.27   Orientability.

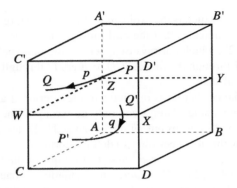

Figure 9.28   A one-sided torus.

Proposition 9.3.9 *A 1-polyhedron presentation is orientable if and only if every orientation of the polyhedron induces opposite orientations on equivalent faces.*

PROOF: If a presentation contains a pair of incoherently oriented equivalent faces, such as $ABCD \equiv A'B'C'D'$ in Figure 9.26, then, as the figure indicates, an in-curl will turn into an out-curl when crossing the so defined pane. This implies the nonorientability of the presentation.

Conversely, if the presentation is such that every pair of equivalent faces is coherently oriented, just like $ABCD \equiv A'B'C'D'$ in Figure 9.27, then, as the figure indicates, crossing any pane will not change the sense of any curl. Consequently the presentation is orientable.   Q.E.D.

It follows that the pseudomanifolds of Examples 3.1–3.5 are orientable whereas that of Example 9.3.6 is nonorientable.

In Chapter 3 we also encountered the notion of a one-sided as opposed to a two-sided surface. In that context, two-sidedness was equivalent to orientability. Such is no longer the case here. Recall that a surface is said to be *two-sided* if it separates its ambient space into two portions such that any arc joining a point of one portion to a point of the other portion must intersect the dividing surface. The sphere and the torus are two-sided surfaces in $\mathbb{R}^3$. Each of the links of the nodes of a pseudomanifold are also two-sided in their ambient pseudomanifolds: one side is the vicinity of the node and the other is the rest of the pseudomanifold. Since the link of the node x of Example 9.3.6 is the nonorientable projective plane, we have here an instance of a two-sided non-orientable surface. The next example comes from Weeks (2002).

Example 9.3.10 In the cube of Figure 9.28, W, X, Y, Z are the midpoints of the sides CC', DD', BB', AA', respectively. The identifications

$$ABDC \equiv A'B'D'C', \qquad ABB'A' \equiv C'D'DC, \qquad ACC'A' \equiv BDD'B' \qquad (6)$$

can be visualized respectively as two straightforward translations and a translation followed by a reflection. This turns the square $WXYZ$ into a torus inside the resulting 3-manifold and identifies the point Q on $C'D'XW$ with the point Q' on $ABYZ$. Consequently, the arbitrary point P in the "top half" of the manifold can be joined by the path $p \cdot q$ to the arbitrary point P' of the "bottom half." Since the path $p \cdot q$ does not intersect the torus $WXYZ$, this torus is one-sided in the ambient nonorientable manifold.

It is not hard to prove that in orientable manifolds such distinctions cannot arise (see Exercise 29).

The computation of the fundamental group of a 3-manifold $\mathscr{S}$ proceeds along the same lines that were used for surfaces. We begin with a presentation Π and assume, without loss of generality, that Π has only one polyhedron. A base point X is selected in the interior of this unique polyhedron. For every pane p, a directed loop p^* is selected that is based at X, crosses p exactly once, and does not intersect the other panes at all. The loop p^* is the *dual* of the pane p. If c is any loop at X, then, for every crossing of c with the interior of a pane p, the *crossing index* ε_p is 1 or -1 according as p^* and c cross p in the same direction or not. Since every such loop c can be perturbed so as to avoid the arcs of the presentation, it follows that

$$c \simeq (p_{i_1}^*)^{\varepsilon_1} \cdot (p_{i_2}^*)^{\varepsilon_2} \cdots (p_{i_k}^*)^{\varepsilon_k},$$

where c crosses the panes $p_{i_1}, p_{i_2}, \ldots, p_{i_k}$ with respective indices $\varepsilon_1, \varepsilon_2, \ldots, \varepsilon_k$. Consequently,

$$[c] = [p_{i_1}^*]^{\varepsilon_1} [p_{i_2}^*]^{\varepsilon_2} \cdots [p_{i_k}^*]^{\varepsilon_k}.$$

In other words, the elements $[p_i^*]$ generate the fundamental group $\pi(\mathscr{S}, X)$. To describe the relations that bind these generators, we introduce the notion of the *book* $\beta(a)$ of an arc a. This consists of a cyclic ordering of the panes that share the arc a, an ordering in which any consecutive pair of panes form a dihedral angle of the polyhedron of Π. If $\beta(a) = (p, q, r, \ldots)$, then the arc a contributes the relator

$$W_a = [p^*]^{\varepsilon_p} [q^*]^{\varepsilon_q} [r^*]^{\varepsilon_r} \cdots$$

where $\varepsilon_p, \varepsilon_q, \varepsilon_r, \ldots$ are the crossing indices of a small loop surrounding a that runs in the same circular direction as the book. The rationale for this is provided by a modification of Figure 9.12. It is only necessary to add a dimension by replacing the node A with an arc a and replacing the arcs p, q, r, s, t with corresponding panes (see Fig. 9.29). Upon the replacement of each $[p_i^*]$ by α_i we then have the following theorem.

Theorem 9.3.11 *Let Π be a 1-polyhedron presentation of the 3-manifold $\mathscr{S}$ with panes $p_1, p_2, \ldots, p_n$, arcs $a_1, a_2, \ldots, a_k$, and a base point X. Then the fundamental group $\pi(\mathscr{S}, X)$ has the presentation*

$$\langle \alpha_1, \alpha_2, \ldots, \alpha_n ; W_1, W_2, \ldots, W_k \rangle$$

where, for each $i = 1, 2, \ldots, k$,

$$W_i = \alpha_{i_1}^{\varepsilon_1} \alpha_{i_2}^{\varepsilon_2} \cdots,$$

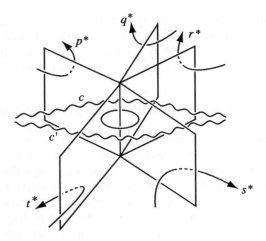

Figure 9.29   The creation of a relator.

$(p_{i_1}, p_{i_2}, \ldots)$ *constitute the book* $\beta(a_i)$, *and the exponents* $\varepsilon_1, \varepsilon_2, \ldots$ *are the corresponding crossing indices of a small loop surrounding* a_i.   □

The books $\beta(a_i)$ are described by a "rotate and switch" process that is very similar to that used in Sections 3.1 and 9.2 for the purpose of describing the star of a node in the presentation of a surface.

Example 9.3.12 To compute the fundamental group of the 3-torus $\mathscr{S}_1$, its arcs and panes are labeled as in Figure 9.30. Begin by selecting duals p^*, q^*, r^* as in the figure. To find $\beta(a)$ we proceed by arbitrarily choosing a point 1 near a on the face $DCAB$ (actually, any of the faces that have a as a side could have furnished the starting point 1). Next rotate through the dihedral angle at this copy of a until you reach the face $A'ACC'$, and, switching to the equivalent $B'BDD'$, mark on it the point 2. Repeat this process using 2 as the starting point: Rotate till you reach $ABDC$, switch to the equivalent $A'B'D'C'$, and then mark the point 3. Starting from 3, rotate to the face $DD'B'B$, switch to $CC'A'A$, and mark the point 4. Starting from 4, rotate to the face $D'C'A'B'$, switch to $DCAB$, and note that you have returned to the point of origin 1. The four arcs that describe the aforementioned four dihedral rotations constitute a small loop circling a. Relative to the choice of duals, the small circling loop has indices -1, 1, 1, -1 at the crossings 1, 2, 3, 4, respectively. Hence a contributes the relator

$$[p^*]^{-1}[r^*][p^*][r^*]^{-1}.$$

The crossings of the small loop circling b contribute the relator

$$[p^*][q^*]^{-1}[p^*]^{-1}[q^*],$$

and the crossings of the small loop circling c contribute the relator

$$[r^*]^{-1}[q^*]^{-1}[r^*][q^*].$$

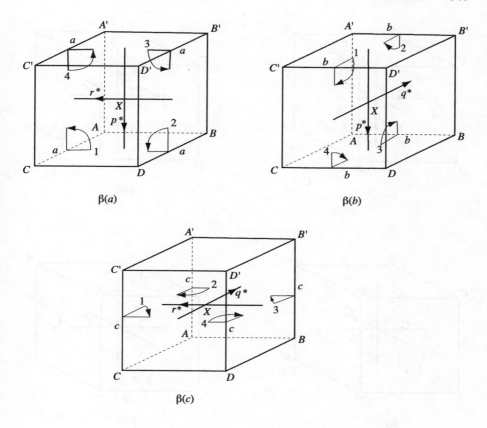

Figure 9.30   Relators for the 3-torus.

If we set $\alpha = [p^*]$, $\beta = [q^*]^{-1}$, $\gamma = [r^*]^{-1}$, then we may conclude that the fundamental group of the 3-torus has the presentation

$$\langle \alpha, \beta, \gamma \,;\; \alpha^{-1}\gamma^{-1}\alpha\gamma, \alpha\beta\alpha^{-1}\beta^{-1}, \gamma\beta\gamma^{-1}\beta^{-1}\rangle$$

and is therefore the free abelian group of width three.

Example 9.3.13 To find the fundamental group of the manifold $\mathscr{S}_3$ of Example 9.3.3, we label the edges, faces, and duals as in Figure 9.31. The arcs a, b, c, d then contribute the relators

$$a: \quad [p^*][r^*]^{-1}[q^*]^{-1},$$
$$b: \quad [p^*][q^*]^{-1}[r^*],$$
$$c: \quad [p^*][q^*][r^*]^{-1},$$
$$d: \quad [p^*][r^*][q^*].$$

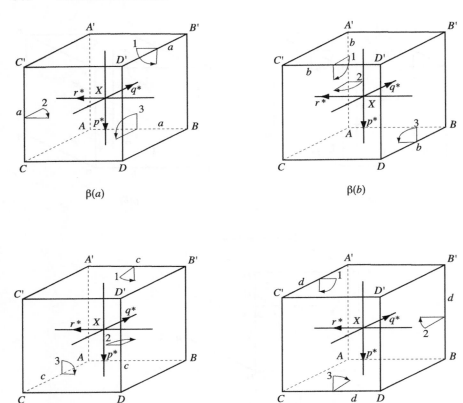

Figure 9.31 Relators for $\mathcal{S}_3$.

The substitution $\alpha = [p^*]$, $\beta = [r^*]^{-1}$, $\gamma = [q^*]^{-1}$ yields the presentation

$$\langle \alpha, \beta, \gamma; \alpha\beta\gamma, \alpha\gamma\beta^{-1}, \alpha\gamma^{-1}\beta, \alpha\beta^{-1}\gamma^{-1} \rangle.$$

It follows from Example B.3.12 that the abelianization of $\pi(\mathcal{S}_3, X)$ is $\mathbb{Z}_2^2$. Since the fundamental group of the 3-torus is $\mathbb{Z}^3$, it follows that the 3-torus $\mathcal{S}_1$ and $\mathcal{S}_3$ are not homeomorphic.

Example 9.3.14 To find the fundamental group of the manifold $\mathcal{S}_4$ of Example 9.3.4, we label the edges and duals as in Figure 9.32. The relators are

$$a: \quad [p^*]^{-1}[q^*]^{-1}[p^*][r^*]^{-1},$$
$$b: \quad [r^*]^{-1}[q^*]^{-1}[r^*][q^*],$$
$$c: \quad [p^*]^{-1}[r^*][p^*][q^*]^{-1}.$$

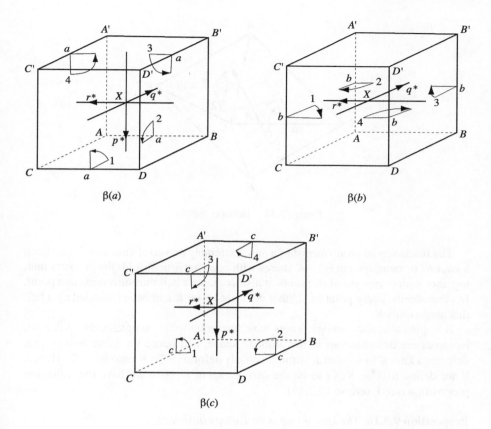

Figure 9.32   Relators for $\mathscr{S}_4$.

The substitutions $\alpha = [p^*]$, $\beta = [q^*]^{-1}$, $\gamma = [r^*]^{-1}$ yield the presentation

$$\langle \alpha, \beta, \gamma; \alpha^{-1}\beta\alpha\gamma, \gamma\beta\gamma^{-1}\beta^{-1}, \alpha^{-1}\gamma^{-1}\alpha\beta \rangle.$$

By Example B.3.13, the abelianization of this group is $\mathbb{Z} \times \mathbb{Z}_2$. It follows that $\mathscr{S}_4$ is homeomorphic to neither $\mathscr{S}_1$ nor $\mathscr{S}_3$.

Example 9.3.15 To find the fundamental group of the manifold $\mathscr{S}_5$ we begin by labeling the duals from the interior point X to the faces *DEA*, *ECA*, *CBA*, *DBA* as, p^*, q^*, r^*, s^*, respectively. It follows that the duals from X to the faces *BCF*, *DBF*, *EDF*, *CEF* are, respectively, p^{*-1}, q^{*-1}, r^{*-1}, s^{*-1}. The book of the arc a has two panes and contributes the relator p^*r^* (see Fig. 9.33). The other five arcs contribute the relators q^*s^*, $p^{*-1}s^*$, $q^{*-1}r^*$, r^*s^{*-1}, q^*p^{*-1}. The last four relators imply that $p^* = q^* = r^* = s^*$, and so it follows that $\pi(\mathscr{S}_5, X) \cong \mathbb{Z}_2$.

It follows from the foregoing discussion that the 3-manifolds $\mathscr{S}_1$, $\mathscr{S}_3$, $\mathscr{S}_4$, $\mathscr{S}_5$ are all topologically distinct.

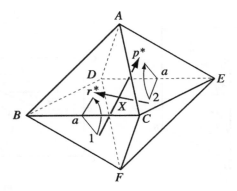

Figure 9.33   Relators for $\mathscr{S}_5$.

The fundamental group furnishes a new, extremely powerful knot invariant. Given a knot K, its complement $\mathbb{R}^3 - K$ shares with compact 3-manifolds the property that, together with every one of its points, it also contains a ball that surrounds that point. In other words, every point of $\mathbb{R}^3$ that does not lie on K can be surrounded by a ball that also avoids K.

It is plausible that isotopic knots have homeomorphic complements. After all, isotopies are deformations of the ambient space, and hence the same isotopy that deforms a knot K to another K' simultaneously deforms $\mathbb{R}^3 - K$ onto $\mathbb{R}^3 - K'$. Hence, if we define $\pi(\mathbb{R}^3 - K, X)$ to be the *knot group* of K, then we have the following proposition (see Exercise 10.2.15).

Proposition 9.3.16 *The knot group is an isotopy invariant.*   □

It would be hard to imagine a presentation of the knot group that is simpler than the one provided next.

Theorem 9.3.17 (W. Wirtinger) *Let D be an oriented diagram of the knot K with strands $a_1, a_2, \ldots, a_n$ and crossings $C_1, C_2, \ldots, C_k$. Then the knot group of K has the presentation*

$$\langle \alpha_1, \alpha_2, \ldots, \alpha_n \ ; \ \alpha_{i_1}^{-1} \alpha_{i_2}^{\eta_i} \alpha_{i_3} \alpha_{i_2}^{-\eta_i}, \ i = 1, 2, \ldots, k \rangle,$$

where η_i is the value of the crossing C_i, a_{i_1} is the understrand into C_i, a_{i_2} is the overstrand at C_i, and a_{i_3} is the understrand leaving C_i.

PROOF: As indicated by the statement of the theorem, the fundamental group of $\mathbb{R}^3 - K$ will be described in terms of a diagram D of the knot K. However, rather than think of this diagram as a figure in a plane, we will treat it as the actual knot (viewed from a certain point "outside" the knot). This will allow us to visualize the elements of $\pi(\mathbb{R}^3 - K, X)$ in relation to K more easily.

Arbitrarily orient the given knot K so that each of the strands of D is a directed arc. Let X be a point off K, and for each $j = 1, 2, \ldots, n$ let a_j^* be a loop that begins at

X, circles a_j, and then returns to X along a path that is very close to its first portion (see Figure 9.34). It is clear that a_j^* can be selected so that both of its crossings with a_j are right-handed with value 1.

While it is reasonable to stipulate that a_j^* avoids all the crossings of D, it may be impossible to prevent a_j^* from crossing the other strands of D. Still, we may assume that whenever a_j^* does cross a strand $a_{j'}$ distinct from a_j, it does so as an overcrossing.

Let c be any circuit of $\mathbb{R}^3 - K$ based at X. If c undercrosses none of the strands of D, then $[c]$ is assigned the empty word. Otherwise, let its ordered list of under-crossings with the strands $a_{j_1}, a_{j_2}, \ldots, a_{j_k}$ be $D_1, D_2, \ldots, D_k$, respectively, with values $\varepsilon_1, \varepsilon_2, \ldots, \varepsilon_k$. We then assign to $[c]$ the word

$$[a_{j_1}^*]^{\varepsilon_1} [a_{j_2}^*]^{\varepsilon_2} \cdots [a_{j_k}^*]^{\varepsilon_k}. \tag{7}$$

We next show that the value of the expression (7) is preserved by homotopies. Suppose c and d are homotopic circuits based at X. If the homotopy does not change any of the crossings, then it is clear that (7) endows $[c]$ and $[d]$ with the same expressions. If the homotopy is of the type displayed in Figure 9.35(a) and w_1, w_2 are the contributions of the portions of c off the diagram, then the corresponding expressions of (7) are

$$w_1 [a_{j_i}^*]^{\varepsilon} [a_{j_i}^*]^{-\varepsilon} w_2 \quad \text{and} \quad w_1 w_2,$$

which are clearly equal. If the homotopy is of the type of Figure 9.35(b), then $[c]$ and $[d]$ are assigned identical expressions by (7).

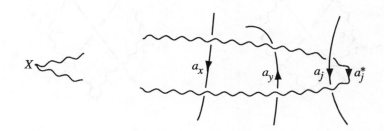

Figure 9.34 A generator for the knot group.

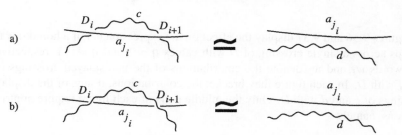

Figure 9.35 Two homotopies.

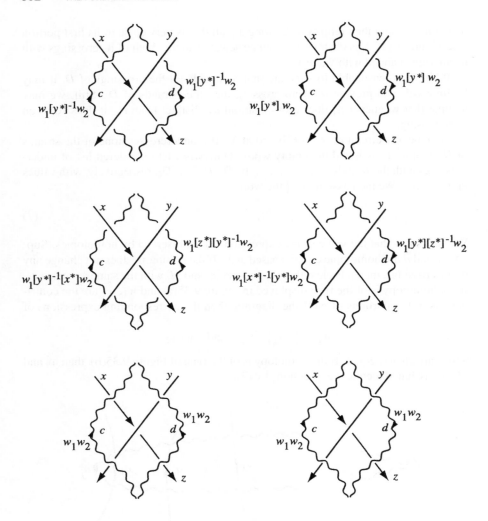

Figure 9.36 Homotopies across a crossing of value 1.

Figures 9.36 and 9.37 display the effect on the expression (7) of a homotopy that sweeps across a single crossing of D with values $\eta = 1$ and $\eta = -1$, respectively. The words w_1 and w_2 denote the contributions of the undisplayed crossings of c and d with D. In each figure they bracket the contributions, if any, of the displayed crossings of c and d with D. Only the middle pairs of each figure require nontrivial relations, namely

$$[y^*]^{-1}[x^*] = [z^*][y^*]^{-1} \qquad \text{in Figure 9.36,} \quad \text{where} \quad \eta = 1,$$

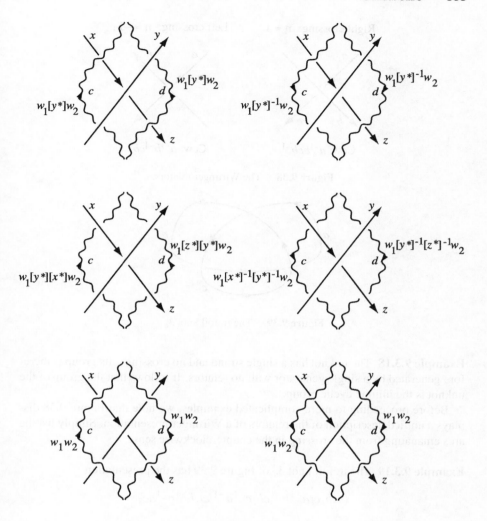

Figure 9.37   Homotopies across a crossing of value -1.

and

$$[y^*][x^*] = [z^*][y^*] \qquad \text{in Figure 9.37, where} \quad \eta = -1.$$

It is easily verified that these two relations can be summarized as the relator

$$[x^*]^{-1}[y^*]^{\eta}[z^*][y^*]^{-\eta}.$$

This accounts for the relators in the theorem's statement, provided we set $\alpha_j = [a_j^*]$.

Q.E.D.

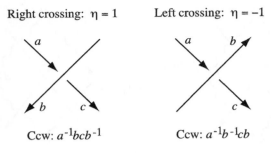

Right crossing: $\eta = 1$ Left crossing: $\eta = -1$

Ccw: $a^{-1}bcb^{-1}$ Ccw: $a^{-1}b^{-1}cb$

Figure 9.38   The Wirtinger relators.

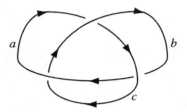

Figure 9.39   The trefoil knot 3_a.

Example 9.3.18 The unknot has a single strand and no crossings. Its group is therefore generated by a single generator with no relators. It follows that the group of the unknot is the infinite cyclic group.

Before proceeding to more complicated examples, we note that Figure 9.38 displays a unified description of the relators of a Wirtinger presentation. Simply list the arcs emanating from the crossing in the counterclockwise sense.

Example 9.3.19 The trefoil knot 3_a of Figure 9.39 has the presentation

$$\langle a,b,c; a^{-1}b^{-1}cb, c^{-1}a^{-1}ba, b^{-1}c^{-1}ac \rangle.$$

It can be shown (see Exercise 9.3.28) that this group also has the simpler presentation

$$\langle a,b \; ; \; abab^{-1}a^{-1}b^{-1} \rangle.$$

Example 9.3.20 The diagram of Figure 9.40 yields the Wirtinger presentation

$$\langle a,b,c \; ; \; a^{-1}ccc^{-1}, c^{-1}cbc^{-1}, b^{-1}c^{-1}ac \rangle.$$

The first two relators imply that $a = b = c$, so that the group of this knot is infinite cyclic. This, of course, also follows from Proposition 9.3.16, Example 9.3.18, and the fact that Figure 9.40 represents the unknot.

It is clear that obverse (mirror image) knots have homeomorphic complements and hence they have isomorphic groups. Thus, the knot group fails to distinguish between nonisotopic obverse knots such as the two trefoil knots of Figure 5.5. Still, for prime knots this is the only exception. In other words, *if two prime knots have isomorphic groups, then they are either isotopic or obverses of each other.*

Exercises 9.3

The identifications of Exercises 1–17 refer to the cube used to define $\mathscr{S}_1$–$\mathscr{S}_5$. For each of the resulting 3-pseudomanifolds, (i) determine its orientability character; (ii) determine whether it is a 3-manifold; (iii) if so, find a presentation for its fundamental group and obtain as much information as possible about the abelianization of this group; (iv) if not, determine the nature of the link of each of its nodes.

1. $ABDC \equiv B'D'C'A', ABB'A' \equiv DD'C'C, ACC'A' \equiv BDD'B'.$
2. $ABDC \equiv B'D'C'A', ABB'A' \equiv DD'C'C, ACC'A' \equiv DD'B'B.$
3. $ABDC \equiv D'C'A'B', ABB'A' \equiv CDD'C', ACC'A' \equiv BDD'B'.$
4. $ABDC \equiv D'C'A'B', ABB'A' \equiv DD'C'C, ACC'A' \equiv BDD'B'.$
5. $ABDC \equiv D'C'A'B', ABB'A' \equiv D'C'CD, ACC'A' \equiv BDD'B'.$
6. $ABDC \equiv D'C'A'B', ABB'A' \equiv D'C'CD, ACC'A' \equiv DD'B'B.$
7. $ABDC \equiv C'A'B'D', ABB'A' \equiv D'C'CD, ACC'A' \equiv DD'B'B.$
8. $ABDC \equiv A'B'D'C', ABB'A' \equiv CDD'C', ACC'A' \equiv DBB'D'.$
9. $ABDC \equiv D'C'A'B', ABB'A' \equiv CDD'C', ACC'A' \equiv DBB'D'.$
10. $ABDC \equiv D'C'A'B', ABB'A' \equiv D'C'CD, ACC'A' \equiv DBB'D'.$
11. $ABDC \equiv D'C'A'B', ABB'A' \equiv DCC'D', ACC'A' \equiv DBB'D'.$
12. $ABDC \equiv B'D'C'A', ABB'A' \equiv CDD'C', ACC'A' \equiv DBB'D'.$
13. $ABDC \equiv B'D'C'A', ABB'A' \equiv DD'C'C, ACC'A' \equiv DBB'D'.$
14. $ABDC \equiv B'D'C'A', ABB'A' \equiv DCC'D', ACC'A' \equiv DBB'D'.$
15. $ACC'A' \equiv BDD'B', CDD'C' \equiv ABB'A', ABC \equiv A'D'C', BCD \equiv B'A'D'.$
16. $ACC'A' \equiv D'B'BD, CDD'C' \equiv ABB'A', ABC \equiv A'D'C', BCD \equiv B'A'D'.$
17. $ACC'A' \equiv DD'B'B, CDD'C' \equiv ABB'A', ABC \equiv A'D'C', BCD \equiv B'A'D'.$
18. Let S be the manifold obtained by applying the identifications $ABC \equiv FBC, ACE \equiv FCE,$ $AED \equiv FED, ADB \equiv FDB$ to the octahedron of Figure 9.33.

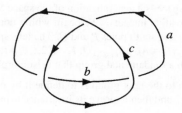

Figure 9.40   A diagram of the unknot.

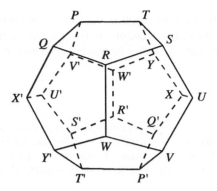

Figure 9.41   Antipodal pairs of points on the dodecahedron.

(a) Use Theorem 9.3.11 to compute a presentation for the fundamental group of S.

(b) Prove that the fundamental group of S is trivial.

(c) Explain why S is homeomorphic to the 3-sphere, which is defined as the graph of the equation $x^2 + y^2 + z^2 + w^2 = 0$ in $\mathbb{R}^4$.

19. (Poincaré dodecahedral space.) On the surface of the dodecahedron (Fig. 9.41) each pentagonal face is identified with the opposite pentagon by a projection followed by a counterclockwise rotation by an angle of $\pi/5$. Some identified faces are $PQRST \equiv S'T'P'Q'R'$ and $PVU'X'Q \equiv XQ'P'V'U$. Show that the resulting space $\mathscr{S}$ is a manifold with five nodes, ten arcs, and six panes. Find a presentation for the fundamental group of $\mathscr{S}$. Show that the abelianization of this group is trivial.

20. (Seifert–Weber dodecahedral space.) On the surface of the dodecahedron (Fig. 9.41) each pentagonal face is identified with the opposite pentagon by a projection followed by a counterclockwise rotation by an angle of $3\pi/5$. Some identified faces are $PQRST \equiv S'T'P'Q'R'$ and $PVU'X'Q \equiv XQ'P'V'U$. Show that the resulting space $\mathscr{S}$ is a manifold with one node, six arcs, and six panes. Find a presentation for the fundamental group of $\mathscr{S}$.

21. Suppose T is the double tetrahedron with equator ABC and north and south poles P and Q. Let S be the pseudomanifold resulting from the identifications $ABP \equiv BCQ, BCP \equiv CAQ, CAP \equiv ABQ$. Show that S is a manifold and that its fundamental group is isomorphic to $\mathbb{Z}_3$.

22. (Lens space.) The lens space is a generalization of the space S of Exercise 21. Let p and q be relatively prime positive integers. We begin with a double pyramid with equator $A_1, A_2, \ldots, A_p$ and north and south poles P and Q. The lens space $L(p, q)$ is the result of the identifications $A_i A_{i+1} P \equiv A_{i+p} A_{i+q+1} Q$, $i = 1, 2, \ldots, p$ (addition modulo p). Show that S is a manifold with a fundamental group that is isomorphic to $\mathbb{Z}_p$.

23. An alternative approach to the knot group is to define it by means of the group presentation of Theorem 9.3.17 and then to prove its invariance using the Reidemeister moves. Do so.

24. Prove that the abelianization of every knot group is the infinite cyclic group.

25.* Derive a presentation of the fundamental group of an arbitrary link. Prove that its abelianization is a free abelian group whose rank equals the number of components of the link.

26. By replacing the generators a, b, c of the trefoil group G of Example 9.3.19 with the transpositions $(1\ 2)$, $(1\ 3)$, $(2\ 3)$ of the symmetric group S_3, prove that 3_a is not an unknot.

27. Obtain the Wirtinger presentation of the diagram in Figure 5.18.

28.* Prove that the group of the knot 3_a of Example 9.3.19 has the presentation

$$\langle a, b \; ; \; abab^{-1}a^{-1}b^{-1} \rangle.$$

29.* Prove that in an orientable 3-manifold a surface is orientable if and only if it is two-sided.

30. (Tetrahedral space.) Let A, B, C, D be the vertices of a tetrahedron, and let $\mathscr{S}$ be obtained by means of the identifications $ACD \equiv CBA, BCD \equiv ADB$. Show that $\mathscr{S}$ is a manifold with fundamental group $\mathbb{Z}_5$.

31. (Octahedral space). A space is obtained by identifying each triangular face of the octahedron with its opposite face after a clockwise 60° turn. Show that this space is a manifold and that the abelianization of its fundamental group is an infinite cyclic group.

9.4   The Poincaré Conjecture

It is generally agreed amongst mathematicians that the Poincaré Conjecture is the most important problem of topology as well as one of the top ten problems of pure mathematics. This question will now be motivated and described.

It is reasonable to ask which manifolds have trivial fundamental groups. We first answer this question for the 2-manifolds, that is, the closed surfaces. Their fundamental groups are described in Theorem 9.2.4. As noted in Example B.3.14, the abelianization of the groups of the orientable surfaces of nonzero genus contains a nontrivial free abelian group and is therefore not a trivial group. A similar conclusion is drawn in Example B.3.15 regarding the groups of the nonorientable surfaces of genus greater than 1. The group of the remaining projective plane is cyclic of order 2, and so it too is nontrivial. It follows that the 2-sphere S_0 is the only closed surface with a trivial fundamental group.

One way to define the *3-sphere* is to modify the Cartesian equation of the 2-sphere. Accordingly,

$$3\text{-sphere} = \{(x, y, z, t) \mid x^2 + y^2 + z^2 + t^2 = 1\}.$$

Alternatively, and more in the spirit of this book, we can view the 2-sphere S_0 as the closed surface obtained by identifying the boundaries of two disks. Analogously, the 3-sphere could be defined as the 3-manifold obtained by identifying the boundaries of two disjoint (solid) balls. This could be visualized by placing the balls symmetrically with respect to a separating plane and then identifying corresponding points on their boundaries. The readers can convince themselves of the equivalence of the two definitions by first examining their two-dimensional analogs. The

2-sphere S_0 can, of course, be visualized as the graph of the equation $x^2 + y^2 + z^2 = 1$. This being said, note that this sphere can be viewed as the result of pasting its northern hemisphere $\{(x,y,z) \mid x^2 + y^2 + z^2 = 1, z \geq 0\}$ to its southern hemisphere $\{(x,y,z) \mid x^2 + y^2 + z^2 = 1, z \leq 0\}$ along their common border, which is the equator. By separating the two hemispheres slightly, one obtains two disks which are situated symmetrically with respect to the xy-plane.

Similarly, the northern and southern hemispheres of the 3-sphere can be defined as the sets

$$\text{NH} = \{(x,y,z,t) \mid x^2 + y^2 + z^2 + t^2 = 1,\, t \geq 0\}$$

and

$$\text{SH} = \{(x,y,z,t) \mid x^2 + y^2 + z^2 + t^2 = 1,\, t \leq 0\}$$

respectively. For each specific value of t, the set

$$S_t = \{(x,y,z,t) \mid x^2 + y^2 + z^2 + t^2 = 1\}$$

can be visualized as a 2-sphere centered at the origin with radius $\sqrt{1 - t^2}$. Keeping this visualization in mind, it is clear that

$$\text{3-sphere} = \text{NH} \cup \text{SH} \approx \left(\bigcup_{0 \leq t \leq 1} S_t \right) \cup \left(\bigcup_{-1 \leq t \leq 0} S_t \right)$$

where NH and SH are both homeomorphic to balls and intersect in their common boundary surface, which is a homeomorph of the 2-sphere S_0.

The fundamental group of the 3-sphere is easily computed. First deform each of NH and SH into a square-based pyramid, and then paste these two pyramids along their bases. This results in a 1-polyhedron presentation of the 3-sphere whose eight faces are identified as in Exercise 9.3.18. As noted in that exercise, the associated fundamental group is trivial. Shortly after inventing the fundamental group and the discipline of algebraic topology, Poincaré made the following conjecture.

Conjecture 9.4.1 (Poincaré, 1904) *Every 3-manifold with a trivial fundamental group is homeomorphic to the 3-sphere.* □

At first Poincaré had conjectured that every 3-manifold with a fundamental group whose *abelianization* was trivial would be homeomorphic to the 3-sphere. However, he soon realized that the dodecahedral space of Exercise 9.3.19 constituted a counterexample to this conjecture, which he then replaced with the version cited above.

Surprisingly, the n-dimensional analogs of Poincaré's conjecture for higher values of n turned out to be easier than the original problem. Stephen Smale proved its validity for all $n \geq 5$ in 1961, and Michael Freedman did the same for the case $n = 4$ in 1982. The validity of the case $n = 2$ is implicit in Theorem 3.2.10, whereas the case $n = 1$ is valid only because there are no 1-manifolds with a trivial fundamental group.

Added in proof: During 2002 and 2003 Grigori Perelman wrote three papers that constitute a proof of the Poincaré Conjecture. The validity of this proof has been verified by several teams of mathematicians.

Chapter Review Exercises

Are the following statements true or false? Justify your answers.

1. Every topological space has a fundamental group.
2. Some topological spaces have several nonisomorphic fundamental groups.
3. The fundamental group of every topological space is infinite.
4. Homotopic arcs are isotopic.
5. Isotopic paths are homotopic.
6. The plane and the 2-sphere have isomorphic fundamental groups.
7. If two closed surfaces are not homeomorphic, then their fundamental groups are not isomorphic.
8. The fundamental group of a surface and that of a 3-manifold cannot be isomorphic.
9. Every polyhedron with an even number of faces can be used to define a 1-polyhedron presentation.
10. If a 3-pseudomanifold contains a singular edge, then it is not a 3-manifold.
11. If a 3-pseudomanifold has Euler characteristic -1, then it is a 3-manifold.
12. A nonorientable surface can be two-sided inside a 3-manifold.
13. The knot group of the knot K is its fundamental group.
14. Isotopic knots have isomorphic knot groups.
15. All knots have isomorphic fundamental groups.
16. All knots have isomorphic knot groups.

CHAPTER 10

GENERAL TOPOLOGY

In this chapter we formalize the various fundamental concepts that were introduced in an informal fashion in the first nine chapters. While this ad hoc procedure may strike some readers as a questionable mathematical practice, it is undeniable that this seemingly inverted order does in fact reflect the evolution of the subject matter. This is particularly true with regard to the notions of continuity, homeomorphism, connectedness, and homotopy. All of these terms were freely used by mathematicians long before the precise language required for their rigorous definitions was developed.

10.1 Metric and Topological Spaces

We begin by defining the objects with which topology deals. A *metric space* (S, d) consists of a set S, whose elements are called *points*, and a real-valued function d of pairs of points of S such that

$$
\begin{array}{llll}
1. & d(P,Q) = d(Q,P) \geq 0 & \text{for any } P, Q \in S; & \\
2. & d(P,Q) = 0 & \text{if and only if } P = Q; & (1)
\end{array}
$$

Introduction to Topology and Geometry, Second Edition.
By Saul Stahl and Catherine Stenson Copyright © 2013 John Wiley & Sons, Inc.

$$3. \quad d(P,Q) \leq d(P,R) + d(R,Q) \qquad \text{for any } P, Q, R \in S.$$

It is customary to refer to the function d as a *metric* on S. Property 3 is also known as the *triangle inequality*.

Example 10.1.1 Let $S = \mathbb{R}^3$, and for any two points P and Q let $d(P,Q)$ be the standard Euclidean (i.e., straight line) distance between P and Q. The first two properties of (1) are clearly satisfied. The third one follows immediately from the triangle inequality of Euclidean geometry (Euclid's Proposition I.20).

Example 10.1.2 Let S be the surface of a sphere in $\mathbb{R}^3$, and for any two points P and Q of S let $d(P,\ Q)$ denote the length of the shortest curve that lies in S and connects P and Q. It is known that $d(P,\ Q)$ is realized by the shorter of the two arcs that join P and Q and lie in the plane determined by P, Q and the center of the sphere. The first two properties of (1) are again clearly satisfied. The validity of the third one is shown as follows. For any two points X and Y of S let $\gamma_{X,Y}$ be a curve joining X to Y whose length equals $d(X,Y)$. Since $\gamma_{P,R} \cup \gamma_{R,Q}$ is a particular curve that joins P and Q, we may conclude that

$$d(P,Q) \leq (\text{length of } \gamma_{P,R} \cup \gamma_{R,Q}) = (\text{length of } \gamma_{P,R}) + (\text{length of } \gamma_{R,Q})$$
$$= d(P,R) + d(R,Q).$$

Example 10.1.3 Let S be any geodesically connected surface in $\mathbb{R}^3$, and for any two points P and Q of S let $d(P,Q)$ denote the length of the shortest curve that lies in S and connects P and Q. The function d is again a metric. As usual, the validity of the first two properties of (1) is immediate. The third property can be shown in exactly the same manner that was used in Example 10.1.2 above. With the exception of the plane and the sphere, the actual value of this metric is quite difficult to compute.

Example 10.1.4 Let S be any subset of $\mathbb{R}^3$, and for any two points P and Q of S let $e(P,Q)$ denote the Euclidean (i.e., straight line) distance between P and Q. The Euclidean triangle inequality implies that e is indeed a metric on S. Note that this example and the previous one demonstrate that the same set S can be subject to two different metrics. These spaces will be termed *Euclidean spaces* and will be denoted by (S,e).

Example 10.1.5 Let $S = \mathbb{R}^3$, and set

$$d(P,Q) = 1 \qquad \text{whenever } P \neq Q,$$
$$d(P,Q) = 0 \qquad \text{whenever } P = Q.$$

If P, Q, and R are three distinct points, then

$$d(P,R) + d(R,Q) = 1 + 1 > 1 = d(P,Q),$$

and so property 3 holds in this case. In fact, this property also holds when the points are not distinct (see Exercise 1), so that d is indeed a metric, albeit a strange and unnatural one.

Example 10.1.6 Let S be any set, and for any points $P, Q \in S$ set

$$D(P,Q) = 1 \qquad \text{whenever } P \neq Q,$$
$$D(P,Q) = 0 \qquad \text{whenever } P = Q.$$

As was the case in the previous example, D constitutes a metric on S. This metric space is termed *discrete* and will be denoted by (S, D).

Let (S, d) be a metric space. The infinite sequence of points $P_1, P_2, P_3, \ldots$ of S is denoted by either $\{P_n\}_{n=1}^{\infty}$ or its abbreviation $\{P_n\}$. If P is yet another point of S, we say that $\{P_n\}$ *converges* to P, or that P is the *limit* of $\{P_n\}$, and write

$$\{P_n\} \to P,$$

provided that

$$\lim_{n \to \infty} d(P_n, P) = 0.$$

If (S, d) is a Euclidean metric space, then the recognition of convergence is in general easy. One need simply examine the coordinates of the points in question. More precisely, if $S = \mathbb{R}^3$, $P = (x, y, z)$, and for each $n = 1, 2, 3, \ldots$

$$P_n = (x_n, y_n, z_n),$$

then $\{P_n\} \to P$ if and only if

$$\lim_{n \to \infty} (x_n - x) = 0, \qquad \lim_{n \to \infty} (y_n - y) = 0, \qquad \text{and} \qquad \lim_{n \to \infty} (z_n - z) = 0.$$

Thus, for example,

$$\left\{ \left(\frac{1}{n}, \frac{1+n}{n}, \frac{1+6n}{2-3n} \right) \right\} \to (0, 1, -2) \qquad \text{in } (\mathbb{R}^3, e).$$

If (S, D) is a discrete metric space, then convergence looks quite different. Note first that if $\{a_n\}$ is a sequence in $\mathbb{R}$ each of whose terms is either 0 or 1, then

$$\lim_{n \to \infty} a_n = 0$$

if and only if all the terms are eventually zero, or, more precisely, there exists an index n' such that

$$a_n = 0 \qquad \text{for all } n \geq n'.$$

Now suppose that we are given a sequence $\{P_n\}$ in some discrete metric space (S, D). Then $\{P_n\} \to P$ if an only if $\lim_{n \to \infty} d(P_n, P) = 0$, which, as noted above, can only happen if there exists an index n' such that for all $n \geq n', d(P_n, P) = 0$. This, however, because of the nature of the discrete metric, is tantamount to saying that

$$P_n = P \qquad \text{for all } n \geq n'. \tag{2}$$

For example, if $S = \mathbb{R}$, then, under the discrete metric,

$$2, 2, 2, 3, 3, 3, 4, 4, 4, 4, 4, 4, 4, \cdots \rightarrow 4,$$

whereas the sequence $\{1/n\}$ does not converge at all, since its terms do not stabilize. Nor, for that matter, does the sequence

$$1, 0, \frac{1}{2}, 0, 0, 0, 0, \frac{1}{3}, 0, 0, 0, 0, 0, 0, 0, 0, 0, \frac{1}{4}, 0, 0, \ldots,$$

whose nth segment consists of $1/n$ followed by n^2 zeros, converge in $(\mathbb{R}, D)$. It will prove convenient to refer to a sequence that possesses the property expressed in (2) as *stable* and to say that it *stabilizes* at P. The following proposition formalizes the above discussion.

Proposition 10.1.7 *A sequence* $\{P_n\}$ *in a discrete metric space is convergent if and only if it is stable.*   □

We now explain the relationship between metric spaces and topology.

Given any set S, a *topology* on S is a family $\mathscr{T} = \{\mathscr{T}_\alpha \mid \alpha \in A\}$, where A is some indexing set, each $\mathscr{T}_\alpha \subseteq S$, and with the following properties:

1. The empty set $\emptyset$ is in $\mathscr{T}$.

2. The given set S is in $\mathscr{T}$.

3. The intersection of any two sets of $\mathscr{T}$ is in $\mathscr{T}$.

4. The union of any number of sets of $\mathscr{T}$ is in $\mathscr{T}$.

The pair $(S, \mathscr{T})$ is called a *topological space*.

Example 10.1.8 Let S be any set, and let $\mathscr{T}$ be the family of all the subsets of S. This is called the *discrete topology* on S. In an obvious sense, this is the largest topology that can be imposed on S. Its relationship to the discrete metric is explained below by Exercise 2.

Example 10.1.9 Let S be any set, and let $\mathscr{T} = \{\emptyset, S\}$. This is the smallest topology that can be imposed on S. By way of contrast with the discrete topology, this is called the *indiscrete topology*.

Two definitions are required in order to associate topologies with metric spaces. Let (S, d) be a metric space, P a point of S, and r a positive real number. The *r-neighborhood* of P, denoted $N_r(P)$, is the set

$$\{Q \in S \mid d(P, Q) < r\}.$$

If $(S,d) = (\mathbb{R}^2, e)$ then the r-neighborhood of P consists of the interior of the circle of radius r centered at P. If $(S,d) = (\mathbb{R}, e)$ then the r-neighborhood of P is the open interval $(P-r, P+r)$. If (S,d) is any discrete metric space, then the r-neighborhood of P consists of either the whole of S or just $\{P\}$, according as r is greater than or at most 1.

If (S,d) is a metric space and U is a subset of S, then U is said to be *open* provided that every point of U has a neighborhood that is completely contained in U. In other words, for every $P \in U$ there is a positive real number r, whose value may depend on P, such that

$$U \supset N_r(P).$$

For example, if $(S,d) = (\mathbb{R}^2, e)$ then the interior of a circle, rectangle, or any loop whatsoever is an open set. For, no matter how close a point P of this interior is to the bounding curve, it has a nonzero distance r from that curve, and the neighborhood $N_r(P)$ is clearly contained in this interior. If $(S,d) = (\mathbb{R}, e)$, then every open interval, as well as every union of open intervals, is an open set. Finally, if (S, D) is a discrete metric space, then every subset U of S is open. The reason for this is that, as noted above, $N_1(P) = \{P\}$ and hence trivially

$$U \supset N_1(P) = \{P\} \qquad \text{whenever } P \in U.$$

Regardless of the metric d, the sets S and $\emptyset$ are always open in (S,d)—the first because it contains all neighborhoods; the second because it contains all the neighborhoods of all the points in it (there aren't any such points). The following proposition shows how to convert a metric space into a topological one.

Proposition 10.1.10 *The open sets of a metric space (S, d) constitute a topology on S.*

PROOF: Let $\mathscr{T}$ be the family of all the open sets of (S,d). As the comments preceding this proposition state, $S, \emptyset \in \mathscr{T}$. Suppose U and V are any open sets and $P \in U \cap V$. Then there exist positive real numbers r and s such that

$$U \supset N_r(P) \quad \text{and} \quad V \supset N_s(P).$$

If t denotes the smaller of r and s, then

$$U \supset N_r(P) \supset N_t(P) \quad \text{and} \quad V \supset N_s(P) \supset N_t(P),$$

and consequently

$$U \cap V \supset N_t(P).$$

Thus, $U \cap V$ is also open. Finally, let $\mathscr{U} = \{U_\beta \mid \beta \in B\}$ be a family of open sets, and let U be the union of all the U_β. If P is any point of U, then there exists an index β_0 such that $P \in U_{\beta_0}$, and hence, since U_{β_0} is open, there exists a positive real number r such that

$$U \supset U_{\beta_0} \supset N_r(P).$$

It follows that the set U is also open.   Q.E.D.

A subset F of the topological space $(S, \mathscr{S})$ is said to be *closed* if its complement $S - F$ is open. It is clear that any property of a topological space that has been expressed in terms of open sets can also be expressed in terms of closed sets and vice versa. In fact, point set topology can be developed in the language of closed sets, and the decision to employ open sets in this context was arrived at on subjective grounds. Some of the properties of closed sets will be listed at the end of each of this chapter's sections, and their proofs will be relegated to the exercises.

Proposition 10.1.11 *The family of closed sets $\mathscr{F}$ of any topological space $(S, \mathscr{S})$ has the following properties:*

(i) $\emptyset, S \in \mathscr{F}$.

(ii) *The union of any finite number of closed sets is closed.*

(iii) *The intersection of any number of closed sets is closed.* $\square$

Proposition 10.1.12 *A subset F of the metric space (S, d) is closed if and only if it has the property that if*

$$F \supseteq \{s_n\}_{n=1}^{\infty} \quad and \quad \{s_n\}_{n=1}^{\infty} \to s,$$

then $s \in F$. In other words, a set is closed if and only if it contains the limit of every one of its convergent sequences. $\square$

Exercises 10.1

1. Let (S, D) be a discrete metric space. Prove that if $P, Q, R \in S$ are not distinct, then $D(P, R) + D(R, Q) \geq D(P, Q)$.

2. Prove that the family of open sets of the discrete metric space (S, D) constitutes the discrete topology on S.

3. Let (S, d) be a metric space. Prove that every r-neighborhood of (S, d) is an open set.

4. Prove that if the set S has more than one point, then there is no metric d on S such that the open sets of d constitute an indiscrete topology of S.

5. Prove that for any two distinct points P and Q of the metric space (S, d) there exist disjoint open sets U and V such that $P \in U$ and $Q \in V$.

6. A subset of the set S is said to be *cofinite* in S if it is obtained by deleting a finite number of elements from S. Prove that the cofinite subsets of S, together with the empty set, constitute a topology on S. Prove that if S is infinite, then this cofinite topology is not the family of open sets of any metric d on S.

7. For any two distinct points z and w of the complex plane $\mathbb{C}$, define $d(z, w) = |z - w|$ if the infinite straight line joining z and w contains 0. Otherwise, set $d(z, w) = |z| + |w|$. Prove that $(\mathbb{C}, d)$ is a metric space.

8. The *taxicab metric* d_t is defined on $\mathbb{R}^2$ by setting $d_t(s,s') = |x-x'| + |y-y'|$ whenever $s = (x,y)$ and $s' = (x',y')$. Prove that $(\mathbb{R}^2, d_t)$ is a metric space. Prove that the topologies associated with $(\mathbb{R}^2, e)$ and $(\mathbb{R}^2, d_t)$ are identical.

9. The *maxi metric* d_M is defined on $\mathbb{R}^2$ by setting $d_t(s,s') = \max\{|x-x'|, |y-y'|\}$ whenever $s = (x,y)$ and $s' = (x',y')$. Prove that $(\mathbb{R}^2, d_M)$ is a metric space. Prove that the topologies associated with $(\mathbb{R}^2, e)$ and $(\mathbb{R}^2, d_M)$ are identical.

10. (Hilbert cube) Let $\mathbb{H}$ be the set of all sequences whose terms come from $[0,1]$. For any two such sequences $s = \{s_n\}_{n=1}^{\infty}$ and $t = \{t_n\}_{n=1}^{\infty}$ define

$$h(s,t) = \sum_{n=1}^{\infty} \frac{1}{2^n} \mid s_n - t_n \mid$$

Show that $(\mathbb{H}, h)$ is a metric space.

11. Let S be any set and let F be the set of all the functions of S into $[0,1]$. For any two such functions f and g define

$$d(f,g) = \sup\{|f(s) - g(s)|\}, \qquad \text{where } s \in S.$$

Prove that (F, d) is a metric space.

12. (Baire space) Let S be the set of all the sequences of positive integers. For any two such sequences $s = \{s_n\}_{n=1}^{\infty}$ and $t = \{t_n\}_{n=1}^{\infty}$, let $d(s,t) = 1/r$, where r is the least index such that $s_r \neq t_r$. Prove that (S,d) is a metric space.

13. Prove Proposition 10.1.11.

14. Prove Proposition 10.1.12.

15. Let $\mathscr{S}^{<} = \{(-\infty,x) \mid x \in \mathbb{R}\} \cup \{\emptyset, \mathbb{R}\}$ and $\mathscr{S}^{\leq} = \{(-\infty,x] \mid x \in \mathbb{R}\} \cup \{\emptyset, \mathbb{R}\}$.

   (a) Prove that $(\mathbb{R}, \mathscr{S}^{<})$ is a topological space.

   (b) Prove that $(\mathbb{R}, \mathscr{S}^{\leq})$ is not a topological space.

   (c) Is $(\mathbb{R}, \mathscr{S}^{<} \cup \mathscr{S}^{\leq})$ a topological space?

16. The family $\mathscr{S}$ consists of the empty set as well as all the subsets of $\mathbb{R}$ that contain 1. Is $(\mathbb{R}, \mathscr{S})$ a topological space?

17. The family $\mathscr{S}$ consists of all the subsets of $\mathbb{R}$ that contain $\mathbb{Q}$ as well as all the subsets of $\mathbb{Q}$. Is $(\mathbb{R}, \mathscr{S})$ a topological space?

18. Let (S,d) and (T,e) be two metric spaces. Their *product space* has underlying set

$$S \times T = \{(s,t) \mid s \in S, t \in T\}$$

and metric

$$m((s_1,t_1),(s_2,t_2)) = \sqrt{(s_1 - s_2)^2 + (t_1 - t_2)^2}.$$

Verify that $(S \times T, m)$ is a metric space.

10.2 Continuity and Homeomorphisms

Next we formalize the notion of equivalence in the context of topology. Given two metric spaces (S,d) and (S',d'), a function $f : S \to S'$ is said to be *continuous* (with respect to d and d') provided that it transforms convergence in (S, d) into convergence in (S', d'). More precisely, for every convergent sequence $\{P_n\} \to P$ of (S, d) we also have $\{f(P_n)\} \to f(P)$ in (S', d'). All of the functions listed below are continuous functions of $(\mathbb{R}, e)$ into $(\mathbb{R}, e)$ (Exercise 16):

1. ax^n (where a is any real number and $n = 0, 1, 2, 3, \ldots$);

2. $|x|, \sin x, \cos x, e^x$;

3. $f + g, f - g, f \cdot g, f \circ g$ (where f and g are already known to be continuous).

On the other hand, the function

$$f(x) = \begin{cases} 1 - x & \text{for } x \le 0. \\ x & \text{for } x > 0 \end{cases}$$

is not continuous with repect to the Euclidean metrics. To see this, note that while 0 is the limit of $\{1/n\}$, it is not the case that $1 = f(0)$ is the limit of $\{f(1/n)\} = \{1/n\}$.

Similarly, the *Dirichlet function*

$$\delta(x) = \begin{cases} 1 & \text{if } x \text{ is rational,} \\ 0 & \text{if } x \text{ is irrational} \end{cases}$$

is not continuous with respect to the Euclidean metrics. Consider, for example, the convergent sequence

$$\left\{ \frac{e}{n} \right\} \to 0,$$

where $e = 2.71828\ldots$ is well known to be irrational. Note that

$$\left\{ \delta \left(\frac{e}{n} \right) \right\} = \{0, 0, 0, 0, \ldots\} \nrightarrow 1 = \delta(0),$$

so that $\delta(x)$ fails to preserve convergence. These examples may seem to be rather artificial, but they cannot be disregarded as advanced mathematics does have uses for such constructs.

The continuity of functions from $(\mathbb{R}^k, e)$ into $(\mathbb{R}, e)$ is similar to that of functions from $(\mathbb{R}, e)$ into $(\mathbb{R}, e)$. Working with the case $k = 3$ for the sake of definiteness, and denoting the typical point P of $\mathbb{R}^3$ by (x, y, z), the following functions are well known to be continuous (Exercise 17):

1. $f(P) = x, f(P) = y, f(P) = z$;

2. $f + g, f - g, f \cdot g$, where f and g are already known to be continuous.

Discontinuous functions from $(\mathbb{R}^k, e)$ into $(\mathbb{R}, e)$ can be produced by the same means that were used to create such functions from $(\mathbb{R}, e)$ into $(\mathbb{R}, e)$.

The continuity of a function f from $(\mathbb{R}^k, e)$ into $(\mathbb{R}^m, e)$ is tantamount to the continuity of its component functions. Thus, the function

$$f(P) = (f_1(P), f_2(P), f_3(P), \ldots, f_m(P)), \qquad \text{where } P \in \mathbb{R}^k,$$

is continuous if and only if each of the component functions is a continuous function $f_i : \mathbb{R}^k \to \mathbb{R}$ with respect to the Euclidean metrics.

We next turn to the discrete metrics. The situation here is rather surprising, as it turns out that in this context *every* function is continuous. For suppose (S, D) and (S', D') are two discrete metric spaces and we are given a function $f : S \to S'$. As noted above, in a discrete topological space any convergent sequence $\{P_n\} \to P$ stabilizes at P, and so its transformation $\{f(P_n)\}$ must also stabilize at $f(P)$. Thus f transforms every convergent sequence of (S, D) into a convergent sequence of (S', D'), and it follows that f is continuous. Hence, the following proposition has been proved.

Proposition 10.2.1 *If (S, D) and (S', D') are discrete metric spaces, then every function $f : (S, D) \to (S', D')$ is continuous.* □

One more general proposition needs to be proved before we pass on to the next topic.

Proposition 10.2.2 *The composition of continuous functions is continuous.*

PROOF: Let

$$f : (S, d) \to (S, d') \quad \text{and} \quad g : (S', d') \to (S', d')$$

be continuous functions, and let $\{P_n\} \to P$ be a convergent sequence in (S, d). The continuity of f guarantees that $\{f(P_n)\} \to f(P)$, and the continuity of g guarantees that $\{g(f(P_n))\} \to g(f(P))$. This is equivalent to saying that

$$\{(g \circ f)(P_n)\} \to (g \circ f)(P),$$

which, in turn, is tantamount to the continuity of $g \circ f$. Q.E.D.

Given any set S, the *identity function* on S, denoted by Id_S, is the unique function such that

$$\mathrm{Id}_S(P) = P \qquad \text{for all } P \text{ in } S.$$

It is clear that this identity function is continuous. Given two sets S and S' and two functions $f : S \to S'$ and $g : S' \to S$ such that

$$f \circ g = \mathrm{Id}_{S'} \quad \text{and} \quad g \circ f = \mathrm{Id}_S,$$

the functions f and g are said to be *inverses* of each other, and each is said to be *invertible*. Under these circumstances it is customary to write $f^{-1} = g$ and $g^{-1} = f$. The functions $x + 1$ and $x - 1$ from $\mathbb{R}$ into $\mathbb{R}$ are inverses of each other, as are x^3 and $x^{1/3}$. Note that if f and g are invertible, then so is their composition (Exercise 1). The following fundamental proposition is easily proven (Exercise 2).

Proposition 10.2.3 *The function $f : S \to S'$ is invertible if and only if it has the following two properties:*

1. *If $a \neq b$ are distinct elements of S, then $f(a) \neq f(b)$ (injectivity).*

2. *For every element y of S' there is an element x of S such that $y = f(x)$ (surjectivity).* □

It follows that the function $f(x) = (x, x+1)$ from $\mathbb{R}$ into $\mathbb{R}^2$ is not invertible, because it is not surjective (the origin is not covered, for one). Similarly, the function $f(x, y) = x - y$ from $\mathbb{R}^2$ into $\mathbb{R}$ also fails to be invertible, this time because it is not injective ($f(1, 0) = f(2, 1) = f(3, 2) = \cdots$).

If $f : (S, d) \to (S', d')$ and $g : (S', d') \to (S, d)$ are both continuous and also inverses of each other, then each is called a *homeomorphism* and the metric spaces (S, d) and (S', d') are said to be *homeomorphic*. The homeomorphism of these spaces is denoted by $(S, d) \approx (S', d')$.

The easiest homeomorphisms to visualize are the rigid motions of Euclidean space. Thus, if S is any subset of Euclidean space, such as a line segment or an arc, and if f is any rigid motion of Euclidean space, then f constitutes a homeomorphism of (S, e) and $(f(S), e)$. The invertibility of f is obvious, and the continuity of both f and its inverse follows immediately from the fact that they preserve the distances between the points they transform. There are many homeomorphisms, however, that do not preserve distances.

Example 10.2.4 The intervals $([0, 1], e)$ and $([0, 2], e)$ are homeomorphic. This is demonstrated by the functions

$$f(x) = 2x, \qquad f : [0, 1] \to [0, 2],$$
$$g(x) = \frac{x}{2}, \qquad g : [0, 2] \to [0, 1],$$

which are clearly continuous and inverses of each other.

In fact, every two intervals in $\mathbb{R}$ are homeomorphic, as are every two line segments in both $\mathbb{R}^2$ and $\mathbb{R}^3$ (see Exercises 3, 4).

Example 10.2.5 Let S denote the arc of the parabola $y = x^2$, $-1 \leq x \leq 1$, in $\mathbb{R}^2$. Then (S, e) is homeomorphic to the interval $([-1, 1], e)$. The homeomorphism is provided by the functions

$$f(x) = (x, x^2), \qquad f : [-1, 1] \to S,$$
$$g(x, x^2) = x, \qquad g : S \to [-1, 1],$$

which are clearly continuous and inverses of each other.

In fact, if $f : \mathbb{R} \to \mathbb{R}$ is any continuous function and $[a, b]$ is any interval in $\mathbb{R}$, then the graph of $y = f(x)$, $a \leq x \leq b$, with the Euclidean metric, is homeomorphic to the interval $([a, b], e)$ (Exercise 5).

Example 10.2.6 Let S denote the graph of $z = x^2 + y^3$, $0 \leq x, y \leq 1$, in $\mathbb{R}^3$. Let S' denote the square $0 \leq x, y \leq 1$ in $\mathbb{R}^2$. Then the functions

$$f(x, y) = (x, y, x^2 + y^3)$$

$$g(x,y,x^2+y^3) = (x,y)$$

are homeomorphisms of (S,d) and (S',d').

In fact, if $f : \mathbb{R}^2 \to \mathbb{R}$ is any continuous function and a,b,c,d are any real numbers, then the graph in $\mathbb{R}^3$ of $z = f(x,y)$, restricted to the rectangular domain $a \leq x \leq b$, $c \leq y \leq d$, is homeomorphic to that rectangular domain, assuming that both sets are endowed with the Euclidean metric (Exercise 6).

Turning to discrete metrics, we remind the reader that every function from a discrete space to a discrete space is continuous. It follows that two discrete metric spaces are homeomorphic to each other if and only if there is an invertible function from one to the other. In particular, two sets of n^2 points, one arranged as a square $n \times n$ grid in $\mathbb{R}^2$, and the other lined up along $\mathbb{R}$, and both endowed with discrete metrics, constitute homeomorphic metric spaces.

It is known that if S is any infinite set, then there is an invertible function from S to

$$S^2 = \{(x,y) \mid x,y \in S\}.$$

In fact, if k is any positive integer and S^k denotes the set of all ordered k-tuples of elements of an infinite set S, then there is an invertible function from S^k to S. It follows that for an infinite set S, the metric spaces (S^k, D) and (S, D) are homeomorphic. In particular we have the following proposition.

Proposition 10.2.7 *The metric spaces* $(\mathbb{R}, D)$ *and* $(\mathbb{R}^k, D)$ *are homeomorphic for each* $k = 1,2,3,\ldots.$ □

The following general proposition is useful when discussing homeomorphism of metric spaces.

Proposition 10.2.8 *Homeomorphism of metric spaces is an equivalence relation in the sense that for any three metric spaces* (S,d), (S',d'), *and* (S'',d''):

1. $(S,d) \approx (S,d)$;

2. *if* $(S,d) \approx (S',d')$ *then* $(S',d') \approx (S,d)$;

3. *if* $(S,d) \approx (S',d')$ *and* $(S',d') \approx (S'',d'')$, *then also* $(S,d) \approx (S'',d'')$.

PROOF: The identity Id_S is a homeomorphism of (S,d) with itself, thus proving part 1. Part 2 follows from the fact that a function is invertible if and only if its inverse is invertible. As for part 3, let $f : (S,d) \to (S',d')$ and $g : (S',d') \to (S'',d')$ be homeomorphisms. It then follows from Propositions 10.1.9 and Exercise 3 that $g \circ f : (S,d) \to (S'',d'')$ is invertible and that both it and its inverse are continuous. Consequently $g \circ f$ is also a homeomorphism. Q.E.D.

It follows from the previous two propositions that for any positive integers k and m the spaces $(\mathbb{R}^k, D)$ and $(\mathbb{R}^m, D)$ are homeomorphic to each other. It should also be mentioned that if S is any surface in $\mathbb{R}^3$ and d is the metric of Example 10.1.3, which employs distances along the surface rather than the straight line Euclidean distance between points, then (S,d) and (S,e) are homeomorphic. While this observation is plausible, its formal proof is too long to be displayed here.

We turn next to the issue of recognizing when spaces are not homeomorphic. Given two finite sets S and S', there is an invertible function $f : S \to S'$ if and only if the two sets have the same number of points (Exercise 7). Consequently, it is easy to decide whether or not two finite discrete metric spaces are homeomorphic.

Proposition 10.2.9 *If the two sets S and S' are finite, then (S,D) and (S',D) are homeomorphic if and only if S and S' have the same number of points.*   □

It is easy to see that when the sets S and S' are finite and d and d' are arbitrary metrics, the spaces (S,d) and (S',d') are homeomorphic if and only if S and S' have the same number of elements (Exercise 18). When the underlying sets are infinite, the problem is more difficult. The following proposition demonstrates that there is more to the issue of homeomorphism than just counting elements in a set.

Proposition 10.2.10 *The metric spaces $(\mathbb{R},e)$ and $(\mathbb{R},D)$ are not homeomorphic.*

PROOF: Suppose $f : \mathbb{R} \to \mathbb{R}$ is an invertible function. Then, because of the injectivity property,

$$f\left(\frac{1}{m}\right) \neq f\left(\frac{1}{n}\right) \qquad \text{whenever } m \neq n,$$

and hence the sequence $\{f(1/n)\}$ does not stabilize. Thus, f transforms the sequence $\{1/n\}$, which is convergent in $(\mathbb{R},e)$, into the sequence $\{f(1/n)\}$, which is not convergent in $(\mathbb{R},D)$ (see Proposition 10.1.7). It follows that the function f is not continuous and therefore cannot constitute a homeomorphism of $(\mathbb{R},e)$ and $(\mathbb{R},D)$.
Q.E.D.

The surprisingly difficult to prove fact that the Euclidean spaces $(\mathbb{R}^k,e)$ and $(\mathbb{R}^m,e)$ are not homeomorphic whenever $k \neq m$ is one of the milestones of topology. This fact, known as the *invariance of dimension*, was demonstrated circa 1910 by L. E. J. Brouwer (1881–1966), who is considered by many to be the founder of topology.

The following lemma and propositions rephrase continuity using open sets and closed sets.

Lemma 10.2.11 *Let (S,d) and (T,ρ) be metric spaces and $f: S \to T$ a function. Then f is continuous at $s \in S$ if and only if for each open set V containing $f(s)$ there is an open set U containing s such that $f(U) \subseteq V$.*

PROOF: Set $t = f(s)$. We begin by assuming that f is continuous at s. By the definition of open sets in a metric space, there is an ε such that $N_\varepsilon(t) \subseteq V$. For each positive integer n, let $U_n = N_{1/n}(s)$. These U_n's are, of course, open sets containing s, and we show that for at least one of them $f(U_n) \subseteq V$.

Suppose, by way of contradiction, that for each n, $f(U_n)$ is not contained in V. It follows that for each n there is an $s_n \in U_n$ such that $f(s_n) \notin V$. In particular, $f(s_n) \notin N_\varepsilon(t)$, implying that

$$\rho(f(s_n), t) \geq \varepsilon, \qquad n = 1, 2, 3, \ldots. \tag{3}$$

Since

$$d(s_n, s) < \frac{1}{n}, \qquad n = 1, 2, 3, \ldots,$$

it follows that $s_n \to s$ and hence, by the definition of continuity,

$$f(s_n) \to f(s) = t.$$

This, however, contradicts (3). Hence, at least one of the $f(U_n)$'s is contained in V.

Conversely, suppose for each open set $V \subseteq T$ containing t there is an open set $U \subseteq S$ containing s such that $f(U) \subseteq V$. Let $s_n \to s$ and let $\varepsilon > 0$. Clearly

$$t \in N_\varepsilon(t) \subseteq T.$$

Consequently there exists an open set U_ε such that

$$s \in U_\varepsilon \subseteq S \quad \text{and} \quad f(U_\varepsilon) \subseteq N_\varepsilon(t).$$

It follows from the definition of open sets that there exists a number $\delta > 0$ such that

$$s \in N_\delta(s) \subseteq U_\varepsilon \subseteq S.$$

Since $s_n \to s$, it follows that there exists an index n' such that

$$d(s_n, s) < \delta \quad \text{or} \quad s_n \in N_\delta(s) \subseteq U_\varepsilon \qquad \text{for all } n \geq n'.$$

It follows that

$$f(s_n) \in f(U_\varepsilon) \subseteq N_\varepsilon(t) \quad \text{or} \quad \rho(f(s_n), t) < \varepsilon \qquad \text{for all } n \geq n'.$$

Since ε is an arbitrary positive number, it follows that $f(s_n) \to t = f(s)$.   Q.E.D.

Proposition 10.2.12 *Given two metric spaces* (S, d) *and* (T, ρ), *the function* $f : S \to T$ *is continuous if and only if*

$$f^{-1}(V) \text{ is open in } S \text{ whenever } V \text{ is open in } T. \tag{4}$$

PROOF: Suppose first that the given function f is continuous and V is an open subset of T. By the lemma, for each $t \in V$ and for each $s \in f^{-1}(t)$ there is an open set U_s such that

$$s \in U_s \subseteq f^{-1}(V).$$

Since each U_s is open and

$$f^{-1}(V) = \cup\{U_s \mid s \in f^{-1}(V)\},$$

it follows that $f^{-1}(V)$ is an open subset of S.

Conversely, suppose $f^{-1}(V)$ is open whenever V is open. Then given any $s \in S$ and V open in (S,d) such that $f(s) \in V$, we may, upon setting $U = f^{-1}(V)$, conclude from the lemma that f is continuous at s. Since s is arbitrary, it follows that f is continuous.   Q.E.D.

It is now possible to extend the notion of continuity to all topological spaces. Let $(S,\mathscr{S})$ and $(T,\mathscr{T})$ be topological spaces. A function $f : S \to T$ is said to be *continuous* provided $f^{-1}(V)$ is in $\mathscr{S}$ whenever V is in $\mathscr{T}$. Or, somewhat less formally, f is continuous if the inverse image of every open set is also open. This can of course also be phrased in the language of closed sets.

Proposition 10.2.13 *Let $(S,\mathscr{S})$ and $(T,\mathscr{T})$ be topological spaces. A function $f : S \to T$ is continuous if and only if $f^{-1}(F)$ is closed in S whenever F is closed in T.*
                                                                                     $\square$

It is now possible to formalize some other terms that were used informally in the previous chapters.

An *(ambient) isotopy* is a continuous function $I(t,P)$, where

$$P \in \mathbb{R}^3, \qquad 0 \le t \le 1, \qquad I(t,P) \in \mathbb{R}^3,$$

and

1. the function $I_t : \mathbb{R}^3 \to \mathbb{R}^3$ defined by $I_t(P) = I(t,P)$ is a homeomorphism for each $t \in [0,1]$;

2. I_0 is the identity function on $\mathbb{R}^3$.

Given a subset S of $\mathbb{R}^3$, such a function is said to be an isotopy from S to $I_1(S)$. It helps to think of t as a time parameter and of $I_t(S)$ as the location of S at time t. In particular, $I_0(S) = S$ should be thought of as the initial position of S, whereas $I_1(S)$ denotes its final position.

Example 10.2.14 Let the coordinates of the typical point $P \in \mathbb{R}^3$ be (x,y,z), and set

$$I(t,P) = I(t,x,y,z) = (1+t)(x,y,z). \tag{5}$$

Then I can be described as expanding space away from the origin by a factor of 2. It is an isotopy of the sphere of radius r centered at the origin onto the sphere of radius $2r$ with the same center. It is also an isotopy of the line segment joining $(1,2,3)$ and $(2,-1,0)$ onto the line segment joining $(2,4,6)$ and $(4,-2,0)$.

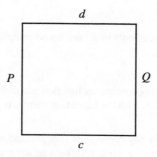

Figure 10.1   A homotopy.

Example 10.2.15 The isotopy

$$I(t,x,y,z) = (x\cos \pi t + y\sin \pi t, -x\sin \pi t + y\cos \pi t, z+t)$$

is the combination of a $180°$ counterclockwise rotation about the z-axis with a unit slide in the direction of the positive z-axis.

Given two points P and Q of a topological space S, a *path* of S from P to Q is a continuous function $c : [0,1] \to S$ such that $c(0) = P$ and $c(1) = Q$.

Example 10.2.16 Both of the functions

$$c(s) = (s,s) \quad \text{and} \quad d(s) = (s^2,s)$$

are paths from $(0,0)$ to $(1,1)$.

Let I denote the unit interval $[0,1]$ and I^2 the unit square $0 \le s,t \le 1$. Given two paths c and d from point P to point Q of a topological space S, a *homotopy* from c to d is a continuous function $H : I^2 \to S$ such that

$$
\begin{aligned}
H(s,0) &= c(s), & H(s,1) &= d(s) & \text{for all } s \in I, \\
H(0,t) &= P, & H(1,t) &= Q & \text{for all } t \in I.
\end{aligned}
$$

Example 10.2.17 The function $H(s,t) = (s^{1+t},s)$ is a homotopy of the two paths in Example 10.2.16.

It is customary to encode the various functions of (6) by means of the diagram of Figure 10.1. The square in the figure is to be understood as the unit square.

Exercises 10.2

1. Prove that if $f : S \to S'$ and $g : S' \to S''$ are invertible functions, then $g \circ f$ is also invertible and $(g \circ f)^{-1} = f^{-1} \circ g^{-1}$.

2. Prove Proposition 10.2.3.

3. Prove that every two line segments in $\mathbb{R}^2$ are homeomorphic provided they are endowed with the Euclidean metric.

4. Repeat Exercise 3 in $\mathbb{R}^3$.

5. Prove that if $f : \mathbb{R} \to \mathbb{R}$ is any continuous function and $[a,b]$ is any interval in $\mathbb{R}$, then the graph of $y = f(x)$, $a \le x \le b$, with the Euclidean metric, is homeomorphic to the interval $([a,b], e)$.

6. Prove that if $f : \mathbb{R}^2 \to \mathbb{R}$ is any continuous function and a, b, c, d are any real numbers, then the graph in $\mathbb{R}^3$ of $z = f(x,y)$, restricted to the rectangular domain $a \le x \le b$, $c \le y \le d$, is homeomorphic to that rectangular domain, assuming that both sets are endowed with the Euclidean metric.

7. Given two finite sets S and S', prove that there is an invertible function $f : S \to S'$ if and only if the two sets have the same number of points.

8. Let $(S, \mathscr{S})$ and $(T, \mathscr{T})$ be two topological spaces. If $t_0 \in T$, prove that the (constant) function $f : S \to T$ defined by $f(s) = t_0$ for all $s \in S$ is continuous.

9. Let $(S, \mathscr{S})$ and $(T, \{\emptyset, T\})$ be two topological spaces (the second one is indiscrete). Prove that every function $f : S \to T$ is continuous.

10. Let (S, D) be a discrete metric space, and let $(T, \mathscr{T})$ be an arbitrary topological space. Is every function $f : S \to T$ necessarily continuous?

11. Let $(\mathbb{R}, \mathscr{S}^<)$ be the topological space of Exercise 10.1.15. Prove that the function $f(x) = 2x$ is a homeomorphism of $(\mathbb{R}, \mathscr{S}^<)$ with itself. Prove that the function $g(x) = x^2$ is not continuous. Is the function $h(x) = x^3$ continuous? Generalize these observations.

12. Let $(S, \mathscr{S})$ be a cofinite topological space. Prove that a function $f : S \to S$ is continuous if and only if $f^{-1}(s)$ is finite for every $s \in S$.

13. Prove that if the sets S and S' are finite and d and d' are arbitrary metrics, the spaces (S, d) and (S', d') are homeomorphic if and only if S and S' have the same number of elements.

14. Let $S \times T$ be the product metric space of Exercise 10.1.18, and prove that the functions $p : S \times T \to S$ and $q : S \times T \to T$ defined by

$$p(s,t) = s, q(s,t) = t$$

are continuous.

15. Prove Proposition 9.3.16 formally.

16. Prove that all the functions listed below are continuous functions of $(\mathbb{R}, e)$:

   (a) ax^n (where a is any real number and $n = 1, 2, 3, \ldots$);

   (b) $|x|$, $\sin x$, $\cos x$, e^x;

   (c) $f + g$, $f - g$, $f \cdot g$, $f \circ g$ (where f and g are already known to be continuous).

17. If $P = (x, y, z)$, prove that the following functions are continuous:

   (a) $f(P) = x$, $g(P) = y$, $h(P) = z$;

   (b) $f + g$, $f - g$, $f \cdot g$ (where f and g are already known to be continuous).

10.3   Connectedness

Many of the objects studied in this text, such as paths, surfaces, and knots, are connected. Others, such as links, are not. This distinction is now formalized. A topological space $(S, \mathscr{S})$ is said to be *connected* if there are no two disjoint open nonempty subsets of S whose union equals S. More generally, a subset A of S is said to be *connected* if there are no two open subsets U and V of X such that

$$U \cap A \neq \emptyset,$$
$$V \cap A \neq \emptyset,$$
$$U \cap V \cap A = \emptyset,$$

and

$$(U \cap A) \cup (V \cap A) = A.$$

When such sets U and V do exist they are said to *separate* $U \cap A$ and $V \cap A$, and the set A is said to be *disconnected*.

Example 10.3.1 On $\mathbb{R}$ let $a < b < c < d$. Then the set $A = [a,b] \cup [c,d]$ is disconnected because the open sets $U = (-\infty, (b+c)/2)$ and $V = ((b+c)/2, \infty)$ separate $[a,b]$ and $[c,d]$.

Example 10.3.2 The punctured line $(\mathbb{R} - \{0\}, e)$ is disconnected because the open sets $\{x \in \mathbb{R} \mid x > 0\}$ and $\{x \in \mathbb{R} \mid x < 0\}$ separate themselves.

The proof of the following plausible proposition is surprisingly intricate.

Proposition 10.3.3 *The metric space* $(\mathbb{R}, e)$ *is connected.*

PROOF: Assume U and V are two disjoint open nonempty subsets of $\mathbb{R}$ such that $U \cup V = \mathbb{R}$. Let $u \in U$ and $v \in V$. Without loss of generality we may assume that $u < v$. Let

$$A = \{x \in \mathbb{R} \mid [u,x) \subseteq U\}.$$

The fact that U is open implies that A is nonempty, and the fact that V is not empty implies that $\alpha = \sup(A)$ exists.

If $\alpha \in U$, then there is an interval (a,b) such that

$$\alpha \in (a,b) \subseteq U,$$

but then, regardless of the relative positions of a and u, we have $b \in A$, contradicting the fact that $\alpha = \sup(A)$.

If $\alpha \notin U$, then $\alpha \in V$, and hence there is an interval (c,d) such that

$$\alpha \in (c,d) \subseteq V.$$

Note that for each $y \in (c,d)$ there is a $z \in (c,d)$ such that $z < y$ and $z \notin U$. It follows that $y \notin A$, so that α cannot be the supremum of A. Either way we have a contradiction, and hence $(\mathbb{R},e)$ is connected.   Q.E.D.

The set of rationals $\mathbb{Q}$ is separated in $(\mathbb{R},e)$ by the sets $U = (-\infty,\pi)$ and $V = (\pi,\infty)$. We now show that the continuous functions preserve connectedness.

Proposition 10.3.4 *Let $(S,\mathscr{S})$ and $(T,\mathscr{T})$ be topological spaces, and suppose $f: S \to T$ is a continuous function. If $(S,\mathscr{S})$ is connected, then $f(S)$ is connected in $(T,\mathscr{T})$.*

PROOF: By contradiction. Suppose $f(S)$ not connected in $(T,\mathscr{T})$. Then there exist open subsets U and V of $(T,\mathscr{T})$ such that.

$$U \cap f(S) \neq \emptyset,$$
$$V \cap f(S) \neq \emptyset,$$
$$U \cap V \cap f(S) = \emptyset,$$
$$(U \cap f(S)) \cup (V \cap f(S)) = f(S).$$

It follows from Proposition 10.2.12 that $f^{-1}(U)$ and $f^{-1}(V)$ are open subsets of $(S,\mathscr{S})$ such that

$$f^{-1}(U) \neq \emptyset,$$
$$f^{-1}(V) \neq \emptyset,$$
$$f^{-1}(U) \cap f^{-1}(V) = \emptyset,$$
$$f^{-1}(U) \cup f^{-1}(V) = S.$$

This, however, contradicts the connectedness of $(S,\mathscr{S})$. Hence, $f(S)$ is connected in $(T,\mathscr{T})$.   Q.E.D.

Corollary 10.3.5 *Let $(S,\mathscr{S})$ and $(T,\mathscr{T})$ be homeomorphic topological spaces. Then $(S,\mathscr{S})$ is connected if and only if $(T,\mathscr{T})$ is connected.*   □

Corollary 10.3.6 *Every arc is connected in its ambient topological space.*

PROOF: We first note that the unit interval (I,e) is connected. The reason for this is that the continuous function $|\sin x|$ maps the connected real line onto this interval. Similarly, the continuity of the function $f(x) = (\cos x, \sin x)$ implies the connectedess of the unit circle. Since every arc is, by definition, homeomorphic to either the unit interval or the unit circle, the desired conclusion follows from Corollary 10.3.5. Q.E.D.

A topological space $(S,\mathscr{S})$ is said to be *arcwise connected* if every two of its points can be joined by an arc that lies entirely in S.

Proposition 10.3.7 *Every arcwise connected topological space is connected in its ambient topological space.*

PROOF: By contradiction. Let S be an arcwise connected space, and let U and V be disjoint nonempty open subsets of S such that $U \cup V = S$. If $u \in U$ and $v \in V$, then there is an arc a that joins u and v. Then

$$U \cap a \neq \emptyset, \qquad V \cap a \neq \emptyset, \quad \text{and} \quad (U \cap a) \cup (V \cap a) = a$$

which contradicts the connectedness of a (see Corollary 10.3.6).   Q.E.D.

Exercises 10.3

1. Prove that every discrete metric space with at least two points is not connected.

2. Prove that every indiscrete topological space is connected.

3. Prove that a cofinite topological space with at least two points is connected if and only if the underlying set is infinite.

4. Prove that the product of two arcwise connected metric spaces is arcwise connected (see Exercise 10.1.18).

5. Let $(S, \mathscr{S})$ and $(T, \mathscr{T})$ be two topological spaces where S and T are disjoint. Prove that their disjoint union $(S \cup T, \mathscr{S} \cup \mathscr{T})$ is a topological space that is not connected.

6. Prove that the topological space of Exercise 10.1.15(a) is connected.

7. (Intermediate Value Theorem) Let $f : [a,b] \to \mathbb{R}$ be a continuous function relative to the Euclidean metric such that $f(a) > 0$ and $f(b) < 0$. Prove that there is a number c such that $f(c) = 0$.

8. Let $f : [0,1] \to [0,1]$ be a continuous function relative to the Euclidean metric. Prove that there is a number c such that $f(c) = c$.

9. Let S^1 denote the unit circle. Prove that if $f : S^1 \to \mathbb{R}$ is continuous relative to the Euclidean metric, then there is a pair of antipodal points $\{z, -z\}$ such that $f(z) = f(-z)$.

10. Prove that every connected subset of $(\mathbb{R}, e)$ is an interval, an infinite ray, or $\mathbb{R}$ itself.

11.* Prove that the Baire space of Exercise 10.1.12 is not connected.

12. A topological space is connected if and only if it has a nonempty proper subset that is both open and closed.

13. Let P be any point of $\mathbb{R}^2$. Is the punctured plane $(\mathbb{R}^2 - P, e)$ connected?

14. Give an example of a disconnected subset of $(\mathbb{R}^2, e)$.

15.* Prove that $(\mathbb{R}, e)$ and $(\mathbb{R}^2, e)$ are not homeomorphic.

10.4   Compactness

If a metric space has two distinct points, say x and y, then it has a sequence, namely

$$x, y, x, y, x, y, \ldots,$$

which does not converge. Consequently, the only nonempty metric space in which every sequence converges is the trivial metric space that consists of a single point. Still, the sequence above does have convergent subsequences, and this motivates our next definition.

A metric space (S,d) in which every sequence has a convergent subsequence is said to be *compact*. We emphasize that the convergent subsequence's limit must be an element of S. For instance, the sequence 1, 1.4, 1.41, 1.414, 1.4142, ... of decimal approximations to $\sqrt{2}$ converges in $(\mathbb{R},e)$ but not in $(\mathbb{Q},e)$. It is clear that every finite metric space is compact, since in such a space every sequence must contain an infinite number of repetitions of some element. Similarly, it follows from the completeness of the real line that for any interval $([a,b])$ the Euclidean metric space $([a,b],e)$ is compact. On the other hand, the space $(\mathbb{R},e)$ is not compact, since the sequence

$$1,2,3,\ldots$$

contains no convergent subsequences. Similarly, no Euclidean open interval of the form $((a,b),e)$ is compact. The reason for this is that every such interval contains a sequence of the form

$$a+\frac{1}{k}, a+\frac{1}{2k}, a+\frac{1}{3k},\ldots$$

for some positive integer k. As this sequence and each of its subsequences converge to the element $a \notin (a,b)$ it follows that none of these subsequences converge to an element of (a,b); in other words, $((a,b),e)$ is not compact.

We now go on to formulate the property of compactness in topological language, i.e., in terms of open subsets of metric or topological spaces. As the resolution of this issue is quite complex, it is broken down into several lemmas.

Given any point s and any nonempty subset X of the metric space (S,d), the *distance* from s to X is

$$d(s,X) = \inf\{d(s,x) \mid x \in X\}.$$

Thus, the distance from the origin to the open interval $(3,4)$ is 3, and the distance from the origin to the straight line $x+y=2$ is $\sqrt{2}$.

Lemma 10.4.1 *Let* (S,d) *be a compact metric space, and let* $\varepsilon > 0$. *Then there exists a finite subset* A_ε *of S such that*

$$d(s,A_\varepsilon) < \varepsilon \qquad \textit{for all } s \in S.$$

PROOF: The set A_ε is defined inductively. Let a_1 be any point of S. Assume that $a_1, a_2, \ldots, a_n$ have all been defined so that

$$d(a_i,a_j) \geq \varepsilon \qquad \text{whenever } 0 \leq i < j \leq n.$$

If

$$d(s, \{a_1, a_2, \ldots, a_n\}) < \varepsilon \qquad \text{for all } s \in S,$$

then $\{a_1, a_2, \ldots, a_n\}$ is the required set A_ε, and we are done. Otherwise, there exists an element $a_{n+1} \in S$ such that

$$d(a_i, a_{n+1}) \geq \varepsilon \qquad \text{for all } i = 1, 2, 3, \ldots, n.$$

Were this process to continue indefinitely, we would obtain a sequence $\{a_n\}_{n=1}^{\infty}$ such that

$$d(a_i, a_j) \geq \varepsilon \qquad \text{whenever } 0 \leq i < j.$$

Since such a sequence has no convergent subsequences, this would contradict the compactness of (S, d). It follows that this process must terminate in a finite number of steps, thus implying the existence of the requisite A_ε.   Q.E.D.

Lemma 10.4.2 *Every compact metric space (S, d) has a sequence $\{x_n\}_{n=1}^{\infty}$ such that every element of S is the limit point of one of its subsequences.*

PROOF: For each positive integer n let $A_{1/n}$ be the finite set defined by Lemma 10.4.1. Let A be the sequence obtained by first listing the elements of A_1, then those of $A_{1/2}$, then those of $A_{1/3}$, and so on. Let s be any point of S, and for each n let b_n be an element of $A_{1/n}$ such that

$$d(s, b_n) < \frac{1}{n}.$$

Then $\{b_n\}_{n=1}^{\infty}$ is a subsequence of A that converges to s.   Q.E.D.

A family $\{S_\lambda \mid \lambda \in \Lambda\}$ of subsets of the set S is said to be a *cover* of S provided

$$S = \bigcup \{S_\lambda \mid \lambda \in \Lambda\}.$$

If $(S, \mathscr{S})$ is a topological space and each of the subsets S_λ is open, then the cover is also said to be *open*.

Suppose $\mathscr{F}$ and $\mathscr{G}$ are two families of subsets of the set S. We say that $\mathscr{G}$ is a *refinement* of $\mathscr{F}$ provided that whenever $s \in F \in \mathscr{F}$, there is a set $G \in \mathscr{G}$ such that

$$s \in G \subset F.$$

For example, let $S = \mathbb{R}$, let $\mathscr{F}$ be the set of all open intervals with rational endpoints, and let $\mathscr{G}$ be the family of all intervals with irrational endpoints. Then $\mathscr{F}$ is a refinement of $\mathscr{G}$, and $\mathscr{G}$ is a refinement of $\mathscr{F}$.

Lemma 10.4.3 *If (S, d) is a compact metric space, then there is a sequence $\{B_n\}_{n=1}^{\infty}$ of open subsets of S that refines every open cover of (S, d).*

PROOF: Let $\{x_n\}_{n=1}^{\infty}$ be the sequence of S whose existence is guaranteed in Lemma 10.4.2. For each positive integer n let A_{n+1} be the *finite* sequence of sets

$$A_{n+1} = N_{1/1}(x_n), N_{1/2}(x_{n-1}), \ldots, N_{1/n}(x_1).$$

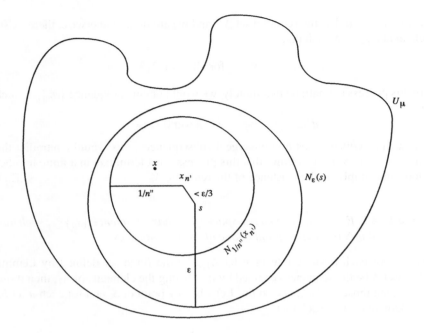

Figure 10.2

Finally, let $\{B_n\}_{n=1}^{\infty}$ be the sequence obtained by first listing the elements of A_2, then those of A_3, then those of A_4, and so on.

Suppose $\{U_\lambda \mid \lambda \in \Lambda\}$ is a cover of (S,d) and s is a point in U_μ for some $\mu \in \Lambda$. Since U_μ is open, there exists a number $\varepsilon, 1 > \varepsilon > 0$, such that

$$s \in N_\varepsilon(s) \subseteq U_\mu.$$

By the definition of the sequence $\{x_n\}$ there exists an index n' such that

$$d(s, x_{n'}) < \frac{\varepsilon}{3}.$$

If n'' is a positive integer such that $\varepsilon/3 < 1/n'' < 2\varepsilon/3$, then (see Fig. 10.2), for each $x \in N_{1/n''}(x_{n'})$,

$$d(s,x) \le d(x, x_{n'}) + d(s, x_{n'}) < \frac{1}{n''} + \frac{\varepsilon}{3} < \frac{2\varepsilon}{3} + \frac{\varepsilon}{3} = \varepsilon,$$

and hence

$$s \in N_{1/n''}(x_{n'}) \subseteq N_s(\varepsilon) \subseteq U_\mu.$$

Since each set $N_{1/n''}(x_{n'})$ appears in $A_{n'+n''}$ and consequently also in $\{B_n\}_{n=1}^{\infty}$, it follows that this latter sequence is indeed a refinement of $\{U_\lambda \mid \lambda \in \Lambda\}$. Q.E.D.

Lemma 10.4.4 *If* $\{U_\lambda \mid \lambda \in \Lambda\}$ *is an open cover of the compact metric space* (X,d), *then there exists a sequence of sets* $\{U_{\lambda_n}\}_{n=1}^\infty$ *that is also a cover of* (X,d).

PROOF: Let $\{U_\lambda \mid \lambda \in \Lambda\}$ be an open cover of the compact metric space (X,d), and let $\{B_n\}_{n=1}^\infty$ be the sequence of sets whose existence is guaranteed by Lemma 10.4.3. Let $\{B_{k_n}\}_{n=1}^\infty$ be the subsequence of those members of $\{B_n\}$ that are contained in some member of the given cover $\{U_\lambda \mid \lambda \in \Lambda\}$. Thus, for each n there exists an index $\lambda_n \in \Lambda$ such that

$$B_{k_n} \subseteq U_{\lambda_n}, \qquad n = 1, 2, 3, \ldots . \tag{7}$$

We now prove that $\{B_{k_n}\}_{n=1}^\infty$ constitutes a cover of S. Let $s \in S$. Since $\{U_\lambda \mid \lambda \in \Lambda\}$ is a cover of S it follows that for some $\lambda' \in \Lambda$

$$s \in U_{\lambda'}.$$

By Lemma 10.4.3 there is a set B_m such that

$$s \in B_m \subseteq U_{\lambda'}.$$

However, this B_m must be one of the B_{k_n}'s, and hence

$$S \subseteq \bigcup\{B_{k_n}, \; n = 1, 2, 3, \ldots\}.$$

In view of (7) it follows that $\{U_{\lambda_n}\}$ is also a cover of S. Q.E.D.

Lemma 10.4.5 *If the sequence of open sets* $\{U_n\}_{n=1}^\infty$ *is a cover of the compact metric space* (X,d), *then there is an index* n' *such that* $\{U_1, U_2, \ldots, U_{n'}\}$ *is also a cover of* (X,d).

PROOF: Set
$$G_n = U_1 \cup U_2 \cup \cdots \cup U_n, \qquad n = 1, 2, 3, \ldots . \tag{8}$$
Each set G_n is open, and moreover,

$$G_1 \subseteq G_2 \subseteq G_3 \subseteq \cdots \quad \text{and} \quad \bigcup\{G_n, \; n = 1, 2, 3, \ldots\} = S. \tag{9}$$

If none of the G_n's equals S then for each n there exists an element $s_n \in S$ such that

$$s_n \notin G_n. \tag{10}$$

Since (S,d) is compact, there exists a convergent subsequence $\{s_{k_n}\}_{n=1}^\infty$ of $\{s_n\}_{n=1}^\infty$ whose limit is, say, s. By (9) there exists an index p such that

$$s \in G_p.$$

Since G_p is open, there exists an $\varepsilon > 0$ such that

$$N_\varepsilon(s) \subseteq G_p.$$

Since $\{s_{k_n}\} \to s$, it follows that there exists a index $k_q > p$ such that

$$d(s_{k_q}, s) < \varepsilon,$$

and hence,

$$s_{k_q} \in N_\varepsilon(s) \subseteq G_p \subseteq G_{k_q},$$

which contradicts (10).   Q.E.D.

Theorem 10.4.6 *A metric space is compact if and only if every open cover has a finite subcover.*

PROOF: Suppose the metric space (S,d) is compact. Then it follows from Lemmas 10.4.4 and 10.4.5 that every open cover of (S,d) has a finite subcover.

Conversely, let every open cover of the metric space (S,d) have a finite subcover. Suppose, by way of contradiction, that $S' = \{s_n\}_{n=1}^{\infty}$ is a sequence in S none of whose subsequences converges. It follows that for each $s \in S$, s is not the limit of a subsequence of S' and so there is an $\varepsilon_s > 0$ such that

$$N_{\varepsilon_s}(s) \cap S' = \emptyset \qquad \text{for all } s \notin S' \tag{11}$$

and

$$N_{\varepsilon_s}(s) \cap S' = \{s\} \qquad \text{for all } s \in S'. \tag{12}$$

It is clear that $\{N_{\varepsilon_s}(s) \mid s \in S\}$ is an open cover of S and hence it has a finite subcover, say

$$\{N_{\varepsilon_{t_1}}(t_1), N_{\varepsilon_{t_2}}(t_2), \dots, N_{\varepsilon_{t_k}}(t_k)\}, \qquad t_1, t_2, \dots, t_k \in S.$$

By (11) and (12) the only elements of S' contained in this cover are at most $t_1, t_2, \dots, t_k$. It follows that, as a set, S' is finite, and consequently one of its elements recurs an infinite number of times. This infinite recurrence constitutes a convergent subsequence of S', which is a contradiction. Hence such a sequence S' cannot exist and (S,d) is compact.   Q. E. D.

A topological space $(S, \mathscr{S})$ is said to be *compact* provided every open cover of $(S, \mathscr{S})$ has a finite subcover. Theorem 10.4.6 says that a metric space is compact if and only if it is compact as a topological space. There do exist nonmetric topological spaces that are compact. For every set S, the indiscrete topological space $(S, \{\emptyset, S\})$ is compact. Similarly, every topological space with only a finite number of open sets is compact. In particular, any topological space whose underlying set S is finite is compact.

In conclusion we note that arcs, surfaces, and 3-manifolds defined via presentations are all compact. On the other hand, the knot complements used to define knot groups are not compact.

Exercises 10.4

1. Prove that the disjoint union of two compact topological spaces (see Exercise 10.3.5) is compact.

2. Prove that the product of two compact metric spaces is compact (Exercise 10.1.18).

3. Prove that every compact metric space (S,d) is bounded in the sense that there exists a number δ such that
$$d(s,t) < \delta \qquad \text{for all } s,t \in S.$$

4. A subset C of the metric space (S,d) is said to be *compact* provided that (C,d) is a compact metric space. Prove that if A and B are disjoint compact subsets of the metric space (S,d), then
$$\inf\{d(a,b) \mid a \in A, b \in B\} > 0.$$

5. Prove that every closed subset of a compact metric space is also compact.

6. Prove that Proposition 10.4.2 holds for the Euclidean metric space $(\mathbb{R}^2, e)$. In other words, this space contains a sequence such that every point of the space is the limit of one of its subsequences.

7. Let T be a finite subset of the infinite set S, and let $\mathscr{S}$ be the topology on S whose members consist of the null set and the complements of the subsets of T in S. Prove that $(S, \mathscr{S})$ is a compact nonmetric topological space.

8*. Prove that the Baire space of Exercise 10.1.12 is not compact.

9. A *Cauchy sequence* $\{s_n\}_{n=1}^{\infty}$ in a metric space (S,d) is a sequence with the property that for every $\varepsilon > 0$ there exists an index n_ε such that
$$d(s_m, s_n) < \varepsilon \qquad \text{whenever } m, n \geq n_\varepsilon.$$
Prove that in a compact space every Cauchy sequence converges.

10. (Banach) Let (S,d) be a compact metric space, and suppose $f : S \to S$ is a function that satisfies the condition
$$d(f(s), f(t)) < kd(s,t) \qquad \text{for all } s,t \in S,$$
where k is some constant, $0 < k < 1$. Prove that there exists a point $s_0 \in S$ such that $f(s_0) = s_0$.

11. (Cantor) In a compact topological space every decreasing sequence of nonempty closed subsets
$$F_1 \supseteq F_2 \supseteq \cdots \supseteq F_n \supseteq \cdots$$
satisfies the inequality
$$\bigcap \{F_n \mid n = 1, 2, 3, \ldots\} \neq \emptyset.$$

12. Let $\{F_\lambda \mid \lambda \in \Lambda\}$ be a family of closed subsets of the compact topological space $(S, \mathscr{S})$ such that for any finite set of indices $\{\lambda_1, \lambda_2, \ldots, \lambda_n\}$,
$$F_{\lambda_1} \cap F_{\lambda_2} \cap \cdots \cap F_{\lambda_n} \neq \emptyset.$$
Then
$$\bigcap \{F_\lambda \mid \lambda \in \Lambda\} \neq \emptyset.$$

13. Let f be a continuous function on the compact topological space (S, d). Prove that if F is a closed subset of S, then $f(F)$ is also closed.

14. Suppose f is a continuous function from the compact space $(S, \mathscr{S})$ onto the space $(T, \mathscr{T})$. Prove that $(T, \mathscr{T})$ is compact.

15. Prove that a continuous real-valued function on a compact metric space assumes both its maximum and its minimum.

Chapter Review Exercises

1. All metric spaces are also topological spaces.

2. All topological spaces are also metric spaces.

3. The triangle inequality holds for all topological spaces.

4. Every discrete topological space is also a metric space.

5. Every indiscrete topological space is also a metric space.

6. If S is any set, then every function from (S, D) into $(S, \{\emptyset, S\})$ is continuous.

7. If S is any set, then every function from $(S, \{\emptyset, S\})$ into (S, D) is continuous.

8. In a compact topological space every sequence converges.

9. Some link with two components is connected.

10. Every compact topological space is also connected.

11. If the set S is finite, then every topological space $(S, \mathscr{S})$ is connected.

12. If the set S is finite, then no topological space $(S, \mathscr{S})$ is connected.

13. If the set S is finite, then every topological space $(S, \mathscr{S})$ is compact.

14. If the set S is finite, then every function $f : (S, \mathscr{S}) \to (S, \mathscr{S})$ is continuous.

15. If the set S is finite, then no function $f : (S, \mathscr{S}) \to (S, \mathscr{S})$ is continuous.

CHAPTER 11

POLYTOPES

11.1   Introduction to Polytopes

In Chapter 1, we saw that the Platonic solids satisfy the Euler relation and claimed that other solids do, too, provided that we give a careful definition of "solid."

The solids we want are *polytopes*. The Platonic solids are examples of 3-dimensional polytopes; convex polygons are 2-dimensional polytopes. Intuitively, then, a polytope should be something with flat sides and sharp corners, no matter what the dimension. The formal definition of a polytope will be given in *d*-dimensional space, but the emphasis in this chapter is on dimensions 2, 3, and 4. We will explore some basic properties of polytopes, consider their graphs, count their faces, and define the 4-dimensional analogs of the Platonic solids. While polytopes are lovely mathematical objects in their own right, they also have applications, particularly in the optimization problems described at the end of Section 11.2.

A polytope can be defined in two ways. The first definition is in terms of points. A set S of points in $\mathbb{R}^d$ is *convex* if, given any two points $\mathbf{x}$ and $\mathbf{y}$ in S, all points on the line segment between $\mathbf{x}$ and $\mathbf{y}$ are in S. This is a geometric property, not a topological property. For example, the cube in Figure 1.1 is convex, but the curved cube in Figure 1.4 is not because the line segment between the top points of the two peaks is not contained within the curved cube.

The idea of "between" is clear for two points. The line segment between **x** and **y** is the set of all points of the form $\lambda\mathbf{x} + (1 - \lambda)\mathbf{y}$, where $0 \leq \lambda \leq 1$. To define a polytope, we need to generalize this idea to an arbitrary set of points.

The *convex hull* of a set A of points in $\mathbb{R}^d$ is the set of all points of the form

$$\mathbf{x} = \sum_{i=1}^{n} \lambda_i \mathbf{x}_i,$$

where

$$\mathbf{x}_i \in A, \ \lambda_i \geq 0, \ \text{and} \ \sum_{i=1}^{n} \lambda_i = 1.$$

Such a point **x** is a *convex combination* of the points $\mathbf{x}_1$ through $\mathbf{x}_n$. Informally, we can think of the convex hull as the result of shrink-wrapping the set of points. A *polytope* is the convex hull of a finite set of points. By this definition, a polytope is convex (Exercise 3).

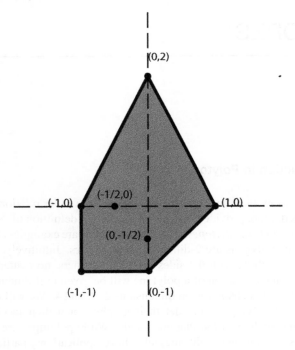

Figure 11.1   The shaded pentagon is the convex hull of $\{(1,0),(0,2),(0,-1/2),$ $(0,-1),(-1/2,0),(-1,0),(-1,-1)\}$.

Figure 11.1 shows the convex hull of

$$A_0 = \{(1,0),(0,2),(0,-1/2),(0,-1),(-1/2,0),(-1,0),(-1,-1)\}.$$

For example, $(0,0)$ is in the convex hull of A_0 because

$$(0,0) = \frac{1}{5}(0,2) + \frac{2}{5}(1,0) + \frac{2}{5}(-1,-1).$$

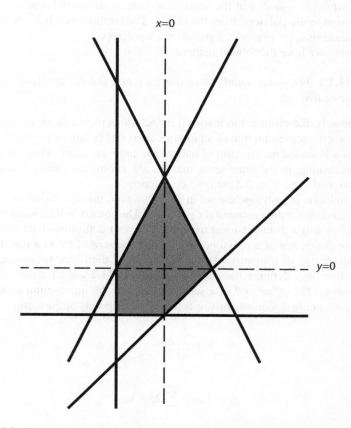

Figure 11.2   The pentagon from Figure 11.1 as the intersection of five closed halfspaces.

The second definition of a polytope is in terms of linear inequalities. Figure 11.2 represents the pentagon from Figure 11.1 as the set of all points satisfying

$$y \geq -1, \qquad x \geq -1, \qquad y \geq x - 1,$$
$$y \leq -2x + 2, \qquad \text{and } y \leq 2x + 2.$$

To generalize to higher dimensions, it helps to write the inequalities in terms of dot products, so that the last inequality, for example, is

$$(-2,1) \cdot (x,y) \leq 2.$$

The points $\mathbf{x}$ in $\mathbb{R}^d$ satisfying a linear inequality

$$\mathbf{a} \cdot \mathbf{x} \leq a_0$$

form a *closed halfspace*. A *polytope* is an intersection of a finite number of closed halfspaces which is bounded in the sense that there is an upper bound on the distance of a point in the polytope from the origin. The requirement that a polytope be bounded means that, for example, a cone is not a polytope.

Fortunately, we have the following theorem.

Theorem 11.1.1 *The point definition of polytope and the inequality definition of polytope are equivalent.* □

Unfortunately, the proof is too involved to include here, but there are algorithms that will convert a representation of a polytope as a convex hull of points to a representation as a bounded intersection of halfspaces and vice versa. These algorithms are time-consuming, in the same sense that the algorithms for finding Hamiltonian cycles mentioned in Section 2.2 are time-consuming.

Both definitions of polytope are set in $\mathbb{R}^d$. However, the dimension of the space does not determine the dimension of the polytope. The convex hull of 3 non-collinear points in $\mathbb{R}^6$ is still a 2-dimensional triangle. The next definitions lead to the definition of the dimension of a polytope. An *affine subspace* of $\mathbb{R}^d$ is a translate of a linear subspace and its dimension is the dimension of the linear subspace; for example, the five lines defining the pentagon in Figure 11.2 are all affine subspaces of dimension 1. The *affine hull* of a set of points A is the intersection of all affine subspaces containing A; equivalently, it is the set of all points of the form

$$\mathbf{x} = \sum_{i=1}^{n} \lambda_i \mathbf{x}_i,$$

where

$$\mathbf{x}_i \in A \text{ and } \sum_{i=1}^{n} \lambda_i = 1.$$

This is like the definition of convex hull, except that the restriction $\lambda_i \geq 0$ has been removed. Thus the affine hull of two distinct points is a line and the affine hull of three non-collinear points is a plane. Just as the linear subspace spanned by k vectors is at most k-dimensional, the affine hull of k points is at most $(k-1)$-dimensional. The dimension of a polytope is the dimension of its affine hull, and a d-dimensional polytope is called a *d-polytope*. A 0-polytope is a point, a 1-polytope is a line segment, and a 2-polytope is a convex polygon. A 2-polytope with n sides is an *n-gon*.

In Chapter 1, we counted the vertices, edges, and (2-dimensional) faces of the Platonic solids. Analogously, a 4-dimensional polytope ought to have a boundary consisting of polytopes of dimension 3 and lower. The next set of definitions leads to a general definition of a face of a polytope.

A *hyperplane* in $\mathbb{R}^d$ is an affine subspace of dimension $d-1$. It breaks $\mathbb{R}^d$ into two closed halfspaces which intersect in the hyperplane itself. If a d-polytope P is entirely contained in one of the closed halfspaces defined by a hyperplane H and $P \cap H$ is nonempty, then H is a *supporting hyperplane* of P and $P \cap H$ is a *face* of P. The dimension of a face is the dimension of its affine hull. For example, in Figure 11.2, the supporting hyperplane $y = 2x + 2$ intersects the pentagon in a line segment, and that line segment is a 1-dimensional face of the pentagon. The supporting hyperplane $y = 2$ (not drawn in the figure) intersects the pentagon in the point $(0, 2)$, and that point is a 0-dimensional face of the pentagon.

A 0-dimensional face of a d-polytope is a *vertex*, a 1-dimensional face is an *edge*, and a $(d - 1)$-dimensional face is a *facet*. In general, a k-dimensional face is called a *k-face*. In addition, both P and $\emptyset$ are considered faces of P, and $\emptyset$ has dimension -1. A face other than P or $\emptyset$ is a *proper face* of P.

Proposition 11.1.2 *Let F be a face of a polytope P. Then F is also a polytope.*

PROOF: If F is either P or $\emptyset$, then F is a polytope. Otherwise, $F = P \cap H$ for some supporting hyperplane H defined by $\mathbf{a} \cdot \mathbf{x} = a_0$. Then F is the intersection of all the closed halfspaces that define P and the closed halfspaces given by

$$\mathbf{a} \cdot \mathbf{x} \le a_0 \text{ and } \mathbf{a} \cdot \mathbf{x} \ge a_0.$$

Because P is bounded and F is a subset of P, F is bounded as well. Q.E.D.

Proposition 11.1.3 *Let F be a face of a polytope P and let G be a face of F. Then G is a face of P.*

PROOF: The proposition is clear if $F = P$, $G = F$, $F = \emptyset$, or $G = \emptyset$. Otherwise, $F = P \cap H_1$ for some supporting hyperplane H_1 of P defined by

$$\mathbf{a} \cdot \mathbf{x} = a_0, \text{ where } \mathbf{a} \cdot \mathbf{x} \le a_0 \text{ for all } \mathbf{x} \in P,$$

and $G = F \cap H_2$ for some supporting hyperplane H_2 of F defined by

$$\mathbf{b} \cdot \mathbf{x} = b_0, \text{ where } \mathbf{b} \cdot \mathbf{x} \le b_0 \text{ for all } \mathbf{x} \in F.$$

Now H_2 is not necessarily a supporting hyperplane of P; it is possible that P is not entirely on one side of H_2. The idea is to perturb H_1 towards H_2 to get a supporting hyperplane of P that intersects P in G. See Figure 11.3. Consider the inequality

$$(\mathbf{b} + \lambda \mathbf{a}) \cdot \mathbf{x} \le b_0 + \lambda a_0 \tag{1}$$

for some positive real number λ. Inequality (1) is satisfied by all $\mathbf{x} \in F$, with equality if and only if $\mathbf{x} \in G$. The goal is to choose λ so that the inequality is satisfied by all $\mathbf{x} \in P$. Because P is a polytope, it is the convex hull of a finite set of points A. Inequality (1) is already satisfied by any points of A in F. The set $A \setminus F$ is finite and

$$a_0 - \mathbf{a} \cdot \mathbf{x} \ne 0 \text{ for } \mathbf{x} \in A \setminus F,$$

so choose λ so that

$$\lambda > \frac{\mathbf{b} \cdot \mathbf{x} - b_0}{a_0 - \mathbf{a} \cdot \mathbf{x}}$$

for all $\mathbf{x} \in A \setminus F$. Because every point in P is a convex combination of points in A, with this choice of λ inequality (1) holds for all $\mathbf{x} \in P$, with equality if and only if $\mathbf{x} \in G$. Therefore, G is a face of P.   Q.E.D.

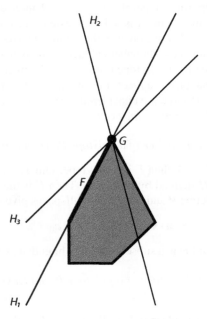

Figure 11.3   The hyperplane H_1 is a supporting hyperplane for the edge F of the pentagon. The hyperplane H_2 is a supporting hyperplane for the vertex G as a face of F, although not as a face of the pentagon. Perturbing H_1 towards H_2 yields H_3, a supporting hyperplane for G as a vertex of the pentagon.

The 2-faces of the Platonic solids intersect nicely in lower-dimensional faces. This property holds for all polytopes.

Proposition 11.1.4 *Let P be a polytope. The intersection of two faces of P is also a face of P.*

PROOF: Let F and G be faces of P. The proposition is clear if either F or G is P or $\emptyset$. Otherwise, $F = P \cap H_1$ for some supporting hyperplane H_1 of P defined by

$$\mathbf{a} \cdot \mathbf{x} = a_0, \text{ where } \mathbf{a} \cdot \mathbf{x} \le a_0 \text{ for all } \mathbf{x} \in P,$$

and $G = P \cap H_2$ for some supporting hyperplane H_2 of P defined by

$$\mathbf{b} \cdot \mathbf{x} = b_0, \text{ where } \mathbf{b} \cdot \mathbf{x} \le b_0 \text{ for all } \mathbf{x} \in P.$$

Then

$$(\mathbf{a} + \mathbf{b}) \cdot \mathbf{x} = a_0 + b_0$$

for $\mathbf{x} \in F \cap G$ and

$$(\mathbf{a} + \mathbf{b}) \cdot \mathbf{x} \leq a_0 + b_0$$

for $\mathbf{x} \in P$.

If $F \cap G$ is the empty set, then it is a face of P. If $F \cap G$ is nonempty, then $(\mathbf{a} + \mathbf{b}) \cdot \mathbf{x} = a_0 + b_0$ defines a supporting hyperplane for P.

It remains to show that the intersection of this hyperplane with P is $F \cap G$. Suppose $\mathbf{x} \in P$ and $(\mathbf{a} + \mathbf{b}) \cdot \mathbf{x} = a_0 + b_0$. Because $\mathbf{a} \cdot \mathbf{x} \leq a_0$ and $\mathbf{b} \cdot \mathbf{x} \leq b_0$, it must be the case that $\mathbf{a} \cdot \mathbf{x} = a_0$ and $\mathbf{b} \cdot \mathbf{x} = b_0$, and therefore $\mathbf{x} \in F \cap G$. Q.E.D.

A given point in a polytope may have more than one representation as a convex combination of points in a set. For example, in Figure 11.1, $(0,0)$ can be written not only as the convex combination given previously, but as $\frac{1}{2}(-1,0) + \frac{1}{2}(1,0)$ and as $\frac{1}{3}(0,2) + \frac{2}{3}(0,-1)$. On the other hand, a given point may have essentially one representation as a convex combination. For example, in Figure 11.1, suppose

$$(1,0) = \sum_{i=1}^{n} \lambda_i \mathbf{x}_i,$$

where

$$\mathbf{x}_i \in A_0, \ \lambda_i \geq 0, \ \text{and} \ \sum_{i=1}^{n} \lambda_i = 1,$$

and suppose the $\mathbf{x}_i$ are distinct. Because the first coordinate of $(1,0)$ is 1 and all other points in A_0 have first coordinate strictly less than 1, one term $\lambda_j \mathbf{x}_j$ must be $1 \cdot (1,0)$, and the other terms must have $\lambda_i = 0$. It is not a coincidence that $(1,0)$ is a vertex of the pentagon, while $(0,0)$ is not.

Proposition 11.1.5 *Let the polytope P be the convex hull of a set of points A and let $\mathbf{v}$ be a vertex of P. Then $\mathbf{v}$ cannot be written as a convex combination of the points in $A \setminus \{\mathbf{v}\}$.*

PROOF: Exercise 8. □

Proposition 11.1.6 *A polytope is the convex hull of its vertices.*

PROOF: The proof is by induction on d, the dimension of the polytope. The proposition is clear for $d = -1, 0$, and 1. Suppose the proposition is true for all d satisfying $-1 \leq d \leq n$. Let P be an $(n+1)$-polytope and let V be the set of its vertices. Let Q be the convex hull of V, and suppose by way of contradiction that $Q \neq P$. Because Q is a polytope, there must be some linear inequality $\mathbf{a} \cdot \mathbf{x} \leq a_0$ that is satisfied by all $\mathbf{x} \in Q$ but not by all $\mathbf{x} \in P$. Let

$$b = \max\{\mathbf{a} \cdot \mathbf{x} : \mathbf{x} \in P\}.$$

Then $b > a_0$ and $\mathbf{a} \cdot \mathbf{x} \leq b$ is satisfied by all $\mathbf{x} \in P$. The intersection of P with the hyperplane H given by $\mathbf{a} \cdot \mathbf{x} = b$ is nonempty (see Exercise 10.4.15), so $P \cap H$ is a proper face of P and therefore a polytope of dimension less than $n + 1$. By the induction hypothesis, $P \cap H$ is the convex hull of its vertices, and by Proposition 11.1.3, the vertices of $P \cap H$ are vertices of P. Thus there are vertices of P that are not in Q, which is a contradiction.   Q.E.D.

The Platonic solids fall into pairs based on their face counts. The cube has 8 vertices, 12 edges, and 6 2-faces, while the octahedron has 6 vertices, 12 edges, and 8 2-faces. Similarly, the dodecahedron has 20 vertices, 30 edges, and 12 2-faces, while the icosahedron has 12 vertices, 30 edges, and 20 2-faces. The numbers of vertices and 2-faces are switched. The tetrahedron seems to be off on its own, but since it has both 4 vertices and 4 2-faces, it is really its own partner. This suggests that there should be some formal way to find a correspondence between the vertices of one Platonic solid and the 2-faces of its partner. Indeed there is.

Consider the cube and the octahedron. The cube can be given by the inequalities

$$x \leq 1, \qquad\qquad -x \leq 1,$$
$$y \leq 1, \qquad\qquad -y \leq 1,$$
$$z \leq 1, \qquad\qquad \text{and } -z \leq 1,$$

and the octahedron can be given by the convex hull of

$$\{(1,0,0),(-1,0,0),(0,1,0),(0,-1,0),(0,0,1),(0,0,-1)\}.$$

With these representations, the correspondence is clear: the coefficients of x, y, and z in the inequality defining a facet of the cube give the coordinates of the corresponding vertex of the octahedron.

The correspondence works in the other direction as well. The vertex set of the cube is

$$\{(1,1,1),(-1,1,1),(1,-1,1),(1,1,-1),$$
$$(-1,-1,1),(-1,1,-1),(1,-1,-1),(-1,-1,-1)\},$$

and the facets of the octahedron are given by

$$x+y+z \leq 1, \qquad\qquad -x+y+z \leq 1,$$
$$x-y+z \leq 1, \qquad\qquad x+y-z \leq 1,$$
$$-x-y+z \leq 1, \qquad\qquad -x+y-z \leq 1,$$
$$x-y-z \leq 1, \qquad\qquad \text{and } -x-y-z \leq 1.$$

Even better, this correspondence can be extended to include the edges. The edge between the two facets of the cube given by $x \leq 1$ and $y \leq 1$ corresponds to the edge of the octahedron between the vertices $(1,0,0)$ and $(0,1,0)$.

In fact, any polytope has such a partner. The *polar set* of a d-polytope P is

$$P^\Delta = \{\mathbf{y} \in \mathbb{R}^d : \mathbf{y} \cdot \mathbf{x} \leq 1 \text{ for all } \mathbf{x} \in P\}.$$

If the origin $\mathbf{0}$ is in the interior of P, then P^Δ is also a polytope containing $\mathbf{0}$ and it is called the *polar dual* of P. For example, the polar dual of the pentagon from Figures

11.1 and 11.2 is shown in Figure 11.4. Rewriting the inequalities from Figure 11.2 in the format $\mathbf{y} \cdot \mathbf{x} \le 1$ gives

$$(0, -1) \cdot \mathbf{x} \le 1,$$
$$(-1, 0) \cdot \mathbf{x} \le 1,$$
$$(1, -1) \cdot \mathbf{x} \le 1,$$
$$(1, 1/2) \cdot \mathbf{x} \le 1,$$

and

$$(-1, 1/2) \cdot \mathbf{x} \le 1,$$

and again the coefficients of the facet-defining inequalities of P give the vertices of P^Δ. Thinking of the vertices of P as vectors, each is normal to one facet of P^Δ. For a more formal development of these ideas, see Exercises 10 to 13.

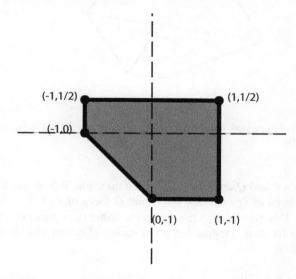

Figure 11.4   The polar dual of the pentagon from Figures 11.1 and 11.2.

So far we have given all definitions in terms of coordinates, which is essential for computational work. However, the lopsided cube in Figure 1.2 shares the same general structure as the cube in Figure 1.1, even though the coordinates are different. Two polytopes P and Q are *combinatorially equivalent* if there is a bijection Φ from the set of faces of P to the set of faces of Q such that for F and G faces of P, $F \subseteq G$ if and only if $\Phi(F) \subseteq \Phi(G)$. The bijection preserves inclusion of faces.

Matching the top face of the regular cube with the top face of the lopsided cube, the bottom face of the regular cube with the bottom face of the lopsided cube, and so forth, gives the bijection that shows these two cubes are combinatorially equivalent. In contrast, the cube is not combinatorially equivalent to the polytope in Figure 11.5, which was created by cutting off two of the vertices of a tetrahedron. This polytope also has 8 vertices, 12 edges, and 6 faces. However, two of its faces are pentagons, and there is no bijection that will map a square face of a cube to a pentagon so that the four edges of the square map to the five edges of the pentagon.

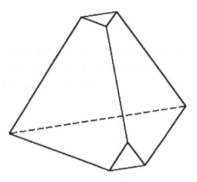

Figure 11.5   A polytope with 8 vertices, 12 edges, and 6 2-faces that is not combinatorially equivalent to a cube.

Two polytopes P and Q are *dual* if there is a bijection Ψ from the set of faces of P to the set of faces of Q such that for F and G faces of P, $F \subseteq G$ if and only if $\Psi(G) \subseteq \Psi(F)$. This type of bijection reverses, rather than preserves, the inclusion of faces. The polar dual construction gives such a bijection and shows that every polytope has a dual.

Proposition 11.1.7 *Let P be a d-polytope in $\mathbb{R}^d$ with $\mathbf{0}$ in the interior. Let F be a face of P, and define*

$$F^* = \{\mathbf{y} \in P^\Delta : \mathbf{y} \cdot \mathbf{x} = 1 \text{ for all } \mathbf{x} \in F\}.$$

Then the mapping Ψ from the faces of P to the faces of P^Δ given by $\Psi(F) = F^*$ is an inclusion-reversing bijection, and therefore P and P^Δ are dual.

PROOF: First we show that F^* is a face of P^Δ. Because every $\mathbf{x} \in F$ is a convex combination of the vertices of F,

$$F^* = \{\mathbf{y} \in P^\Delta : \mathbf{y} \cdot \mathbf{v} = 1 \text{ for each } \mathbf{v} \text{ a vertex of } F\}.$$

For each vertex $\mathbf{v}$ of F, there is some supporting hyperplane H of P such that $P \cap H = \mathbf{v}$, so there is some $\mathbf{z} \in P^{\Delta}$ such that $\mathbf{z} \cdot \mathbf{v} = 1$. (See Exercise 11.) Thus $\mathbf{y} \cdot \mathbf{v} \leq 1$ for all $\mathbf{y} \in P^{\Delta}$ and the hyperplane given by $\mathbf{y} \cdot \mathbf{v} = 1$ is a supporting hyperplane of P^{Δ} and defines a face of P^{Δ}. Therefore F^{*} is the intersection of faces of P^{Δ} and, by Proposition 11.1.4, is itself a face of P^{Δ}. To show that Ψ is an inclusion-reversing bijection, see Exercises 17 and 18.   Q.E.D.

So that we have examples of polytopes beyond polygons and the Platonic solids, we close this section by defining some families of polytopes and describing operations that create new polytopes from old ones.

The *d-simplex* T^{d} is the convex hull of $d + 1$ points whose affine hull is d-dimensional. The 0-simplex is a point, the 1-simplex is a line segment, the 2-simplex is a triangle, and the 3-simplex is a tetrahedron. A polytope whose facets are all simplices is called *simplicial*. For example, the octahedron is a simplicial 3-polytope. The d-simplex itself is simplicial. If it is the convex hull of the set A of $d + 1$ points, then any d points in A have a $(d - 1)$-dimensional affine hull, which is a supporting hyperplane of T^{d}, and hence they are all on a facet of T^{d}. A translate of this supporting hyperplane is a supporting hyperplane for the remaining vertex of T^{d}. Thus every point of A is a vertex of T^{d} and every subset of d points of A is the vertex set of a $(d - 1)$-simplex that is a facet of T^{d}. Therefore every k-face of T^{d} is a k-simplex, and all proper faces of a simplicial polytope are simplices.

The dual of a simplicial d-polytope is a *simple* d-polytope; every vertex is contained in exactly d facets. The polytope in Figure 11.5 is simple.

If the coordinates of $\mathbb{R}^{d}$ are given by x_{i}, where $1 \leq i \leq d$, then the *d-cube* C^{d} is given by the inequalities $-1 \leq x_{i} \leq 1$. Equivalently, it is the convex hull of the 2^{d} points in $\mathbb{R}^{d}$ for which each x_{i} is either 1 or -1. The 0-cube is a point, the 1-cube is a line segment, the 2-cube is a square, and the 3-cube is the usual cube. Every k-face of a d-cube is a k-cube. A polytope whose facets are all combinatorially equivalent to cubes is called *cubical*. An example is in Figure 11.6.

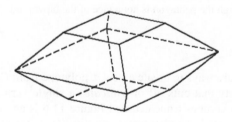

Figure 11.6   A cubical 3-polytope.

Let P be a d-polytope. Here are three operations that create a $(d+1)$-polytope from P. The *pyramid* over P is the convex hull of P and a point outside the affine hull of P. The *prism* over P is the convex hull of P and a translate of P in a direction outside its affine hull; it is the Cartesian product of P and a line segment not contained in its affine hull. The *bipyramid* over P is the convex hull of P and a line segment L not contained in the affine hull of P such that $L \cap P$ is a single point in the interior of both L and P. The bipyramid is two pyramids glued together along P in a way that guarantees that the result is convex. See Figure 11.7 for examples. In particular, the d-simplex T^d is the pyramid over T^{d-1} and the d-cube C^d is the prism over C^{d-1}.

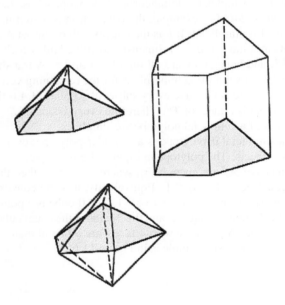

Figure 11.7   The pyramid, prism, and bipyramid over a pentagon. The pentagon is shaded in each polytope, although the pentagon is not a face of the bipyramid.

In addition, there are ways to create a new d-polytope from P. One way is to add a new linear inequality that cuts off or *truncates* a single vertex. Figure 11.5 is a tetrahedron with two vertices truncated and Figure 11.6 is an octahedron with two opposite vertices truncated. Another way to modify P is to take the convex hull of P and a point outside P but near a facet F; in particular, the point should strictly satisfy all the linear inequalities that define facets of P except the one that defines F. This operation is the *stellar subdivision* of P at F. See Figure 11.8. It has the effect of creating a very short pyramid over F and gluing that pyramid to P along F.

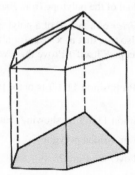

Figure 11.8   The stellar subdivision of the prism over the pentagon from Figure 11.7 at the top pentagon.

Exercises 11.1

1. Draw the convex hull of the set $A_0 = \{(1,0,1),(0,0,1),(0,0,-1),(0,-1,-1),(0,1,-1),(-1,0,1)\}$.

2. Confirm that $(0,0,0)$ is in the convex hull of the set A_0 from Exercise 1 by writing it as a convex combination of points in A_0.

3. Prove that the convex hull of a set A satisfies the definition of convexity.

4. Prove that the intersection of two convex sets is convex.

5. Write the polytope from Exercise 1 as an intersection of closed halfspaces.

6. Draw the polytope defined by the inequalities $y \leq x+1$, $y \leq 2$, $y \geq 2x-2$, $y \geq -2x-2$, $y \leq -x+1$, and $x \geq -1$.

7. Write the polytope from Exercise 6 as the convex hull of a finite set of points.

8. Prove Proposition 11.1.5.

9. Both definitions of polytope include the word "finite." Draw some 2-dimensional figures that are not polytopes but would satisfy the definition(s) if the word "finite" were removed.

10. Prove that if the origin $\mathbf{0}$ is in the interior of a d-polytope P in $\mathbb{R}^d$, then P^Δ is a polytope. Formally, a point $\mathbf{x}$ is in the interior of P if there exists some positive real number ε such that every point $\mathbf{y}$ in $\mathbb{R}^d$ whose distance from $\mathbf{x}$ is less than ε is also in P.

11. Prove that if the origin $\mathbf{0}$ is in the interior of a d-polytope P, any inequality of the form $\mathbf{a} \cdot \mathbf{x} \leq a_0$ that is satisfied by all $\mathbf{x}$ in P can be written in the form $\mathbf{b} \cdot \mathbf{x} \leq 1$.

12. Prove that if a d-polytope P in $\mathbb{R}^d$ contains $\mathbf{0}$ in its interior, then $(P^\Delta)^\Delta = P$.

13. Let P be a d-polytope in $\mathbb{R}^d$ containing $\mathbf{0}$ in its interior, and let $\mathbf{z}$ be a point in P. Prove that if $F = \{\mathbf{y} \in P^\Delta : \mathbf{z} \cdot \mathbf{y} = 1\}$ is a facet of P^Δ, then $\mathbf{z}$ is a vertex of P.

14. Compute and draw the polar dual of the polytope from Exercise 1. Exercise 13 may help.

15. Compute and draw the polar dual of the polytope from Exercise 6. Exercise 13 may help.

16. A point $\mathbf{z}$ is in the *relative interior* of a face F of a polytope P if whenever $\mathbf{a} \cdot \mathbf{x} \leq a_0$ for all $\mathbf{x} \in P$ and $\mathbf{a} \cdot \mathbf{z} = a_0$, it is also true that $\mathbf{a} \cdot \mathbf{x} = a_0$ for all $\mathbf{x} \in F$. (Equivalently, $\mathbf{x}$ is not contained in any face of F except F itself.) Prove that every point of P is in the relative interior of some face of P.

17. Prove that the mapping Ψ in Proposition 11.1.7 is onto. It may help to use Exercises 11 and 16.

18. Complete the proof of Proposition 11.1.7 by showing that $\Psi(\Psi(F)) = F$.

19. What is the dual of an n-gon (an n-sided polygon)?

20. What is the dual of a pyramid over an n-gon?

21. Let P be a polytope and let Q be the pyramid over P created by taking the convex hull of P and a point $\mathbf{v}$ outside the affine hull of P. Prove that the vertices of Q are $\mathbf{v}$ and the vertices of P.

22. Let P be a d-polytope and let Q be the prism over P created by taking the convex hull of all points in $\mathbb{R}^{d+1}$ of the form $(\mathbf{x}, 0)$ and $(\mathbf{x}, 1)$, where $\mathbf{x}$ is a point in P. (Place a copy of P at height 0 in $\mathbb{R}^{d+1}$ and another at height 1 in $\mathbb{R}^{d+1}$ and then take the convex hull.) Prove that the vertices of Q are the points of the form $(\mathbf{v}, 0)$ and $(\mathbf{v}, 1)$, where $\mathbf{v}$ is a vertex of P.

23. Let P be a d-polytope containing the origin $\mathbf{0}$ in its interior. Let Q be the bipyramid over P created by taking the convex hull of all points in $\mathbb{R}^{d+1}$ of the form $(\mathbf{x}, 0)$, where $\mathbf{x}$ is a point in P, and the points $(0, \ldots, 0, 1)$ and $(0, \ldots, 0, -1)$. Prove that the vertices of Q are $(0, \ldots, 0, 1)$, $(0, \ldots, 0, -1)$, and the points of the form $(\mathbf{v}, 0)$, where $\mathbf{v}$ is a vertex of P.

11.2   Graphs of Polytopes

The vertices and edges of a polytope form a graph, where the vertices are the nodes and the edges are the arcs. In this section, we will use "nodes" and "arcs" when discussing graphs in general, as in Chapter 2, and "vertices" and "edges" when discussing graphs of polytopes.

One particularly nice way to draw a d-polytope P, and therefore also its graph, is as a *Schlegel diagram*. Choose a supporting hyperplane H that defines a facet F of P and a point $\mathbf{x} \in \mathbb{R}^d$ outside P very close to F, so that H is the only facet-defining supporting hyperplane separating $\mathbf{x}$ from P, and $\mathbf{x}$ is not on any facet-defining supporting hyperplane of P. The Schlegel diagram is the projection of the polytope onto H by rays through $\mathbf{x}$. Because of the location of $\mathbf{x}$, the image of P in H is contained in F. The dimension of the Schlegel diagram is $d - 1$. Thus Schlegel diagrams offer a way to draw the graph of a 3-polytope in the plane and also give us some hope of visualizing the regular 4-polytopes in the next section. Figures 11.9 and 11.10 show two different Schlegel diagrams for a prism over a triangle.

If P is a 3-polytope, then the images of two edges of P in the Schlegel diagram do not intersect unless they share an endpoint, in which case they intersect in the image of that endpoint. (See Exercise 3.) Thus, graphs of 3-polytopes are planar. To completely characterize graphs of 3-polytopes, we need one other concept. In

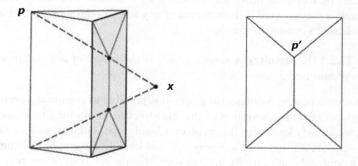

Figure 11.9   Here is the construction of a Schlegel diagram for a triangular prism with the point **x** placed near a shaded rectangular face F. In the left figure, the dashed lines show the projection of two vertices onto F. The gray lines in the interior of F are the projections of the edges not on F. On the right is the Schlegel diagram, where the point $\mathbf{p'}$ is the projection of the point **p**.

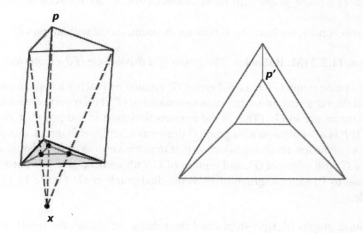

Figure 11.10   Here is the construction of a different Schlegel diagram for a triangular prism. Now the point **x** is placed near a shaded triangular face F. In the left figure, the dashed lines show the projection of three vertices onto F. The gray lines in the interior of F are the projections of the edges not on F. On the right is the Schlegel diagram, where the point $\mathbf{p'}$ is the projection of the point **p**.

Section 2.2, a graph in which every pair of nodes can be joined by a trail was said to be *connected*. More generally, a graph is *k-connected* if every pair of nodes can be connected by k disjoint trails. Equivalently, by a theorem of Hassler Whitney, a simple graph is k-connected if the removal of any $k - 1$ nodes and the arcs incident with them leaves a connected graph.

Theorem 11.2.1 (E. Steinitz) *A simple graph is the graph of a 3-polytope if and only if it is planar and 3-connected.*

PROOF: Given a planar, 3-connected graph, it is possible to construct a corresponding 3-polytope, but the description of that construction is too long to include here. However, we already know that the graph of a 3-polytope is planar, and we can show that it is 3-connected. Let P be a 3-polytope and let G be its graph, and consider G drawn as a Schlegel diagram. We need to show that the removal of any two vertices leaves a connected graph. First, suppose that the two vertices are not connected by an edge. Removal of a vertex of degree k and its incident edges combines k regions into one. Thus the Euler characteristic $v - e + f$ changes by $(-1) + k - (k - 1) = 0$. Because the two vertices are not connected by an edge, they share at most one region. Thus the same argument applies to the second vertex, and the Euler characteristic does not change. Now suppose that the two vertices are connected by an edge and have degrees k_1 and k_2. Removal of the two vertices and their incident edges combines $k_1 + k_2 - 2$ regions into one, and the Euler characteristic changes by $(-2) + (k_1 + k_2 - 1) - (k_1 + k_2 - 3) = 0$. In either case, the deletion of two vertices leaves a plane graph with Euler characteristic 2. By Exercise 4, this graph is connected.   Q.E.D.

More generally, we have the following theorem, stated without proof.

Theorem 11.2.2 (M. Balinski) *The graph of a d-polytope is d-connected.*   ☐

Every plane graph G has a *dual graph* G^* created by placing a node in each region of G and then drawing an arc between two nodes of G^* if their corresponding regions in G share an arc in G. (This is the construction used in the proof of Proposition 2.4.9.) If P is a 3-polytope with graph G drawn as a Schlegel diagram, this procedure creates an inclusion-reversing bijection that matches vertices of G with regions of G^*, edges of G with edges of G^*, and regions of G with vertices of G^*. If the 3-polytope P^* is dual to P, then the graph of P^* is the dual graph of G. Figure 11.11 gives an example.

Because graphs of 3-polytopes and their duals are planar, the results of Section 2.4 have some immediate consequences:

Proposition 11.2.3 *Let P be a 3-polytope with v vertices, e edges, and f 2-faces. Then*

1. $e \leq 3v - 6$.

2. $e \leq 3f - 6$.

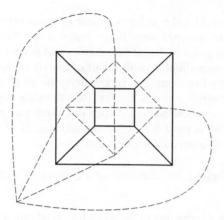

Figure 11.11   The solid lines are a Schlegel diagram of the cube. The dashed lines are its dual graph, which is the graph of the octahedron.

PROOF: Apply Corollary 2.4.4 to P and to its dual.   Q.E.D.

Proposition 11.2.4 *Every 3-polytope contains a vertex with degree no more than 5 and a 2-face with no more than 5 sides.*

PROOF: See the proof of Lemma 2.4.8.   Q.E.D.

Proposition 11.2.5 *The graph of a 3-polytope is 4-colorable.*

PROOF: This is a consequence of the Four-Color Theorem.   Q.E.D.
In addition, the following theorem is a consequence of the Euler relation.

Proposition 11.2.6 *Every 3-polytope has a vertex of degree 3 or a triangular face.*

PROOF: See Exercise 5.   □
Other topics from Chapter 2 apply to graphs of polytopes as well. For colorability and Eulerian trails, see Exercises 6 and 7. Here is one result about homeomorphisms.

Theorem 11.2.7 (B. Grünbaum) *Every d-polytope contains a subgraph homeomorphic to the complete graph K_{d+1}.*

PROOF:The proof of this theorem for $d \leq 3$ is Exercise 8.   □
Questions about Hamiltonian paths are difficult for graphs of polytopes as well as for graphs in general. However, here is one important result, presented without proof.

Theorem 11.2.8 (W. Tutte) *If the graph of a 3-polytope is 4-connected, then it is Hamiltonian.*   □

Graphs of high-dimensional polytopes appear in linear programming problems. These problems seek to optimize some linear function of several variables; the domain of the function is given by linear inequalities and is thus a polytope provided it is bounded. The famous simplex method of G. Dantzig finds the optimal point in the polytope by walking from vertex to vertex along the graph of the polytope. The shorter the walk to the optimal vertex, the less time the algorithm takes. The *diameter* of a graph G is the smallest number δ such that every pair of nodes in G can be connected by a path with at most δ arcs. Motivated by study of the simplex method, W. Hirsch made the following conjecture in 1957.

Conjecture 11.2.9 (Hirsch) *The diameter of a d-polytope with n facets is at most* $n - d$. $\qquad\qquad\qquad\qquad\qquad\qquad\qquad\qquad\qquad\qquad\qquad\qquad\qquad\qquad\qquad$ $\square$

The Hirsch Conjecture was one of the great unsolved problems about polytopes. It was proven for many types of polytopes, but in 2010 F. Santos found a counterexample: a 43-dimensional polytope with 86 vertices and diameter more than 43. Since then, lower-dimensional counterexamples have been found. The larger question still remains: what is the best upper bound for the diameter of a d-polytope with n facets?

Exercises 11.2

1. Draw the Schlegel diagram of a dodecahedron and draw the dual graph.

2. Draw two different Schlegel diagrams of a pyramid with a square base, one by placing the point **x** near a triangular face, and one by placing **x** near the square face. In each case, draw the dual graph.

3. Prove that images of the vertices and edges of a 3-polytope in a Schlegel diagram form a planar graph. Exercise 4 in Section 11.1 may be useful.

4. Determine the Euler characteristic of a plane graph with k connected components and use this to prove that a plane graph is connected if and only if its Euler characteristic is 2. (See also Exercise 2.4.15.)

5. Prove Proposition 11.2.6 by applying Lemma 2.4.3 to both the polytope and its dual.

6. Prove that the graph of a 3-polytope P is 2-colorable if and only if each 2-face of P has an even number of edges.

7. Use Exercise 6 to prove that the graph of a 3-polytope P is 2-colorable if and only if the graph of the dual of P is Eulerian.

8. Prove Theorem 11.2.7 for dimensions 1, 2, and 3. Proposition 11.2.6 may be useful.

11.3 Regular Polytopes

The five Platonic solids are highly symmetric. They are the 3-dimensional analogs of regular polygons and are also known as *regular 3-polytopes*. By definition, the 2-faces of a regular 3-polytope are all congruent regular p-gons, and q of them meet at each vertex. The tetrahedron, cube, octahedron, dodecahedron, and icosahedron clearly satisfy this definition, but are these the only such 3-polytopes?

Euclid says they are and gives the following argument in a remark on the last proposition of the last book of his *Elements*. This is the 1908 translation by Thomas L. Heath. [1]

> I say next that *no other figure, besides the said five figures, can be constructed which is contained by equilateral and equiangular figures equal to one another.*
>
> For a solid angle cannot be constructed with two triangles, or indeed planes.
>
> With three triangles the angle of the pyramid is constructed, with four the angle of the octahedron, and with five the angle of the icosahedron; but a solid angle cannot be formed by six equilateral and equiangular triangles placed together at one point, for, the angle of the equilateral triangle being two-thirds of a right angle, the six will be equal to four right angles: which is impossible, for any solid angle is contained by angles less than four right angles.
>
> For the same reason, neither can a solid angle be constructed by more than six plane angles.
>
> By three squares the angle of the cube is contained, but by four it is impossible for a solid angle to be contained, for they will again be four right angles.
>
> By three equilateral and equiangular pentagons the angle of the dodecahedron is contained; but by four such it is impossible for any solid angle to be contained, for, the angle of the equilateral pentagon being a right angle and a fifth, the four angles will be greater than four right angles: which is impossible.
>
> Neither again will a solid angle be contained by other polygonal figures by reason of the same absurdity.
>
> Therefore etc. Q. E. D.

The statement "any solid angle is contained by angles less than four right angles" means that the sum of the angles of the 2-faces that meet at a vertex must be less than 2π. Euclid says this condition can be met by 3, 4, or 5 equilateral triangles, 3 squares, or 3 pentagons. Since each vertex must be incident with at least three 2-faces, " for a solid angle cannot be constructed with two triangles, or indeed planes," these are all the possibilities. They are satisfied by the tetrahedron, the octahedron, the icosahedron, the cube, and the dodecahedron, respectively.

Euclid does not explicitly say that the same number of 2-faces must meet at each vertex. However, without this assumption, more 3-polytopes would be classified as regular. For example, a bipyramid over an equilateral triangle can be constructed with equilateral triangular 2-faces, but some vertices are incident with 4 2-faces, while others are incident with 3.

The *Schläfli symbol* for a regular 3-polytope whose 2-faces are p-gons meeting q at a vertex is $\{p,q\}$. Despite the use of curly brackets, this is not a set, and $\{p,q\}$ is different from $\{q,p\}$. These symbols are named for Ludwig Schläfli, who discovered higher-dimensional regular polytopes in the 1850s.

[1] The authors wish to acknowledge the kindness of the Dover Publishing Co. for granting them permission to reproduce this proof from *The Thirteen Books of the Elements*, vol. 3, (pp. 507–8) by Euclid, edited by Sir Thomas L. Heath and published by Dover Publications.

Euler's relation gives another argument that the only possible choices for $\{p,q\}$ are $\{3,3\},\{3,4\},\{4,3\},\{3,5\}$, and $\{5,3\}$. If each 2-face is a p-gon, then $2e = pf$, since each edge is in two 2-faces and each 2-face contains p edges. If q 2-faces meet at each vertex, then q edges also meet at each vertex. Therefore $2e = qv$, since each edge is incident with two vertices and each vertex is incident with q edges. For a regular 3-polytope, then, $v - e + f = 2$ becomes

$$\frac{2e}{q} - e + \frac{2e}{p} = 2,$$

and therefore

$$e = \frac{1}{\frac{1}{p} + \frac{1}{q} - \frac{1}{2}}.$$

The denominator of this expression for e must be positive, which means that it is not possible to have both $p > 3$ and $q > 3$. Because $p \geq 3$ and $q \geq 3$, either $p = 3$ or $q = 3$. Given this, a quick check shows that the possible values of $\{p,q\}$ are indeed $\{3,3\}$, $\{3,4\}$, $\{4,3\}$, $\{5,3\}$, and $\{3,5\}$. Furthermore, the above expression for e and the equations $2e = pf$ and $2e = qv$ show that p and q determine v, e, and f.

The values of p and q also determine the graph of the polytope. It is worth drawing a Schlegel diagram for each $\{p,q\}$ to convince yourself of this.

One subtlety is not addressed in Euclid's argument. Could two regular polytopes with the same p and q have different geometry? For example, if the definition of "regular" for 2-polytopes said that the side lengths should be equal and that the same number of edges should meet at a vertex, which is a 2-dimensional analog of the definition of a regular 3-polytope, any rhombus would be regular. To avoid this, a regular 2-polytope, or regular polygon, is instead defined to have equal sides and equal angles.

In 1813, A. Cauchy argued that two combinatorially equivalent 3-polytopes with congruent 2-faces are themselves congruent. (Steinitz found and corrected a flaw in the argument in 1934.) In particular, this means that 3-polytopes are *rigid*: a 3-polytope made with inflexible 2-faces and with hinges for edges will not flex. In contrast, a square with inflexible rods for edges and with hinges for vertices can be deformed into a family of rhombi. As a consequence of Cauchy's work, p and q do determine the geometry of $\{p,q\}$.

What are the 4-dimensional analogs of the Platonic solids? To answer this, we need a definition of regularity for 4-polytopes. The facets of a regular 4-polytope should of course be congruent regular 3-polytopes, and in a regular d-polytope, the facets should be congruent regular $(d-1)$-polytopes. The harder part is finding the right condition on the vertices so that the neighborhood of each vertex is the same. For any polytope P, regular or not, the *vertex figure* of P at $\mathbf{v}$, denoted $P/\mathbf{v}$, is the polytope given by the intersection of P and a hyperplane H that separates $\mathbf{v}$ from all the other vertices of P. Different choices of H give combinatorially equivalent polytopes, so in general it does make sense to call $P/\mathbf{v}$ "the" vertex figure. The i-faces of $P/\mathbf{v}$ correspond to the $(i+1)$-faces of P that contain $\mathbf{v}$. (See Exercise 1.)

However, in order to define a regular d-polytope, we need to specify not just the combinatorics but the geometry at a vertex. For the purposes of this section, then, the vertex figure of a regular d-polytope P at vertex $\mathbf{v}$ is the intersection of P with the hyperplane H that contains the midpoints of all the edges incident with $\mathbf{v}$. Figure 11.12 shows the vertex figure of a 3-cube; the edges and vertices of this triangle are slices of the 2-faces and edges of the cube that contain the top front right vertex. If d edges are incident with $\mathbf{v}$, then the midpoints automatically lie on a hyperplane. If more than d edges are incident with $\mathbf{v}$, then the requirement that the midpoints lie on a hyperplane does restrict the geometry of P. Now we can complete the definition: a d-polytope is *regular* if its facets are congruent regular $(d-1)$-polytopes and its vertex figures are congruent regular $(d-1)$-polytopes.

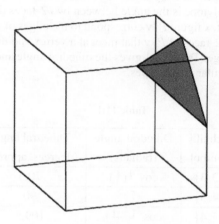

Figure 11.12   The shaded triangle is the vertex figure of the cube. It is the intersection of the cube with the hyperplane containing the midpoints of the three edges incident with the top right front vertex of the cube.

If P is a regular 4-polytope, the 2-faces of $P/\mathbf{v}$ are the vertex figures of the 3-faces of P that meet at $\mathbf{v}$. The vertex figure of $\{p,q\}$ is a q-gon, since q 2-faces meet at a vertex $\mathbf{v}$ of $\{p,q\}$, and each 2-face containing $\mathbf{v}$ corresponds to an edge of the vertex figure. Thus if the facets of P are $\{p,q\}$'s, then its vertex figures are $\{q,r\}$'s. The Schläfli symbol for such a 4-polytope is $\{p,q,r\}$.

Proposition 11.3.1 *There are six regular 4-polytopes.*   □

The rest of this section is devoted to constructing the six regular 4-polytopes and proving they are the only ones possible.

Two regular 4-polytopes are familiar already. The 4-cube C^4 with coordinates given in Section 11.1 has 8 facets given by $\mathbf{x_i} = \pm 1$, and each facet is a regular 3-cube. The 16 vertex figures are all regular tetrahedra. (See Exercise 3.) It is $\{4,3,3\}$.

If the vertices of the 4-simplex T^4 are $(1,0,0,0), (0,1,0,0), (0,0,1,0), (0,0,0,1)$, and $\frac{1-\sqrt{5}}{4} \cdot (1,1,1,1)$, then T^4 is regular. It has 5 facets, each a regular tetrahedron, and 5 vertex figures, each also a regular tetrahedron. (See Exercise 4.) It is $\{3,3,3\}$.

Dual to C^d is the *crosspolytope* β^d, which is the convex hull of $\pm e_i$, $1 \leq i \leq d$, where e_i is the standard basis vector with 1 in the ith coordinate and 0 elsewhere. If $d = 3$, this is the octahedron. The 16 facets of β^4 are regular tetrahedra, and the vertex figures are regular octahedra. (See Exercise 6.) It is $\{3,3,4\}$.

Euclid's proof that only five choices of $\{p,q\}$ correspond to regular 3-polytopes used the fact that the sum of the angles of the p-gons meeting at a vertex must be less than 2π. The analog for a regular 4-polytope P is that the sum of the dihedral angles of the 3-polytopes meeting at an edge must be less than 2π. (The dihedral angle of a regular 3-polytope is the angle between two 2-faces that meet at an edge.) The vertices of the vertex figure $P/\mathbf{v}$ correspond to the edges of P that meet at $\mathbf{v}$, so r is both the number of 2-faces of $P/\mathbf{v}$ that meet at a vertex and the number of 3-faces of P that meet at an edge. Table 11.1 gives the dihedral angle and the possible values of r for the each of regular 3-polytopes.

Table 11.1

Regular 3-polytope	Schläfli symbol	Dihedral angle (radians)	Dihedral angle (degrees, approx.)	Possible values of r
tetrahedron	$\{3,3\}$	$\cos^{-1}(\frac{1}{3})$	70.5	3, 4, 5
cube	$\{4,3\}$	$\frac{\pi}{2}$	90	3
octahedron	$\{3,4\}$	$\cos^{-1}(\frac{-1}{3})$	109.5	3
dodecahedron	$\{5,3\}$	$\cos^{-1}(\frac{-\sqrt{5}}{5})$	116.6	3
icosahedron	$\{3,5\}$	$\cos^{-1}(\frac{-\sqrt{5}}{3})$	138.2	(none)

We have accounted for $\{3,3,3\}$, $\{4,3,3\}$, and $\{3,3,4\}$, while $\{3,4,3\}$, $\{3,3,5\}$, and $\{5,3,3\}$ remain to be described.

The 4-polytope $\{3,4,3\}$ is known as the *24-cell*. It can be constructed by truncating each vertex of β^4 by cutting with the hyperplane through the midpoint of each edge incident with that vertex. This is the hyperplane used to define the vertex figure, so slicing off the 8 vertices creates 8 octahedra. In the process, each of the 16 original tetrahedral facets of β^4 is cut as well. As shown in Figure 11.13, the result of cutting off each vertex of a tetrahedron at the midpoints of its edges is also an octahedron. Thus the 24-cell has 24 octahedral facets, which matches the initial $\{3,4\}$ of $\{3,4,3\}$. The 2-faces of the vertex figure of the 24-cell are the vertex figures of the octahedral facets of the 24-cell, which means they are squares. Each edge of the 24-cell is created by the truncation of β^4. As shown in Figure 11.13, each edge of the 24-cell was originally in the interior of a triangular 2-face of β^4. That 2-face is incident with 2 tetrahedral facets of β^4, and thus an edge of the 24-cell is incident with the two octahedra created by truncating those two facets of β^4 and with one additional octahedron that is a vertex figure of β^4. Therefore an edge of the 24-cell is incident with 3 octahedra, and the 24-cell is indeed $\{3,4,3\}$.

The remaining two regular 4-polytopes are more difficult to construct, so we simply give coordinates. The 4-polytope $\{3,3,5\}$ is called the *600-cell*. It has 600 tetrahedral facets and 120 vertices. Eight of its vertices can be given by the permutations of $(\pm2,0,0,0)$ and another 16 by $(\pm1,\pm1,\pm1,\pm1)$. The remaining 96 vertices are the even permutations of $(\pm\phi,\pm1,\pm\frac{1}{\phi},0)$, where ϕ is the golden ratio. (An even permutation is the result of an even number of swaps of pairs of elements. For example, *badc* and *bcad* are even permutations of *abcd*, while *bacd* is not.)

The 4-polytope $\{5,3,3\}$ is called the *120-cell* and is dual to the 600-cell. It has 120 dodecahedral facets and 600 vertices. Twenty-four of its vertices are given by the permutations of $(\pm2,\pm2,0,0)$, another 64 by the permutations of $(\pm\sqrt{5},\pm1,\pm1,\pm1)$, another 64 by the permutations of $(\pm\phi,\pm\phi,\pm\phi,\pm\frac{1}{\phi^2})$, and another 64 by the permutations of $(\pm\phi^2,\pm\frac{1}{\phi},\pm\frac{1}{\phi},\pm\frac{1}{\phi})$. The remaining vertices are the 96 even permutations of $(\pm\phi,\pm\frac{1}{\phi^2},\pm1,0)$, the 96 even permutations of $(\pm\sqrt{5},\pm\frac{1}{\phi},\pm\phi,0)$, and the 192 even permutations of $(\pm2,\pm1,\pm\phi,\pm\frac{1}{\phi})$.

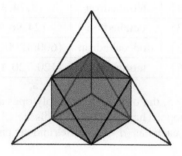

Figure 11.13 The shaded polytope is the octahedron created by cutting off the vertices of a tetrahedron. The hyperplane cutting off a vertex goes through the midpoints of the edges incident with that vertex.

So far we have used v, e, and f to represent the number of vertices, edges, and 2-faces of a 3-polytope. Once mathematicians began looking at dimensions 4 and beyond, it became more common to use f_k to denote the number of k-faces of a polytope. The vector $(f_0, f_1, \ldots, f_{d-1})$ is the *f-vector* of a d-polytope and will be studied in more detail in the next section. Given the number of facets f_3 of a regular

4-polytope $\{p,q,r\}$, it is possible to determine the rest of the f-vector. For example, the 4-simplex T^4 has 5 3-faces. Because each 2-face is in 2 3-faces and each of the 5 3-faces has 4 2-faces, $2f_2 = 5 \cdot 4$ and $f_2 = 10$. Because $r = 3$, each edge is incident with 3 tetrahedra. Each tetrahedron has 6 edges and there are 5 tetrahedra, so $3f_1 = 5 \cdot 6$ and $f_1 = 10$. Because the vertex figure of T^4 is a tetrahedron with 4 2-faces, each vertex of T^4 is incident with 4 3-faces. Each of the five 3-faces has 4 vertices, so $4f_0 = 5 \cdot 4$ and $f_0 = 5$.

The f-vectors of the remaining five regular 4-polytopes are computed in Exercises 7 – 11. Table 11.2 summarizes our information about the regular 4-polytopes. Just as $\{p,q\}$ is dual to $\{q,p\}$, $\{p,q,r\}$ is dual to $\{r,q,p\}$.

Table 11.2

Regular 4-polytope	Schläfli symbol	Type of facet	f-vector	Dual polytope
simplex	$\{3,3,3\}$	tetrahedron	$(5,10,10,5)$	simplex
cube	$\{4,3,3\}$	3-cube	$(16,32,24,8)$	cross-polytope
crosspolytope	$\{3,3,4\}$	tetrahedron	$(8,24,32,16)$	cube
24-cell	$\{3,4,3\}$	octahedron	$(24,96,96,24)$	24-cell
120-cell	$\{5,3,3\}$	dodecahedron	$(600,1200,720,120)$	600-cell
600-cell	$\{3,3,5\}$	tetrahedron	$(120,720,1200,600)$	120-cell

It turns out that for $d \geq 5$, the only regular d-polytopes are the d-simplex, the d-cube, and the d-crosspolytope. Their Schläfli symbols are $\{3,3,\ldots,3\}$, $\{4,3,3,\ldots,3\}$, and $\{3,\ldots,3,3,4\}$, where each has $(d-1)$ entries and the entries indicated by $\ldots$ are all 3.

While we have been able to describe the regular 4-polytopes, it would be nice to actually see them. One approach is through a Schlegel diagram. In the previous section, we saw that a Schlegel diagram of a 3-polytope is a projection of the polytope onto one of its facets. The Schlegel diagram is 2-dimensional, and facets of the 3-polytope correspond to polygons of the Schlegel diagram. Following the definition in the previous section, a Schlegel diagram of a 4-polytope is a projection of the polytope onto one of its facets. The Schlegel diagram is 3-dimensional, and the facets of the 4-polytope correspond to 3-polytopes in the Schlegel diagram. In the next three figures, the Schlegel diagrams of T^3, C^3, and β^3 will help us understand the Schlegel diagrams of T^4, C^4, and β^4.

Figure 11.14 shows Schlegel diagrams of the 3-simplex and the 4-simplex. The Schlegel diagram of T^3 looks like a triangle with a point in its interior connected to the vertices of the triangle. But because we are used to perspective drawing, we can also see the diagram as a tetrahedron with one bold triangular face close to us and one vertex further away in a direction perpendicular to the plane of the bold triangle.

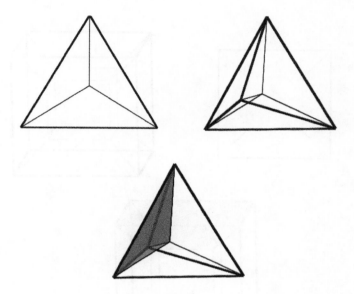

Figure 11.14   Schlegel diagrams of the 3-simplex and the 4-simplex. In the bottom diagram, one of the 3-faces of the 4-simplex is shaded.

The four triangular faces of T^3 are the outer bold triangle and the three triangles created by connecting the central vertex to an edge of the bold triangle.

Similarly, in the Schlegel diagram of T^4 we can think of the bold tetrahedron as close to us and the point in the middle as further away in a direction perpendicular to the 3-dimensional affine hull of the bold tetrahedron. The five tetrahedral faces of T^4 are the bold tetrahedron and the four tetrahedra created by connecting the central vertex to a triangular face of the bold tetrahedron. One of these four tetrahedra is shaded in the third diagram of Figure 11.14.

The Schlegel diagram of the 3-cube C^3 (Figure 11.15) looks like a square inside a square, with corresponding corners connected. Thinking of it as a perspective drawing, we see the smaller square as further away than the larger bold square. The square facets of C^3 are the outer square, the inner square, and the four squares created by connecting an edge of the bold square to the corresponding edge of the smaller square. These four squares are distorted in the 2-dimensional Schlegel diagram, but not in the 3-dimensional polytope.

Similarly, the Schlegel diagram of C^4 looks like a cube inside a cube, with corresponding corners connected. We can think of the smaller cube as further away in a direction perpendicular to the 3-dimensional affine hull of the bold 3-cube. The eight facets of C^4 are the outer bold cube, the inner cube, and the six cubes created by connecting a 2-face of the bold cube to the corresponding 2-face of the smaller

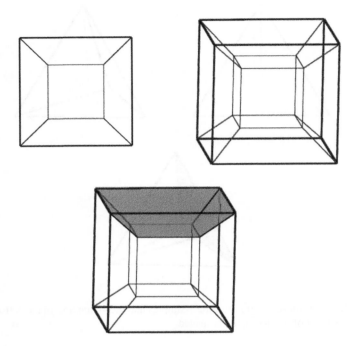

Figure 11.15   Schlegel diagrams of the 3-cube and the 4-cube. In the bottom diagram, one of the 3-faces of the 4-cube is shaded.

cube. One of these six cubes is shaded in the third diagram of Figure 11.15. It is distorted in the Schlegel diagram, but not in the actual polytope.

The Schlegel diagram of the octahedron β^3 (Figure 11.16) looks like a rotated triangle inside another triangle with each vertex of the outer triangle connected to the two closest vertices of the inner triangle. The triangular facets of β^3 are the outer triangle, the inner triangle, the three triangles created by connecting an edge of the outer triangle to the closest vertex of the inner triangle, and the three triangles created by connecting an edge of the inner triangle to the closest vertex of the outer triangle.

Similarly, the Schlegel diagram of β^4 looks like a rotated tetrahedron (shown in grey in Figure 11.16) inside another tetrahedron. Each vertex of the outer tetrahedron is connected to the three closest vertices of the inner tetrahedron. The 16 facets of β^4 are the outer tetrahedron, the inner tetrahedron, the four tetrahedra created by connecting a triangle of the outer tetrahedron to the closest vertex of the inner tetrahedron, the six tetrahedra created by connecting an edge of the outer tetrahedron to the closest edge of the inner tetrahedron, and the four tetrahedra created by connecting a triangle of the inner tetrahedron to the closest vertex of the outer tetrahedron. One tetrahedron of this last type is shaded in the third diagram of Figure 11.16.

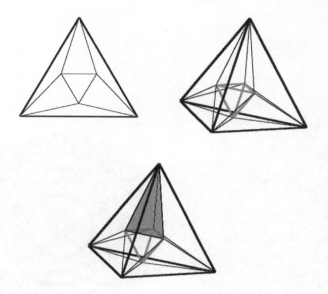

Figure 11.16   Schlegel diagrams of the octahedron and the 4-dimensional crosspolytope. In the bottom diagram, one of the 3-faces of the crosspolytope is shaded.

It is possible to draw Schlegel diagrams of the remaining regular 4-polytopes as well. Because they are quite complicated, we will instead draw the graphs of their vertices and edges in Figure 11.17. In the 24-cell, each vertex has degree 8, since each vertex figure is a 3-cube, which has 8 vertices. In the 120-cell, each vertex has degree 4, since each vertex figure is a tetrahedron. In the 600-cell, each vertex has degree 12, since each vertex figure is an icosahedron.

Finally, it is impossible to write about regular polytopes without mentioning H. S. M. Coxeter's classic text *Regular Polytopes*, which is the best source for further information about these objects.

Exercises 11.3

1. Explain why the i-faces of $P/\mathbf{v}$ correspond to the $(i+1)$-faces of P that contain $\mathbf{v}$. Use this to prove that if H_1 and H_2 are two hyperplanes that separate $\mathbf{v}$ from all the other vertices of P, the vertex figure created by H_1 and the vertex figure created by H_2 are combinatorially equivalent.

2. What are the vertex figures of a pyramid with a square base?

3. Confirm that the vertex figure of C^4 at the vertex $(1,1,1,1)$ is a regular tetrahedron.

4. Confirm that the vertex figure of T^4 at the vertex $(1,0,0,0)$ is a regular tetrahedron.

5. What geometric operation creates β^d from β^{d-1}?

Figure 11.17   Graphs of the 24-cell, the 120-cell, and the 600-cell, created by Tom Ruen and available at http://en.wikipedia.org/wiki/File:24-cell_graph_F4.svg, ../File:120-cell_graph_H4.svg., and ../File:600-cell_t0_p20.svg.

6. Confirm that the vertex figure of β^4 at the vertex $(0,0,0,1)$ is a regular octahedron. Exercise 5 may be useful.

7. Given that the 4-cube has 8 facets and Schläfli symbol $\{4,3,3\}$, find its f-vector.

8. Given that the crosspolytope β^4 has 16 facets and Schläfli symbol $\{3,3,4\}$, find its f-vector.

9. Given that the 24-cell has 24 facets and Schläfli symbol $\{3,4,3\}$, find its f-vector.

10. Given that the 120-cell has 120 facets and Schläfli symbol $\{5,3,3\}$, find its f-vector.

11. Given that the 600-cell has 600 facets and Schläfli symbol $\{3,3,5\}$, find its f-vector.

12. Draw two Schlegel diagrams of a prism over a tetrahedron, one with a tetrahedron as the large facet and one with a prism over a triangle as the large facet.

13. Draw a Schlegel diagram of a pyramid over a pyramid with a square base.

11.4 Enumerating Faces

We know the Euler relation for 3-polytopes. What else can we say about the number of faces of a polytope? Because we will eventually consider dimensions beyond three, we will use the f-vector introduced in the previous section, rather than v, e, and f.

In addition to the Euler relation, the f-vectors of 3-polytopes satisfy the inequalities of Proposition 11.2.3. Surprisingly, these restrictions completely characterize the face counts of 3-polytopes.

Theorem 11.4.1 (Steinitz) *The integer vector* (f_0, f_1, f_2) *is the f-vector of a 3-polytope if and only if it satisfies (i)* $f_0 - f_1 + f_2 = 2$, *(ii)* $f_0 \leq 2f_2 - 4$, *(iii)* $f_2 \leq 2f_0 - 4$.

PROOF: The inequalities are the inequalities of Proposition 11.2.3 rewritten using the Euler relation to eliminate f_1. Rewriting the inequalities this way makes it easier to see how to construct a 3-polytope with a given f-vector. A pyramid over an n-gon has $n + 1$ vertices, $2n$ edges, and $n + 1$ faces, so all of the integer points on the dashed line $f_0 = f_2$ of Figure 11.18 correspond to 3-polytopes. The n vertices of the n-gon all have degree 3, and all faces except the n-gon are triangles. To construct 3-polytopes corresponding to the other integer points in the cone of Figure 11.18 we use two operations. First, cutting off a vertex of degree 3 increases the number of vertices by 2 and the number of 2-faces by 1. (The polytope in Figure 11.5 is the result of cutting off two vertices of a tetrahedron.) The new vertices have degree 3, so this operation can be repeated. Second, the stellar subdivision of a 3-polytope at a triangular facet increases the number of vertices by 1 and the number of 2-faces by 2. The new 2-faces are triangular, so the operation can be repeated. Any point in the cone in Figure 11.18 can be reached by starting at a point $(n + 1, n + 1)$ for $n \geq 3$ and then taking steps of the form $(2, 1)$ or $(1, 2)$ along the grey lines.   Q.E.D.

A simplicial 3-polytope satisfies $3f_2 = 2f_1$, since each 2-face has three edges and each edge is in two 2-faces. Thus if a 3-polytope is simplicial, its f-vector is on the

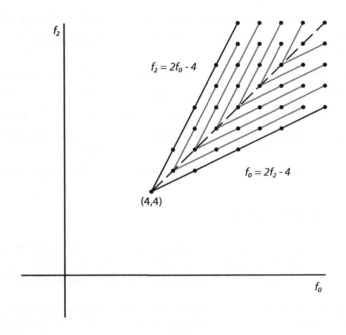

Figure 11.18 The integer points in this cone correspond to f-vectors of 3-polytopes.

line $f_0 = 2f_2 - 4$. Since a 2-face must have at least 3 edges, this is an if and only if condition. Dually, a polytope satisfies $f_2 = 2f_0 - 4$ if and only if it is simple.

There is no equivalent of Theorem 11.4.1 for dimensions 4 and higher. Characterizing the f-vector of d-polytopes is still a big open problem. However, there has been much progress on many pieces of this problem. First, there is the higher-dimensional version of the Euler relation.

Theorem 11.4.2 (Euler, Poincaré) *If P is a 4-polytope, its f-vector satisfies $f_0 - f_1 + f_2 - f_3 = 0$. More generally, if P is a d-polytope, its f-vector satisfies $f_0 - f_1 + f_2 - \cdots + (-1)^{d-1} f_{d-1} = 1 - (-1)^d$.* $\square$

Before giving a formal proof for dimension 4, we will work through the general idea in dimension 3. Let P be a 3-polytope. The plan is to track the changes in the quantity $f_0 - f_1 + f_2$ while building up the boundary of P facet by facet. The first facet (any one will work) is an n-gon for some n, so it contributes $n - n + 1 = 1$ to $f_0 - f_1 + f_2$. The next facet, an m-gon, is chosen so that it intersects the first facet in an edge. Two of its vertices and one of its edges were already counted when the first facet was added, so the new facet contributes $(m-2)-(m-1)+1=0$ to $f_0 - f_1 + f_2$. At every step (and this is the hard part), a new facet is chosen so that it intersects the set of previously chosen facets in a connected sequence of edges which is homeomorphic to a line segment. It will contribute 0 to $f_0 - f_1 + f_2$. The last facet of course cannot be chosen this way. All of its edges and vertices are already present in previously chosen facets, so it contributes $0 - 0 + 1$ to $f_0 - f_1 + f_2$, and thus $f_0 - f_1 + f_2 = 2$.

Poincaré proved Theorem 11.4.2 in the 1890s using algebraic topology. A proof using only the language of polytopes took much longer. L. Schläfli and others gave proofs of the Euler–Poincaré relation that assumed that the facets of any polytope could be ordered so that each facet intersected the previous facets in a particularly nice way, as in the dimension 3 argument above. In the language of Chapter 9, the intersection should be simply connected. Schläfli's work was done in the 1850s and published in 1901. The proof that such an ordering is possible did not come until 1971 in a paper by H. Bruggesser and P. Mani.

Here is how they found that ordering. Let P be a d-polytope, and let l be an oriented line that passes through the interior of P, intersects each facet-defining hyperplane of P exactly once, and intersects distinct facet-defining hyperplanes in distinct points.

Bruggesser and Mani order the facets by starting in the interior of P, moving in the positive direction along l, and recording the facets in the order in which l intersects their supporting hyperplanes. Then they start far on the negative side of l, past all the facet-defining hyperplanes of P, and again move in the positive direction along l, recording the facets in the order in which l intersects their supporting hyperplanes until they return to the interior of P. See Figure 11.19. This ordering of the facets is a *line shelling* of P.

A line shelling of a polytope gives exactly the kind of ordering of the facets needed to prove the Euler–Poincaré relation. We are going to build up to the Euler–Poincaré relation in dimension 4 by working through the lower dimensions. To do that, we want to understand how a line shelling of P induces a line shelling on some of its facets.

Say a point $\mathbf{z}$ outside of P is on the positive side of the line l. Then the facets whose supporting hyperplanes intersect l before $\mathbf{z}$ are exactly those facets such that $\mathbf{z}$ and the interior of P are on opposite sides of the supporting hyperplane. These are the facets that are *visible* from $\mathbf{z}$, and they form the beginning of a line shelling. Figure 11.20 shows the same line shelling as Figure 11.19. The facets F_1 and F_2 are visible from $\mathbf{z}_1$, and they are the first two facets of the line shelling. In addition, the facets not visible from $\mathbf{z}$ are also the beginning of a line shelling, namely the shelling that uses the opposite orientation of l. In Figure 11.20, the facets not visible from $\mathbf{z}_1$ are F_5, F_4, and F_3, which are the first three facets in the shelling using the opposite orientation of the line.

Conversely, given a point $\mathbf{z}$ outside of P, any line l through $\mathbf{z}$ that goes through the interior of P, is oriented so $\mathbf{z}$ is on the positive side, and intersects the facet-defining hyperplanes appropriately, gives a line shelling that begins with exactly the facets visible from $\mathbf{z}$.

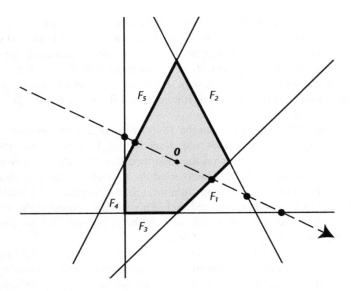

Figure 11.19   A line shelling of a pentagon. Starting at the origin and moving in the positive direction, the line first intersects the hyperplane defining F_1 and then the hyperplanes defining F_2 and F_3. Starting on the negative side of the line far away from the pentagon, the line intersects the hyperplane defining F_4 and finally the hyperplane defining F_5 before returning to the origin.

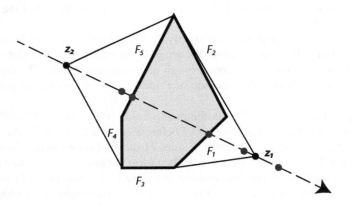

Figure 11.20   This is the line shelling from Figure 11.19 shown without the supporting hyperplanes. The intersection points of the dashed line and the supporting hyperplanes are shown in grey. Facets F_1 and F_2 are visible from the point $\mathbf{z}_1$, and they are the first two facets of this line shelling. Facets F_5 and F_4 are visible from the point $\mathbf{z}_2$, and they are the first two facets of the line shelling given by the opposite orientation of the dashed line.

Proposition 11.4.3 *Let P be a d-polytope, and let $F_1, F_2, \ldots, F_n$ be a line shelling of P by the line l. Then for any F_j satisfying $2 \leq j \leq n - 1$, the intersection of F_j with the previous facets $F_1, F_2, \ldots, F_{j-1}$ is the beginning of a line shelling of F_j.*

PROOF: Let H_j be the hyperplane that defines the facet F_j and let **z** be the intersection point of H_j and l. If **z** is on the positive side of l, then the faces of F_j that are already contained in $F_1, F_2, \ldots, F_{j-1}$ are exactly the faces that are visible from **z** within H_j, and thus they are the beginning of a line shelling of F_j. If **z** is on the negative side of l, then the faces of F_j that are already contained in $F_1, F_2, \ldots, F_{j-1}$ are exactly the faces that are not visible from **z** within H_j, and thus they are the beginning of a line shelling of F_j. See Figure 11.21.   Q.E.D.

It seems a bit excessive to use line shellings to prove that an n-gon has the same number of vertices as edges, but the proof of the next proposition sets the stage for similar arguments in dimensions 3 and 4.

Proposition 11.4.4 *Let P be an n-gon with f-vector (f_0, f_1), and let $F_1, F_2, \ldots, F_n$ be a line shelling of P by the line l. As the boundary of P is built up facet by facet, F_1 contributes 1 to $f_0 - f_1$, F_j contributes 0 to $f_0 - f_1$ for $2 \leq j \leq n - 1$, and F_n contributes -1 to $f_0 - f_1$. As a consequence, $f_0 - f_1 = 0$.*

PROOF: The facet F_1 is an edge and has 2 vertices, so it contributes $2 - 1 = 1$ to $f_0 - f_1$. If $2 \leq j \leq n - 1$, then by Proposition 11.4.3, the intersection of F_j with $F_1 \cup F_2 \cup \cdots \cup F_{j-1}$ is the beginning of a line shelling of F_j. Since F_j is an edge, this intersection must be a single vertex of F_j, and this vertex was already counted in the addition of a previous facet. Thus the addition of F_j contributes one new edge and one new vertex, and therefore contributes 0 to $f_0 - f_1$. Because F_n is the last facet added, its two vertices are already present in $F_1 \cup F_2 \cup \cdots \cup F_{n-1}$. Only the edge F_n itself is new, so it contributes -1 to $f_0 - f_1$. As a consequence, $f_0 - f_1 = 1 + 0 + 0 + \cdots + 0 + (-1) = 0$.   Q.E.D.

The Euler relation for 3-polytopes can also be proven using line shellings.

Proposition 11.4.5 *Let P be a 3-polytope with f-vector (f_0, f_1, f_2), and let $F_1, F_2, \ldots, F_n$ be a line shelling of P by the line l. As the boundary of P is built up facet by facet, F_1 contributes 1 to $f_0 - f_1 + f_2$, F_j contributes 0 to $f_0 - f_1 + f_2$ for $2 \leq j \leq n - 1$, and F_n contributes 1 to $f_0 - f_1 + f_2$. As a consequence, $f_0 - f_1 + f_2 = 2$.*

PROOF: The facet F_1 is an m-gon for some m. By Proposition 11.4.4, it contributes $m - m + 1 = 1$ to $f_0 - f_1 + f_2$. If $2 \leq j \leq n - 1$, then by Proposition 11.4.3, the intersection of F_j with $F_1 \cup F_2 \cup \cdots \cup F_{j-1}$ is the beginning of a line shelling of F_j. By Proposition 11.4.4, the beginning of a line shelling of F_j has one more vertex than edge, and these vertices and edges were counted in the addition of previous facets. Because F_j has the same number of vertices and edges, the part of the boundary of F_j that does not intersect the previous facets must have one more edge than vertex, and it contributes -1 to $f_0 - f_1$. Because F_j itself is a 2-face of P, it contributes 1 to f_2, and the addition of F_j contributes $-1 + 1 = 0$ to $f_0 - f_1 + f_2$. Because F_n is the last facet added, all its vertices and edges are already present in $F_1 \cup F_2 \cup \cdots \cup F_{n-1}$.

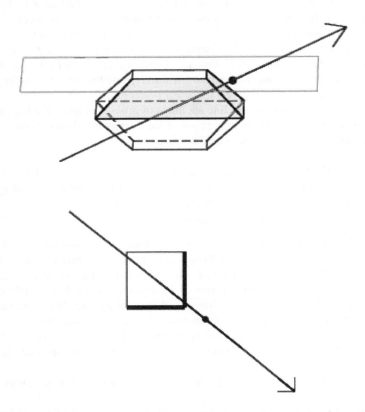

Figure 11.21   The upper figure is the beginning of a line shelling of a 3-polytope P. The two shaded faces start the shelling. The top right face is first, followed by the top front face. The next face will be the top face. Its supporting hyperplane is indicated in grey, and the intersection of the line with the supporting hyperplane is marked. The lower figure is the top face within its supporting hyperplane. The two thicker edges of the square are the two that belong to the two previous (shaded) facets in the shelling of P, and they are also the beginning of a line shelling of the square. Here the line happens to be the projection of the line from the shelling of P. The intersection point of the line and the hyperplane is marked again, and the two thicker edges of the square are exactly the facets of the square visible from this point.

Only the 2-face F_n itself is new, so it contributes 1 to $f_0 - f_1 + f_2$. As a consequence, $f_0 - f_1 + f_2 = 2$.   Q.E.D.

Finally, the same type of argument proves the Euler–Poincaré relation for 4-polytopes.

Proposition 11.4.6 *Let P be a 4-polytope with f-vector (f_0, f_1, f_2, f_3), and let $F_1, F_2,$ $\ldots, F_n$ be a line shelling of P by the line l. As the boundary of P is built up facet by facet, F_1 contributes 1 to $f_0 - f_1 + f_2 - f_3$, F_j contributes 0 for $2 \leq j \leq n - 1$, and F_n contributes -1. As a consequence, $f_0 - f_1 + f_2 - f_3 = 0$.*

PROOF: The facet F_1 is a 3-polytope. By Proposition 11.4.5, it contributes 2 to $f_0 - f_1 + f_2$. It also contributes 1 to f_3, so its total contribution to $f_0 - f_1 + f_2 - f_3$ is $2 - 1 = 1$. If $2 \leq j \leq n - 1$, then by Proposition 11.4.3, the intersection of F_j with $F_1 \cup F_2 \cup \cdots \cup F_{j-1}$ is the beginning of a line shelling of F_j. By Proposition 11.4.5, if the facets forming the beginning of a line shelling of F_j contain v vertices, e edges, and f 2-faces, then $v - e + f = 1$. These faces were already counted in the addition of facets before F_j. Because F_j satisfies the Euler relation for 3-polytopes, the part of the boundary of F_j that does not intersect the previous facets must contribute 1 to $f_0 - f_1 + f_2$. Because F_j itself is a 3-face of P, it contributes 1 to f_3, and the addition of F_j contributes $1 - 1 = 0$ to $f_0 - f_1 + f_2 - f_3$. Because F_n is the last facet added, all its vertices, edges, and 2-faces are already present in $F_1 \cup F_2 \cup \cdots \cup F_{n-1}$. Only the 3-face F_n itself is new, so it contributes -1 to $f_0 - f_1 + f_2 - f_3$. As a consequence, $f_0 - f_1 + f_2 - f_3 = 0$.   Q.E.D.

The proofs of Propositions 11.4.4 – 11.4.6 have exactly the same structure, and the proof of Proposition 11.4.5 relies on Proposition 11.4.4, while the proof of Proposition 11.4.6 relies on Proposition 11.4.5. This sounds like induction, and in fact, it is possible to construct a proof by induction that proves the Euler–Poincaré relation for all values of d.

The Euler–Poincaré relation is the only linear equation satisfied by the f-vectors of d-polytopes. Figure 11.18 makes this clear for $d = 3$, since the affine hull of the points (f_0, f_2) is two-dimensional. The case $d = 4$ is Exercise 13.

In dimension 3, the f-vectors of simplicial and simple polytopes satisfy additional linear equations not satisfied by all 3-polytopes. The same is true in dimension 4. In a simplicial 4-polytope, every 3-face has 4 2-faces, and every 2-face is contained in 2 3-faces, so $2f_2 = 4f_3$. Thus $f_3 = f_0 - f_1 + f_2$ and $f_2 = 2f_1 - 2f_0$, and the f-vector of a simplicial 4-polytope is determined by f_0 and f_1. The space of f-vectors of simplicial 4-polytopes is 2-dimensional. (See Exercise 14). M. Dehn ($d = 5$) and D. Sommerville ($d > 5$) generalized this reasoning to show that the affine hull of f-vectors of simplicial d-polytopes is $\lfloor \frac{d}{2} \rfloor$-dimensional.

In fact, the f-vectors of simplicial (and therefore also simple) d-polytopes have been completely characterized. The characterization was proposed by P. McMullen and proved necessary by R. Stanley and sufficient by L. J. Billera and C. W. Lee. The details of the characterization are too much to include here, but we can at least state (although not prove) two theorems that preceded the characterization. These theorems give us a chance to introduce two more families of polytopes.

A simplicial d-polytope is *stacked* if it is the result of starting with a d-simplex and repeatedly applying stellar subdivision. (The subdivision does not have to be at one of the facets of the original d-simplex, but may be at a facet created by a previous subdivision.) Figure 11.22 shows a stacked 3-polytope created by stellar subdivision at two facets of a tetrahedron. Each stellar subdivision has the effect of gluing a new d-simplex to the previous stacked polytope. It adds one new vertex and only increases f_{d-1} by $d - 1$, since one old facet is removed and d new ones are added. This suggests that among all simplicial polytopes with a given number of vertices, a stacked polytope should have the fewest facets. D. Barnette proved the following Lower Bound Theorem. Here $\binom{d}{k}$ is the binomial coefficient $\frac{d!}{k!(d-k)!}$, the number of ways to choose a subset of k objects from a set of d objects.

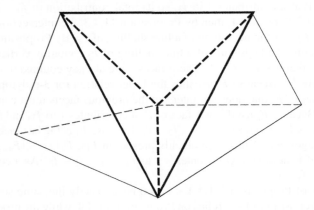

Figure 11.22   A stacked 3-polytope. The edges of the original tetrahedron are drawn in bold. The remaining edges come from stellar subdivision at two of the facets of the original tetrahedron.

Theorem 11.4.7 (D. Barnette) *Let P be a simplicial d-polytope. Then*

$$f_k \geq \binom{d}{k}f_0 - \binom{d+1}{k+1}k$$

for all $1 \leq k \leq d - 2$ and

$$f_{d-1} \geq (d-1)f_0 - (d+1)(d-2).$$

Furthermore, if $d \geq 4$ and equality holds for some f_k, then P must be stacked. $\square$

The *cyclic polytope* $C_d(n)$ is the convex hull of n distinct points on the moment curve $(t, t^2, t^3, \ldots, t^d)$, where $d \geq 2$ and $n \geq d + 1$. Different choices of the n points give combinatorially equivalent polytopes. In particular, the n points are all vertices and $C_d(n)$ is simplicial. For $d \geq 4$ any pair of vertices forms an edge, so the graph of $C_d(n)$ is K_n. More generally, any subset of k vertices forms a $(k - 1)$-face for

$1 \leq k \leq \frac{d}{2}$. Because $C_d(n)$ has as many k-faces as possible for a d-polytope with n vertices for $1 \leq k \leq \frac{d}{2}$, it seems plausible that $C_d(n)$ might have as many k-faces as possible for all k, and in fact it does. The following is the Upper Bound Theorem.

Theorem 11.4.8 (McMullen) *Let P be a d-polytope with n vertices. Then $f_k(P) \leq f_k(C_d(n))$ for $1 \leq k \leq d-1$.* $\qquad\square$

The Upper Bound Theorem applies to all d-polytopes. If a polytope P is not simplicial, it is possible to show that a slight perturbation of its vertices will yield a new simplicial polytope with the same number of vertices as P and at least as many k-faces as P. Therefore, it was enough for McMullen to prove the theorem for simplicial polytopes.

We now turn to 4-polytopes. Pairs of entries of the f-vector are understood.

Theorem 11.4.9 (Grünbaum) *There is a 4-polytope with f_0 vertices and f_3 3-faces if and only if $5 \leq f_0 \leq \frac{1}{2}f_3(f_3-3)$ and $5 \leq f_3 \leq \frac{1}{2}f_0(f_0-3)$.* $\qquad\square$

Theorem 11.4.10 (Grünbaum) *There is a 4-polytope with f_0 vertices and f_1 edges if and only if $10 \leq 2f_0 \leq f_1 \leq \frac{1}{2}f_0(f_0-1)$ and (f_0,f_1) is not $(6,12)$, $(7,14)$, $(8,17)$, or $(10,20)$.* $\qquad\square$

The necessity of these conditions is Exercises 17 and 18. The construction of a 4-polytope with given values of f_0 and f_3 (or f_0 and f_1) is in the same spirit as the proof of Theorem 11.4.1. It begins with an appropriate 4-polytope (the 4-simplex, a cyclic 4-polytope, or a pyramid over one of several 3-polytopes) and then proceeds by either stellar subdivision or truncation of vertices. The pairs (f_0,f_2) and (f_1,f_2) have been characterized by Barnette and Reay, and duality takes care of the other pairs.

Many of the proofs about f-vectors have not counted faces individually, but instead have counted the number of times a face of some dimension is incident with a face of some other dimension. For example, the inequalities in Theorem 11.4.1 came from Corollary 2.4.4, which was proven by counting the edges within each 2-face and then summing over all 2-faces. Because each edge of a 3-polytope is contained in two facets, this sum is simply $2f_1$. Thus the incidences of faces of different dimensions give insight into the structure of a polytope. This information is encoded in the *flag f-vector*. We start with an example before giving the full definition. Let P be a 3-polytope with p_3 triangular facets, p_4 quadrilateral facets, p_5 pentagonal facets, and so forth. Then the sum $3p_3 + 4p_4 + 5p_5 + \cdots$ from the proof of Corollary 2.4.4 is f_{12}, the number of edge-(2-face) incidences of P. The exact same sum is also f_{02}, the number of vertex-(2-face) incidences of P, since of course every 2-face has the same number of vertices as edges. In general, if $S = \{i_1, i_2, \ldots, i_k\}$ is a subset of $\{0, 1, \ldots, d-1\}$, then f_S is the number of $(i_1\text{-face})$-$(i_2\text{-face})$-$\cdots$-$(i_k\text{-face})$ incidences in P. (Here we omit the brackets and commas from the subscript S when there is no danger of confusion; a 3-polytope has no 12-faces.) The flag f-vector has entries f_S, one for each subset S of $\{0, 1, \ldots, d-1\}$. If S contains only a single element, then f_S is also an entry of the f-vector.

For a 3-polytope P, the flag f-vector is determined by the ordinary f-vector. Because each edge has two vertices, $f_{01} = 2f_1$. This argument applies the Euler relation $f_0 = 2$ from dimension 1 to each 1-face of P. By applying $f_{01} = 2f_1$ to the dual of P, we get $f_{12} = 2f_1$. Applying the Euler relation $f_0 - f_1 = 0$ to all 2-faces of P gives $f_{12} = f_{02}$. See Exercise 19 for f_{012}. M. M. Bayer and L. J. Billera generalized this idea to d-polytopes. They applied the Euler relation to faces of all dimensions, to faces of vertex figures, and so forth, to find all linear equations satisfied by the flag f-vectors of d-polytopes. In particular, all elements of the flag f-vector of a 4-polytope are determined by f_0, f_1, f_2, and f_{02}. (In dimension 4, it is still true that $f_{02} = f_{12}$, but the equation $f_{12} = 2f_1$ from dimension 3 no longer holds. In dimension 4, the dual of $f_{01} = 2f_1$ is $f_{23} = 2f_3$.)

Here are the known linear inequalities satisfied by the flag f-vectors (f_0, f_1, f_2, f_{02}) of 4-polytopes:

1. $f_{02} - 3f_2 \geq 0$

2. $f_{02} - 3f_1 \geq 0$

3. $f_{02} - 3f_2 + f_1 - 4f_0 + 10 \geq 0$

4. $6f_1 - 6f_0 - f_{02} \geq 0$

5. $f_0 - 5 \geq 0$

6. $f_2 - f_1 + f_0 - 5 \geq 0$.

Inequality (1) holds because each 2-face has at least 3 vertices, and inequality (2) is its dual, using the fact that $f_{02} = f_{13}$ for 4-polytopes. G. Kalai proved inequality (3) with an argument about rigidity, the same concept used by Cauchy in the previous section. Bayer's proof of inequality (4) is Exercise 20. Inequalities (5) and (6) just say that a 4-polytope must have at least 5 vertices, and dually must have at least 5 facets. In dimension 3, the inequalities $f_0 - 4 \geq 0$ and $f_2 - 4 \geq 0$ were implied by the two inequalities in Theorem 11.4.1, but that is not the case in dimension 4.

The flag f-vectors of 4-polytopes also satisfy some nonlinear inequalities. We have already seen that $f_1 \leq \binom{f_0}{2}$ in Theorem 11.4.10. Bayer proved several other nonlinear inequalities, including a strengthening of $f_1 \leq \binom{f_0}{2}$ which is Exercise 21. Her work was generalized by J. Ling.

Returning to dimension 3, we recall that if P is a 3-polytope with p_3 triangular facets, p_4 quadrilateral facets, p_5 pentagonal facets, and so forth, then $3p_3 + 4p_4 + 5p_5 + \cdots = f_{12}$. We can ask, then, not only what f-vectors are possible for 3-polytopes, but what p-vectors $(p_3, p_4, p_5, p_6, \ldots)$ are possible. To answer this question, we also consider the v-vector $(v_3, v_4, v_5, v_6, \ldots)$ of P, where v_i is the number of vertices of degree i. The v-vector of P is the p-vector of P^Δ. Both the p-vector and the v-vector have only a finite number of nonzero entries. The Euler relation leads to the following results.

Proposition 11.4.11 *If P is a 3-polytope, then*

$$\sum_{i\geq 3}(6-i)p_i + 2\sum_{i\geq 3}(3-i)v_i = 12.$$

In particular, if P is simple, then

$$\sum_{i\geq 3}(6-i)p_i = 12.$$

PROOF: See Exercise 24. □

Proposition 11.4.12 *If P is a 3-polytope, then*

$$\sum_{i\geq 3}(4-i)(p_i+v_i) = 8.$$

PROOF: See Exercise 25. □

Because of the factors of $(6-i)$ and $(3-i)$, Proposition 11.4.11 does not say anything about the values of p_6 and v_3. Similarly, Proposition 11.4.12 does not say anything about p_4 and v_4. While the p-vectors and v-vectors of 3-polytopes have not been completely characterized, quite a bit is known. Eberhard's Theorem below is from 1891. Interest in the problem was revived by B. Grünbaum's classic 1967 text *Convex Polytopes*, and there has been much work on p-vectors and v-vectors since then. Here are some of the key results.

Theorem 11.4.13 (V. Eberhard) *Given values of p_i for $i = 3, 4, 5$ and $i \geq 7$ satisfying the second equation of Proposition 11.4.11, there exists a value of p_6 such that $(p_3, p_4, p_5, p_6, \ldots)$ is the p-vector of some simple 3-polytope P.* □

Theorem 11.4.14 (Grünbaum) *Given values of p_i and v_i for $i = 3$ and $i \geq 5$ satisfying the equation in Proposition 11.4.12 such that $\sum_i i p_i$ and $\sum_i i v_i$ are both even, there exist values of p_4 and v_4 such that $(p_3, p_4, p_5, p_6, \ldots)$ and $(v_3, v_4, v_5, v_6, \ldots)$ are the p-vector and v-vector of some 3-polytope P.* □

Theorem 11.4.15 (S. Jendrol') *Let p_i be given for $i = 3, 4, 5$ and $i \geq 7$ and v_i be given for $i \geq 4$ so that the first condition of Proposition 11.4.11 is satisfied. If $\sum_i p_i \neq 0$, where the sum is over all odd values of i, and $\sum_i v_i = 1$, where here the sum is over all i that are not divisible by 3, then there is no corresponding 3-polytope. Otherwise, there are infinitely many choices of p_6 and v_3 such that $(p_3, p_4, p_5, p_6, \ldots)$ and $(v_3, v_4, v_5, v_6, \ldots)$ are the p-vector and v-vector of some 3-polytope P.* □

These three theorems are proven by constructing planar 3-connected graphs with appropriate p-vectors and v-vectors; the constructions are too lengthy to include here.

Exercises 11.4

1. Using the constructions described in the proof of Theorem 11.4.1, describe a 3-polytope with $(f_0, f_2) = (6, 7)$ and another 3-polytope with $(f_0, f_2) = (8, 7)$.

2. What is the f-vector of a prism over an n-gon? Where are the corresponding points in Figure 11.18?

3. What is the f-vector of a bipyramid over an n-gon? Where are the corresponding points in Figure 11.18?

4. Find a linear equation satisfied by the f-vectors of cubical 3-polytopes.

5. Give a different line shelling of the pentagon in Figure 11.19.

6. Confirm that the regular 4-polytopes satisfy the Euler–Poincaré relation.

7. Given the f-vector of a polytope P, explain how to find the f-vector of the pyramid over P. Exercise 21 in Section 11.1 may help.

8. Given the f-vector of a polytope P, explain how to find the f-vector of the prism over P. Exercise 22 in Section 11.1 may help.

9. Given the f-vector of a polytope P, explain how to find the f-vector of the bipyramid over P. Exercise 23 in Section 11.1 may help.

Part (b) of each of Exercises 10 – 12 requires some familiarity with either recurrence relations or the binomial coefficients.

10. Find the f-vector of (a) the 5-simplex (b) the d-simplex. One approach is to use Exercise 7; other approaches are possible.

11. Find the f-vector of (a) the 5-cube (b) the d-cube. One approach is to use Exercise 8; other approaches are possible.

12. Find the f-vector of (a) the 5-crosspolytope (b) the d-crosspolytope. One approach is to use Exercise 9; other approaches are possible.

13. Find four 4-polytopes whose f-vectors show that the affine hull of the f-vectors of 4-polytopes is 3-dimensional.

14. Find three simplicial 4-polytopes whose f-vectors show that the affine hull of the f-vectors of simplicial 4-polytopes is 2-dimensional.

15. Prove that stacked 4-polytopes satisfy the conditions of the Lower Bound Theorem with equality.

16. Draw or describe two stacked 3-polytopes that have the same number of vertices but are not combinatorially equivalent.

17. Prove the necessity of the inequalities in Theorem 11.4.9.

18. Prove the necessity of the inequalities in Theorem 11.4.10.

19. Express f_{012} of a 3-polytope in terms of the entries of the f-vector.

20. Prove inequality (4) for 4-polytopes. Start with the fact that for any 3-polytope, $2f_1 \geq 3f_0$. In the course of the proof, you will probably want to explain why $f_{03} - f_{02} + f_{01} = 2f_0$ for 4-polytopes. You may also use the fact that $f_{13} = f_{02}$ for 4-polytopes without giving a proof.

21. Prove that $2(f_{02} - 3f_2) + f_1 \leq \binom{f_0}{2}$ for 4-polytopes by counting the pairs of vertices that form edges and the pairs of vertices that are in the same 2-face but do not form an edge.

22. Describe two 3-polytopes with the following p-vectors, which differ only in the value of p_6: $(0, 6, 0, 0)$ and $(0, 6, 0, 2)$. Both polytopes have $p_i = 0$ for $i \geq 7$.

23. Describe two 3-polytopes with the following p-vectors, which differ only in the value of p_6: $(6, 0, 0, 0)$ and $(6, 0, 0, 1)$. Both polytopes have $p_i = 0$ for $i \geq 7$.

24. Prove Proposition 11.4.11.

25. Prove Proposition 11.4.12.

APPENDIX A

CURVES

A.1 Parametrization of Curves and Arclength

For $n = 2$ or 3, a *smooth parametrized curve* is a subset of $\mathbb{R}^n$ of the form

$$\{\mathbf{C}(t) \mid t \in (a,b)\}$$

where $\mathbf{C}(t) : (a,b) \to \mathbb{R}^n$ is assumed to be at least twice differentiable, and $\mathbf{C}'(t) \neq 0$ for any t in (a,b). It is quite common to think of the function $\mathbf{C}(t)$ itself as the curve, and such will be the case here as well. Thus, $\mathbf{C}_1(t) = (t^2, \cos t)$ is a curve in $\mathbb{R}^2$, and $\mathbf{C}_2(t) = (\cos t, \sin t, t)$ is a curve in $\mathbb{R}^3$ for any domain (a,b), but $\mathbf{C}_3(t) = (t^2, |t|)$ and $\mathbf{C}_4(t) = (t^3, t^2)$ fail to be curves for any domain (a,b) that contains 0. Of course, both $\mathbf{C}_3(t)$ and $\mathbf{C}_4(t)$ are smooth parametrized curves over any domain (a,b) that does not contain 0. In general, the domain (a,b) will be left unspecified, and it will be assumed that every curve is in fact smooth and parametrized.

An *arc* of the curve $\mathbf{C}(t)$ is any portion that corresponds to a closed subinterval $[a,b]$ of the curve's domain. The length of such an arc is defined as

$$\int_a^b |\mathbf{C}'(t)| \, dt.$$

Introduction to Topology and Geometry, Second Edition.
By Saul Stahl and Catherine Stenson Copyright © 2013 John Wiley & Sons, Inc.

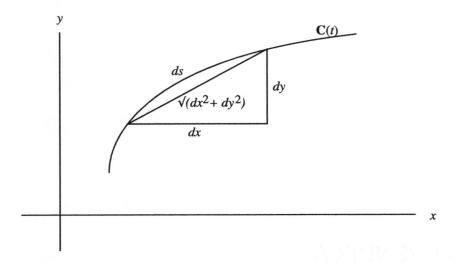

Figure A.1   Arclength.

This definition is justified by the following informal argument. In Figure A.1 let ds denote the length of a small arc of the curve $\mathbf{C}(t) = (x(t), y(t))$ in $\mathbb{R}^2$. Then

$$ds \approx \sqrt{dx^2 + dy^2} \approx \sqrt{[x'(t)dt]^2 + [y'(t)]^2}\, dt$$

$$= \sqrt{[x'(t)]^2 + [y'(t)]^2}\, dt = |\mathbf{C}'(t)|\, dt. \tag{1}$$

The justification of this definition for curves in $\mathbb{R}^3$ is relegated to Exercise 7.

Example A.1.1 Find the lengths of the arcs of the curves $\mathbf{C}_1(t) = (t, t^2, t^3)$ and $\mathbf{C}_2(t) = (\cos t, \sin t, t)$ for $1 \le t \le 2$.

These lengths are respectively

$$\int_1^2 \sqrt{1 + 4t^2 + 9t^4}\, dt = 7.70755\ldots$$

and

$$\int_1^2 \sqrt{\sin^2 t + \cos^2 t + 1}\, dt = \int_1^2 \sqrt{2}\, dt = \sqrt{2}.$$

Every curve of the form $y = f(x)$ can be converted into the parametrized curve $\mathbf{C}(t) = (t, f(t))$. Its arc over the interval $[c, d]$ has length

$$\int_c^d |\mathbf{C}'(t)|\, dt = \int_c^d \sqrt{1 + [f'(t)]^2}\, dt = \int_c^d \sqrt{1 + y'^2}\, dx.$$

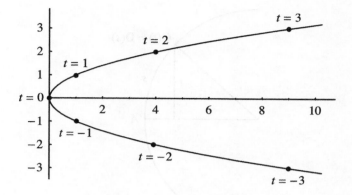

Figure A.2 A parametrized curve.

Theorem A.1.2 *Let $\mathbf{C}(t)$ be a curve. Then $\mathbf{C}'(t_0)$ is parallel to the tangent to $\mathbf{C}(t)$ at $\mathbf{C}(t_0)$, and $|\mathbf{C}'(t_0)|$ is the instantaneous rate with which this parametrization traces the curve at $\mathbf{C}(t_0)$.*

PROOF: By definition

$$\mathbf{C}'(t_0) = \lim_{\Delta t \to 0} \frac{\mathbf{C}(t_0 + \Delta t) - \mathbf{C}(t_0)}{\Delta t}.$$

Since $(1/\Delta t)[\mathbf{C}(t_0 + \Delta t) - \mathbf{C}(t_0)]$ is parallel to the chord from $\mathbf{C}(t_0)$ to $\mathbf{C}(t_0 + \Delta t)$, and since the tangent line at $\mathbf{C}(t_0)$ is the limiting position of this chord, it follows that $\mathbf{C}'(t_0)$ is parallel to the tangent at $\mathbf{C}(t_0)$.

Let Δs denote the length of the arc of $\mathbf{C}(t)$, $t_0 \le t \le t_0 + \Delta t$. The instantaneous speed with which the curve is traced out by the paramter t at $\mathbf{C}(t_0)$ is

$$\lim_{\Delta t \to 0} \frac{\Delta s}{\Delta t} = \lim_{\Delta t \to 0} \frac{\int_{t_0}^{t_0 + \Delta t} |\mathbf{C}'(t)|\, dt}{\Delta t},$$

which limit, by the Mean Value Theorem for integrals, has value $|\mathbf{C}'(t_0)|$.   Q.E.D.

Example A.1.3 The curve $\mathbf{C}(t) = (t^2, t)$ of Figure A.2 has the points corresponding to $t = -3, -2, -1, 0, 1, 2, 3$ marked on it. The different lengths of the arcs determined by these points are roughly proportional to the corresponding values of

$$|\mathbf{C}'(t)| = \sqrt{1 + 4t^2}.$$

The parameter s of the curve $\mathbf{C}(s)$ is said to be an *arclength parameter* if $|\mathbf{C}'(s)| = 1$ for all s in the domain.

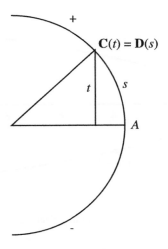

Figure A.3   The arclength parameter of a semicircle.

Example A.1.4 In the curve $\mathbf{C}(s) = (1/\sqrt{2})(\cos s, \sin s, s), s$ is an arclength parameter, because

$$|\mathbf{C}'(s)| = \frac{1}{\sqrt{2}}\sqrt{\sin^2 s + \cos^2 s + 1} = \frac{1}{\sqrt{2}}\sqrt{2} = 1.$$

Theorem A.1.5 *Every curve has an arclength parametrization.*

PROOF: Let $\mathbf{C}(t)$ be a curve with a point A on it. The point A divides the curve into two parts of which one (it doesn't matter which one) is designated as positive and the other as negative. If s is a positive (negative) number, let $\mathbf{D}(s)$ denote that point P on the positive (negative) part such that the arc of $\mathbf{D}(s)$ from A to P has length $|s|$. Define the function $t = t(s)$ by the equation

$$\mathbf{C}(t(s)) = \mathbf{D}(s),$$

and let $s = s(t)$ be its inverse.
    Then

$$|\mathbf{D}'(s)| = \left|\mathbf{C}'(t)\frac{dt}{ds}\right| = |\mathbf{C}'(t)|\left|\frac{1}{ds/dt}\right| = |\mathbf{C}'(t)|\frac{1}{|\mathbf{C}'(t)|} = 1. \qquad \text{Q.E.D.}$$

Example A.1.6 For the semicircle $\mathbf{C}(t) = (\sqrt{1-t^2}, t), -1 < t < 1$, set $A = (1, 0)$, and designate the upper part as the positive direction (Fig. A.3). Then

$$|\mathbf{C}'(t)| = \sqrt{\left(\frac{-2t}{2\sqrt{1-t^2}}\right)^2 + 1^2} = \frac{1}{\sqrt{1-t^2}},$$

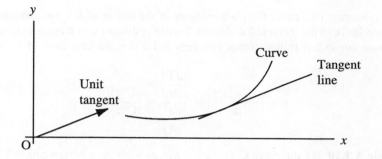

Figure A.4   Tangents.

and hence

$$s(t) = \int_0^t |\mathbf{C}'(\tau)|\,d\tau = \int_0^t \frac{d\tau}{\sqrt{1-\tau^2}} = \sin^{-1} t.$$

Consequently, $t(s) = \sin s$, and hence

$$\mathbf{D}(s) = \mathbf{C}(t(s)) = (\sqrt{1-\sin^2 s}, \sin s) = (\cos s, \sin s).$$

For any curve $\mathbf{C}(t)$, the *unit tangent* is

$$\mathbf{T}(t) = \frac{\mathbf{C}'(t)}{|\mathbf{C}'(t)|}.$$

We note in passing that while the tangent line to a curve is visualized at its point of contact with the curve, the unit tangent is anchored at the origin (Fig. A.4).

Example A.1.7 If $\mathbf{C}(t) = (a_1 + b_1 t, a_2 + b_2 t, a_3 + b_3 t)$, then

$$\mathbf{C}'(t) = (b_1, b_2, b_3) \quad \text{and} \quad \mathbf{T}(t) = \frac{(b_1, b_2, b_3)}{\sqrt{(b_1^2 + b_2^2 + b_3^2)}},$$

which, as expected, is a constant.

Example A.1.8 If $\mathbf{C}(t) = (\cos t, \sin t, t)$, then

$$\mathbf{C}'(t) = (-\sin t, \cos t, 1) \quad \text{and} \quad \mathbf{T}(t) = \frac{1}{\sqrt{2}}(-\cos t, \sin t, 1).$$

Here $\mathbf{C}$ is a helix, and $\mathbf{T}$ rotates about the origin while keeping a constant angle to the xy-plane.

Example A.1.9 If $\mathbf{C}(t) = (t, t^2, t^3)$, then

$$\mathbf{C}'(t) = (1, 2t, 3t^2) \quad \text{and} \quad \mathbf{T}(t) = \frac{(1, 2t, 3t^2)}{\sqrt{1 + 4t^2 + 9t^4}}.$$

The *curvature* of a curve $\mathbf{C}(t)$ is a measure of the rate at which it is changing direction (relative to the traversed distance). Since the direction of a curve is specified by its unit tangent, it follows from Theorem A.1.2 that the curvature of $\mathbf{C}$ should equal

$$\kappa = \left| \frac{d\mathbf{T}(t(s))}{ds} \right| = \frac{\left| \dfrac{d\mathbf{T}}{dt} \right|}{\left| \dfrac{ds}{dt} \right|} = \frac{|\mathbf{T}'(t)|}{|\mathbf{C}'(t)|}.$$

Example A.1.10 For the curve $\mathbf{C}(t) = (a_1 + b_1 t, a_2 + b_2 t, a_3 + b_3 t)$, we have $\mathbf{T}'(t) = \mathbf{0}$, whereas $\mathbf{C}'(t) \neq \mathbf{0}$ and hence $\kappa = 0$.

Example A.1.11 For the curve $\mathbf{C}(t) = (\cos t, \sin t, t)$, we have $\mathbf{C}'(t) = (-\sin t, \cos t, 1)$, $\mathbf{T}'(t) = (-\cos t, \sin t, 0)$, and hence

$$\kappa = \frac{|\mathbf{T}'(t)|}{|\mathbf{C}'(t)|} = \frac{1}{\sqrt{2}}.$$

Example A.1.12 For the curve $\mathbf{C}(t) = (t, t^2, t^3)$,

$$\mathbf{T}'(t) = \frac{(-4t - 18t^3, 2 - 18t^4, 6t + 12t^3)}{(1 + 4t^2 + 9t^4)^{3/2}},$$

$$|\mathbf{T}'(t)| = \frac{\sqrt{4 + 52t^2 + 216t^4 + 468t^6 + 324t^8}}{(1 + 4t^2 + 9t^4)^{3/2}},$$

$$|\mathbf{C}'(t)| = |(1, 2t, 3t^2)| = \sqrt{1 + 4t^2 + 9t^4},$$

$$\kappa = \frac{\sqrt{4 + 52t^2 + 216t^4 + 468t^6 + 324t^8}}{(1 + 9t^2 + 9t^4)^2}.$$

In the interest of brevity the derivative $d\mathbf{T}(t(s))/ds$ will be abbreviated to $d\mathbf{T}/ds$, where s denotes the arclength parameter of $\mathbf{C}(t)$. Note that since $\mathbf{T}$ is a unit vector,

$$\mathbf{T} \cdot \mathbf{T} = 1.$$

When this equation is differentiated with respect to s, we get

$$\frac{d\mathbf{T}}{ds} \cdot \mathbf{T} + \mathbf{T} \cdot \frac{d\mathbf{T}}{ds} = 0$$

and hence

$$\frac{d\mathbf{T}}{ds} \cdot \mathbf{T} = 0.$$

It follows that the vector $d\mathbf{T}/ds$ is orthogonal to the direction of $\mathbf{C}$. For this reason, the *principal normal* to $\mathbf{C}(t)$ is defined as the unit vector

$$\mathbf{N} = \frac{\dfrac{d\mathbf{T}}{ds}}{\left| \dfrac{d\mathbf{T}}{ds} \right|} = \frac{1}{\kappa} \frac{d\mathbf{T}}{ds}. \tag{2}$$

Example A.1.13 For the curve $\mathbf{C}(t) = (\cos t, \sin t, t)$, set $A = (1,0,0)$, so that

$$s(t) = \int_0^t \sqrt{\cos^2 \tau + \sin^2 \tau + 1}\, d\tau = \sqrt{2}\, t,$$

Hence

$$\mathbf{D}(s) = \left(\cos \frac{s}{\sqrt{2}}, \sin \frac{s}{\sqrt{2}}, \frac{s}{\sqrt{2}} \right)$$

is the arclength parametrization of $\mathbf{C}(t)$. Consequently,

$$\mathbf{T}(s) = \mathbf{D}'(s) = \frac{1}{\sqrt{2}} \left(-\sin \frac{s}{\sqrt{2}}, \cos \frac{s}{\sqrt{2}}, \frac{s}{\sqrt{2}} \right),$$

$$\frac{d\mathbf{T}}{ds} = \frac{1}{2} \left(-\cos \frac{s}{\sqrt{2}}, -\sin \frac{s}{\sqrt{2}}, 0 \right),$$

and

$$\mathbf{N} = \left(-\cos \frac{s}{\sqrt{2}}, -\sin \frac{s}{\sqrt{2}}, 0 \right).$$

Theorem A.1.14 *Let $\mathbf{C}(t)$ be a curve, and let $'$ and $''$ denote first and second derivatives with respect to t. Then*

$$\kappa = \frac{|\mathbf{C}' \times \mathbf{C}''|}{|\mathbf{C}'|^3}.$$

PROOF: Let $\mathbf{D}(s)$ be the arclength parametrization of $\mathbf{C}(t)$. By the chain rule,

$$\mathbf{C}' = \frac{d\mathbf{C}}{dt} = \frac{d\mathbf{D}}{ds}\frac{ds}{dt} = s'\mathbf{T},$$

$$\mathbf{C}'' = s''\mathbf{T} + s'\mathbf{T}' = s''\mathbf{T} + s'\frac{d\mathbf{T}}{ds}\frac{ds}{dt} = s''\mathbf{T} + s'^2\kappa\mathbf{N},$$

and hence

$$\mathbf{C}' \times \mathbf{C}'' = s's''\mathbf{T} \times \mathbf{T} + s'^3\kappa\mathbf{T} \times \mathbf{N}.$$

However, since $\mathbf{T}$ and $\mathbf{N}$ are othogonal unit vectors, it follows that $\mathbf{T} \times \mathbf{N}$ is also a unit vector. Hence

$$|\mathbf{C}' \times \mathbf{C}''| = s'^3\kappa,$$

from which it follows that

$$\kappa = \frac{|\mathbf{C}' \times \mathbf{C}''|}{s'^3} = \frac{|\mathbf{C}' \times \mathbf{C}''|}{|\mathbf{C}'|^3}. \qquad \text{Q.E.D.}$$

Example A.1.15 For the curve $\mathbf{C}(t) = (t, t^2, t^3)$ we have

$$\mathbf{C}'(t) = (1, 2t, 3t^2),$$

$$\mathbf{C}''(t) = (0, 2, 6t),$$

and hence

$$\mathbf{C}' \times \mathbf{C}'' = (6t^2, -6t, 2).$$

Consequently,

$$\kappa = \frac{|\mathbf{C}' \times \mathbf{C}''|}{|\mathbf{C}'|^3} = \frac{\sqrt{36t^4 + 36t^2 + 4}}{(1 + 4t^2 + 9t^4)^{3/2}} = 2\sqrt{\frac{9t^4 + 9t^2 + 1}{(9t^4 + 4t^2 + 1)^3}}.$$

The reader should take note of the fact that both the derivation and the final answer here are simpler than those in Example A.1.12.

Example A.1.16 The plane graph of $y = f(x)$ can be placed in $\mathbb{R}^3$ as the curve $\mathbf{C}(t) = (t, f(t), 0)$ with $\mathbf{C}'(t) = (1, f'(t), 0)$ and $\mathbf{C}''(t) = (0, f''(t), 0)$. It follows that its curvature is

$$\kappa = \frac{|\mathbf{C}' \times \mathbf{C}''|}{|\mathbf{C}'|^3} = \frac{|(0, 0, f'')|}{(\sqrt{1 + f'^2})^3} = \frac{|f''|}{(1 + f'^2)^{3/2}}.$$

The *signed curvature* of a plane curve is $\pm\kappa$, where the positive value is used if and only if the curve is turning counterclockwise.

For a given curve $\mathbf{C}(t)$, the vector

$$\mathbf{B}(t) = \mathbf{T}(t) \times \mathbf{N}(t)$$

is called the *binormal* of $\mathbf{C}(t)$. It is clear that the vectors $\mathbf{T}, \mathbf{N}, \mathbf{B}$ form an orthonormal triple. The arclength derivatives of these vectors are used to describe how the curve twists through space. Let s be an arclength parameter of $\mathbf{C}$. Since $\mathbf{B} \cdot \mathbf{B} = 1$, it follows by differentiation with respect to the arclength s that

$$\frac{d\mathbf{B}}{ds} \cdot \mathbf{B} = 0.$$

In addition,

$$\frac{d\mathbf{B}}{ds} \cdot \mathbf{T} = \frac{d\mathbf{T}}{ds} \times \mathbf{N} \cdot \mathbf{T} + \mathbf{T} \times \frac{d\mathbf{N}}{ds} \cdot \mathbf{T} = \kappa \mathbf{N} \times \mathbf{N} \cdot \mathbf{T} + \mathbf{0} = \mathbf{0}.$$

The two equations above imply that $d\mathbf{B}/ds$ is orthogonal to both $\mathbf{B}$ and $\mathbf{T}$. Since $\mathbf{N}$ is also orthogonal to these two vectors, it follows that there exists a scalar function τ, called the *torsion* of $\mathbf{C}$, such that

$$\frac{d\mathbf{B}}{ds} = -\tau \mathbf{N}.$$

We will describe how this function is computed and used after the above considerations are summarized.

Theorem A.1.17 (Frenet–Serret formulas) *Let s be an arclength parameter of the curve* **C**. *Then*

$$\frac{d\mathbf{T}}{ds} = \kappa\mathbf{N},$$

$$\frac{d\mathbf{N}}{ds} = -\kappa\mathbf{T} + \tau\mathbf{B},$$

$$\frac{d\mathbf{B}}{ds} = -\tau\mathbf{N}.$$

PROOF: It only remains to prove the second of the three equations. This, however, follows by differentiating the equation $\mathbf{N} = \mathbf{B} \times \mathbf{T}$ with respect to s:

$$\frac{d\mathbf{N}}{ds} = \mathbf{B} \times \frac{d\mathbf{T}}{ds} + \frac{d\mathbf{B}}{ds} \times \mathbf{T}$$

$$= \mathbf{B} \times \kappa\mathbf{N} - \tau\mathbf{N} \times \mathbf{T} = -\kappa\mathbf{T} + \tau\mathbf{B}. \qquad \text{Q.E.D.}$$

Proposition A.1.18 *The curve* **C** *is a straight line if and only if* $\kappa \equiv 0$. *It is a plane curve if and only if* $\tau \equiv 0$.

PROOF: If **C** is a straight line, then $\mathbf{T} \equiv 0$, and it follows from the first of the Frenet–Serret equations that $\kappa \equiv 0$.

If **C** is a plane curve, then both **T** and **N** lie in a plane parallel to that of **C**. The binormal **B** must then be one of the two normals to this plane, and it is therefore constant. It follows from the third equation that $\tau \equiv 0$.

If s is arclength and $\kappa \equiv 0$, then by the first equation $d\mathbf{T}/ds = 0$ and hence $\mathbf{T} = d\mathbf{C}/ds$ is constant. It follows that

$$\mathbf{C}(s) = s\mathbf{E} + \mathbf{F}$$

for some constant vectors **E** and **F**. In other words, **C** is a straight line.

If $\tau \equiv 0$, then, by the third equation, **B** is constant. Set

$$f(s) = (\mathbf{C}(s) - \mathbf{C}(0)) \cdot \mathbf{B} \qquad \text{for all } s.$$

Then

$$\frac{df}{ds} = \frac{d\mathbf{C}}{ds} \cdot \mathbf{B} = \mathbf{T} \cdot \mathbf{B} = 0.$$

In other words, $f(s)$ is constant. Since $f(0) = 0$, it follows that

$$(\mathbf{C}(s) - \mathbf{C}(0)) \cdot \mathbf{B} = 0 \qquad \text{for all } s$$

implying that $\mathbf{C}(s)$ is contained in the plane orthogonal to **B** and containing the point $\mathbf{C}(0)$. \qquad Q.E.D.

Proposition A.1.18 makes it reasonable to regard the torsion as a measure of the nonplanarity of the curve. There is a straightforward analog of Theorem A.1.14 for the computation of the torsion.

Theorem A.1.19 *If* ′ *denotes differentiation with respect to t, then the torsion of* **C**(*t*) *is given by*

$$\tau = \frac{\mathbf{C}' \times \mathbf{C}'' \cdot \mathbf{C}'''}{|\mathbf{C}' \times \mathbf{C}''|^2}.$$

PROOF: The following equations are easily verified, and the details are relegated to Exercise 7:

$$\mathbf{C}'' = s''\mathbf{T} + \kappa(s')^2\mathbf{N}, \tag{3}$$

$$\mathbf{C}' \times \mathbf{C}'' = \kappa(s')^3\mathbf{B}, \tag{4}$$

$$\mathbf{C}''' = [s''' - \kappa^2(s')^3]\mathbf{T} + [3\kappa's'' + \kappa'(s')^2]\mathbf{N} + \kappa\tau(s')^3\mathbf{B}, \tag{5}$$

$$\tau = \frac{\mathbf{C}' \times \mathbf{C}'' \cdot \mathbf{C}'''}{|\mathbf{C}' \times \mathbf{C}''|^2}, \qquad\qquad \square \ (6)$$

A curve **C** is said to be a *cylindrical helix* provided there is a fixed unit vector **A** such that the angle from **T** to **A** is a constant θ. Equivalently, $\mathbf{T} \cdot \mathbf{A} = \cos\theta$ is constant. For example, the helix

$$\mathbf{C}(t) = (a\cos t, a\sin t, bt)$$

is a cylindrical helix with $\mathbf{A} = (0, 0, 1)$.

Proposition A.1.20 *The curve* **C** *with* $\kappa > 0$ *is a cylindrical helix if and only if* τ/κ *is a constant.*

PROOF: Let ′ denote differentiation with respect to an arclength parameter of the cylindrical helix **C** with $\mathbf{T} \cdot \mathbf{A} = \cos\theta$. Then

$$0 = (\mathbf{T} \cdot \mathbf{A})' = \mathbf{T}' \cdot \mathbf{A} = \kappa\mathbf{N} \cdot \mathbf{A}.$$

Since $\kappa > 0$, it follows that $\mathbf{N} \cdot \mathbf{A} = 0$, and consequently **A** lies in the plane spanned by **T** and **B**. Since all these vectors have unit length,

$$\mathbf{A} = \cos\theta\,\mathbf{T} + \sin\theta\,\mathbf{B}.$$

Differentiation and the first and third Frenet–Serret formulas yield

$$0 = \kappa\cos\theta\,\mathbf{N} - \tau\sin\theta\,\mathbf{N}.$$

or

$$\tau/\kappa = \cot\theta = \text{constant}$$

Conversely, suppose $\kappa > 0$ and τ/κ is constant. Let $\theta = \cot^{-1}(\tau/\kappa)$. If we set

$$\mathbf{A} = \cos\theta\,\mathbf{T} + \sin\theta\,\mathbf{B},$$

then, again by the Frenet–Serret formulas,

$$\mathbf{A}' = (\kappa\cos\theta - \tau\sin\theta)\mathbf{N} = \mathbf{0}.$$

It follows that $\mathbf{A}$ is a constant unit vector such that $\mathbf{T}\cdot\mathbf{A}$ is also constant.   Q.E.D.

Proposition A.1.21 *The curve* $\mathbf{C}$ *is part of a circle if and only if it has constant nonzero curvature and its torsion vanishes everywhere.*

PROOF: Suppose $\mathbf{C}$ is part of a circle. After a rotation and/or a translation it may be assumed that $\mathbf{C}$ lies in the xy-plane and is centered at the origin, so that it has arclength parametrization

$$\mathbf{C}(s) = r(\cos(s/r), \sin(s/r), 0),$$

where r is the radius of the circle. It is then easily verified by means of Theorems A.1.14 and A.1.18 that $\kappa = 1/r = $ constant, and τ is identically 0.

Conversely, suppose $\kappa \neq 0$ is constant and τ vanishes everywhere. It follows from Proposition A.1.18 that $\mathbf{C}$ is planar. We next show that if s is an arclength parameter of $\mathbf{C}$, then the point

$$\mathbf{D}(s) = \mathbf{C}(s) + \frac{1}{\kappa}\mathbf{N}(s)$$

is constant. When the equation above is differentiated with respect to s, we get

$$\mathbf{D}'(s) = \mathbf{C}'(s) + \frac{1}{\kappa}\mathbf{N}'(s) = \mathbf{T} + \frac{1}{\kappa}(-\kappa\mathbf{T} + \tau\mathbf{B}) = \mathbf{T} - \mathbf{T} = \mathbf{0}.$$

Hence, $\mathbf{D}(s)$ is a constant point, say $\mathbf{D}$. Since the distance of $\mathbf{C}(s)$ from this point $\mathbf{D}$ is the constant $1/\kappa$, and since $\mathbf{C}$ is known to be planar, it follows that it is also part of a circle.   Q.E.D.

Exercises A

1. Suppose

$$\mathbf{D}(s) = \left(\frac{(1+s)^{3/2}}{3}, \frac{(1-s)^{3/2}}{3}, \frac{s}{\sqrt{2}}\right), \qquad -1 < s < 1.$$

   (a) Show that this is an arclength parametrization.
   (b) Show that $\kappa = 1/\sqrt{8(1-s^2)}$.
   (c) Show that

   $$\mathbf{N}(s) = \left(\frac{\sqrt{2(1-s)}}{2}, \frac{\sqrt{2(1+s)}}{2}, 0\right).$$

2. Compute the curvature of the curve

   $$\mathbf{C}(s) = \left(\frac{1}{\sqrt{2}}\cos t, \sin t, \frac{1}{\sqrt{2}}\cos t\right).$$

3. Let $(x(t), y(t))$ be a parametrized plane curve. Show that

$$\kappa = \frac{|x'y'' - x''y'|}{(x'^2 + y'^2)^{3/2}}.$$

4. The curve $\mathbf{C}(t) = (\cosh t, \sinh t, t)$ is known as the *hyperbolic helix*. Show that its curvature is $1/(2\cosh^2 t)$.

5. Let θ denote the inclination of the tangent line to the graph of $y = f(x)$ to the x-axis. Prove that $d\theta/ds$ equals the signed curvature.

6. Suppose $f(x) = Ax^2 + g(x)$, where $0 = g'(0) = g''(0)$. Then the signed curvature of $y = f(x)$ at $x = 0$ is $2A$.

7. Prove Equations (3)–(6).

8. Suppose $\mathbf{C}(t)$ is a curve in $\mathbb{R}^3$, $\mathbf{A}$ is a fixed 3-vector, and $\mathbf{B}$ is a fixed 3×3 rotation matrix. Set $\mathbf{D}(t) = \mathbf{C}(t) + \mathbf{A}$, $\mathbf{E}(t) = \mathbf{B}\mathbf{C}(t)$. Prove that

$$\kappa_E = \kappa_D = \kappa_C, \qquad \tau_E = \tau_D = \tau_C.$$

9. Evaluate the curvature and torsion of the following curves:

   (a) $\mathbf{C}(t) = (3t - t^3, 3t^2, 3t + t^3)$.

   (b) $\mathbf{C}(t) = (t - \sin t, 1 - \cos t, t)$.

   (c) $\mathbf{C}(t) = (1 + \cos 2t, \sin 2t, 2\sin t)$.

   (d) $\mathbf{C}(t) = (t, t^2, t^3)$.

APPENDIX B

A BRIEF SURVEY OF GROUPS

B.1 The General Background

A *group* $(G, \cdot)$ consists of a set G and a binary operation '$\cdot$', called the *product*, on the elements of G, that satisfy the constraints:

1. $a \cdot b \in G$ for all $a, b \in G$.

2. $a \cdot (b \cdot c) = (a \cdot b) \cdot c$ for all $a, b, c \in G$.

3. There exists an *identity element* 1_G in G such that $a \cdot 1_G = 1_G \cdot a = a$ for all a in G.

4. For every $a \in G$ there exists an element a^{-1} (the *inverse* of a) such that $a \cdot a^{-1} = a^{-1} \cdot a = 1_G$.

The following are all well-known groups:

Introduction to Topology and Geometry, Second Edition.
By Saul Stahl and Catherine Stenson Copyright © 2013 John Wiley & Sons, Inc.

$(\mathbb{Z}, +)$	— the integers with addition;
$(\mathbb{Q}, +)$	— the rational numbers with addition;
$(\mathbb{R}, +)$	— the real numbers with addition;
$(\mathbb{C}, +)$	— the complex numbers with addition;
$(\mathbb{Z}_n, +)$	— addition modulo n;
$(\mathbb{Q} - \{0\}, \times)$	— the nonzero rationals with multiplication;
$(\mathbb{R} - \{0\}, \times)$	— the nonzero reals with multiplication;
$(\mathbb{C} - \{0\}, \times)$	— the nonzero complex numbers with multiplication;
$(\mathscr{M}_{m,n}, +)$	— the $m \times n$ real matrices with matrix addition;
$(S_n, \circ)$	— the permutations of $\{1, 2, \ldots, n\}$ with composition (see Appendix C);
$(\mathscr{I}(\mathbb{R}^2), \circ)$	— the isometries of the Euclidean plane with composition;
$(\mathrm{GL}_n(\mathbb{R}), \cdot)$	— the nonsingular $n \times n$ real matrices with matrix mutliplication;
$(\mathrm{SL}(2, \mathbb{R}), \circ)$	— the real 2×2 matrices of determinant 1 with matrix multiplication.

Figures B.1 and B.2 display several small groups as multiplication tables. It is easily proved that the identity element of any group G is unique, that the inverse of any element a is unique, and that $(ab)^{-1} = b^{-1}a^{-1}$ for any two elements a and b (Exercises 15, 16). It follows from the existence of inverses that each column and each row of the multiplication table is necessarily a permutation of the group's elements (Exercise 22).

In referring to specific groups it is customary to delete the operation, so that, for example, the additive group of integers is denoted simply by $\mathbb{Z}$. A group in which $a \cdot b = b \cdot a$ for any two elements a, b, is said to be *abelian*, or *commutative*. The first nine groups listed above are abelian, whereas the rest are not. The only nonabelian group in Figure B.1 is S_3, whereas both of the groups of Figure B.2 are nonabelian. It is customary to refer to the binary operation of an abelian group as a sum and to denote it by '$+$'. Note that the multiplication table of an abelian group is symmetric relative to its main diagonal.

The number of elements of a group is its *order*. Groups of order 1 are said to be *trivial*. The iterated product of n g's is denoted by g^n. (In abelian groups this is written as ng.) If $g \neq 1_G$ is an element of G, then the least positive integer k, if any, such that $g^k = 1_G$ is the *order* of g. In abelian groups this equation is replaced by $kg = 0$, where 0 denotes the identity element. If no such k exists, then g is said to have *infinite order*. The identity element always has order 1 and is the only element of that order.

Example B.1.1 The group K has one element of order 1 and three elements of order 2. The group S_3 has one element of order 1, three elements of order 2, and two elements of order 3.

Orders of elements and orders of groups are related by the following fundamental observation. The proof is relegated to Exercise 13.

Proposition B.1.2 *The order of every element of a finite group divides that group's order.*   □

Given two groups G and H, a function $f : G \rightarrow H$ is said to be a *homomorphism* provided

$$f(ab) = f(a)f(b) \qquad \text{for all } a, b \in G.$$

The function $f(m) = 2m$ is a homomorphism of $\mathbb{Z}$ into itself. If $[m]$ denotes the residue class of m modulo n, then the function $f(m) = [m]$ is a homomorphism of $\mathbb{Z}$ into $\mathbb{Z}_n$. If tr(A) denotes the trace (sum of the diagonal elements) of the $n \times n$ matrix A, then this function is a homomorphism of $\mathcal{M}_{n,n}$ into $\mathbb{R}$. If $\det(A)$ denotes the determinant of the matrix A, then this function is a homomorphism of GL$_n(\mathbb{R})$ into the group $(\mathbb{R} - \{0\}, \times)$. For any such group homomorphism f, one has $f(1_G) = 1_H$ and $f(g^{-1}) = [f(g)]^{-1}$ for all $g \in G$ (see Exercise 17).

A homomorphism $f : G \rightarrow H$ that is both surjective (onto) and injective (one-to-one) is an *isomorphism* of G and H, and we write $G \cong H$. It is clear that isomorphic groups have the same order, but the converse is not true (Exercise 9). Moreover, if $f : G \rightarrow H$ is an isomorphism, then g and $f(g)$ have the same order for each $g \in G$ (Exercise 20). It is customary to consider isomorphic groups as virtually identical, as the homomorphism property $f(ab) = f(a)f(b)$ ensures that, after an appropriate relabeling, isomorphic groups have identical multiplication tables (see Figure B.3).

Example B.1.3 The order 2 group of Figure B.4 is isomorphic to the group $\mathbb{Z}_2$ of Figure B.1. In fact, it is clear that every two groups of order 2 are isomorphic. The same goes for groups of order 3. To see this, note that the multiplication table of any group of order 3 must contain the partial information of Figure B.5(a). Since, by Proposition B.1.2, no element of such a group can have order 2, and since the rows and columns of a multiplication table are permutations of the elements of the group, it follows that the table of Figure B.5(a) can only be completed into the table of Figure B.5(b).

A subset H of the group G that is closed with respect to the binary operation of G as well as the taking of inverses is called a *subgroup* of G. Each of the groups $\mathbb{Z}$, $\mathbb{Q}$, $\mathbb{R}$, $\mathbb{C}$ is a subgroup of the next. On the other hand, if m and n are distinct positive integers, then $\mathbb{Z}_m$ is generally not a subgroup of $\mathbb{Z}_n$. The intersection of any number of subgroups of the group G is again a subgroup of G (Exercise 21).

Example B.1.4 The group K has the subgroups $\{1\}, \{1, a\}, \{1, b\}, \{1, c\}, K$.

The order of a subgroup is related to the order of the ambient group by an observation of Lagrange (see Exercise 14).

Proposition B.1.5 *The order of a subgroup of a group divides the order of the group.*   □

Given any subset S of the group G, the *subgroup of G generated by S*, $\langle S \rangle$, is the intersection of all the subgroups of G that contain S. Alternatively, this is the set of all the elements of G that are expressible as finite products (or sums) of elements of S and their inverses.

Example B.1.6 The subgroup of $\mathbb{Z}$ generated by $\{2\}$ consists of all the even integers. The subgroup of $\mathbb{Z}$ generated by $\{6, 8\}$ also consists of all the even integers. The subgroup of $\mathbb{Q}$ generated by $\{1/2, 1/3\}$ consists of all the fractions with denominator 1, 2, 3, or 6.

A group that is generated by a single element is said to be *cyclic*. The groups $\mathbb{Z}$ and $\mathbb{Z}_m$, $m = 1, 2, \ldots$, are all cyclic, and we have the following important result (Exercise 18).

Proposition B.1.7 *Every cyclic group is isomorphic to exactly one of the groups* $\mathbb{Z}$, $\mathbb{Z}_m$, $m = 1, 2, \ldots$ $\quad\square$

Given two groups G and H, their *direct product $G \times H$* is a group whose underlying set is

$$G \times H = \{(g,h) \mid g \in G, h \in H\}$$

and whose binary operation is

$$(g,h)(g',h') = (gg',hh').$$

For any three groups G, H, K the iterated products $G \times (H \times K)$ and $(G \times H) \times K$ are isomorphic, and they are both denoted by $G \times H \times K$. A similar convention holds for longer iterated direct products of groups. The iterated direct product of n copies of G is denoted by G^n.

We will mostly be interested in products of cyclic groups, in which case the resulting products are also abelian. It is convenient to describe the elements of such products of cyclic groups as character strings. By this convention the elements of $\mathbb{Z}_2^2 \times \mathbb{Z}_3$ can be listed as

$$000, 100, 010, 110, \qquad 001, 101, 011, 111, \qquad 002, 102, 012, 112$$

where, by way of examples,

$$012 + 111 = 100, \qquad 102 + 012 = 111, \qquad 111 + 111 = 002.$$

Exercises B.1

1. Find the orders of all the elements of the following groups: (a) $\mathbb{Z}_3$, (b) $\mathbb{Z}_4$, (c) $\mathbb{Z}_5$, (d) $\mathbb{Z}_6$, (e) $\mathbb{Z}_7$, (f) $\mathbb{Z}_8$, (g) D_4, (h) $\mathbb{Z}_2 \times \mathbb{Z}_4$, (i) $\mathbb{Z}_3 \times \mathbb{Z}_4$, (j) $\mathbb{Z}_4 \times \mathbb{Z}_4$.

2. List all the subgroups of each of the following groups: (a) $\mathbb{Z}_3$, (b) $\mathbb{Z}_4$, (c) $\mathbb{Z}_5$, (d) $\mathbb{Z}_6$, (e) $\mathbb{Z}_7$, (f) $\mathbb{Z}_8$, (g) D_4, (h) $\mathbb{Z}_2 \times \mathbb{Z}_4$, (i) $\mathbb{Z}_3 \times \mathbb{Z}_4$, (j) $\mathbb{Z}_4 \times \mathbb{Z}_4$.

3. Find all the subgroups of the quaternion group of Figure B.2.

4. Describe all the subgroups of $(\mathbb{Z}, +)$.

5. Let n be a positive integer. Describe all the subgroups of $(\mathbb{Z}_n, +)$.

6. Let g be a group element of order n. Prove that the element g^k has order $n/(k, n)$, where (k, n) denotes the greatest common divisor of k and n.

7. Prove that the function $f(x) = 5x$ is a homomorphism of $\mathbb{Z}$ into $\mathbb{Z}$.

8. Let x be a fixed element of the group G. Prove that the function

$$f_x(g) = xgx^{-1}$$

is an isomorphism of G onto itself.

9. Prove that $\mathbb{Z}_2^2$ and $\mathbb{Z}_4$ are nonisomorphic groups of order 4.

10. Prove that every group of order 4 is isomorphic to either K or $\mathbb{Z}_4$.

11. Prove that every subgroup of a cyclic group is also cyclic.

12. Prove that the groups $G \times H$ and $H \times G$ are isomorphic to each other.

13. Prove Proposition B.1.2.

14. Prove Proposition B.1.5.

15. Prove that the identity element of any group G is unique.

16. Prove that the inverse of any element a of a group G is unique, and that $(ab)^{-1} = b^{-1}a^{-1}$ for any $a, b \in G$.

17. Prove that if $f : G \to H$ is a group homomorphism, then $f(1_G) = 1_H$ and $f(g^{-1}) = [f(g)]^{-1}$ for all $g \in G$.

18. Prove Proposition B.1.7.

19. Construct multiplication tables akin to Table B.1 for the following groups: (a) $\mathbb{Z}_2^2$, (b) $\mathbb{Z}_6$, (c) $\mathbb{Z}_2 \times \mathbb{Z}_3$, (d) $\mathbb{Z}_2^3$, (e) $\mathbb{Z}_2 \times \mathbb{Z}_4$.

20. Suppose $f : G \to H$ is an isomorphism. Prove that for every $g \in G$, g and $f(g)$ have the same orders.

21. Prove that the intersection of any number of subgroups of a group G is also a subgroup of G.

22. If $f : G \to H$ is a group homomorphism, then the set

$$\ker f = \{g \in G \mid f(g) = 1_H\}$$

is the *kernel* of f. Prove:

(a) Ker f is a subgroup of G;

(b) f is injective (one-to-one) if and only if Ker f is trivial.

23. Prove that each column and each row of a group's multiplication table is a permutation of that group's elements.

B.2   Abelian Groups

Given an abelian group G, a set S of elements of G is said to generate G provided that every element of G is expressible as the sum of a finite number of elements of S and their inverses. For example, the set $\{1\}$ generates the group $\mathbb{Z}_n$, as does the set $\{1, 2\}$ (provided that $n \geq 2$). The set $\{p^{-n}\}$, where p varies over all the positive primes and n is any positive integer, generates $\mathbb{Q}$. The direct product $\mathbb{Z}_k \times \mathbb{Z}_m \times \mathbb{Z}_n$ is generated by the set

$$\{100, 010, 001\}.$$

A group is said to be *finitely generated* if it has a finite generating set. Any iterated direct product of groups selected from $\mathbb{Z}$ and $\mathbb{Z}_n$, $n = 1, 2, \ldots$, is a finitely generated abelian group, because it is generated by the set

$$\{100\cdots00, 010\cdots00, \ldots, 000\cdots10, 000\cdots01\}.$$

On the other hand, the group $\mathbb{Q}$ is not finitely generated (Exercise 10). The structure of finitely generated groups is well known.

Theorem B.2.1 *Let G be a finitely generated abelian group. Then there exist integers $r, s \geq 0$, and $t_1, t_2, \ldots, t_s > 1$ such that*

a. *t_i divides t_{i+1} for $i = 1, 2, \ldots, s - 1$;*

b. *$G \cong \mathbb{Z}^r \times \mathbb{Z}_{t_1} \times \mathbb{Z}_{t_2} \times \cdots \times \mathbb{Z}_{t_s}$.*   □

The integer r of this theorem is the *rank*, or *Betti number*, of G, and $t_1, t_2, \ldots, t_s$ are its *torsion coefficients*. The sum $r + s$ is the *width* of G.

Theorem B.2.2 *Two finitely generated abelian groups are isomorphic if and only if they have the same Betti numbers and torsion coefficients. The width of G is the minimum cardinality of its generating sets.*   □

The abelian groups of width 1 are the cyclic groups. The direct product of two cyclic groups may have width 1 or 2. The product $\mathbb{Z}_2 \times \mathbb{Z}_3$ has width 1, because it is generated by the single element 11. On the other hand, $\mathbb{Z}_2^2$ has width 2, because it is not generated by any of its four elements $\{00, 10, 01, 11\}$.

Exercises B.2

1. Prove that the groups $\mathbb{Z}_2 \times \mathbb{Z}_3$ and $\mathbb{Z}_6$ are isomorphic.

2. Prove that for any two relatively prime positive integers m, n, one has $\mathbb{Z}_m \times \mathbb{Z}_n \cong \mathbb{Z}_{mn}$.

3. Prove that no two of the groups $\mathbb{Z}_2^3, \mathbb{Z}_2 \times \mathbb{Z}_4, \mathbb{Z}_8$ are isomorphic.

4. Suppose the elements a and b of the abelian group G have orders m and n, respectively. Show that if m and n are relatively prime, then $a + b$ has order mn. Generalize this to arbitrary m and n.

5. Let G be an abelian group of order mn where m and n are relatively prime. Prove that there exist two groups M and N, of orders m and n, respectively, such that $G \cong M \times N$.

6. Describe all the subgroups of $\mathbb{Z}_4^2$, and specify their widths.

7. Prove that every group of order at most 5 is abelian.

8. Every element of a group G has order at most 2. Prove that G is abelian.

9. Prove that every subgroup of an abelian group of width 1 also has width 1.

10. Prove that the group $\mathbb{Q}$ is not finitely generated.

11. Prove directly, without using Theorem B.2.1, that the group $\mathbb{Z}_2^n$ cannot be generated by fewer that n of its elements.

B.3 Group Presentations

Given a finite set of symbols $A = \{a, b, c, \ldots\}$, a *word* w in A is a finite character string $w = x_1^{d_1} x_2^{d_2} x_3^{d_3} \cdots x_n^{d_n}$ where each x_i is in A and each d_i is an integer. The empty word is denoted by 1. As is customary, the exponent 1 is deleted, and letters can be combined or deleted according to the usual rules of arithmetic: For any integers m, n and word w,

$$w^m w^n = w^{m+n}, \qquad w^0 = 1, \qquad w1 = 1w = w. \tag{1}$$

Thus,

$$a^1 c^{-1} bbb^{-1} c^1 c^1 b^{-1} b^{-1} a^{-1} a \quad \text{and} \quad ac^{-1} bc^2 b^{-3}$$

are considered as equal words in A. The *free group* $F(A)$ consists of all the words in A, under the operation of juxtaposition, with the understanding that two words are equal if the rules of (1) can be used to modify one into the other. Inverses in $F(A)$ are given by the equation

$$(x_1^{d_1} x_2^{d_2} \cdots x_n^{d_n})^{-1} = x_n^{-d_n} \cdots x_2^{-d_2} x_1^{-d_1}.$$

In particular, if $A = \{a\}$, then

$$F(A) = \{a^n \mid n \in \mathbb{Z}\},$$

which is an infinite cyclic group. On the other hand, if $A = \{a, b\}$, then

$$F(A) = \{a^{i_1} b^{j_1} a^{i_2} b^{j_2} \cdots a^{i_k} b^{j_k} \mid i_1, j_1, \ldots, i_k, j_k \in \mathbb{Z}\}. \tag{2}$$

Given a finite set $A = g_1, g_2, \ldots, g_m$ of symbols and a finite set $R = \{R_1, R_2, \ldots, R_n\}$ of words in A, we refer to the double list

$$\langle g_1, g_2, \ldots, g_m \; ; \; R_1, R_2, \ldots, R_n \rangle \tag{3}$$

as a *presentation* of the group obtained from $F(A)$ by stipulating that

$$R_i = 1, \qquad i = 1, 2, \ldots, n.$$

While the definition of this group may look forbidding, it is in fact very easy to work with. We simply manipulate words in A as elements of the free group $F(A)$, with the added stipulation that whenever a relator R_i appears as a subword of a word w, R_i can be replaced by the empty word. In other words (no pun intended), for any two words w_1 and w_2 in A, and any relator R_i, we have

$$w_1 R_i w_2 = w_1 w_2$$

in the group of (3).

Even though identical groups may have different presentations, it is customary to identify a group with its presentation. The symbols $g_1, g_2, \ldots, g_m$ are the *generators* of the group of (3), whereas the words $R_1, R_2, \ldots, R_n$ are its *relators*.

Example B.3.1 Suppose $A = \{a\}$ and $R_1 = \{a^2\}$. Then every word a^n equals either 1 or a in the group $\langle a; a^2 \rangle$. It follows that this group is cyclic of order 2. A similar argument demonstrates that the presentation $\langle a; a^m \rangle$ defines a cyclic group of order m for every positive integer m.

Example B.3.2 Suppose $A = \{a, b\}$ and $R = \{b\}$. Then the presentation $\langle a, b; b \rangle$ defines a group that is clearly isomorphic to $F(\{a\})$.

Example B.3.3 The presentation $\langle a, b; a^2, b^2 \rangle$ defines a group whose nonidentity elements are all the finite alternating words

$$ababa \cdots, \qquad babab \cdots.$$

Example B.3.4 The presentation $\langle a, b; a^2, ab \rangle$ defines a group whose only two elements can be written as 1 and a, because from the equation

$$ab = 1$$

we can deduce that

$$b = a^{-1}.$$

Note that if $a_1 a_2 \cdots a_k$ is a relator of a presentation of a group G, then we also have

$$a_2 \cdots a_k a_1 = a_1^{-1} a_1 a_2 \cdots a_k a_1 = a_1^{-1} 1 a_1 = 1.$$

In other words:

The words obtained by cycling the symbols of a relator also have value 1_G.

Example B.3.5 In the group $\langle a, b, c; abca \rangle$ the equation

$$aca^2 ba = a^2$$

holds.

The following observation is also useful:

If $w_1 w_2$ is a relator in a presentation of a group G, then $w_1 = w_2^{-1}$ in G.

Example B.3.6 In the group $\langle a, b, c; a^2 bc \rangle$ the equations

$$cbca^3 = ca^{-2}a^3 = ca$$

hold.

Equations of the form $w_1 = w_2$ that can be derived from the relators of a presentation are known as *relations*. A particularly important class of relators are the *commutators* $aba^{-1}b^{-1}$. It is clear that this relator is equivalent to the relation

$$ab = ba.$$

Example B.3.7 The elements of the group $\langle a, b; aba^{-1}b^{-1} \rangle$ can be listed as

$$a^m b^n, \qquad m, n \in \mathbb{Z}.$$

This group is easily seen to be isomorphic to $\mathbb{Z}^2$ via the map

$$f(a^h, b^k) = (h, k).$$

Example B.3.8 The group

$$\langle g_1, g_2, \ldots, g_r; g_i g_j g_i^{-1} g_j^{-1}, \ 1 \le i < j \le r \rangle$$

is the *free abelian group of rank r*. It is clear that these groups also have width r. The reader should not be misled by the terminology. Free abelian groups are *not* free groups in the sense previously defined. They are, however, abelian, and the given free abelian group is in fact isomorphic to $\mathbb{Z}^r$.

The following plausible presentations of the finitely generated abelian groups classified in Theorem B.2.1 are obtained by setting

$$g_1 = 100\cdots00, g_2 = 010\cdots00, \ldots, g_{r+s} = 000\cdots01.$$

Theorem B.3.9 *Suppose*

$$G \cong \mathbb{Z}^r \times \mathbb{Z}_{t_1} \times \mathbb{Z}_{t_2} \times \cdots \times \mathbb{Z}_{t_s}.$$

Then

$$G = \langle g_1, g_2, \ldots, g_{r+s}; g_i g_j g_i^{-1} g_j^{-1}, \ 1 \le i < j \le r, g_k^{t_k}, \ r < k \le r+s \rangle.$$

$\square$

The *abelianization* of a presented group is both easily described and easy to work with.

Proposition B.3.10 *Let* $G = \langle g_1, g_2, \ldots, g_m; R_1, R_2, \ldots, R_n \rangle$. *Then*

$$\langle g_1, g_2, \ldots, g_m; R_1, R_2, \ldots, R_n, g_i g_j g_i^{-1} g_j^{-1}, \ 1 \le i < j \le m \rangle$$

is a presentation of the abelianization of G.

This proposition is tantamount to saying that the abelianization of a presented group is obtained by pretending that the generators of G commute.

Example B.3.11 Let $A = a_1, \ldots, a_m$. Then the abelianization G of the free group $F(A)$ has the presentation $\langle a_1, a_2, \ldots, a_n; a_i a_j a_i^{-1} a_j^{-1}, \ 1 \le i < j \le n \rangle$. The elements of G can be listed as $a_1^{d_1} a_2^{d_2} \cdots a_n^{d_n}$. The function

$$f(a_1^{d_1} a_2^{d_2} \cdots a_n^{d_n}) = (d_1, d_2, \ldots, d_n)$$

is an isomorphism of G with $\mathbb{Z}^n$. Thus, the abelianization of a free group is a free abelian group.

Example B.3.12 Let $G = \langle a, b, c; abc, acb^{-1}, ac^{-1}b, ab^{-1}c^{-1} \rangle$. In the abelianization of G, written additively, we have

$$a + b + c = 0,$$
$$a - b + c = 0,$$
$$a + b - c = 0,$$
$$a - b - c = 0.$$

It is easy to see that this system of equations is equivalent to

$$2a = 2b = 2c = a + b + c = 0.$$

It follows that the abelianization of G has only the four elements

$$0, \ a, \ b, \ a + b,$$

This group is isomorphic to $\mathbb{Z}_2^2$.

Example B.3.13 Let $G = \langle a, b, c; a^{-1}bac, cbc^{-1}b^{-1}, a^{-1}c^{-1}ab \rangle$. Then the abelianization H of G has the presentation

$$\langle a, b, c; bc, c^{-1}b, aba^{-1}b^{-1}, aca^{-1}c^{-1}, cbc^{-1}b^{-1} \rangle.$$

It is clear that, in H, the only constraint on a is that it must commute with all the other elements. On the other hand, the generators b and c satisfy the additional constraints

$$b = c, \qquad b = -c.$$

Thus, $2b = 0$ and the elements of H can be listed as

$$\{ka + lb \mid k \in \mathbb{Z}, \ l \in \mathbb{Z}\}.$$

Hence H is is isomorphic to $\mathbb{Z} \times \mathbb{Z}_2$.

Example B.3.14 As noted in Theorem 9.2.4, the fundamental group of the closed orientable surface S^n, $n \geq 1$, has the presentation

$$\langle a_1, b_1, \ldots, a_n, b_n; a_1 b_1 a_1^{-1} b_1^{-1} a_2 b_2 a_2^{-1} b_2^{-1} \cdots a_n b_n a_n^{-1} b_n^{-1} \rangle.$$

Since the single relator of this presentation is a product of commutators, it follows that the abelianization of this group is the free abelian group of rank $2n$ which is isomorphic to $\mathbb{Z}^{2n}$.

Example B.3.15 As noted in Theorem 9.2.4, the fundamental group of the closed nonorientable surface $\tilde{S}^m$ has the presentation

$$\langle a_1, a_2, \ldots, a_m; a_1^2 a_2^2 \cdots a_m^2 \rangle.$$

Its abelianization G therefore has width at most m. If its rank were m, G would be isomorphic to $\mathbb{Z}^m$, a group which contains no elements of finite order. This, however, would contradict the fact that the abelianized element $a_1 + a_2 + \cdots + a_n$ has order 2. (If it had order 1, then G would have had width at most $m - 1$.) Consequently, this group has rank at most $m - 1$.

Exercises B.3

1. Show that the group $\langle a, b; a^3, b^2, aba^{-2}b^{-1} \rangle$ has order at most 6.

2. Show that the group $\langle a, b, c; a^3, b^2, aba^{-2}b^{-1}, c^2, aca^{-1}c^{-1}, bcb^{-1}c^{-1} \rangle$ has order at most 12.

3. Show that the group $\langle a, b; a^n, b^2, abab^{-1} \rangle$ has order at most $2n$. Show that the abelianization of this group has order at most 4.

4. Show that the group $\langle a, b; a^4, a^2b^{-2}, abab^{-1} \rangle$ has order at most 8. Show that the abelianization of this group has order at most 4.

5. Show that the group $\langle a, b; a^2, b^2, (ab)^3 \rangle$ has order at most 11. Show that its abelianization has order at most 4.

6. Show that the group $\langle a, b; a^3, b^4, abab \rangle$ has order at most 12. Show that its abelianization has order at most 2.

7. Show that the group $\langle a, b; a^5, b^4, aba^{-2}b^{-1} \rangle$ has order at most 20. Show that its abelianization has order at most 4.

8. Show that the group $\langle a, b, c; a^3, b^3, c^4, acac^{-1}, aba^{-1}bc^{-1}b^{-1} \rangle$ is trivial.

9. Show that the group $\langle a, b; a^4, a^2b^{-3} \rangle$ has a finite abelianization.

10. Show that the group $\langle a, b; a^2, b^2, (ab)^r \rangle$ has order less than $4r$. Show that its abelianization has order at most 4.

11. Prove that the group $\langle a, b; a^3b^2, a^5b^3 \rangle$ is trivial.

12. Prove that the abelianization of the group $\langle a, b, c; a^3b^2c^5, a^2b^2c^3 \rangle$ has width at most 2.

13. Prove that the abelianization of the group $\langle a, b, c; a^3 b^2 c^5, a^2 b^2 c^3, abc \rangle$ has width 1.

14. Show that the presentation $\langle a, b; a^2, b^2, (ab)^2 \rangle$ defines the group K.

15. Show that the presentation $\langle a, b, c; a^2, b^2, c^2, (ab)^2, (bc)^2, (ac)^2 \rangle$ defines a group isomorphic to $\mathbb{Z}_2 \times \mathbb{Z}_2 \times \mathbb{Z}_2$.

16. Show that the presentation $\langle a, d; a^3, d^2, (da)^2 \rangle$ defines the group S_3.

17. Show that the presentation $\langle a, d; a^4, d^2, (da)^2 \rangle$ defines the group D_4.

+	0
0	0

The trivial group

+	0	1
0	0	1
1	1	0

$\mathbb{Z}_2$

+	0	1	2
0	0	1	2
1	1	2	0
2	2	0	1

$\mathbb{Z}_3$

+	0	1	2	3
0	0	1	2	3
1	1	2	3	0
2	2	3	0	1
3	3	0	1	2

$\mathbb{Z}_4$

·	1	a	b	c
1	1	a	b	c
a	a	1	c	b
b	b	c	1	a
c	c	b	a	1

The Klein 4-group

+	0	1	2	3	4
0	0	1	2	3	4
1	1	2	3	4	0
2	2	3	4	0	1
3	3	4	0	1	2
4	4	0	1	2	3

$\mathbb{Z}_5$

·	1	a	b	c	d	e
1	1	a	b	c	d	e
a	a	1	d	e	b	c
b	b	e	1	d	c	a
c	c	d	e	1	a	b
d	d	c	a	b	e	1
e	e	b	c	a	1	d

S_3

Figure B.1

·	1	a	b	c	d	e	f	g
1	1	a	b	c	d	e	f	g
a	a	b	c	1	g	f	d	e
b	b	c	1	a	e	d	g	f
c	c	1	a	b	f	g	e	d
d	d	f	e	g	1	b	a	c
e	e	g	d	f	b	1	c	a
f	f	e	g	d	c	a	1	b
g	g	d	f	e	a	c	b	1

The dihedral group D_4

·	1	a	b	c	d	e	f	g
1	1	a	b	c	d	e	f	g
a	a	d	c	f	e	1	g	b
b	b	g	d	a	f	c	1	e
c	c	b	e	d	g	f	a	1
d	d	e	f	g	1	a	b	c
e	e	1	g	b	a	d	c	f
f	f	c	1	e	b	g	d	a
g	g	f	a	1	c	b	e	d

The quaternion group

Figure B.2

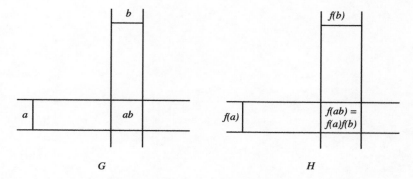

Figure B.3 The multiplication tables of isomorphic groups.

	1	a
1	1	a
a	a	1

Figure B.4

·	1	a	b
1	1	a	b
a	a		
b	b		

(a)

·	1	a	b
1	1	a	b
a	a	b	1
b	b	1	a

(b)

Figure B.5

APPENDIX C

PERMUTATIONS

A *permutation* of a finite set S is a function $\sigma : S \to S$ that is both injective (one-to-one) and surjective (onto). The *cyclic form* of the permutation σ dislays it in terms of its iterations. Thus, σ is written as a collection of cycles each of which has the form

$$(a_1 \, a_2 \, \ldots \, a_k)$$

where

$$a_{i+1} = f(a_i) \quad \text{for } i = 1, 2, \ldots, k-1$$

and

$$a_1 = f(a_k).$$

Example C.1.1 The permutation $\sigma = (1\ 6)(5\ 3\ 7\ 2\ )(4)$ is the function that maps 1, 2, 3, 4, 5, 6, 7 onto 6, 5, 7, 4, 3, 1, 2, respectively.

It is clear that the order of the cycles within a permutation is immaterial, as is the starting point of each cycle. Thus,

$$(1\ 6)(5\ 3\ 7\ 2\ )(4) = (5\ 3\ 7\ 2\ )(4)(1\ 6) = (7\ 2\ 5\ 3\ )(6\ 1)(4).$$

Introduction to Topology and Geometry, Second Edition.
By Saul Stahl and Catherine Stenson Copyright © 2013 John Wiley & Sons, Inc.

If σ and ρ are permutations of the set S, so is their composition $\sigma \circ \rho$, where

$$(\sigma \circ \rho)(x) = \sigma(\rho(x)) \qquad \text{for all } x \in S.$$

Example C.1.2

$$(1\ 6)(5\ 3\ 7\ 2)(4) \circ (1\ 2\ 3\ 4\ 5\ 6\ 7) = (1\ 5)(2\ 7\ 6)(3\ 4),$$
$$(1\ 2\ 3\ 4\ 5\ 6\ 7) \circ (1\ 6)(5\ 3\ 7\ 2)(4) = (1\ 7\ 3)(2\ 6)(4\ 5).$$

A set P of permutations of S is said to act *transitively* on S provided that for any two elements $s, t \in S$, there is a sequence $\sigma_1, \sigma_2, \ldots, \sigma_k$ of (not necessarily distinct) permutations in P such that

$$(\sigma_1 \circ \sigma_2 \circ \cdots \circ \sigma_k)(s) = t.$$

Example C.1.3 The permutations $\{(1\ 3\ 5\ 7\ 9)(2\ 4\ 6\ 8), (3\ 9)(7\ 1\ 5)(2\ 6)(4\ 8)\}$ do not act transitively, because no sequence of permutations will transform an odd number to an even number. On the other hand, $\{(1\ 3\ 5\ 7\ 8)(2\ 4\ 6\ 9), (3\ 9)(7\ 1\ 5)(2\ 6)(4\ 8)\}$ do act transitively.

It is clear that a single permutation acts transitively if and only if it has a single cycle.

If π is any permutation, then $\|\pi\|$ denotes the number of cycles in its cyclic form. Thus, $\|(1)(2)(3)\| = 3$ and $\|(1\ 2\ 3)(4\ 5\ 6\ 7)\| = 2$.

Proposition C.1.4 *Suppose $a, b \in S$, and π, ρ, σ, τ are permutations of S such that*

$$a, b \text{ belong to the same cycle of } \pi;$$
$$a, b \text{ belong to different cycles of } \rho.$$

Then

1. $\|(a\ b) \circ \pi\| = \|\pi \circ (a\ b)\| = \|\pi\| + 1;$

2. $\|(a\ b) \circ \rho\| = \|\rho \circ (a\ b)\| = \|\rho\| - 1;$

3. $\|\rho \circ \sigma \circ \rho^{-1}\| = \|\sigma\|;$

4. $\|\sigma \circ (a\ b) \circ \tau\| = \|\sigma \circ \tau\| \pm 1.$

PROOF: 1: Suppose π contains the cycle $(a\ A\ b\ B)$ where A and B are character strings. Then the assertion follows from the observation that $(a\ b) \circ (a\ A\ b\ B) = (a\ A)(b\ B)$.

2: Suppose ρ contains the cycles $(a\ A)(b\ B)$ where A and B are character strings. Then the assertion follows from the observation that $(a\ b) \circ (a\ A)(b\ B) = (a\ A\ b\ B)$.

3: The assertion follows from the fact that $(a\ b\ c \cdots x)$ is a cycle of σ if and only if $(\rho(a)\ \rho(b)\ \rho(c) \cdots \rho(x))$ is a cycle of $\rho \circ \sigma \circ \rho^{-1}$.

4: It follows from parts 1, 2, and 3 that

$$\|\sigma \circ (a\ b) \circ \tau\| = \|\tau \circ \sigma \circ (a\ b) \circ \tau \circ \tau^{-1}\| = \|\tau \circ \sigma \circ (a\ b)\|$$
$$= \|\tau \circ \sigma\| \pm 1 = \|\tau^{-1} \circ \tau \circ \sigma \circ \tau\| \pm 1 = \|\sigma \circ \tau\| \pm 1.$$

Q.E.D.

Exercises C.1

1. Given $\rho = (1\ 2\ 3\ 4)(5)(6\ 7\ 8\ 9), \sigma = (1\ 2)(3\ 4)(5\ 9\ 8\ 7\ 6), \tau = (1\ 9\ 8\ 7\ 6\ 5\ 4\ 3\ 2)$, compute:

   (a) $\rho \circ \sigma$,

   (b) $\sigma \circ \rho$,

   (c) $\rho \circ \tau$,

   (d) $\tau \circ \rho$,

   (e) $\sigma \circ \tau$,

   (f) $\tau \circ \sigma$,

2. Decide which of the following sets of permutations act transitively:

   (a) $\{(1\ 2\ 3)(4\ 5\ 6)(7\ 8)(9), (1\ 3\ 2\ 8\ 7)(4\ 5\ 6\ 9)\}$,

   (b) $\{(1\ 2\ 3)(4\ 5\ 6)(7\ 8)(9), (1\ 3\ 2\ 8\ 6)(4\ 5\ 7\ 9)\}$,

   (c) $\{(1\ 2\ 3)(4\ 5\ 6)(7\ 8)(9), (1\ 3\ 2\ 9)(4\ 5\ 6\ 7\ 8)\}$,

   (d) $\{(1\ 2\ 3)(4\ 5\ 6)(7\ 8)(9), (1\ 3\ 2\ 4)(9\ 5\ 6\ 7\ 8)\}$,

   (e) $\{(1\ 2\ 3\ 4\ 5\ 6\ 7\ 8\ 9), (9\ 8\ 7\ 6\ 5\ 4\ 3\ 2\ 1)\}$,

   (f) $\{(1\ 2\ 3\ 4\ 5\ 6\ 7)(8\ 9), (9\ 8)(7\ 6\ 5\ 4\ 3\ 2\ 1)\}$.

APPENDIX D

MODULAR ARITHMETIC

Let n, a, b be any integers. We define

$$a \equiv b \quad (\text{mod } n)$$

(pronounced "a is congruent to b modulo n") provided $a - b$ is divisible by n (i.e., $(a - b)/n$ is an integer). For instance,

$8 \equiv 3 \ (\text{mod } 5), \qquad 7 \equiv 3 \ (\text{mod } 4), \qquad 7 \equiv 1 \ (\text{mod } 3), \qquad 7 \equiv 1 \ (\text{mod } 2),$

$12 \equiv 0 \ (\text{mod } 4), \qquad 17 \equiv -3 \ (\text{mod } 5), \qquad -7 \equiv 1 \ (\text{mod } 4), \qquad -7 \equiv -4 \ (\text{mod } 3),$

$7x \equiv 3x \ (\text{mod } 4), \qquad 7x \equiv 2x \ (\text{mod } 5), \qquad 7x \equiv x \ (\text{mod } 3), \qquad -17x \equiv -x \ (\text{mod } 16),$

where x is also any integer.

If n is any positive integer, then *arithmetic modulo n* consists of the integers 0, 1, 2, ..., $n - 1$, subject to the standard operations of addition, subtraction, and multiplication, with the added stipulation that equality is to be replaced by congruence. Thus, in arithmetic modulo 6,

$$3 + 5 \equiv 2, \qquad 3 \cdot 5 \equiv 3, \qquad 3 - 5 \equiv 4, \qquad 4x + 5x \equiv 3x,$$

Introduction to Topology and Geometry, Second Edition.
By Saul Stahl and Catherine Stenson Copyright © 2013 John Wiley & Sons, Inc.

while in arithmetic modulo 7,

$$3+4 \equiv 0, \qquad 3 \cdot 4 \equiv 5, \qquad 3-4 \equiv 6, \qquad 4x + 4x \equiv x.$$

Division is problematic. In arithmetic modulo 6 it could be argued that $2 \div 5 \equiv 4$ because $4 \cdot 5 \equiv 2$. However, what about $3 \div 4$? Since there is no integer x such that $4x \equiv 3$, it is clear that, modulo 6, division by 4 is not well defined. In general, divisibility by m is only well defined provided m and n are relatively prime. Consequently, if p is a prime number, then divisibilty by each of the nonzero elements $1, 2, 3, \ldots, p-1$ is well defined.

Arithmetic modulo n is very similar to the standard arithmetic. The operations of addition and subtraction are commutative, associative, and distributive. The additive inverse of a is $n - a$. Multiplicative inverses, when they exist, are harder to describe. For the purposes of this text they can be found by trial and error. For instance, in arithmetic modulo 5, the multiplicative inverses of 1, 2, 3, and 4 are 1, 3, 2, and 4 respectively.

Exercises D

1. Evaluate the following in modulo 6 arithmetic:

   (a) $2+4$

   (b) $2 \cdot 4$

   (c) $2-4$

   (d) $3+4$

   (e) $3-4$

   (f) $3 \cdot 4$

   (g) $3 \cdot 3$

   (h) $4 \cdot 4$

2. Evaluate the following in modulo 7 arithmetic:

   (a) $4+5$

   (b) $4-5$

   (c) $4 \cdot 5$

   (d) $3+5$

   (e) $3-5$

   (f) $3 \cdot 5$

   (g) $4 \cdot 4$

   (h) $5 \cdot 5$

3. Evaluate the following in modulo 11 arithmetic:

   (a) $4+8$

(b) $4 - 8$

(c) $4 \cdot 8$

(d) $3 + 10$

(e) $3 - 10$

(f) $3 \cdot 10$

(g) $4 \cdot 4$

(h) $5 \cdot 5$

4. Simplify the following expressions in modulo 5 arithmetic:

   (a) $2x + 4x$

   (b) $2x + 3y - 3x - y$

   (c) $2x - 3y - (4x - y)$

   (d) $3x - y + 4z + 2(x - 2y + 3z)$

   (e) $3(2x - 3y + 4z) - 2(4x - y + 3z)$

5. Simplify the following expressions in modulo 7 arithmetic:

   (a) $2x + 6x$

   (b) $2x + 3y - 3x - y$

   (c) $2x - 3y - (4x - y)$

   (d) $3x - y + 4z + 3(x - 2y + 3z)$

   (e) $4(2x - 3y + 4z) - 3(4x - y + 3z)$

6. Simplify the following expressions in modulo 3 arithmetic:

   (a) $2x + 2x$

   (b) $2x + y - x - 2y$

   (c) $2x - y - (2x - y)$

   (d) $2x - y + 2z + 2(x - 2y + z)$

   (e) $2(2x - y + 2z) - (2x - y + 2z)$

APPENDIX E

SOLUTIONS AND HINTS TO SELECTED EXERCISES

Exercises 1.1

1. Letting $a, b, c, d, \ldots, z$ refer to the 26 elements of Figure 1.18, we have

$$c \approx i \approx j \approx l \approx m \approx n \approx s \approx u \approx v \approx w \approx z,$$
$$d \approx o, \qquad e \approx f \approx g \approx t \approx y, \qquad q \approx r, x \approx k.$$

Exercises 2.1

4. (a) Proposition 2.1.1 entails $4q = n(n+1)$, and hence the sequence is not graphical if $n \equiv 1, 2 \pmod 4$. Moreover, if G_n realizes $(n, n-1, \ldots, 1)$, then a graph G_{n+4} realizing $(n+4, n+3, \ldots, 1)$ is obtained by adding two Hamiltonian cycles to the arcs of G_n as well as a completely disjoint copy of the easily constructed G_4. Since G_3 and G_4 are easily constructed, G_n exists for all $n \equiv 3, 4 \pmod 4$.

Introduction to Topology and Geometry, Second Edition.
By Saul Stahl and Catherine Stenson Copyright © 2013 John Wiley & Sons, Inc.

(b) Proposition 2.1.1 entails $4q = n(n+3)$, and hence the sequence is not graphical if $n \equiv 2, 3 \pmod 4$. If $n \equiv 0$, then add to G_{n-1} (see part (a) above) two disjoint copies of the bouquet on $n/2$ loops. If $n \equiv 1$, then add to G_{n-1} a disjoint copy of the connected graph that consists of two bouquets on $(n-1)/2$ loops joined by a single arc.

5. (b) Only for $n = 1$. Otherwise, the two nodes of degree n must be adjacent to every other node. This means that no node can have degree 1.

(e) Let H_0 consist of the isolated nodes u_0, v_0. Assume that H_n has been defined so that it has nodes $u_0, v_0, \ldots, u_n, v_n$ such that $\deg u_i = \deg v_i = i$. Define H_{n+1} by adding to H_n the arcs $u_i v_{n-i}, i = 0, 1, 2, \ldots, n$, as well as two new isolated nodes u_{-1}, v_{-1}. Of course, it is now necessary to add 1 to each index. Note that each new generation of arcs is centered in a new location and so this is indeed a simple graph.

8. If the sequence is graphical, then the evenness of the sum follows from Proposition 2.1.1. Conversely, suppose the sum is even. It follows that the number of odd a_i's is even. Each of the even a_i's is realized by a bouquet of loops at one vertex. Two odd terms, say a_i and a_j, can be realized by joining a bouquet of $(a_i - 1)/2$ loops to a bouquet of $(a_j - 1)/2$ loops by a single arc.

Exercises 2.2

6. By Theorem 2.2.2, if and only if $n_1 + n_2, n_1 + n_3, n_2 + n_3$ are all even, which happens if and only if n_1, n_2, n_3 all have the same parity.

9. Let G be traversable. If G is Eulerian, then it has no odd nodes. If G is not Eulerian, connect the endpoints of its traversing walk by a new arc to obtain a graph G'. Since G' is Eulerian, it follows from Theorem 2.2.2 that G has exactly two odd nodes, namely, the said endpoints. Conversely, if G has no odd nodes, we are done by Theorem 2.2.2. If G has exactly two odd nodes, add to G a new arc joining these two odd nodes. The resulting graph G' is, by Theorem 2.2.2, Eulerian. By deleting the new arc from the Eulerian walk of G' we obtain a traversing walk of G.

16. Label the five outside nodes O and the five inside nodes I. A Hamilton cycle would then consist of a cyclic string S of 10 O's and I's. It is easy to see that S cannot contain any of the strings OOOOO, IIIII, IOI, OIO, OIIIIO, IOOOOI as substrings. It follows that it may be assumed without loss of generality that $S = $ OOIIOOOIII. This possibility, however, can be eliminated by trial and error.

19. If a is a loop, then $G - a$ is connected. Otherwise, let u and v be the distinct endpoints of a. Let G_u denote the subgraph of G spanned by the set of nodes that can be connected to u by a path that does not contain v. Let G_v denote the subgraph of G spanned by the set of nodes that can be connected to u by a path that does not contain v. Both G_u and G_v are connected. If they are disjoint, then they are the two components of $G - a$; otherwise, $G - a$ is connected.

21. As Figure E.1 indicates, this graph is really the Petersen graph in disguise, and hence, by Exercise 14, it is not Hamiltonian.

Exercises 2.3

2. The endpoints of a path of maximum length in the graph must have degree 1 in the graph. Since every subgraph of an acyclic graph is also acyclic, it follows that every subgraph of an acyclic graph is 1-degenerate. By either Proposition 2.3.2 or Theorem 2.3.1, it is 2-colorable.

8. The subdivision graph $S(G)$ is clearly homeomorphic to G. It is 2-colorable with all the old vertices colored 1 and the new ones colored 2.

Exercises 2.4

2. See Figures E.2 and E.3.
a is nonplanar, as demonstrated by the cycle 123456789 with chords 16, 208, 49.
b is planar.
c is planar.
d is nonplanar, because it has $p = 8$, $q = 20 > 3p - 6 = 18$.
e is nonplanar, because it has $p = 6$, $q = 14 > 3p - 6 = 12$.
f is planar.
g is planar.
h is nonplanar with cycle 12348765 and chords 14, 27, 36.
i is planar.

3. Let P be Hamiltonian cycle. The regions in the interior of P are formed by diagonal arcs, and hence they are 2-colorable. Ditto for the regions outside P. Thus, the regions are 4-colorable.

8. It is easy to see that $K_{n,1,1}$, $K_{2,2,1}$, and $K_{2,2,2}$ are planar. If $n_1 \geq 3$, then $K_{n_1,2,n_3}$ contains $K_{3,3}$. It follows that these three graphs are the only K_{n_1,n_2,n_3}, $n_1 \geq n_2 \geq n_3 \geq 1$, that are planar.

12. For each straight line, select one of its half-planes as its preferred half-plane. Note that for any two abutting regions the numbers of preferred regions that contain them differ by 1. Hence, coloring each region by the parity of the number of preferred regions that contain it is a 2-coloring of the map.

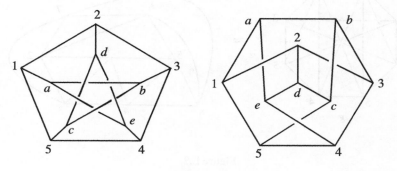

Figure E.1

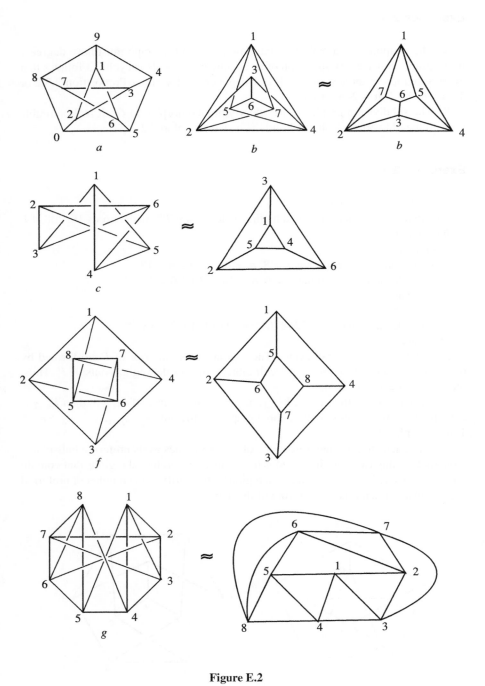

Figure E.2

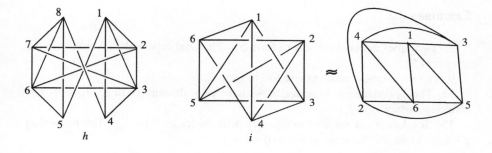

Figure E.3

Exercises 2.5

1. $a \approx d, b \approx c \approx e$.
4. $c \approx d \approx e$.

Exercises 3.1

1. Sphere on left: $db^{-1}, bc, c^{-1}a, a^{-1}d^{-1}$. Sphere on right: $de^{-1}b^{-1}, ef^{-1}c^{-1}$, $fd^{-1}a^{-1}, bh^{-1}g, ci^{-1}h, ag^{-1}i$.
4. Torus on left: $adfa^{-1}eb, cd^{-1}f^{-1}c^{-1}e^{-1}b^{-1}$. Torus on right: $ada^{-1}b, cd^{-1}c^{-1}b^{-1}$.

Exercises 3.2

2. (a) $\tilde{S}_1$, (b) S_0, (c) $\tilde{S}_2$, (d) $\tilde{S}_1$, (e) S_0, (f) $\tilde{S}_3$.
5. For this surface $2 \geq \chi = r \geq 1$. Since the surface is orientable, we must have $\chi = 2$, so this is S_0.
9. Since $2q = pd = r\rho$, it then follows that $2q/d - q + 2q/\rho = 2$, from which it may be concluded that $1/d + 1/\rho > 1/2$. Hence the only possibilities for (d, ρ) are $(3,3), (3,4), (3,5), (4,3), (5,3)$. The desired triples follow from the equation $q = (2/d + 2/\rho - 1)^{-1}$.
14. That $3r = 2q$ follows from the fact that each region is bounded by three distinct arcs and each arc is on the boundary of two distinct regions. The second equation follows from the substitution of the first one into $p - q + r = \chi$.
16. In any such triangulation we must, by Exercise 4, have $q = 3p$. Since the skeleton is simple, it follows that $q \leq p(p-1)/2$. Hence, $p \geq 7$. Figure E.4 displays a triangulation of the torus in which all the triangles have been pasted to form a single polygon. The nodes are labeled with numbers.

Exercises 3.3

5. Two tubes connect the tori to form S_3. The final tube adds a handle. So the answer is S_4.

9. The $n-1$ tubes connect these tori to form S_n.

11. This is tantamount to adding $n-1$ handles to the surface of Exercise 9. The resulting surface is S_{2n-1}.

15. It takes $p-1$ fat arcs to connect p fat nodes to form S_0. The remaining $q-(p-1)$ fat arcs raise the genus to $q-p+1$.

Exercises 3.4

2. (f) No such surfaces exist for $n > 2$. Let $n \leq 2$. For $\beta \equiv n \pmod 2$ we have the surfaces $S_{(2-n-\beta)/2,\beta}$ and $\tilde{S}_{2-n-\beta,\beta}$ whenever $0 \leq \beta \leq 2-n$. For $\beta \equiv n+1 \pmod 2$ we have $\tilde{S}_{2-n-\beta,\beta}$ whenever $0 \leq \beta \leq 2-n$.

3. (a) $S \approx S_{0,2}$. (b) $S \approx \tilde{S}_{1,1}$. (c) $S \approx S_{0,3}$. (d) $S \approx \tilde{S}_{2,1}$.

4. (a) See Figure E.5. $\beta = 2$, orientable, $\chi(S) = 10 - 15 + 3, \chi(S^c) = -2 + 2 = 0, \gamma(S^c) = 1, S \approx S_{1,2}$.

(c) See Figure E.6. $\beta = 2$, nonorientable, $\chi(S) = 10 - 15 + 3 = -2, \chi(S^c) = 0, \gamma(S^c) = 2, S \approx \tilde{S}_{2,2}$.

6. No. $2 - 2\gamma(S_{0,1}) = 2 - 0 = 2$, whereas $\chi(S_{0,1}) = \chi(S_0) - 1 = 2 - 1 = 1$.

11. By Proposition 3.3.1, $\chi(S^c) = \chi(S'^c) + \chi(S''^c) - 2$. Hence $\chi(S) = \chi(S^c) - \beta(S) = \chi(S'^c) + \chi(S''^c) - 2 - (\beta(S') + \beta(S'') - 2) = \chi(S') + \chi(S'')$.

16. If $\beta \geq 1$, then $\tilde{S}_{n,\beta}$ is obtained by excising $\beta - 1$ holes in $\tilde{S}_{n,1}$, and hence, by the above exercise, it is embeddable in $\mathbb{R}^3$. The converse is the content of Proposition 5.3.6.

Exercises 3.5

1. $\qquad O^1: \qquad a_1^{1,-} a_4^{2,-}, a_3^{3,-}, a_2^{4,-}, a_1^{1,-}$

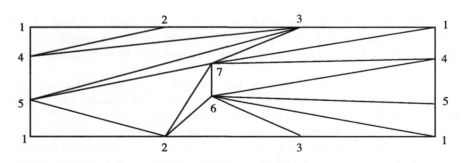

Figure E.4

$$O^2: \quad a_1^{2,-} a_4^{3,-}, a_3^{4,-}, a_2^{1,-}, a_1^{2,-}$$

$$O^3: \quad a_1^{3,-} a_4^{4,-}, a_3^{1,-}, a_2^{2,-}, a_1^{3,-}$$

$$O^4: \quad a_1^{4,-} a_4^{3,-}, a_3^{2,-}, a_2^{1,-}, a_1^{4,-}$$

$$A_{1,(1234)}: \quad (a_1^{1,-})^{-1}, (a_1^{2,-})^{-1}, (a_1^{3,-})^{-1}, (a_1^{4,-})^{-1}, (a_1^{1,-})^{-1}$$

$$A_{2,(1234)}: \quad (a_2^{1,-})^{-1}, (a_2^{2,-})^{-1}, (a_2^{3,-})^{-1}, (a_2^{4,-})^{-1}, (a_2^{1,-})^{-1}$$

$$A_{3,(1234)}: \quad (a_3^{1,-})^{-1}, (a_3^{2,-})^{-1}, (a_3^{3,-})^{-1}, (a_3^{4,-})^{-1}, (a_3^{1,-})^{-1}$$

$$A_{4,(1234)}: \quad (a_4^{1,-})^{-1}, (a_4^{2,-})^{-1}, (a_4^{3,-})^{-1}, (a_4^{4,-})^{-1}, (a_4^{1,-})^{-1}$$

The characteristic of the capped surface is $8 - 16 + 4 = -4$, and its genus is 3. The Riemann surface is therefore $S_{3,4}$.

5. Since the genus is an integer, the numerator of the expression of Theorem 3.5.1 must be even. The desired equality now follows from the fact that $2 - 2n$ is an even integer.

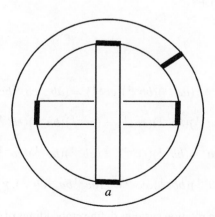

Figure E.5

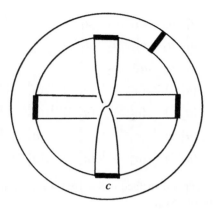

Figure E.6

Exercises 4.1

2. (a) Two regions. One has genus 0 with the single border $cb^{-1}c^{-1}a$. The other has genus 1 with two borders b and a^{-1}.

(b) Two regions. One has genus 0 with two borders a, b^{-1}. The other has genus 1 with one border $cbc^{-1}a^{-1}$.

(c) Two regions. One of genus 0 with border $bc^{-1}a^{-1}c$. One of genus 1 with border $adb^{-1}d^{-1}$.

(d) Two regions. One of genus 0 with borders c^{-1}, d. One of genus 0 with perimeter $ab^{-1} \cup cbd^{-1}a^{-1}$.

6. By Proposition 4.1.2, a given graph G can be embedded on some orientable surface S. If a crosscap is added to S, this results in an embedding of G on the nonorientable surface $S + c$.

Exercises 4.2

1. $(abc)(a^{-1}b^{-1}c^{-1}) \circ (aa^{-1})(bb^{-1})(cc^{-1}) = (ab^{-1}ca^{-1}bc^{-1}), \chi = 2 - 3 + 1 = 0$, genus = 1.
$(abc)(a^{-1}c^{-1}b^{-1}) \circ (aa^{-1})(bb^{-1})(cc^{-1}) = (ac^{-1})(ba^{-1})(cb^{-1}), \chi = 2 - 3 + 3 = 2$, genus = 0.
$(acb)(a^{-1}b^{-1}c^{-1}) \circ (aa^{-1})(bb^{-1})(cc^{-1}) = (ab^{-1})(bc^{-1})(ca^{-1}), \chi = 2 - 3 + 3 = 2$, genus = 0.
$(acb)(a^{-1}c^{-1}b^{-1}) \circ (aa^{-1})(bb^{-1})(cc^{-1}) = (ac^{-1}ba^{-1}cb^{-1}), \chi = 2 - 3 + 1 = 0$, genus = 1.

6. Two of the sixteen rotation systems define embeddings of genus 0, and fourteen of genus 1.

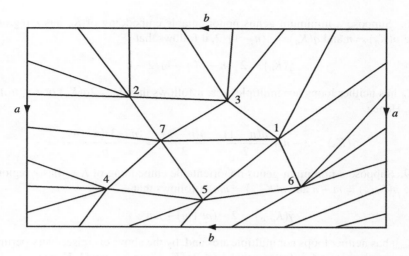

Figure E.7

11. Suppose G has a polygonal orientable embedding with genus m and r regions. Then $p(G) - q(G) + r = 2 - 2m$, or $1 \leq r = 2 - 2m - p(G) + q(G)$. The desired inequality now follows immediately.

16. Proof by contradiction. Suppose the disconnected graph G has a polygonal embedding, which is also a polygonal presentation, say Π. Then there is a region R of the embedding which contains arcs that belong to different components of G. These arcs must be contained in different borders of R. This, however, contradicts the fact that R, being a topological disk, can have only one border.

Exercises 4.3

1. Figure E.7 illustrates an embedding of K_7 on the torus. It follows that $\gamma(K_5) \leq \gamma(K_6) \leq \gamma(K_7) \leq 1$. However, since K_5 is nonplanar (Proposition 2.4.5), it now follows that equality holds.

3. The nodes on the perimeter of each region must alternate between the two colors, and hence these perimeters must have even lengths. Since G has no multiple arcs, it follows that each such perimeter has length at least 4.

10. For each $i = 1, 2$, let R_i be a region of an embedding of G_i on $S^i = S_{\gamma(G_i)}$ whose perimeter contains a. Connect the interiors of R_1 and R_2 by means of a tube T to obtain an embedding of $G_1 \cup G_2$ on $S^1 + S^2$ in which the two copies of a appear on the perimeter of the same region (which contains T). Now push the two copies of A towards each other along T until they coalesce. This results in an embedding of $G_1 +_a G_2$ on the surface $S^1 + S^2 \approx S_{\gamma(G_1)} + S_{\gamma(G_2)}$.

15. Figure E.8 illustrates an embedding of $K_{3,3}$ on the projective plane $\tilde{S}_1$. Since $K_{3,3}$ is nonplanar (see Proposition 2.4.5), this proves that $\tilde{\gamma}(K_{3,3}) = 1$.

16. Suppose a minimum genus nonorientable embedding of K_n has r regions. Since $p(K_n) = n$ and $q(K_n) = n(n-1)/2$, it follows that

$$\gamma(K_n) = 2 - n + n(n-1)/2 - r.$$

As K_n has neither loops nor multiple arcs, it follows that $3r \leq 2q(K_n)$ or $r \leq n(n-1)/3$. Hence,

$$\gamma(K_n) \geq 2 - n + \frac{n(n-1)}{2} - \frac{n(n-1)}{3} = \frac{(n-3)(n-4)}{6}.$$

19. Suppose a minimum-genus nonorientable embedding of $K_{m,n}$ has r regions. Since $p(K_{m,n}) = m + n$ and $q(K_{m,n}) = mn$, it follows that

$$\tilde{\gamma}(K_{m,n}) = 2 - (m+n) + mn - r.$$

As $K_{m,n}$ has neither loops nor multiple arcs and, by the above exercise, every perimeter has length at least 4, it follows that $4r \leq 2q(K_{m,n})$, or $r \leq mn/2$. Hence,

$$\tilde{\gamma}(K_{m,n}) \geq 2 - m - n + mn - \frac{mn}{2} = \frac{(m-2)(n-2)}{2}.$$

23. In Figure E.9 we modify Figure E.7 by adding a crosscap, which allows for the embedding of an additional copy of K_5 (with node set 1, 8, 9, 10, 11.) Thus, Figure E.9 demonstrates that $\tilde{\gamma}(K_5 +_v K_7) \leq 3 < 4 = \tilde{\gamma}(K_5) + \tilde{\gamma}(K_7)$.

Exercises 4.4

2. (a) See Figure E.10. (b) K_5. (c)–(i) See Figure E.10. (j) There are 15 nodes $v_1, v_2, \ldots, v_{15}$, and each v_i is adjacent to $v_i \pm 3$ and $v_i \pm 5$ (indices mod 15). (k) There are 221 nodes $v_1, v_2, \ldots, v_{221}$, and each v_i is adjacent to $v_i \pm 13$ and $v_i \pm 17$ (indices modulo 221). (l) $K_{3,3}$.

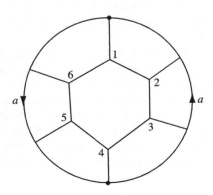

Figure E.8

3. This is the rotation system of the actual embedding defined by Figure 4.41, and its regions have perimeters as follows: R_1, $ad^{-1}b^{-1}$; R_2, bec^{-1}; R_3, dfe^{-1}; R_4, $a^{-1}cf^{-1}$. Hence $\varphi(R_1) = -3, \varphi(R_2) = 0, \varphi(R_3) = 2, \varphi(R_4) = 1$. If $\Gamma = \mathbb{Z}_3$, then $k_1 = k_2 = 1$, $k_3 = k_4 = 3$. The derived embedding therefore has six triangles, two nonagons, and genus 0. If $\Gamma = \mathbb{Z}_4$, then $k_1 = 4$, $k_2 = 1$, $k_3 = 2$, $k_4 = 4$. The derived embedding therefore has four triangles, two hexagons, two dodecagons, and genus 1. If $\gamma = \mathbb{Z}_n$ then four cases must be distinguished. If n is divisible by 6, then $k_1 = n/3, k_2 = 1, k_3 = n/2, k_4 = n$. The derived embedding therefore has thre n-gons, n triangles, two $(3n/2)$-gons, one $(3n)$-gon, and genus $n/2 - 2$. If n is divisible by 3 but not by 2, then $k_1 = n/3, k_2 = 1, k_3 = n, k_4 = n$. The derived embedding therefore has three n-gons, n triangles, two $3n$-gons, and genus $(n-3)/2$. If n is divisible by two but not by 3, then $k_1 = n$, $k_2 = 1$, $k_3 = n/2$, $k_4 = n$. The derived embedding therefore has two $(3n)$-gons, n triangles, two $(3n/2)$-gons, and genus $n/2 - 1$. If n is divisble by neither 2 nor 3, then $k_1 = n$, $k_2 = 1$, $k_3 = n$, $k_4 = n$. The derived embedding therefore has three $3n$-gons, n triangles, and genus $(n-1)/2$.

5. The perimeters of the regions of the given rotation system are

$$abcd, a^{-1}, b^{-1}, c^{-1}, d^{-1},$$

and the genus of the embedding is 0. The perimeters of the lifted regions are

$$a_0b_1c_2d_3, a_1b_2c_3d_0, a_2b_3c_0d_1, a_3b_0c_1d_2$$

and

$$a_0^{-1}a_1^{-1}a_2^{-1}a_3^{-1}, b_0^{-1}b_1^{-1}b_2^{-1}b_3^{-1}, c_0^{-1}c_1^{-1}c_2^{-1}c_3^{-1}, d_0^{-1}d_1^{-1}d_2^{-1}d_3^{-1}.$$

The ambient surface of the lifted embedding has genus 3. Moreover, each of these eight perimeters is a Hamiltonian cycle through the nodes 0, 1, 2, 3. Hence, if new nodes $v_1, v_2, \ldots, v_8$ are selected in the interiors of these eight regions and joined to the four nodes 0, 1, 2, 3, we obtain an embedding of $K_{8,4}$ on S_3. On the other hand, by Exercise 4.3.4, $\gamma(K_{8,4}) \geq (8-2)(4-2)/2 = 3$.

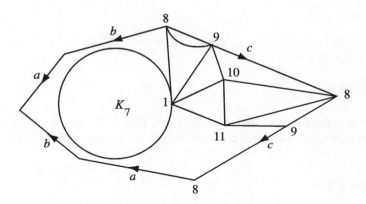

Figure E.9

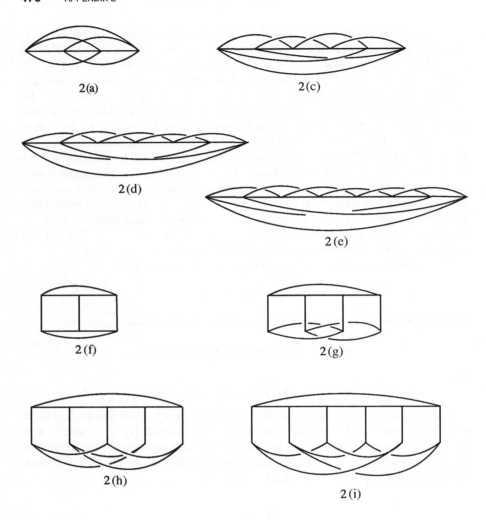

2(a)

2(c)

2(d)

2(e)

2(f)

2(g)

2(h)

2(i)

Figure E.10

Exercises 5.1

1. Up to plane isotopies, the crossing A (see Fig. E.11) can be completed to a knot diagram without additional crossings only in the manners depicted in B and C.

Exercises 5.2

4. (a) 5_1; see Figure E.12. A: $2x - w - y \equiv 0$; B: $2y - x - z \equiv 0$; C: $2z - y - t \equiv 0$;

D: $2t - w - z \equiv 0$; E: $2w - x - t \equiv 0$. If $p = 5$, then $x = 0, y = 4, z = 3, w = 1, t = 2$ is a solution. No labelings exist for any other p.

(b) 5_2; see Figure E.12. A: $2x - y - z \equiv 0$; B: $2y - x - w \equiv 0$; C: $2w - x - t \equiv 0$; D: $2z - y - t \equiv 0$; E: $2t - w - z \equiv 0$. If $p = 7$, then $x = 6, y = 3, z = 2, w = 0, t = 1$ is a solution. No labelings exist for any other p.

(f) 7_1 : $p = 7$. (g) 7_2 : $p = 11$. (h) 7_3 : $p = 13$. (i) 7_4 : $p = 3, 5$.

6. Any $n \geq 3$ and any odd prime divisor p of $2n - 3$.

10. Suppose a p-labeling of a knot uses exactly two distinct labels, say a and b. It follows from the connectedness of the diagram of a knot that some crossing equation involves both a and b and hence $a \equiv b$, contradicting their distinctness. Every split link diagram has a p-labeling with two distinct labels: Label each strand on one side of the splitting line 0 and each strand on the other side 1.

14. For $k = 2$ use 2^2 of Figure 5.8. For $k = 3$ use text Figure 5.20. For $k \geq 4$ see Figure E.13.

Exercises 5.3

1. If n is even, there are two components. The linking numbers are $\pm n/2$.

5. Fix k, and for each $j = 1, 2, \ldots, k$ let C_j be the counterclockwise circle of radius 1 centered at the point with polar coordinates $(0.5, 2\pi/j)$, beginning and ending at the point $(1.5, 2p/j)$. At each crossing choose the overstrand and understrand so that the value is 1.

8. (a) (136, 245), (145, 236), (146, 235).

9. (a) (146, 235).

Exercises 5.4

1. For a, b, c, d: $[\vec{D}] = A^4 d^3 + 4A^3 Bd^2 + 6A^2 B^2 d + 4AB^3 + B^4 d$ and $\langle \vec{D} \rangle = -A^{10} + A^6 - A^2 - A^{-6}$.

For a, d: $w(\vec{D}) = -4, V(\vec{L}) = (-A)^{-3(-4)} \langle \vec{D} \rangle = -A^{22} + A^{18} - A^{14} - A^6$.

For b, c: $w(\vec{D}) = 4, V(\vec{L}) = (-A)^{-3(4)} \langle \vec{D} \rangle = -A^{-2} + A^{-6} - A^{-10} - A^{-18}$.

9. $\vec{L}$ and $\vec{L}^*$ have identical bracket and Kauffman polynomials. It follows from the definition of the Jones polynomials that $V(\vec{L})/V(\vec{L}^*)$ is a power of A.

14. $(-A^{-2} - A^{-10})^n$.

21. One of the two distinct polynomials is $-A^{-6} - 2A^{-14} + A^{-18} - A^{-22} + A^{-26}$.

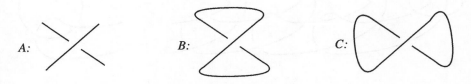

Figure E.11

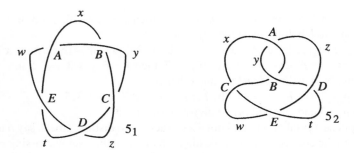

Figure E.12

Exercises 5.5

1. (a) Use the fact that max deg $V(K) = -3w(K) + v(D) + 2(r_B(D) - 1))$, min deg $V(K) = -3w(K) - v(D) - 2(r_A(D) - 1))$, and, for achiral knots, max deg $= -$min deg.

(b) Suppose first that $w(K) \geq 0$. If D is achiral, then $v \leq 3w = r_B - r_A$ and $v = r_A + r_B - 2$. It follows that $r_A \leq 1$ and hence $r_A = 1$. This implies that every crossing is an isthmus, a contradiction. A similar proof works for negative w.

Exercises 5.6

1. The presentation consisting of the Seifert disks with the identified antiparallel arcs has two nodes for each crossing, making for a total of $2v$ nodes. It has $2v$ arcs that join distinct crossings as well as v arcs at the crossings, making for a total of $3v$ arcs. Hence this presentation has characteristic $\chi(S) = 2v - 3v + s = s - v$ and $\beta = 1$. It follows that $\chi(S^c) = s - v + 1$. Hence $\gamma(S) = \gamma(S^c) = (2 - \chi(S^c))/2 = (v - s + 1)/2$.

11. Because 3_a is not an unknot, its genus is not 0. It has two Seifert circles and three crossings. Hence, by Exercise 1, it has genus 1.

19. The eleven regions labeled A in Figure E.14 can be used to create an orientable

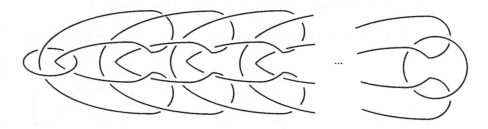

Figure E.13

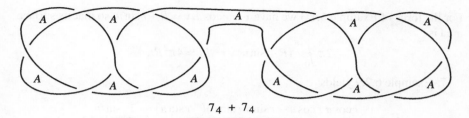

$$7_4 + 7_4$$

Figure E.14

spanning surface S of $7_4 + 7_4$. This surface has characteristic $2 \cdot 14 - 3 \cdot 14 + 11 = -3$. It follows that $\chi(S^c) = -3 + 1 = -2$ and hence $\gamma(S^c) = 2$. By Exercise 18, $\tilde{\gamma}(7_4 + 7_4) \leq 2 \cdot 2 + 1 = 5 < 3 + 3 = \tilde{\gamma}(7_4) + \tilde{\gamma}(7_4)$.

Exercises 6.1

4. The typical point on the said line segment is (ka, kb, c). When this point is rotated about the x-axis into the xz-plane, it falls onto the point $(ka, 0, \sqrt{k^2 b^2 + c^2})$. This point lies on the hyperbola

$$\frac{z^2}{b^2} - \frac{x^2}{a^2} = \frac{c^2}{b^2}$$

in the xz-plane.

Exercises 6.2

3.
$$\mathbf{X}_u = R(-\sin u, \cos u \cos v, \cos u \sin v),$$
$$\mathbf{X}_v = R(0, -\sin u \sin v, \sin u \cos v),$$
$$\mathbf{n} = (\cos u, \sin u \cos v, \sin u \sin v),$$
$$\mathbf{X}_{uu} = R(-\cos u, -\sin u \cos v, -\sin u \sin v),$$
$$\mathbf{X}_{uv} = R(0, -\cos u \sin v, \cos u \cos v),$$
$$\mathbf{X}_{vv} = R(0, -\sin u \cos v, -\sin u \sin v),$$
$$E = R^2, \quad F = 0, \quad G = R^2 \sin^2 u, \quad L = -R, \quad M = 0, \quad N = -R \sin^2 u.$$

6.
$$\mathbf{X}_u = (x_u, y_u \cos v, y_u \sin v),$$
$$\mathbf{X}_v = (0, -y \sin v, y \cos v),$$
$$\mathbf{X}_u \times \mathbf{X}_v = y(y_u, -x_u \cos v, -x_u \sin v),$$
$$|\mathbf{X}_u \times \mathbf{X}_v| = y \sqrt{x_u^2 + y_u^2}$$

Hence the surface area is

$$\int_0^{2\pi} \int_a^b y \sqrt{x_u^2 + y_u^2} \, du \, dv = 2\pi \int_a^b y \sqrt{x_u^2 + y_u^2} \, du$$

For the torus of Example 6.1.3 we have $x = r\cos u, y = R + r\sin u$, so that its surface area is

$$2\pi \int_0^{2\pi} (R + \sin u)\sqrt{r^2}\, du = 4\pi^2 Rr.$$

7. Example 6.2.9 yields

$$K = \frac{(-r\cos u\, r\cos u - r\sin u\, r\sin u)(-r\sin u)}{r^2(R + r\sin u)} = \frac{\sin u}{R + r\sin u}.$$

It follows from the computations of the previous exercise that the total curvature of the torus is

$$\int_0^{2\pi} \int_0^{2\pi} (R + r\sin u)\sqrt{r^2}\, \frac{\sin u}{R + r\sin u}\, du\, dv = 0.$$

Exercises 6.3

1. (a)

$$s = \int_1^2 \sqrt{(2 + 4t^4)(2t)^2 + 2(2t^3 - 1)(2t)(3t^2) + (4t^6 + 1)(3t^2)^2}\, dt \approx 67.946.$$

5. (a) Here $t = 2, \tau = -1$,

$$\cos\theta = \frac{(18)(4)(-2) + 7[(4)(-2) + (1)(-2)] + 6(1)(-2)}{\sqrt{18 \cdot 16 + 14 \cdot 4 \cdot 1 + 6 \cdot 1}\sqrt{18 \cdot 4 + 14 \cdot 4 + 6 \cdot 4}} = \frac{-226}{\sqrt{350}\sqrt{152}},$$

with $\theta \approx 168.5°$.

10. Let α be the angle from the u-parameter curve to the bisector. Then

$$\cos\alpha = \frac{E \cdot 1 \cdot f' + G \cdot 0 \cdot g'}{\sqrt{E \cdot 1^2 + G \cdot 0^2}\sqrt{E \cdot f'^2 + G \cdot g'^2}} = \frac{\sqrt{E}f'}{\sqrt{E \cdot f'^2 + G \cdot g'^2}}.$$

Let β be the angle from the bisector to the v-parameter curve. Then

$$\cos\beta = \frac{E \cdot f' \cdot 0 + G \cdot g' \cdot 1}{\sqrt{E \cdot f'^2 + G \cdot g'^2}\sqrt{E \cdot 0^2 + G \cdot 1^2}} = \frac{\sqrt{G}g'}{\sqrt{E \cdot f'^2 + G \cdot g'^2}}.$$

The equality of α and β yields the desired equation.

13. We work with the standard parametrization of surfaces of revolution whose first fundamental form is specified in Example 6.3.5. The latitudes are the v-parameter curves, and hence $u' = 0, v' = 1$. Since this parametrization is orthogonal, we have $\theta \equiv \pi/2$ and $\theta' = 0$.

Gauss's Equation becomes

$$0 = \frac{0 \cdot 0 - (2yy'(u)) \cdot 1}{2\sqrt{EG}},$$

which is clearly satisfied if and only if $y'(u) = 0$.

19. For this cylinder we have the first fundamental form

$$ds^2 = du^2 + r^2 dv^2$$

Since $E_u = 0 = G_v$ Gauss's Equation becomes $\theta' = 0$. For the given helix, $u' = k$, $v' = 1$ and so, by Lemma 6.3.11, we have

$$\cos\theta = \frac{\sqrt{1} \cdot k}{\sqrt{1 \cdot k^2 + r^2 \cdot 1^2}},$$

so that θ is a constant and has derivative 0.

Exercises 6.4

1. For this surface,

$$E = 1 + f_x^2, \quad F = f_x f_y, \quad G = 1 + f_y^2,$$

$$L = \frac{f_{xx}}{\sqrt{1 + f_x^2 + f_y^2}}, \quad M = \frac{f_{xy}}{\sqrt{1 + f_x^2 + f_y^2}}, \quad N = \frac{f_{yy}}{\sqrt{1 + f_x^2 + f_y^2}}.$$

At the point $P = (a, b, f(a,b))$, the normal section is parametrized by $x = a + t\cos\alpha$, $y = b + t\sin\alpha$, so that $x' = \cos\alpha$ and $y' = \sin\alpha$. The substitution of all these value into the formula of Lemma 6.4.1 yields

$$\kappa_{\mathbf{d}(\alpha)} = \frac{1}{\sqrt{1 + f_x^2 + f_y^2}} \frac{f_{xx}\cos^2\alpha + f_{xy}\sin(2\alpha) + f_{yy}\sin^2\alpha}{(1 + f_x^2)\cos^2\alpha + f_x f_y \sin(2\alpha) + (1 + f_y^2)\sin^2\alpha}.$$

8. The point is an umbilic if and only if the two solutions of the quadratic equation of Proposition 4.7 are equal. This is tantamount to the discriminant

$$\Delta = (EN + GL - 2FM)^2 - 4(EG - F^2)(LN - M^2)$$

vanishing. However, this discriminant equals

$$4\left(\frac{EG - F^2}{E^2}\right)(EM - FL)^2 + \left[EN - GL - \frac{2F}{E}(EM - FL)\right]^2,$$

which, since $EG - F^2 > 0$, vanishes if and only if E, F, G are proportional to L, M, N.

Exercises 6.5

2. $K = -1$.

6. Since $\mathbf{X}(u,v) = (x(u), y(u)\cos v, y(u)\sin v)$, it follows that $\mathbf{X}_u = (x', y'\cos v, y'\sin v)$ and $\mathbf{X}_v = (0, -y\sin v, y\cos v)$, and hence

$$E = x'^2 + y'^2, \quad F = 0.$$

Since u is an arclength parameter of the generating curve, it follows that $E = 1$.

Exercises 6.7

3. Proof by contradiction. Let Π be the digon formed by two such geodesics. Then

$$0 > \int\int_{\Pi} K \, dS = \alpha_1 + \alpha_2 \geq 0,$$

which is impossible.

Exercises 6.8

4.
$$\sqrt{(x+1.5)^2 + (y-0.5)^2} \quad \text{if } x+y \leq 1 \text{ and } x \leq y;$$
$$\sqrt{(x-0.5)^2 + (y-2.5)^2} \quad \text{if } x+y \geq 1 \text{ and } x \leq y;$$
$$\sqrt{(x-2.5)^2 + (y-0.5)^2} \quad \text{if } x+y \geq 1 \text{ and } x \geq y;$$
$$\sqrt{(x-0.5)^2 + (y+1.5)^2} \quad \text{if } x+y \leq 1 \text{ and } x \geq y.$$

Exercises 7.1

1. The path in question is parametrized as $(u, v) = (t \cos\theta, t \sin\theta), 0 \leq t \leq r$.
(a)
$$\text{Length} = \int_0^r \frac{dt}{\sqrt{1-t^2}} = \sin^{-1} r \to \frac{\pi}{2}.$$

(b)
$$\text{Length} = \int_0^r \frac{dt}{\sqrt{1-t^2}} = \sin^{-1} r \to \frac{\pi}{2}.$$

6.
$$[Df] = \frac{1}{[(u+1)^2 + v^2]^2} \begin{pmatrix} (u+1)^2 - v^2 & -2v(u+1) \\ 2v(u+1) & (u+1)^2 - v^2 \end{pmatrix},$$

$$M_{\mathscr{H}} = \frac{1}{v^2} \begin{pmatrix} 1 & 0 \\ 0 & 1 \end{pmatrix} \qquad \text{(on domain } f\text{)},$$

$$M_{\mathscr{H}} = \frac{1}{\bar{v}^2} \begin{pmatrix} 1 & 0 \\ 0 & 1 \end{pmatrix} = \left[\frac{(u+1)^2 + v^2}{v} \right]^2 \begin{pmatrix} 1 & 0 \\ 0 & 1 \end{pmatrix} \qquad \text{(on range } f\text{)},$$

and the required Equation (7.14) is tantamount to the easily verified matrix multiplication

$$\begin{pmatrix} (u+1)^2 - v^2 & 2v(u+1) \\ -2v(u+1) & (u+1)^2 - v^2 \end{pmatrix} \begin{pmatrix} (u+1)^2 - v^2 & -2v(u+1) \\ 2v(u+1) & (u+1)^2 - v^2 \end{pmatrix}$$

$$= [(u+1)^2 + v^2]^2 \begin{pmatrix} 1 & 0 \\ 0 & 1 \end{pmatrix}.$$

15. Let $\Delta = 1 + (c/4)(u^2 + v^2)$. Then

$$K = -\frac{1}{2\dfrac{1}{\Delta^2}} \left[\frac{-c + \dfrac{c^2u^2}{4} - \dfrac{c^2v^2}{4}}{\Delta^2} + \frac{-c - \dfrac{c^2u^2}{4} + \dfrac{c^2v^2}{4}}{\Delta^2} \right] = c.$$

20. Here $(\tilde{u}, \tilde{v}) = (au, av)$. Consequently $[Df] = a\mathrm{Id}$. We have $M_{\mathscr{H}} = v^{-2}\mathrm{Id}$, for the domain, and $M_{\mathscr{H}} = \tilde{v}^{-2}\mathrm{Id}$ for the range of D_a, and Equation (7.14) is easily verified.

26. Since $a\mathbb{P}$ has curvature $-1/a$ whereas $\mathbb{P}_b$ has curvature $-b^2$, it follows that $a = b^{-2}$ is a necessary condition for the isometry. To show that it is also sufficient, it suffices to show that the function $f(u,v) = (\tilde{u}, \tilde{v}) = (u/\sqrt{a}, v/\sqrt{a})$ is an isometry from $a\mathbb{P}$ to $\mathbb{P}_{1/\sqrt{a}}$. This is indeed the case, because Equation (7.14) reduces to

$$\begin{pmatrix} \dfrac{4a}{(1-u^2-v^2)^2} & 0 \\ 0 & \dfrac{4a}{(1-u^2-v^2)^2} \end{pmatrix}$$

$$= \begin{pmatrix} \dfrac{1}{\sqrt{a}} & 0 \\ 0 & \dfrac{1}{\sqrt{a}} \end{pmatrix} \begin{pmatrix} \dfrac{4}{(a^{-1}-\tilde{u}^2-\tilde{v}^2)^2} & 0 \\ 0 & \dfrac{4}{(a^{-1}-\tilde{u}^2-\tilde{v}^2)^2} \end{pmatrix} \begin{pmatrix} \dfrac{1}{\sqrt{a}} & 0 \\ 0 & \dfrac{1}{\sqrt{a}} \end{pmatrix}.$$

Exercises 8.1

3. Let $A_0A_1B_1B_0$ be a rectangle (see Fig. E.15). Then construct quadrilaterals $A_iA_{i+1}B_{i+1}B_i$, $i = 1,2,\ldots,n-1$, each of which is congruent to $A_0A_1B_1B_0$. Then $A_0A_nB_nB_0$ is a rectangle in which the sides $A_0A_n = nA_0A_1, B_0B_n = nB_0B_1$ are arbitrarily large. Next apply the same process to $A_0A_nB_nB_0$ to produce a rectangle A_0A_nXY in which A_0Y and A_nX are also arbitrarily large.

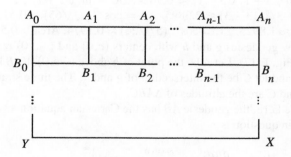

Figure E.15

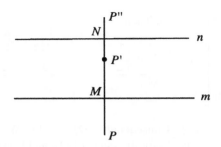

Figure E.16

8. If one triangle has angle sum π then by Exercise 7 a rectangle exists, and so, by Exercise 6, every triangle has angle sum π.

15. Note that if the altitude from a vertex is external, then, by Theorem 8.1.1, one of the two interior angles at the other vertices is obtuse. Since, by Euclid's Proposition 17, at least two of the interior angles are acute, it follows that the altitude from the vertex of the third angle is internal.

21. Let $m \parallel n$, and let P be a point such that $P' = \rho_m(P)$ falls between m and n (Fig. E.16). Set $P'' = \rho_n(P')$. Note that

$$PP'' = PP' + P'P'' = 2P'M + 2P'N = 2MN \ (= \text{constant}).$$

It follows that $\rho_n \circ \rho_m$ acts on such P like a translation in a direction perpendicular to m and n. It follows from Theorem 8.1.8 that this is true for all P.

26. By Theorem 1.11, every isometry f is the composition of $n \leq 3$ reflections. If $n = 0$, f is the identity. If $n = 1$, f is a reflection. If $n = 2$, then, by Proposition 8.1.12 and Exercise 21, f is either a rotation or a translation. If $n = 3$, then, by Exercises 24, 25, f is a glide reflection.

Exercises 8.2

1. $AB = \ln(\sqrt{5}+1/\sqrt{5}-1) \approx 0.9624$, $BC = \ln(\sqrt{2}+1/\sqrt{2}-1) \approx 1.7627$, $AC = \ln(\sqrt{13}+3/\sqrt{13}-3) \approx 2.3895$. $\angle A = \cos^{-1}(7/\sqrt{65}) \approx 0.5191$, $\angle B = \pi - \cos^{-1}(1/\sqrt{10}) \approx 1.8925$, $\angle C = \cos^{-1}(5/\sqrt{26}) \approx 0.1974$. Area ≈ 0.5325.

6. Yes. Draw geodesics g and h with centers $(c,0)$ and $(-c,0)$ respectively and radius $r > 2c$ (Fig. E.17). Let A be the point of h above $(-c,0)$, let B be the point of g above $(c,0)$, and let C be the intersection of g and h. The three straight geodesics through A, B, and C are the altitudes of $\triangle ABC$.

13. In Figure E.18, the geodesic AB has the Cartesian equation $u^2 + v^2 = r^2$, and hence the area in question is

$$\int_{r\cos(\pi-\beta)}^{r\cos\alpha} \int_{\sqrt{r^2-u^2}}^{\infty} \frac{du\,dv}{v^2} = \int_{r\cos(\pi-\beta)}^{r\cos\alpha} \left. -v^{-1} \right]_{\sqrt{r^2-u^2}}^{\infty} du$$

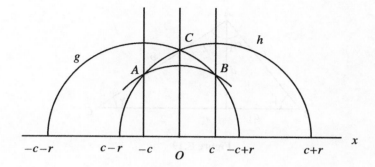

Figure E.17

$$= \int_{r\cos(\pi-\beta)}^{\cos\alpha} \frac{du}{\sqrt{r^2 - u^2}} = \sin^{-1}\frac{u}{r}\bigg]_{r\cos(\pi-\beta)}^{r\cos\alpha}$$
$$= \sin^{-1}(\cos\alpha) - \sin^{-1}(\cos(\pi-\beta))$$
$$= \frac{\pi}{2} - \alpha - \left[\frac{\pi}{2} - (\pi-\beta)\right] = \pi - \alpha - \beta.$$

16. By Proposition 8.2.5, the inversion $I_{(b,0),|a-b|}$ reduces this configuration to that of the previous exercise.

Exercises 8.3

1. Note that in each of (a)–(d) the two equations are symmetrical in a and b, and so only one of them needs to be proved. All of the following proofs refer to Figure E.19 and make use of some of the identities established in the proof of Theorem 8.3.2.

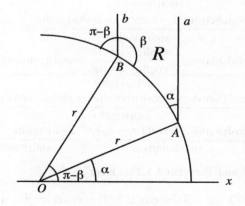

Figure E.18

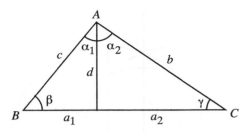

Figure E.19

(a)

$$\tanh b = \cos(\alpha + \alpha_1) = s, \qquad \sinh a \tan \beta = \frac{k^2 - 1}{2k} \cdot \frac{k}{u} = su \cdot \frac{1}{u} = s,$$

(b)

$$\sinh b = \cot(\alpha + \alpha_1) = s/t$$

$$\sinh c \sin \beta = \frac{\cos(\pi - \beta) + \cos \alpha_1}{\sin(\pi - \beta) \sin \alpha_1} \cdot \frac{k}{r} = \frac{\dfrac{-u}{r} + \dfrac{u+s}{r}}{\dfrac{k}{r} \cdot \dfrac{t}{r}} \cdot \frac{k}{r} = \frac{s}{t}.$$

6. (a) The proof below applies to the half-plane triangle of Figure E.19, whose altitude d is internal. Parts (b) and (c) of Exercise 1, the half-plane Theorem of Pythagoras, and the formulas for $\cos(A + B)$ and $\cosh(A + B)$ yield

$$
\begin{aligned}
\cos \alpha &= \cos \alpha_1 \cos \alpha_2 - \sin \alpha_1 \sin \alpha_2 \\
&= \frac{\tanh d}{\tanh c} \cdot \frac{\tanh d}{\tanh b} - \frac{\sinh a_1}{\sinh c} \cdot \frac{\sinh a_1}{\sinh b} \\
&= \frac{\cosh b \cosh c \tanh^2 d - \sinh a_1 \sinh a_2}{\sinh b \sinh c} \\
&= \frac{\cosh b \cosh c (1 - \operatorname{sech}^2 d) - \sinh a_1 \sinh a_2}{\sinh b \sinh c} \\
&= \frac{\cosh b \cosh c - \dfrac{\cosh b}{\cosh d} \cdot \dfrac{\cosh c}{\cosh d} - \sinh a_1 \sinh a_2}{\sinh b \sinh c} \\
&= \frac{\cosh b \cosh c - (\cosh a_1 \cosh a_2 + \sinh a_1 \sinh a_2)}{\sinh b \sinh c} \\
&= \frac{\cosh b \cosh c - \cosh(a_1 + a_2)}{\sinh b \sinh c} = \frac{\cosh b \cosh c - \cosh a}{\sinh b \sinh c}
\end{aligned}
$$

11. By Exercise 1 and Theorem 8.3.2 we have

$$
\begin{aligned}
\sin S &= \sin[\pi - (\pi/2 + \alpha + \beta)] = \cos(\alpha + \beta) = \cos \alpha \cos \beta - \sin \alpha \sin \beta \\
&= (\cosh a \sin \beta)(\cosh b \sin \alpha) - \sin \alpha \sin \beta = (\cosh a \cosh b - 1) \sin \alpha \sin \beta
\end{aligned}
$$

$$= (\cosh c - 1)\frac{\sinh a}{\sinh c} \cdot \frac{\sinh b}{\sinh c} = \frac{(\cosh c - 1)\sinh a \sinh b}{\cosh^2 c - 1}$$

$$= \frac{\sinh a \sinh b}{\cosh c + 1} = \frac{\sinh a \sinh b}{1 + \cosh a \cosh b}.$$

17. First use Taylor expansions to verify that $\sin(x/i) = -i\sinh x$. Then

$$\text{RHS}(11) = \frac{\cos\frac{a}{ri} - \cos\frac{b}{ri}\cos\frac{c}{ri}}{\sin\frac{b}{ri}\sin\frac{c}{ri}} = \frac{\cosh\frac{a}{r} - \cosh\frac{b}{r}\cosh\frac{c}{r}}{i^2\sinh\frac{b}{r}\sinh\frac{c}{r}} = \text{RHS}(8),$$

$$\cos\frac{a}{ri} = \cosh\frac{a}{r},$$

$$\frac{\sin\frac{a}{ri}}{\sin\alpha} = \frac{-i\sinh\frac{a}{r}}{\sin\alpha}, \qquad \text{etc.}$$

Exercises 8.4

1. The isometry T^{-1} of Corollary 8.4.7 is a Möbius transformation, and hence it maps the geodesics of $\mathscr{H}$, which are circles of $\mathbb{C}^*$, into circles of $\mathbb{C}^*$, which are necessarily the geodesics of $\mathscr{P}$. Since T^{-1} transforms the x-axis into the unit circle, and since the geodesics of $\mathscr{H}$ are orthogonal to this axis, the geodesics of $\mathscr{P}$ must be orthogonal to the unit circle.

5. Let $\triangle ABC$ be a geodesic triangle in such a geometry. Because this geometry is isometric to $\mathscr{H}$, it follows that $\text{area}_{\mathscr{H}}(\triangle ABC) = \pi - \alpha - \beta - \gamma$. Because this geometric is conformal and its geodesics are Eucidean straight lines, it follows that $\alpha + \beta + \gamma = \pi$, implying that the triangle has area 0, which is impossible.

9. Because of Corollary 8.4.7 it suffices to prove the exercise for $\mathbb{P}$. There it may be assumed that the circle is centered at the origin. Example 7.1.3 implies that if this circle has Euclidean radius r, then $R = \ln[(1+r)/(1-r)]$ and hence $r = \tanh(R/2)$. If the circle is parametrized as $u = r\cos t$, $v = r\sin t$, then its perimeter relative to $\mathscr{P}$ is

$$\int_{\text{circle}} \frac{2\sqrt{du^2 + dv^2}}{1 - u^2 - v^2} = \int_0^{2\pi} \frac{2r\,dt}{1 - r^2} = \frac{4\pi r}{1 - r^2} = \frac{4\pi\tanh(R/2)}{1 - \tanh^2(R/2)}$$

$$= 4\pi \frac{\dfrac{\sinh(R/2)}{\cosh(R/2)}}{\dfrac{1}{\cosh^2(R/2)}} = 2\pi\sinh(R).$$

14. This follows easily from the fact that

$$f \circ f = \frac{(a^2 + bc)z + b(a + d)}{c(a + d)z + (bc + d^2)}.$$

Exercises 9.1

2. We use the notation of Section 10.2. Let $a(s)$, $b(s)$, $c(s) : I^2 \to S$ be such that $a(0) = b(0) = c(0) = P$, $a(1) = b(1) = c(1) = Q$. The function $H_a : I^2 \to S$ defined by $H_a(s,t) = a(s)$ is a homotopy of $a(s)$ with itself. If $H(s,t)$ is a homotopy of $a(s)$ and $b(s)$, then $G(s,t) = H(s, 1-t)$ is a homotopy of $b(s)$ and $a(s)$. If $H(s,t)$ is a homotopy of $a(s)$ and $b(s)$, and if $G(s,t)$ is a homotopy of $b(s)$ and $c(s)$, then

$$K(s,t) = \begin{cases} H(s, 2t), & 0 \leq t \leq 1/2, \\ G(s, 2t-1), & 1/2 \leq t \leq 1, \end{cases}$$

is a homotopy of $a(s)$ and $c(s)$.

8. This is the free group generated by $n-1$ generators.

14. It is trivial.

18. It is the free group on n generators.

Exercises 9.2

2. See Figure E.20. There is only one node (or star) corresponding to the loop 1, 2, 3, 4, 5, 6, 7, 8. Therefore this is the surface $\tilde{S}^4$. This loop yields the relator $[a^*]^{-1}[d^*]^{-1}[a^*]^{-1}[b^*][c^*]^{-1}[d^*][c^*]^{-1}[b^*]$. The substitutions $\alpha = [a^*]^{-1}$, $\beta = [b^*]$, $\gamma = [c^*]^{-1}$, $\delta = [d^*]$ yield the relator $\alpha\delta^{-1}\alpha\beta\gamma\delta\gamma\beta$ which is cyclically equivalent to the proposed relator. The bound on the rank follows from Theorem 9.2.4 and Example B.3.15.

Exercises 9.3

1. Orientable. $\chi = 1 - 1 + 3 - 1 = 2$, and the link is S^2.

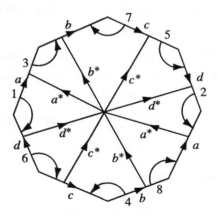

Figure E.20

6. Orientable. $\chi = 1 - 3 + 3 - 1 = 0$,

$$\pi(\mathscr{S},X) = \langle \alpha,\beta,\gamma; \alpha\beta^{-1}, \alpha\beta, \alpha\gamma\beta\gamma\alpha^{-1}\gamma\beta^{-1}\gamma \rangle,$$

and its abelianization is $\mathbb{Z}_2 \times \mathbb{Z}_4$.

11. Nonorientable. $\chi = 2 - 4 + 3 - 1 = 0$.

$$\pi(\mathscr{S},X) = \langle \alpha,\beta,\gamma; \gamma^{-1}\alpha^{-1}\gamma^{-1}\alpha, \alpha\beta\alpha^{-1}\beta, \beta\gamma^{-1}, \gamma\beta \rangle,$$

and its abelianization is $\mathbb{Z} \times \mathbb{Z}_2$.

13. Nonorientable. $\chi = 2 - 2 + 3 - 1 = 2$. The links are S^1 and $\tilde{S}^2$.

19. The identifications are:

$$
\begin{aligned}
\alpha: & \quad PQRST & \equiv S'T'P'Q'R', \\
\beta: & \quad PV'U'X'Q & \equiv XQ'P'VU, \\
\gamma: & \quad QX'Y'WR & \equiv W'R'Q'XY, \\
\delta: & \quad RWVUS & \equiv U'S'R'W'V' \\
\varepsilon: & \quad SUXYT & \equiv Y'T'S'U'X', \\
\phi: & \quad TYW'V'P & \equiv VP'T'Y'W,
\end{aligned}
$$

where α denotes the dual of $PQRST$ directed from $PQRST$ to $S'T'P'Q'R'$, and β, γ, δ, ε, ϕ are similarly defined. By symmetry, a typical node is $\{P \equiv S' \equiv X \equiv W\}$, and a typical arc is $\{PQ \equiv S'T' \equiv XU\}$. Hence

$$\chi = \frac{20}{4} - \frac{30}{3} + \frac{12}{2} - 1 = 0.$$

The crossings 1, 2, 3 of Figure E.21 yield the relator $\beta\delta\alpha^{-1}$, and symmetry adds the relators $\gamma\varepsilon\alpha^{-1}, \delta\phi\alpha^{-1}, \varepsilon\beta\alpha^{-1}, \phi\gamma\alpha^{-1}$. The crossings 4, 5, 6 of Figure E.21 yield the relator $\phi\beta^{-1}\gamma$ and symmetry adds the relators $\beta\gamma^{-1}\delta, \gamma\delta^{-1}\varepsilon, \delta\varepsilon^{-1}\phi, \varepsilon\phi^{-1}\beta$. The fundamental group therefore has the presentation

$$\langle \alpha,\beta,\gamma,\delta,\varepsilon,\phi; \alpha = \beta\delta = \gamma\varepsilon = \delta\phi = \varepsilon\beta = \phi\gamma,$$
$$\beta = \gamma\phi, \gamma = \delta\beta, \delta = \varepsilon\gamma, \varepsilon = \phi\delta\phi = \beta\varepsilon \rangle.$$

To argue that this fundamental group is nontrivial it suffices to find a nontrivial group some of whose elements satisfy the above relations. For each $i = 1, 2, \ldots, 6$ let R_i denote the $72°$ rotation of the icosahedron of Figure E.22 about the vertex i. Then it is easily verified that

$$R_2 \circ R_4 = R_3 \circ R_5 = R_4 \circ R_6 = R_5 \circ R_2 = R_6 \circ R_3 = R_1$$

and

$$R_2 = R_3 \circ R_6, \quad R_3 = R_4 \circ R_2, \quad R_4 = R_5 \circ R_3, \quad R_5 = R_6 \circ R_4, \quad R_6 = R_2 \circ R_5.$$

Finally, note that in the abelianization of this group, the first five relators yield the equations

$$\beta = \gamma = \delta = \varepsilon = \phi, \qquad \alpha = 2\beta.$$

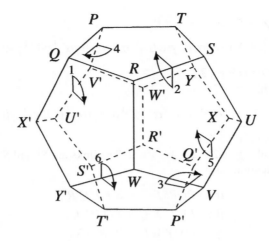

Figure E.21

The sixth relator now permits us to conclude that all the generators are zero, so that the abelianization is trivial.

26. $(1\,2)(1\,3)(2\,3)(1\,3) = (2\,3)(1\,2)(1\,3)(1\,2) = (1\,3)(2\,3)(1\,2)(2\,3) = \mathrm{Id}$. Since these transpositions do not commute, it follows that the knot group is not abelian and, in particular, is not $\mathbb{Z}$. Hence this knot is not trivial.

31. $\chi = 1 - 4 + 4 - 1 = 0$. The fundamental group has presentation

$$\langle \alpha, \beta, \gamma, \delta; \alpha\delta^{-1}\gamma, \beta\alpha^{-1}\delta, \gamma\beta^{-1}\alpha, \delta\gamma^{-1}\beta \rangle.$$

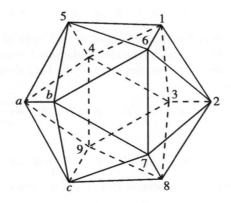

Figure E.22

Exercises 10.1

1. If $P = Q \neq R$, we have $0 + 1 \geq 1$. If $P \neq Q = R$, we have $1 + 0 \geq 1$. If $P = R \neq Q$, we have $1 + 1 \geq 0$. If $P = Q = R$, we have $0 + 0 \geq 0$.

6. Suppose S is infinite and contains the two distinct points P and Q, and suppose its cofinite topology comes from a metric d. Let $\delta = d(P, Q)$. The neighborhoods $N_{\delta/2}(P)$ and $N_{\delta/2}(Q)$ are then disjoint open sets each of which has a finite complement. This contradicts the infinitude of S.

11. The positivity and symmetry of d are immediate. Let f, g, h be three elements of F. Then

$$
\begin{aligned}
d(f, h) = \sup\{|f(s) - h(s)|\} &= \sup\{|f(s) - g(s) + g(s) - h(s)|\} \\
&\leq \sup\{|f(s) - g(s)| + |g(s) - h(s)|\} \\
&\leq \sup\{|f(s) - g(s)|\} + \sup\{|g(s) - h(s)|\} \\
&= d(f, g) + d(g, h).
\end{aligned}
$$

16. Yes.

Exercises 10.2

1. Note that $(g \circ f) \circ (f^{-1} \circ g^{-1}) = g \circ (f \circ f^{-1}) \circ g^{-1} = g \circ \mathrm{Id}_{S'} \circ g^{-1} = g \circ g^{-1} = \mathrm{Id}_{S''}$. Similarly, $(f^{-1} \circ g^{-1}) \circ (g \circ f) = f^{-1} \circ (g^{-1} \circ g) \circ f = f^{-1} \circ \mathrm{Id}_{S''} \circ f = f^{-1} \circ f = \mathrm{Id}_S$.

6. Let D denote the given domain, $f : D \to \mathbb{R}$ be the given continuous function, and $G = \{(x, y, f(x, y)) \mid (x, y) \in D\}$ be its graph. Let $\phi : D \to G$ be defined by $\phi(x) = (x, y, f(x, y))$. Then ϕ is clearly both injective and bijective, and so it is invertible. The continuity of f implies that of ϕ. The continuity of ϕ^{-1} follows from the fact that it is distance-reducing.

11. The reason $f(x) = x^2$ is not continuous is that $(-\infty, 1)$ is an open set but $f^{-1}((-\infty, 1)) = \{x \mid x^2 < 1\} = (-1, 1)$ is not an open set.

Exercises 10.3

1. Let P and Q be two distinct points of S. Then the sets $U = \{P\}$ and $V = S - \{P\}$ are disjoint nonempty open subsets whose union is S.

6. There are no two disjoint nonempty subsets of S.

13. Yes.

Exercises 10.4

1. An open cover $\mathscr{U}$ of $S \cup T$ contains a finite subcover $\mathscr{U}_S$ of S and a finite subcover $\mathscr{U}_{\mathscr{T}}$ of T. It is clear that $\mathscr{U}_{\mathscr{S}} \cup \mathscr{U}_{\mathscr{T}}$ is a finite subcover of $S \cup T$.

5. Let C be a closed subset of the compact space S, and suppose $\mathscr{U}$ is an open cover of C. Then $\mathscr{U} \cup \{S - C\}$ is an open cover of S. If follows that $\mathscr{U} \cup \{S - C\}$ has a finite subcover of S, say $\mathscr{U}'$. Then $\mathscr{U}' - \{S - C\}$ is the required finite subcover of C.

11. Let S be the underlying compact space, and let $G_i = S - F_i, i = 1, 2, 3, \ldots$ Then

$$G_1 \subset G_2 \subset G_3 \subset \cdots,$$

and if $\cap F_i = \emptyset$, then $\{G_i\}$ is an open cover of S. Since S is compact, there is an index α such that $G_\alpha = S$. But then $F_\alpha = \emptyset$, which contradicts the fact that the F_i are nonempty.

Exercises 11.1

10. For any $\mathbf{y}$ in P^Δ, $\varepsilon \frac{\mathbf{y}}{\|\mathbf{y}\|}$ is in P.

13. Write $\mathbf{z}$ as a convex combination of the vertices of P. Show that if the coefficient λ_i of some vertex $\mathbf{v_i}$ in this convex combination is nonzero, then F is contained in the face of P^Δ defined by the supporting hyperplane $\mathbf{v}_i \cdot \mathbf{y} = 1$. What faces of P^Δ can contain a facet?

16. Consider the intersection of all faces containing the point.

18. Show that if $\mathbf{z} \in P$ and $\mathbf{z} \notin F$, then $\mathbf{z} \notin \Psi(\Psi(F))$.

21 − 23. This hint is for all three exercises. Let Q be a polytope that is the convex hull of a set A. Explain why any vertex of Q must be an element of A.

Exercises 11.2

3. Let e_1 and e_2 be distinct edges of a 3-polytope P and let $S(e_1)$ and $S(e_2)$ be their images in the Schlegel diagram. Suppose there is some point $\mathbf{p}$ in both $S(e_1)$ and $S(e_2)$. Then the ray from $\mathbf{x}$ (the point used to define the Schlegel diagram) through $\mathbf{p}$ must intersect both e_1 and e_2. Explain why this means that e_1 and e_2 intersect in a vertex of P on this ray. Then explain why this means that the Schlegel diagram is a planar graph.

6. Use Theorem 2.3.1. A Schlegel diagram and Theorem 2.4.1 may help.

8. Let P be a 3-polytope. If P has a vertex $\mathbf{v}$ of degree 3, consider the 2-faces of P that contain $\mathbf{v}$. If P has a triangular face, let G be the graph from the Schlegel diagram of P that uses that triangle as the outer face. Let the vertices of that triangle be $\mathbf{a}$, $\mathbf{b}$, and $\mathbf{c}$. Let $\mathbf{x}$ be a vertex of the other 2-face of P that contains $\mathbf{a}$ and $\mathbf{b}$. Use the fact that G is 3-connected to find an appropriate path from $\mathbf{c}$ to $\mathbf{x}$.

Exercises 11.3

6. By Exercise 5, β^4 is the bipyramid over β^3. The edges of β^4 containing $(0,0,0,1)$ connect that vertex to the vertices of β^3, namely $(1,0,0,0)$, $(0,1,0,0)$,

$(0,0,1,0)$, $(-1,0,0,0)$, $(0,-1,0,0)$, and $(0,0,-1,0)$. The midpoints of these six edges are $(\frac{1}{2},0,0,\frac{1}{2})$, $(0,\frac{1}{2},0,\frac{1}{2})$, $(0,0,\frac{1}{2},\frac{1}{2})$, $(-\frac{1}{2},0,0,\frac{1}{2})$, $(0,-\frac{1}{2},0,\frac{1}{2})$, and $(0,0,-\frac{1}{2},\frac{1}{2})$. These are the vertices of the vertex figure. They all lie in the hyperplane $x_4 = \frac{1}{2}$. Their first three coordinates are the coordinates of the vertices of β^3 multiplied by $\frac{1}{2}$. Thus the vertex figure is a regular octahedron.

9. Each of the 24 3-faces of the 24-cell is an octahedron. Each 2-face is in 2 3-faces and each 3-face has 8 2-faces, so $2f_2 = (24)(8)$ and $f_2 = 96$. Because the vertex figure of the 24-cell is a 3-cube, which has 6 2-faces, each vertex of the 24-cell is in 6 3-faces. Each 3-face has 6 vertices, so $6f_0 = (24)(6)$, and $f_0 = 24$. In the 3-cube, each vertex is in 3 2-faces. Because the 3-cube is the vertex figure of the 24-cell, each edge of the 24-cell is in 3 3-faces. Each 3-face has 12 edges, so $3f_1 = (24)(12)$, and $f_1 = 96$.

13. Make the large outer face a pyramid with a square base. There is only one other vertex. Place that in the middle of the pyramid with a square base and then connect it appropriately.

Exercises 11.4

7. Let P be a d-polytope with f-vector $(f_0, f_1, f_2, \ldots, f_{d-1})$. Let A be the affine hull of P. Let Q be the pyramid over P created by taking the convex hull of P and a point $\mathbf{v}$ outside A. The faces of Q either contain $\mathbf{v}$ or do not contain $\mathbf{v}$. We will consider these cases separately.

Suppose a face F of Q does not contain $\mathbf{v}$. The face F is the convex hull of its vertices, which are also vertices of Q. Because the vertices of Q are $\mathbf{v}$ and the vertices of P (Exercise 21 in Section 11.1), F must be a face of P. Also, because P is a face of Q (with supporting hyperplane A), every face of P is a face of Q.

Now let F' be a face of Q that contains $\mathbf{v}$ and at least one point of P, and let H' be its supporting hyperplane. The intersection of H' and A is a supporting hyperplane H of P. Let $F = H \cap P$. Then F' is the pyramid over F, and if F' is a $(k+1)$-face of Q, then F is a k-face of P.

Conversely, let F be a face of P with supporting hyperplane H in A. Then H can be extended to a supporting hyperplane H' of Q by taking the affine hull of H and $\mathbf{v}$. The hyperplane H' intersects Q in the face F', which is the convex hull of F and $\mathbf{v}$. The face F' is the pyramid over F, so if F is a k-face of P, then F' is a $(k+1)$-face of Q.

Thus the faces of Q are $\mathbf{v}$, the faces of P, and the pyramids over nonempty faces of P. Thus Q has $f_0 + 1$ vertices. For $k \geq 1$, Q has $f_k + f_{k-1}$ k-faces, f_k of them in P and f_{k-1} of them from pyramids over $(k-1)$-faces of P. Therefore, the f-vector of Q is $(f_0 + 1, f_1 + f_0, f_2 + f_1, \ldots, f_{d-1} + f_{d-2}, 1 + f_{d-1})$.

10. (a) We already know that the f-vector of the 4-simplex is $(5, 10, 10, 5)$. The 5-simplex is the pyramid over the 4-simplex, so by Exercise 7, the f-vector of the 5-simplex is $(6, 15, 20, 15, 6)$. Alternatively, we know that any d of the $d + 1$ vertices of a d-simplex define a facet of the d-simplex, and that facet is a $(d - 1)$-simplex.

Thus any set of k vertices of the 5-simplex defines a $(k-1)$-face, and the f-vector is $\left(\binom{6}{1}, \binom{6}{2}, \binom{6}{3}, \binom{6}{4}, \binom{6}{5}\right) = (6, 15, 20, 15, 6)$.

13. Try some of the regular 4-polytopes. The Euler-Poincaré relation says that f_3 is determined by f_0, f_1, and f_2. It remains to show that the affine hull of the vectors (f_0, f_1, f_2) of the four chosen 4-polytopes is 3-dimensional. Recall that the affine hull is a translate of a linear subspace. Translate this affine hull so that it goes through the origin by choosing one of the four 4-polytopes and subtracting its vector (f_0, f_1, f_2) from that of the other three 4-polytopes. Now show that these three new vectors are linearly independent.

17. Here is an outline of the proof that $f_3 \le \frac{1}{2} f_0(f_0 - 3)$. Each 2-face is in 2 3-faces. Each 3-face must have at least 4 2-faces. Thus $2f_2 \ge 4f_3$. Apply the Euler-Poincaré relation to get $f_1 - f_0 \ge f_3$. Now use the fact that $f_1 \le \binom{f_0}{2}$, since there can be at most one edge for each pair of vertices. Once the inequality $f_3 \le \frac{1}{2} f_0(f_0 - 3)$ is established, apply it to the dual of a polytope P in order to get $f_0 \le \frac{1}{2} f_3(f_3 - 3)$.

Exercises A

1. (a)
$$\mathbf{D}(s) = \left(\frac{\sqrt{1+s}}{2}, -\frac{\sqrt{1-s}}{2}, \frac{1}{\sqrt{2}} \right), \qquad |\mathbf{D}'(s)| = 1,$$

so
$$\mathbf{T}(s) = \mathbf{D}'(s), \mathbf{T}'(s) = \left(\frac{1}{4\sqrt{1+s}}, \frac{1}{4\sqrt{1-s}}, 0 \right).$$

So
$$\kappa = |\mathbf{T}'(S)| = \frac{1}{8\sqrt{1-s^2}}.$$

9. (a) $1/[3(t^2+1)^2]$.

Exercises B.1

3. The subgroups are $\{1\}, \{1,d\}, \{1,a,e,d\}, \{1,b,f,d\}, \{1,c,g,d\}, \{1,a,b,c,d, e,f,g\}$.

7. For any $x, y \in \mathbb{Z}, f(x+y) = 5(x+y) = 5x+5y = f(x)+f(y)$. Hence f is a homeomorphism.

10. If the group contains an element of order 4, then it is necessarily isomorphic to $\mathbb{Z}_4$. If not, then every nonidentity element has order 2. Once this information is placed in the multiplication table, it becomes clear that the group must be isomorphic to $\mathbb{Z}_2^2$.

15. Suppose both 1 and $1'$ are identities of G. Then $1 = 1 \cdot 1' = 1'$.

21. Suppose $\{G_\lambda\}, \lambda \in \Lambda$, is a family of subgroups of the group G, and
$$H = \bigcap G_\lambda, \qquad \lambda \in \Lambda.$$

Group property 2 holds for H simply because $H \subset G$. If $a, b \in H$ then $a, b \in G_\lambda$ for each $\lambda \in \Lambda$. It follows that $ab \in G_\lambda$ for each $\lambda \in \Lambda$, and hence $ab \in H$. This establishes group property 1 for H. The validity of group properties 3 and 4 for H is similarly established.

Exercises B.2

1. The element 11 of $\mathbb{Z}_2 \times \mathbb{Z}_3$ has order 6, and so this group is cyclic. By Proposition B.1.7 we are done.

5. Let $M(N)$ be the set of all the elements of G whose order is a divisor of $m(n)$. Since m and n are relatively prime, it follows that $M \cap N = \{0\}$. Let g be an element of G of order $k = m'n'$ where $m'|m$ and $n'|n$. Define $g_M = n'g, g_N = m'g$. It is clear that $g_M \in M, g_N \in N$. Let A, B be integers such that $An' + Bm' = 1$. Then $Ag_M \in M, Bg_N \in N$ are such that $g = (An' + Bm')g = Ag_M + Bg_N$. In other words, $M + N = G$. We now show that the function $f : M \times N \to G$ defined by $f(a, b) = a + b$ is an isomorphism. Since $M + N = G$, this function is surjective. If $f(a, b) = f(a', b')$ then $a - a' = b' - b \in M \cap N = \{0\}$ and hence $(a, b) = (a', b')$. Thus f is also injective. Finally,

$$f[(a, b) + (a', b')] = f(a + a', b + b') = a + a' + b + b' = a + b + a' + b'$$
$$= f(a, b) + f(a', b'),$$

and so f is also a homomorphism.

10. Let $S = \{m_i/n_i, i = 1, 2, \ldots, k\}$ be a finite set of elements of $\mathbb{Q}$. If p is a prime number that does not divide the product $n_1 n_2 \cdots n_k$, then $1/p$ is not in the subgroup of $\mathbb{Q}$ generated by S. Hence S does not generate $\mathbb{Q}$.

Exercises B.3

1. Since $ab = ba^2$, it follows that every element of the group is expressible in the form $b^m a^n$. Since $b^3 = a^2 = 1$, there are at most six different elements of this form.

6. Since $ab = b^{-1}a^{-1} = b^2 a^3$, it follows that every element can be written in the form $b^n a^m$, $0 \le m \le 2$, $0 \le n \le 3$, and so there are at most 12 different elements. In the abelianization we have $1 = a^2 b^2 = a^2 b^{-2}$ and hence $a^2 = b^2$. It follows that here $a^4 = b^4 = 1$. Since $a^3 = 1$, we have $a = 1$ and $b^2 = 1^2 = 1$. Hence the abelianization consists of at most $\{1, b\}$.

11. From $a^3 b^2 = 1 = a^5 b^3$ we conclude that $a^3 = a^5 b$ and hence $a^{-2} = b$. Substitution into the first relator yields $a^{-1} = 1$ from which it follows that the group is trivial.

Exercises C

1. (a) $(1\ 3)(2)(4)(5\ 6)(7)(8)(9)$ b) $(1)(2\ 4)(3)(5\ 9)(6)(7)(8)$.

Exercises D

1. (a) 0,   (b) 2,   (c) 4.
2. (a) 2,   (b) 6,   (c) 6.
4. (a) x,   (b) 4x + 2y.
5. (a) x,   (b) 6x + 2y.

REFERENCES AND RESOURCES

Adams, C. C., *The Knot Book: An Elementary Introduction to the Mathematical Theory of Knots,* W. H. Freeman, New York, 1994.

Archdeacon, D., "A Kuratowski Theorem for the Projective Plane," *J. Graph Theory,* 5 (1981), 243–246.

Archdeacon, D., and Huneke, P., "A Kuratowski Theorem for Non-orientable Surfaces," *J. Combinatorial Theory Ser. B,* 46 (1989), 173–231.

Anderson, J. W., *Hyperbolic Geometry,* Springer, New York, 1999.

Archibald, T., "Connectivity and Smoke Rings: Green's Second Identity in Its First Fifty Years," *Mathematics Magazine,* 62 (1989), 219–232.

Arnold, B. H., *Intuitive Concepts in Elementary Topology,* Prentice Hall, Upper Saddle River, NJ, 1962.

Banchoff, T. F., "Critical Points and Curvature for Embedded Polyhedral Surfaces," *Amer. Math. Monthly,* 77 (1970), 475–485.

Battle, J., Harary, F., Kodama, Y., and Youngs, J. W. T., "Additivity of the Genus of a Graph," *Bull. Amer. Math. Soc.* 68 (1962), 565-568.

Bayer, M. M., "The Extended f-Vectors of 4-Polytopes," *J. Combinatorial Theory Ser. A*, 44 (1987), 141–151.

Bayer, M. M., and Lee, C. W., "Combinatorial Aspects of Convex Polytopes," in *Handbook of Convex Geometry*, P. M. Gruber and J. M. Wills, editors, Elsevier, Amsterdam, 1993.

Beardon, A. F., *The Geometry of Discrete Groups*, Springer-Verlag, New York, 1983.

Blackett, D. W., *Elementary Topology*, Academic Press, New York, 1967.

Bondy, J. A., and Murty, U. S. R., *Graph Theory with Applications*, North Holland, New York, 1979.

Bonola, R., *Non-Euclidean Geometry*, Dover Publications, New York, 1955.

Bruggesser, H., and Mani, P., "Shellable Decompositions of Cells and Spheres," *Math. Scand.*, 29 (1971), 197–205.

Cairns, S. S., *Introductory Topology*, Ronald Press, New York, 1961.

Carlson, S. C., *Topology of Surfaces, Knots, and Manifolds: A First Undergraduate Course*, Wiley, New York, 2001.

Conway, J. H., and Gordon, C. M., "Knots and Links in Spatial Graphs," *J. Graph Theory*, 7 (1983), 445–453.

Coxeter, H. S. M., *Regular Polytopes*, Dover, New York, 1973.

Coxeter, H. S. M., *Non-Euclidean Geometry*, 6th ed., Mathematical Association of America, Washington, DC, 1998.

Cromwell, P. R., *Polyhedra*, Cambridge University Press, Cambridge, 1997.

Dieudonné, J., *A History of Algebraic and Differential Topology*, Birkhäuser, Boston, 1989.

Do Carmo, M., *Differential Geometry of Curves and Surfaces*, Prentice Hall, Upper Saddle River, NJ, 1976.

Euclid, *The Elements*, Sir Thomas L. Heath, editor, Dover Publications, New York, 1956.

Farris, F. A., "The Edge of the Universe," *Math Horizons*, September 2001, 16–23.

Fenchel, W., *Elementary Geometry in Hyperbolic Space*, Walter de Gruyter, Berlin, 1989.

Firby, P. A., and Gardiner, C. F., *Surface Topology*, Halsted Press, Wiley, New York, 1982.

Flegg, H. G., *From Geometry to Topology*, Crane, Russak, New York, 1974.

Fort, M. K., Jr., editor, *Topology of 3-Manifolds*, Prentice Hall, Englewood Cliffs, NJ, 1962.

Gauss, C. F., *General Investigations of Curved Surfaces,* Raven Press, Hewlett, NY, 1965.

Gilbert, N. D., and Porter, T., *Knots and Surfaces,* Oxford University Press, Oxford, 1994.

Goodman, J. E., and O'Rourke, J., editors, *Handbook of Discrete and Computational Geometry,* Chapman and Hall/CRC, Boca Raton, FL, 2004.

Gramain, A., *Topology of Surfaces,* BCS Associates, Moscow, ID, 1988.

Graustein, W. C., *Differential Geometry,* Dover Publications, New York, 1966.

Greenberg, M. J., *Euclidean and Non-Euclidean Geometries: Development and History,* 3rd ed., W. H. Freeman, New York, 1993.

Gross, J. L., and Tucker, T. W., *Topological Graph Theory,* Wiley Interscience, New York, 1987.

Grünbaum, B., *Convex Polytopes*, 2nd ed., Springer-Verlag, New York, 2003.

Harary, F., *Graph Theory,* Perseus Publ., 1995.

Heffter, L., "Über das Problem der Nachbargebiete," *Math. Ann.,* 38 (1891), 477–508.

Henderson, D. W., *Differential Geometry: A Geometric Introduction,* Prentice Hall, Upper Saddle River, NJ, 1998.

Henderson, D. W., *Experiencing Geometry in Euclidean, Spherical, and Hyperbolic Spaces,* Prentice Hall, Upper Saddle River, NJ, 2001.

Hilbert, D., and Cohn-Vossen, S., *Geometry and the Imagination,* American Mathematical Society, Providence, RI, 1999.

Hoste, J., Thistlethwaite, M., and Weeks, J., "The First 1,701,936 Knots," *Math. Intelligencer,* 20 (1998), 33–48.

Jendrol', S., "On Face Vectors and Vertex Vectors of Convex Polyhedra," *Discrete Math.,* 118 (1993), 119–144.

Kauffman, L. H., "State Models and the Jones Polynomial," *Topology,* 26 (1987a), 395–407.

Kauffman, L. H., *On Knots,* Annals Studies No. 15, Princeton University Press, Princeton, 1987b.

Kauffman, L. H., "New Invariants in the Theory of Knots," *Amer. Math. Monthly,* 95 (1988), 195–242.

Kauffman, L. H., *Knots and Physics,* World Scientific, 1991.

Kinsey, L. C., *Topology of Surfaces,* Springer-Verlag, New York, 1993.

Kuratowski, K., "Sur le Problème des Courbes Gauche en Topologie," *Fund. Math.*, 15 (1930), 271–283.

Lang, S., *Calculus of Several Variables*, 3d ed., Springer-Verlag, New York, 1987.

Lefschetz, S., *Topology*, Colloq. Publ., No. 12, American Mathematical Society, Providence RI, 1942.

Lickorish, W. B. R., *An Introduction to Knot Theory*, Graduate Texts in Mathematics, No. 175, Springer, New York, 1997.

Lickorish, W. B. R., and Millett, K. C., "The New Polynomial Invariants of Knots and Links," *Mathematics Magazine*, 61 (1988), 3–23.

Ling, J. M., "New Non-Linear Inequalities for Flag-Vectors of 4-Polytopes," *Discrete Comput. Geom.*, 37 (2007), 455–469.

Livingston, C., *Knot Theory*, Carus Mathematical Monographs, No. 24, The Mathematical Association of America, 1993.

Lyndon, R. C., *Groups and Geometry*, Cambridge University Press, New York, 1985.

Lyndon, R. C., and Schupp, P. E., *Combinatorial Group Theory*, Springer-Verlag, New York, 1976.

Maehara, H., "Why is P^2 Not Embeddable in $\mathbb{R}^3$?" *Amer. Math. Monthly*, 100 (1993), 862–864.

Magnus, W., Karrass, A., and Solitar, D., *Combinatorial Group Theory: Presentations of Groups in Terms of Generators and Relators*, Wiley Interscience, New York, 1966.

Martin, G. E., *Transformation Geometry: An Introduction to Symmetry*, Springer-Verlag, New York, 1982.

Massey, W. S., *Algebraic Topology: An Introduction*, Harcourt, Brace & World, New York, 1967.

McCleary, J., *Geometry from a Differentiable Viewpoint*, Cambridge University Press, New York, 1994.

McCleary, J., "Trigonometries," *Amer. Math. Monthly*, 109 (2002), 623–639.

Mendelson, B., *Introduction to Topology*, 3rd ed., Dover Publications, New York, 1990.

Munkres, J. R., *Topology: A First Course*, Prentice Hall, Upper Saddle River, NJ, 1975.

Murasugi, K., *Knot Theory & Its Applications*, Birkhäuser, Boston, 1996.

Nanyes, O., "An Elementary Proof that the Borromean Rings Are Non-splittable," *Amer. Math. Monthly*, 100 (1993), 786–789.

Oprea, J., *Differential Geometry and Its Applications,* Prentice Hall, Upper Saddle River, NJ, 1997.

Ore, O., "Note on Hamilton Circuits," *Amer. Math. Monthly,* 67 (1960), 55.

Picard, E., and Simart, G., *Theorie des Fonctions Algébriques de Deux Variables Independants,* Gauthier-Villars, Paris, Vol. I, 1897; Vol II, 1906.

Poincaré, H., "Cinquième Complément a l'Analysis Situs," *Rendiconti del Circolo matematico di Palermo,* 18 (1904), 45–110.

Poincaré, H., *Oeuvres,* Vol. 6, Gauthier-Villars, Paris, 1953.

Poincaré, H., *Papers on Fuchsian Functions* (J. Stillwell, translator), Springer-Verlag, New York, 1985.

Polya, G., "An Elementary Analog to the Gauss–Bonnet Theorem," *Amer. Math. Monthly,* 61 (1954), 601–603.

Pont, J.-C., *La Topologie Algébrique,* Presses Universitaire de France, Paris, 1974.

Ramsay, A., and Richtmyer, R. D., *Introduction to Hyperbolic Geometry,* Springer-Verlag, New York, 1995.

Reidemeister, K., *Knotentheorie,* Ergebnisse der Mathematic, Vol 1, Springer-Verlag, Berlin, 1932; Boron, L. F., Christenson, C. O., and Smith, B. A., (English translation), BCA Associates, Moscow, Idaho, 1983.

Ringel, G., *Map Color Theorem,* Springer-Verlag, New York, 1974.

Robertson, N., and Seymour, P. D., "Graph Minors. VIII: A Kuratowski Theorem for General Surfaces," *J. Combinatorial Theory, Series B,* 48 (1990), 255–288.

Robertson, N., Seymour, P. D., and Thomas, R., "Linkless Embeddings of Graphs in 3-Space," *Bull. Amer. Math. Soc.,* 28 (1993), 84–89.

Rolfsen, D., *Knots and Links,* Publish or Perish, Berkeley, 1976.

Sachs, H., "On a Spatial Analogue of Kuratowski's Theorem on Planar Graphs— An Open Problem," *Graph Theory,* Lagow 1981, Lecture Notes in Mathematics, No. 1018, Springer-Verlag, Berlin, 1982, 230–241.

Santos, F., "A Counterexample to the Hirsch Conjecture," *Ann. of Math.*, to appear.

Schubert, H., "Die eindeutige Zelegbarkeit eines Knotens in Primknoten," *Sitzungsberichte der Heidelberger Akademie der Wissenschaften Mathematische-Naturewissenschftliche Klasse*, No. 3, 1949, 57–104.

Seifert, H., "Über das Geschlecht von Knoten," *Math. Ann.,* 110 (1934), 571–592.

Seifert, H., and Threlfall, W., *A Textbook of Topology* (Goldman, M. A., translator), Academic Press, New York, 1980.

Singer, D. A., *Geometry: Plane and Fancy,* Springer, New York, 1998.

Sommerville, D. M. Y., *An Introduction to the Geometry of N Dimensions*, Dover, New York, 1958.

Sossinsky, A., *Knots: Mathematics with a Twist,* Harvard Univesity Press, Cambridge, MA, 2002.

Spivak, M., *A Comprehensive Introduction to Differential Geometry,* Publish or Perish, Willmington, DE, 1979.

Stahl, S., *The Poincaré Half-Plane: A Gateway to Modern Geometry,* Jones and Bartlett, Boston, 1993.

Steen, L. A., and Seebach, J. A., Jr., *Counterexamples in Topology,* Holt, Rinehart and Winston, New York, 1970.

Thurston, W. P., *Three-Dimensional Geometry and Topology, Vol. 1,* Princeton University Press, Princeton, 1997.

Todhunter, I., *Spherical Trigonometry,* MacMillan, London, 1886.

Weeks, J., *The Shape of Space,* Marcel Dekker, New York, 2002.

West, D. B., *Introduction to Graph Theory,* Prentice Hall, Upper Saddle River, NJ, 1996.

White, A. T., *Graphs of Groups on Surfaces,* Elsevier, Amsterdam, 2001.

Willmore, T. J., *An Introduction to Differential Geometry,* Oxford University Press, London, 1961.

Wolfe, H. E., *Introduction to Non-Euclidean Geometry,* Holt, Rinehart and Winston, New York, 1945.

Ziegler, G. M., *Lectures on Polytopes*, Springer-Verlag, New York, 1995.

Index

Introduction to Topology and Geometry, Second Edition.
By Saul Stahl and Catherine Stenson Copyright © 2013 John Wiley & Sons, Inc.

PURE AND APPLIED MATHEMATICS

A Wiley Series of Texts, Monographs, and Tracts

Founded by RICHARD COURANT

Editors Emeriti: MYRON B. ALLEN III, DAVID A. COX, PETER HILTON, HARRY HOCHSTADT, PETER LAX, JOHN TOLAND

*Now available in a lower priced paperback edition in the Wiley Classics Library.
†Now available in paperback.

---

*Now available in a lower priced paperback edition in the Wiley Classics Library.
†Now available in paperback.

SENDOV and POPOV—The Averaged Moduli of Smoothness
SEWELL—The Numerical Solution of Ordinary and Partial Differential Equations,
Second Edition
SEWELL—Computational Methods of Linear Algebra, Second Edition
SHICK—Topology: Point-Set and Geometric
SHISKOWSKI and FRINKLE—Principles of Linear Algebra With *Maple*™
SHISKOWSKI and FRINKLE—Principles of Linear Algebra With *Mathematica*®
*SIEGEL—Topics in Complex Function Theory
Volume 1—Elliptic Functions and Uniformization Theory
Volume 2—Automorphic Functions and Abelian Integrals
Volume 3—Abelian Functions and Modular Functions of Several Variables
SMITH and ROMANOWSKA—Post-Modern Algebra
ŠOLÍN–Partial Differential Equations and the Finite Element Method
STADE—Fourier Analysis
STAHL and STENSON—Introduction to Topology and Geometry, Second Edition
STAHL—Real Analysis, Second Edition
STAKGOLD and HOLST—Green's Functions and Boundary Value Problems,
Third Edition
STANOYEVITCH—Introduction to Numerical Ordinary and Partial Differential
Equations Using MATLAB®
*STOKER—Differential Geometry
*STOKER—Nonlinear Vibrations in Mechanical and Electrical Systems
*STOKER—Water Waves: The Mathematical Theory with Applications
WATKINS—Fundamentals of Matrix Computations, Third Edition
WESSELING—An Introduction to Multigrid Methods
†WHITHAM—Linear and Nonlinear Waves
ZAUDERER—Partial Differential Equations of Applied Mathematics, Third Edition

*Now available in a lower priced paperback edition in the Wiley Classics Library.
†Now available in paperback.

Printed and bound by CPI Group (UK) Ltd, Croydon, CR0 4YY

16/04/2025

14658366-0005